1931 Carl Anderson discovers the positron (antielectron).

1931 Wolfgang Pauli suggests existence of neutral particle (neutrino) emitted in beta decay.

1932 James Chadwick discovers the neutron.

1932 John Cockcroft and Ernest Walton produce the first nuclear reaction using a high-voltage accelerator.

1934 Irène and Frédéric Joliot-Curie discover artificially induced radioactivity.

1935 Hideki Yukawa proposes existence of medium-mass particles (mesons).

1938 Otto Hahn, Fritz Strassmann, Lise Meitner, and Otto Frisch discover nuclear fission.

1938 Hans Bethe proposes thermonuclear fusion reactions as the source of energy in stars.

1940 Edwin McMillan, Glenn Seaborg, and colleagues produce first synthetic transuranic elements.

1942 Enrico Fermi and colleagues build first nuclear fission reactor.

1945 Detonation of first fission bomb in New Mexico desert.

1946 George Gamow proposes big-bang cosmology.

1948 John Bardeen, Walter Brattain, and William Shockley demonstrate first transistor.

1952 Detonation of first thermonuclear fusion bomb at Eniwetok atoll.

1956 Frederick Reines and Clyde Cowan demonstrate experimental evidence for existence of neutrino.

1958 Rudolf L. Mössbauer demonstrates recoilless emission of gamma rays.

1960 Theodore Maiman constructs first ruby laser; Ali Javan constructs first helium-neon laser.

1964 Allan R. Sandage discovers first quasar.

1964 Murray Gell-Mann and George Zweig independently introduce three-quark model of elementary particles.

1965 Arno Penzias and Robert Wilson discover cosmic microwave background radiation.

1967 Jocelyn Bell and Anthony Hewish discover first pulsar.

1967 Steven Weinberg and Abdus Salam independently propose a unified theory linking the weak and electromagnetic interactions.

1974 Burton Richter and Samuel Ting and co-workers independently discover first evidence of fourth quark (charm).

1974 Joseph Taylor and Russell Hulse discover first binary pulsar.

1977 Leon Lederman and colleagues discover new particle showing evidence for fifth quark (bottom).

1981 Gerd Binnig and Heinrich Rohrer invent scanning-tunneling electron microscope.

1983 Carlo Rubbia and co-workers at CERN discover $W^{\pm}$ and Z^0 particles.

1986 J. Georg Bednorz and Karl Alex Müller produce first high-temperature superconductors.

1994 Investigators at Fermilab discover evidence for sixth quark (top).

MODERN PHYSICS

MODERN PHYSICS

Second Edition

Kenneth S. Krane

DEPARTMENT OF PHYSICS
OREGON STATE UNIVERSITY

John Wiley & Sons, Inc.
New York   Chichester   Brisbane   Toronto   Singapore

COVER PHOTO   Ken Biggs/Tony Stone Images
ACQUISITIONS EDITOR   Clifford Mills
MARKETING MANAGER   Cathy Faduska
PRODUCTION MANAGEMENT   Jennifer Knapp
DESIGN SUPERVISOR   Harry Nolan
MANUFACTURING MANAGER   Mark Cirillo
ILLUSTRATION COORDINATOR   Jaime Perea
OUTSIDE PRODUCTION MANAGER   Suzanne Ingrao of Ingrao Associates

This book was set in Times Roman by Bi-Comp, Inc., and printed and bound by R. R. Donnelley and Sons, Willard. The cover and insert were printed by Lehigh Press, Inc.

Recognizing the importance of preserving what has been written, it is a
policy of John Wiley & Sons, Inc. to have books of enduring value published
in the United States printed on acid-free paper, and we exert our best
efforts to that end.

The paper on this book was manufactured by a mill whose forest management programs include sustained
yield harvesting of its timberlands. Sustained yield harvesting principles ensure that the number of trees
cut each year does not exceed the amount of new growth.

Library of Congress Cataloging-in-Publication Data

Krane, Kenneth S.
    Modern physics/Kenneth S. Krane.—2nd ed.
        p.   cm.
    Includes index.
    ISBN 0-471-82872-6 (cloth)
    1. Physics. I. Title.
QC21.2.K7   1996
539—dc20
                                            95-6382
                                            CIP

Printed in the United States of America

10 9 8 7 6 5 4

This textbook is meant to serve a first course in modern physics, including relativity, quantum physics, and applications. Such a course generally follows the standard introductory course in calculus-based physics.

The major goal of this text, and of the course for which it is intended, is to instill in the student an appreciation of the concepts and methods of twentieth-century physics. The text has grown from a course at Oregon State University, taken generally during the sophomore year, which serves two functions for two different audiences. (1) Physics majors, who will later take a more rigorous course in quantum physics, find an introductory modern course helpful to broaden their perspective before undertaking the rigors of the traditional junior-year studies in classical mechanics and electromagnetism. (2) Nonmajors, who likely will take no further physics, are increasingly finding need for modern physics in their disciplines; an introductory classical course is hardly sufficient for chemists, computer scientists, nuclear and electrical engineers, and so forth.

Necessary prerequisites for undertaking the text include any standard introductory calculus-based course covering mechanics, electromagnetism, thermal physics, and optics. Calculus is used extensively, but no previous knowledge of differential equations, complex variables, or partial derivatives is assumed (although some familiarity with these topics is helpful).

Chapters 1–8 comprise the core of the text. They cover relativity and quantum theory through atomic structure. At that point the reader may continue with Chapters 9–11 (molecules, statistics, and solids) or branch to Chapters 12–14 (nuclei and particles). The final two chapters (astrophysics and cosmology) can be viewed as the capstone modern physics experience, integrating concepts from relativity and from atomic, molecular, statistical, solid-state, nuclear, and particle physics. Sections of the text that can be omitted without loss of continuity are marked with asterisks.

To maintain the introductory nature of the text and to keep its length within reasonable limits, many topics have been omitted: spacetime diagrams, four-vectors, parity, symmetric and antisymmetric wave functions, total angular momentum, isotopic spin, hypercharge, and nuclear models, among others. My experience in teaching this course for nearly twenty years suggests that the essentials of modern physics can be appreciated without these topics. To do justice to, for example, parity or nuclear models requires a greater exposition than the limited space in this introductory text can allow, and a few offhand paragraphs on these difficult subjects are more likely to lead to confusion than enlightenment.

The unifying theme of the text is the empirical basis of modern physics. Experimental tests of phenomena are discussed throughout. Examples include the latest experimental tests of special and general relativity and studies of wave-particle duality of photons and material particles. Applications of the basic phenomena are used extensively, including barrier penetration, lasers, radioactive dating, nuclear activation analysis, nuclear medicine, and semiconductor devices.

Readers familiar with the first edition will notice a number of changes in its appearance, content, and organization. Much of the text has been rewritten to improve clarity and aid comprehension. The ordering of some topics has been changed to reflect a more logical arrangement or to lower the barrier for entry into a new subject. Color has been added to the artwork to improve the pedagogy, and a new insert contains numerous color photographs that illustrate phenomena discussed in the text. The number of worked examples has been increased by 25%, and the sets of end-of-chapter questions and problems have been expanded by (respectively) 23% and 15%. An effort has been made for the problems to require a greater range of abilities, and some "open-ended" problems have been included. The end-of-chapter reading lists have been updated with recent references to the literature.

Although the approach taken in the text is largely analytical, it should be complemented by computational methods. Computer demonstrations and exercises based on topics from the text are available through the Modern Physics Simulations developed by the Consortium for Upper Level Physics Software (CUPS) and distributed by John Wiley & Sons. A few selected computer exercises from the CUPS materials are included with the end-of-chapter problems.

Specific changes in content and organization from the first edition including the following:

Chapter 2 has been extensively rewritten. Relativistic mass has been eliminated,* and the role of the Lorentz transformation has been deemphasized,** so that it does not appear to occupy quite the central role in relativity that it did in the first edition. The discussion of the experimental tests of special relativity has been updated to include examples of recent elegant and increasingly precise studies.

In Chapter 3, I have interchanged the ordering so that the discussion of the photoelectric effect now precedes that of blackbody radiation. While this reverses the chronological order of the discoveries, the new ordering provides a less challenging introduction to the photon concept, which appears directly in the explanation of the photoelectric effect but is buried under abstract mathematical and statistical concepts in the blackbody radiation formalism. Continuing the emphasis on experimental tests, the chapter features examples of new studies of wave-particle duality for light.

In Chapter 4, the section on the uncertainty principle has been rewritten. Examples of new experimental studies of the wave nature of particles (electrons, neutrons, helium atoms) have been added.

---

* See the discussion by Lev B. Okun, *Physics Today*, June 1989, p. 31.

** See, for example, the discussions by A. John Mallinckrodt, *American Journal of Physics* **61**, 760 (1993) and N. David Mermin, *American Journal of Physics* **62**, 11 (1994), who argue for complete elimination of the Lorentz transformation in a first introduction to relativity.

Chapter 5 now includes examples of the wave behavior of particles revealed using scanning tunneling electron microscopes.

Chapter 7 now has a greater emphasis on quantum-mechanical orbits and spatial probabilities and less emphasis on Bohr-type orbits in atoms.

Chapters 9–11 in the first edition (nuclear and particle physics) have been interchanged with Chapters 12–14 (molecular, statistical, and solid-state physics). In the new ordering, the chapter on molecular physics directly follows those covering atomic physics and is in turn followed by the chapters on statistical physics and solids. The chapters on nuclear and particle physics then lead directly into the last chapters on astrophysics and cosmology. As in the previous edition, an abbreviated course may skip either block of three chapters, although two concepts from statistics are needed in the astrophysics and cosmology chapters: the photon distribution in thermal radiation (for which the material in Chapter 3 is a sufficient introduction) and the relationship between the Fermi energy and the particle density.

Chapter 10 offers a new introduction to statistical concepts and the origin of the difference between classical and quantum statistics, based on a simple counting problem.

Chapter 11 contains a new section on the justification for band theory, a greater focus on the quantum theory of electrical conduction, and a new section on superconductivity.

New features in Chapter 12 include examples of "exotic" nuclear decays and a revised (and simplified) discussion of the calculation of alpha-decay probabilities.

In Chapter 13, the section on the compound nuclear model has been eliminated. Instead, a section on applications of nuclear physics has been added.

Chapter 14 features new introductory sections on particles, forces, anti-particles, and classification schemes. The discussion of the quark model has been updated, and a discussion of the standard model and its implications has been added.

Chapters 15 and 16 have been substantially rewritten. Sections of Chapter 15 dealing with general relativity and black holes have been revised, and the cosmological applications of general relativity that were previously in Chapter 15 have been moved to Chapter 16. Chapter 16 includes the latest results on the cosmic microwave background radiation, a new section on cosmology and general relativity, a new section on dark matter, and a rewritten section on the Big Bang cosmology.

I am grateful to many individuals for help in preparing this new edition. Many users of the first edition have taken the time to write with comments and criticisms; the list is too long to acknowledge them individually here, but their influence extends throughout the text and I am grateful for their participation in this effort. Many thoughtful suggestions have been offered by a loyal cadre of reviewers of the manuscript for this edition:

John Campbell, U.S. Military Academy

John Gagliardi, Rutgers University—Camden

Robert Rasera, University of Maryland Baltimore County

Sharon Rosell, Central Washington University

Morton Sternheim, University of Massachusetts, Amherst

Thor F. Stromberg, New Mexico State University

Carol Thompson, Brooklyn Polytechnic University

Their advice has helped in innumerable ways to improve the text; even in cases in which I have stubbornly refused to follow their wise counsel, it has guided my reasoning and caused me to rethink many of the book's expositions. I am indebted to these reviewers for their contributions to this project.

The staff at John Wiley & Sons has continued to provide strong support and cooperation for this project. I am grateful for the assistance of: Physics Editor Clifford Mills, Production Coordinator Suzanne Ingrao of Ingrao Associates, Photo Researcher Hilary Newman, Illustration Coordinator Jaimé Perea, Editorial Program Assistants Cathy Donovan and Monica Stipanov, Marketing Manager Cathy Faduska, and Designer Harry Nolan.

In my research and other activities over the past decade, I have encountered many students (some of whom are now professional colleagues) who used the first edition of this text. It has been most gratifying to learn how much they enjoyed using the text; some have even reported that their decision to become physicists resulted from their first introduction to modern physics. For most students, the introductory modern physics course offers their first insights into what physicists really do and what is exciting, perplexing, and challenging about our discipline. I hope users of this new edition will continue to find these insights.

Corvallis, Oregon   Kenneth S. Krane
May 1995   kranek@physics.orst.edu

CONTENTS

Contents **xiii**

MODERN PHYSICS

This illustration is from Galileo's Dialogue Concerning the Two Chief World Systems, *published in 1632. The dialogue was a debate between those (represented by Aristotle and Ptolemy, the figures on the left) who supported the accepted view that the Earth was fixed at the center of the universe, and those (like Copernicus, the figure on the right, and like Galileo himself) who supported the view that the Earth moves about the Sun. Galileo had been prohibited by the Church from teaching such "controversial" views, but by presenting both sides of the issue he tried (unsuccessfully, as it turned out) to circumvent the Church's prohibition. Galileo hoped his readers would draw their own conclusions, based on the evidence; for this reason, Galileo is often regarded as the father of modern science.*

INTRODUCTION

At the end of the nineteenth century it seemed that most of what there was to know about physics had already been learned. Newton's dynamics had been carefully and repeatedly tested, and its success had provided a framework for a deep and consistent understanding of nature. Electricity and magnetism had been unified by Maxwell's theoretical work, and the electromagnetic waves predicted by Maxwell's equations had been discovered and investigated in the experiments conducted by Hertz. The laws of thermodynamics and kinetic theory had been particularly successful in providing a unified explanation of a wide variety of phenomena. More generally, the Industrial Revolution had introduced a measure of technological sophistication that would have profound influence on the lives and standard of living of people everwhere. After a period of economic and geographical expansion, the United States was beginning to assert its role as a world power. In Europe, strong monarchies had provided an environment in which industrialization could proceed at a rapid pace. However, beneath this apparent air of stability and optimism there were strong undercurrents, which, in a few years, would plunge the world into the brutal conflict of World War I; the rising tide of militarism, the forces of nationalism and revolution, and the gathering strength of Marxism would soon upset the established order. The fine arts were similarly in the middle of revolutionary change, as new ideas began to dominate the fields of painting, sculpture, and music. The understanding of even the very fundamental aspects of human behavior was subject to serious and critical modification by the Freudian psychologists.

In the world of physics, too, there were undercurrents that would soon cause revolutionary changes. Even though the overwhelming majority of experimental evidence agreed with classical physics, several experiments gave results that were not explainable in terms of the otherwise successful classical theories of mechanics, electromagnetism, and thermodynamics. Classical electromagnetic theory suggested that a medium is needed to propagate electromagnetic waves, but precise experiments failed to detect this medium. Experiments to study the emission of electromagnetic waves by hot, glowing objects gave results that could not be explained by the classical theories of thermodynamics and electromagnetism. Experiments on the emission of electrons from surfaces illuminated with light also could not be understood using classical theories.

These few experiments may not seem significant, especially when viewed against the background of the many successful and well-understood experiments of the nineteenth century. However, these experiments were to have a profound and lasting effect, not only on the world of physics, but on all of science, on the political structure of our world, and on the way we view ourselves and our place in the universe. Within the short span of two decades, the results of these experiments were to lead to the special theory of relativity and to the quantum theory; soon after the revolutions inspired by these new theories came the development of atomic physics, nuclear physics, and solid-state physics, with the monumental impact that applications of research in these fields have had on our daily lives.

Modern physics   The designation *modern physics* usually refers to the developments that began with the relativity and quantum theories, including the applications of those theories to understanding the atom, the atomic nucleus and the particles of which it is composed, collections of atoms in molecules and solids, and, on a cosmic scale, the origin and evolution of the universe. Our discussion of modern physics in this text touches on each of these fields. We begin with the *relativity theory,* exploring its assumptions, implications, and experimental verification. After reviewing the experiments that signaled the inadequacy of classical concepts of particles and waves, we discuss the success of the *quantum theory* (or *wave mechanics,* as it is sometimes known) in resolving those failures. A complete discussion of wave mechanics requires mathematical proficiency beyond the level of this text; therefore we undertake only a superficial introduction to the techniques and applications of wave mechanics. The remainder of the text deals with applications of these principles, first to the study of the structure and properties of the atom, next to the study of groups of atoms, both small groups in molecules and large groups in solids, and then to the study of the structure and properties of the atomic nucleus and the elementary particles. Finally, we turn from the microscopic to the cosmic and discuss the applications of modern physics to the understanding of some problems of astrophysics and cosmology.

As you undertake this study, keep in mind that the details of the story of modern physics have been written only during this century, and that many of the discoveries have been made during our lifetimes. This means that the story of modern physics is not yet complete and will continue to evolve. Often we find that each time we look deeper or refine our techniques we learn something new and must revise our theories to account for the new discoveries. It is this process of refinement and revision that advances the frontiers of science.

1.1 REVIEW OF CLASSICAL PHYSICS

Although we will find many areas in which modern physics differs radically from classical physics, we frequently find the need to refer to concepts of classical physics. Here is a brief review of some of the concepts of classical physics that we may need.

Mechanics

A particle of mass m moving with velocity v has a *kinetic energy* defined by

$$K = \tfrac{1}{2}mv^2 \tag{1.1}$$

and a *linear momentum* **p** defined by

$$\mathbf{p} = m\mathbf{v} \tag{1.2}$$

In terms of the linear momentum, the kinetic energy can be written

$$K = \frac{p^2}{2m} \tag{1.3}$$

When one particle collides with another, we analyze the collision by applying two fundamental conservation laws:

I. **Conservation of Energy.** The total energy of an isolated system (on which no net external forces act) remains constant. This means (in the case of a collision) that the total energy of the two particles *before* the collision is equal to the total energy of the two particles *after* the collision.

II. **Conservation of Linear Momentum.** The total linear momentum of an isolated system remains constant. For the collision, the total linear momentum of the two particles *before* the collision is equal to the total linear momentum of the two particles *after* the collision. Since linear momentum is a vector, application of this law usually gives us two equations, one for the *x* components and another for the *y* components.

These two conservation laws are of the most basic importance to understanding and analyzing a wide variety of problems in classical physics. Problems 1–3, 15, and 16 at the end of this chapter review the use of these laws.

The importance of these conservation laws is both so great and so fundamental that, even though in Chapter 2 we learn that the special theory of relativity modifies Equations 1.1, 1.2, and 1.3, the laws of conservation of energy and linear momentum remain valid.

Another application of the principle of conservation of energy occurs when a particle moves subject to an external force *F*. Corresponding to that external force there is often a potential energy *U*, defined such that (for one-dimensional motion)

$$F = -\frac{dU}{dx} \tag{1.4}$$

The total energy *E* is the sum of the kinetic and potential energies:

$$E = K + U \tag{1.5}$$

As the particle moves, *K* and *U* may change, but *E* remains constant. (In Chapter 2, we find that the special theory of relativity gives us a new definition of total energy.)

When a particle moving with linear momentum **p** is at a displacement **r** from the origin *O*, its *angular momentum* **L** about the point *O* is defined (see Figure 1.1) by

$$\mathbf{L} = \mathbf{r} \times \mathbf{p} \tag{1.6}$$

There is a conservation law for angular momentum, just like that for linear momentum. In practice this has many important applications. For example, when a charged particle moves near, and is deflected by, another charged particle, the total angular momentum of the system (the two particles) remains constant if no net external torques act on the system. If the second particle is so much heavier than the first that its motion is unchanged by the influence of the first particle, the angular momentum of the first particle remains constant (because the second particle acquires no angular momentum). A similar situation occurs when a comet moves in the gravitational

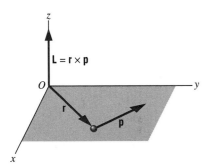

FIGURE 1.1 A particle of mass *m*, located with respect to the origin *O* by position vector **r** and moving with linear momentum **p**, has angular momentum **L** about *O*.

field of the Sun. As it approaches the Sun, **r** decreases, and so **p** must increase if **L** is to remain constant; the comet therefore accelerates as it approaches the Sun.

Electricity and Magnetism

The electrostatic force (Coulomb force) between two charged particles q_1 and q_2 has magnitude

$$F = \frac{1}{4\pi\varepsilon_0} \frac{q_1 q_2}{r^2} \tag{1.7}$$

The direction of **F** is along the line joining the particles (Figure 1.2). In the SI system of units, the constant $1/4\pi\varepsilon_0$ has the value

$$\frac{1}{4\pi\varepsilon_0} = 8.988 \times 10^9 \, \text{N·m}^2/\text{C}^2$$

The corresponding potential energy is

$$U = \frac{1}{4\pi\varepsilon_0} \frac{q_1 q_2}{r} \tag{1.8}$$

In all equations derived from Equation 1.7 or 1.8 as starting points, *the quantity $1/4\pi\varepsilon_0$ must appear.* In some texts and reference books, you may find electrostatic quantities in which this constant does not appear. In such cases, the centimeter-gram-second (cgs) system has probably been used, in which the constant $1/4\pi\varepsilon_0$ is *defined* to be 1. You should always be very careful in making comparisons of electrostatic quantities from different references and check that the units are identical.

A magnetic field **B** can be produced by an electric current i. For example, the magnitude of the magnetic field at the center of a circular current loop of radius r is (see Figure 1.3a)

$$B = \frac{\mu_0 i}{2r} \tag{1.9}$$

The SI unit for magnetic field is the tesla (T), which is equivalent to a newton per ampere-meter. The constant μ_0 is

$$\mu_0 = 4\pi \times 10^{-7} \, \text{N·s}^2/\text{C}^2$$

Be sure to remember that i is in the direction of the conventional (*positive*) current, opposite to the actual direction of travel of the negatively charged electrons that typically produce the current in metallic wires. The direction of **B** is chosen according to the right-hand rule: if you hold the wire in the right hand with the thumb pointing in the direction of the current, the fingers point in the direction of the magnetic field.

It is often convenient to define the *magnetic moment* **μ** of a current loop: *Magnetic moment*

$$|\boldsymbol{\mu}| = iA \tag{1.10}$$

where A is the geometrical area enclosed by the loop. The direction of **μ** is perpendicular to the plane of the loop, according to the right-hand rule.

FIGURE 1.2 Two charged particles experience equal and opposite electrostatic forces along the line joining their centers. If the charges have the same sign (both positive or both negative), the force is repulsive; if the signs are different, the force is attractive.

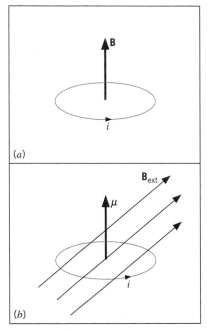

FIGURE 1.3 (*a*) A circular current loop produces a magnetic field **B** at its center. (*b*) A current loop with magnetic moment **μ** in an external magnetic field **B**ₑₓₜ. The field exerts a torque on the loop that will tend to rotate it so that **μ** lines up with **B**ₑₓₜ.

When a current loop is placed in a uniform *external* magnetic field $\mathbf{B}_{\text{ext}}$, there is a torque $\boldsymbol{\tau}$ on the loop that tends to line up $\boldsymbol{\mu}$ with $\mathbf{B}_{\text{ext}}$:

$$\boldsymbol{\tau} = \boldsymbol{\mu} \times \mathbf{B}_{\text{ext}} \tag{1.11}$$

Figure 1.3*b* shows a current loop in an external magnetic field. Another way to interpret this is to assign a potential energy to the magnetic moment $\boldsymbol{\mu}$ in the external field $\mathbf{B}_{\text{ext}}$:

$$U = -\boldsymbol{\mu} \cdot \mathbf{B}_{\text{ext}} \tag{1.12}$$

When the field $\mathbf{B}_{\text{ext}}$ is applied, $\boldsymbol{\mu}$ rotates so that its energy tends to a minimum value, which occurs when $\boldsymbol{\mu}$ and $\mathbf{B}_{\text{ext}}$ are parallel.

It is important for us to understand the properties of magnetic moments, because we will find that particles such as electrons or protons have magnetic moments. Although we don't imagine these particles to be tiny current loops, their magnetic moments do obey Equations 1.11 and 1.12.

A particularly important aspect of electromagnetism is *electromagnetic waves*. In Chapter 3 we discuss some properties of these waves in more detail; in particular, *interference* and *diffraction* are fundamental to our discussions in this text. Electromagnetic waves travel in free space with speed c (the speed of light), which is related to the electromagnetic constants ε_0 and μ_0

$$c = (\varepsilon_0 \mu_0)^{-1/2} \tag{1.13}$$

The waves have a frequency ν and wavelength λ, which are related by

$$c = \lambda \nu \tag{1.14}$$

The wavelengths range from the very short (nuclear gamma rays) to the very long (radio waves). Figure 1.4 shows the electromagnetic spectrum with the conventional names assigned to the different ranges of wavelengths.

Kinetic Theory of Matter

The mean kinetic energy of the molecules of an ideal gas at temperature T is

$$K = \tfrac{3}{2}kT \tag{1.15}$$

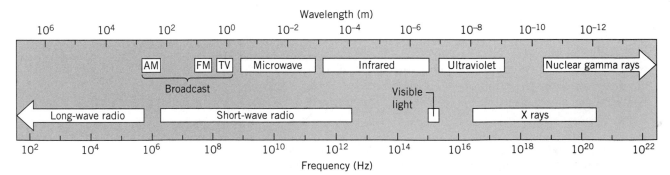

FIGURE 1.4 The electromagnetic spectrum. The boundaries of the regions are not sharply defined.

where k is the Boltzmann constant

$$k = 1.381 \times 10^{-23} \text{ J/K}$$

The SI unit of temperature is the kelvin (K), *not* "degree kelvin." Be careful not to confuse the symbol K for kelvin with the symbol K for kinetic energy; also be careful about confusing the Boltzmann constant k with the wave number $k = 2\pi/\lambda$.

For rough estimates, the quantity kT is often used as a measure of the mean kinetic energy per particle. For example, at room temperature (20°C = 293 K), the mean kinetic energy per particle is about 4×10^{-21} J, while in the interior of a star where $T \sim 10^7$ K, the mean energy is about 10^{-16} J.

1.2 UNITS AND DIMENSIONS

Nearly all of the physical constants and variables we will be using have both *units* and *dimensions*. The dimensions of a constant or variable tell us something about the kind of constant or variable it is; a quantity that in one frame of reference has the dimension of length, for example, will have the dimension of length in *every* frame of reference, although its magnitude and the units in which we express it may vary. We often refer to mass, length, and time as the *fundamental dimensions*.

In working problems, it is *always* necessary to be sure that your equations are dimensionally consistent; if, for example, you have an equation that contains the term $(v + m)$ where $v = $ velocity and $m = $ mass, it is a sure bet that you have made a mistake somewhere—two quantities can *never* be added unless they have the same dimensions. [However, if your equation contains the term $(av + m)$, where a is a constant, it may be dimensionally correct if a has the proper dimensions.]

Checking your results for dimensional consistency is a good habit to acquire. It is one of the simplest techniques to apply and can almost always be done quickly by inspection. If your results check for dimensional consistency, it doesn't guarantee that they are correct; the reverse, however, is true—lack of dimensional consistency *always* indicates an error.

It is sometimes possible for a quantity to have units, but no dimensions! Suppose your watch runs slow and loses 6.0 seconds each day. Its rate of loss is therefore $R = 6.0$ s/day. R is a dimensionless quantity—it has dimensions of t/t—but it has units, and its value changes as its units change; we could also express R as 0.10 min/day or 0.25 s/hour, or even in a unitless form as 6.9×10^{-5}, which gives the fractional loss of time in any interval. As another example, all conversion factors such as 25.4 mm/inch or 1000 g/kg are dimensionless.

We generally use SI units to express physical quantities. However, in modern physics our calculations often yield results that turn out to be very small or very large when expressed in SI units. For example, typical energies associated with atomic or nuclear processes may be 10^{-19} to 10^{-12} J, and typical sizes of atomic and nuclear systems range from 10^{-10} to 10^{-15} m.

Although there is no reason that we could not use these SI units in our study of modern physics, we frequently give way to convenience and use different units.

Length

The SI unit for length is the *meter* (m), but we will need lengths much smaller than the meter for atomic and nuclear systems. We use the following units for lengths:

micrometer $= \mu$m $= 10^{-6}$ m

nanometer $=$ nm $= 10^{-9}$ m

femtometer $=$ fm $= 10^{-15}$ m

Wavelengths are conveniently measured in nanometers—visible light has wavelengths in the range 400 to 700 nm. Atomic sizes are typically 0.1 nm, and nuclear sizes are about 1 to 10 fm. (The unit fm is sometimes known as the *fermi* in honor of Enrico Fermi, who was a pioneer experimental and theoretical nuclear physicist.)

Energy

The electron-volt

The SI unit of energy is the *joule* (J), which is too large to be of convenience for atomic and nuclear physics. A much more convenient unit is the *electron-volt* (eV), defined as the energy gained or lost by a particle of charge equal in magnitude to that of the electron in moving through a potential difference of one volt. (The volt is an SI unit, so we have not gone too far astray!) With the electronic charge of 1.602×10^{-19} C and with 1 V = 1 J/C, we have the equivalence

$$1 \text{ eV} = 1.602 \times 10^{-19} \text{ J}$$

Some convenient multiples of the electron-volt are

keV $=$ kilo electron-volt $= 10^3$ eV

MeV $=$ mega electron-volt $= 10^6$ eV

GeV $=$ giga electron-volt $= 10^9$ eV

(In some older works you may find reference to the BeV, for billion electron-volts; this is a source of confusion, for in the United States a billion is 10^9 while in Europe a billion is 10^{12}.)

Electric Charge

The standard unit of electric charge is the *coulomb* (C), and the basic quantity of charge, that of the electron, is $e = 1.602 \times 10^{-19}$ C. Very frequently we wish to find the potential energy of two basic charges separated by typical atomic or nuclear dimensions, and we wish to have the result expressed in electron-volts. Here is a convenient way of doing this. Let us find the potential energy of two electrons at a separation of $r = 1.00$ nm:

$$U = \frac{1}{4\pi\varepsilon_0} \frac{e^2}{r}$$

The quantity $e^2/4\pi\varepsilon_0$ can be expressed in a convenient form:

$$\frac{e^2}{4\pi\varepsilon_0} = \left(8.988 \times 10^9 \frac{\text{N} \cdot \text{m}^2}{\text{C}^2}\right)(1.602 \times 10^{-19}\,\text{C})^2$$

$$= 2.307 \times 10^{-28}\,\text{N} \cdot \text{m}^2$$

$$= 2.307 \times 10^{-28}\,\text{J} \cdot \text{m} \frac{1}{1.602 \times 10^{-19}\,\text{J/eV}} \cdot \frac{10^9\,\text{nm}}{\text{m}}$$

$$= 1.440\,\text{eV} \cdot \text{nm}$$

With this useful combination of constants it becomes very easy to calculate electrostatic potential energies. For two unit charges separated by 1.00 nm,

$$U = \frac{1}{4\pi\varepsilon_0} \frac{e^2}{r} = \frac{e^2}{4\pi\varepsilon_0} \frac{1}{r} = 1.440\,\text{eV} \cdot \text{nm} \frac{1}{1.00\,\text{nm}}$$

$$= 1.44\,\text{eV}$$

For calculations using typical nuclear sizes, the femtometer is a more convenient unit of distance, so

$$\frac{e^2}{4\pi\varepsilon_0} = 1.440\,\text{eV} \cdot \text{nm} \frac{1\,\text{m}}{10^9\,\text{nm}} \frac{10^{15}\,\text{fm}}{\text{m}} \frac{1\,\text{MeV}}{10^6\,\text{eV}}$$

$$= 1.440\,\text{MeV} \cdot \text{fm}$$

It is remarkable (and convenient to remember) that the quantity $e^2/4\pi\varepsilon_0$ has the same value of 1.440 whether we use typical atomic energies and sizes (eV·nm) or typical nuclear energies and sizes (MeV·fm).

Mass

The *kilogram* (kg) is the basic SI unit of mass, but it is too large a unit to be particularly useful for work in atomic and nuclear physics. Another difficulty, as we discuss in Chapter 2, is that we are frequently interested in using Einstein's equation $E = mc^2$ to express the mass of a particle in terms of its rest energy. Using c^2, a very large number, to do this conversion is inconvenient and can lead to mistakes. We avoid this problem by keeping the factor of c^2 with the mass units, and remembering that $m = E/c^2$. For example, you will often find in tables of the masses of the elementary particles that an electron has a mass of 0.511 MeV/c^2. Although this may look like an energy unit, it is really a measure of mass—the factor of c^2 does the conversion. What is important is that we can keep the factor of c^2 without bothering to put in its numerical value. You can check this result by converting MeV to joules and putting in the numerical value of c^2, and you should obtain 9.11×10^{-31} kg, the mass of the electron. (To avoid round-off errors, use four significant figures for the quantities in this calculation and then round off the final result to three significant figures.)

Atomic mass unit   Another mass unit that we find convenient to use is the *unified atomic mass unit,* u. This is most convenient in calculating atomic and nuclear binding energies. The atomic mass unit is *defined* so that the mass of an atom of the most abundant isotope of carbon (^{12}C) is exactly equal to 12 u. All other atomic masses are measured relative to this value. In Appendix B you will find a table of atomic masses.

Another useful mass unit based on ^{12}C is the *molar mass.* One *mole* of any substance is the quantity of that substance that contains the same number of elementary entities (atoms, molecules, or whatever) as 0.012 kg of ^{12}C. This number is called the Avogadro constant and has the value

$$N_A = 6.022 \times 10^{23} \text{ per mole}$$

That is, one mole of ^{12}C contains N_A atoms and has a molar mass of (exactly) 0.012 kg, one mole of nitrogen gas contains N_A molecules of N_2 and has a molar mass of 0.028 kg, and one mole of water contains N_A molecules of H_2O and has a molar mass of 0.018 kg.

The Speed of Light

One of the fundamental constants of nature is the speed of light, c, which you will be using quite frequently throughout your study of modern physics. Its value is

$$c = 2.998 \times 10^8 \text{ m/s}$$

It will often be convenient for us to measure speeds by comparing them with the speed of light; in Chapter 2 you will find many examples of problems in which speeds are given as a fraction of c, for example, $v = 0.6c$. Fortunately, many of the equations of special relativity involve not v but v/c, and thus it is often not necessary to convert $0.6c$ to a numerical speed in meters per second.

Planck's Constant

Another of the fundamental constants of nature is Planck's constant, h, with a value of

$$h = 6.626 \times 10^{-34} \text{ J·s}$$

Planck's constant obviously has dimensions of energy $\times$ time, but with a little bit of manipulation, you can show that it also has dimensions of linear momentum $\times$ displacement and also of angular momentum. Planck's constant will appear in many applications when we begin our study of quantum physics, and each different way of expressing its dimensions represents an important application.

We have already mentioned our desire to use electron-volts rather than joules for energy, and so it is useful to express Planck's constant using eV:

$$h = 4.136 \times 10^{-15} \text{ eV·s}$$

We also encounter the product hc in many calculations. In our units of convenience, you should be able to derive its value:

$$hc = 1240 \text{ eV·nm}$$
$$= 1240 \text{ MeV·fm}$$

It is interesting to note that hc and $e^2/4\pi\varepsilon_0$ have the same dimensions, and we have in fact calculated both quantities in the same units, eV·nm. The ratio of these two quantities is a pure number, independent of the system of units we have chosen, and we will learn that this ratio is of fundamental importance in atomic physics. The dimensionless constant α, called the *fine structure constant*, is actually 2π times the ratio:

Fine structure constant

$$\alpha = 2\pi \frac{e^2/4\pi\varepsilon_0}{hc}$$
$$= 2\pi \frac{1.440 \text{ eV·nm}}{1240 \text{ eV·nm}}$$
$$= 0.007297 \qquad (1.16)$$

This number is often expressed as $\alpha = 1/137.0$.

1.3   SIGNIFICANT FIGURES

The number of *significant figures* in a quantity tells us something about its precision—a quantity that is measured to eight significant figures is known more precisely than one that is measured to four significant figures. In calculations, the number of significant figures in the result is determined by the number of significant figures in the input quantities by applying two rules:

1. When adding or subtracting, the least significant digit of the number being added or subtracted determines the least significant digit of the sum or difference. *The number of significant figures does not matter.*
2. When multiplying or dividing, count the number of significant figures in the quantities being multiplied or divided. The number of significant figures in the product or quotient is determined by the factor with the smallest number of significant figures. *The location of the least significant digit does not matter.*

Here are some examples that illustrate the use of these rules.

EXAMPLE 1.1

Find the mass difference between a proton and a neutron. Express the result in u and in MeV/c^2.

SOLUTION

The proton and neutron masses are (to seven significant figures):

$$m_n = 1.008665\,u$$
$$m_p = 1.007276\,u$$

The difference is

$$1.008665\,u - 1.007276\,u = 0.001389\,u$$

In all cases, the last digit (indicated in boldface) is the least significant. In addition or subtraction, we do not count the number of significant figures. It is of *no importance* for the subtraction that each of the two masses has seven significant figures while the difference has only four.

The conversion factor from u to MeV/c^2 is

$$1\,u = 931.50\,MeV/c^2$$

The mass difference is therefore

$$0.001389\,u \times \frac{931.50\,MeV/c^2}{1\,u} = 1.294\,MeV/c^2$$

The mass difference in u has four significant figures, the conversion factor has five significant figures, and the product can therefore have only four significant figures, according to the second rule given above.

EXAMPLE 1.2

A proton and an electron can combine to form an atom of hydrogen. Find the total mass of a proton and an electron.

SOLUTION

The values of the masses are

$$m_p = 1.007276\,u$$
$$m_e = 5.4858 \times 10^{-4}\,u$$

The combined mass is

$$1.007276\,u + 0.00054858\,u = 1.007825\,u$$

Notice that the *position* of the least significant digit (the 6 in m_p) determines the *position* of the least significant digit in the sum.

EXAMPLE 1.3

The value of *hc* was given in the last section as 1240 eV·nm. Compute the value of *hc* to four significant figures and determine if the zero in the last digit is significant.

SOLUTION

The values given previously for *h* and *c* contain four significant figures. To avoid round-off errors, we use the values of the constants to five significant figures, and then round off the final result to four figures.

$$h = 6.6261 \times 10^{-34}\,\text{J·s}$$

$$c = 2.9979 \times 10^{8}\,\text{m/s}$$

$$1\,\text{eV} = 1.6022 \times 10^{-19}\,\text{J}$$

Therefore,

$$hc = \frac{(6.6261 \times 10^{-34}\,\text{J·s})(2.9979 \times 10^{8}\,\text{m/s})(10^{9}\,\text{nm/m})}{1.6022 \times 10^{-19}\,\text{J/eV}}$$

$$= 1239.8\,\text{eV·nm}$$

Rounding off to four figures, we obtain *hc* = 1240 eV·nm, so the zero *is* significant. (Of course, it is good practice not to leave zeros on the end of numbers such as this, where we don't know whether or not they are significant. It would be better to express this result as 1.240×10^{3} eV·nm.)

Proper attention to significant figures is a matter of habit, and the sooner you form this good habit the less trouble you will have in expressing the results of calculations. Simply because your calculator display shows eight digits does not make them all significant, and only the significant ones should be written down as the answer to a problem.

1.4 THEORY, EXPERIMENT, LAW

When you first began to study science, perhaps in your elementary or high school years, you may have learned about the "scientific method," which was supposed to be a sort of procedure by which scientific progress was achieved. The basic idea of the "scientific method" was that, on reflecting over some particular aspect of nature, the scientist would invent a *hypothesis* or *theory,* which would then be tested by *experiment* and if successful would be elevated to the status of *law.* This procedure is meant to emphasize the importance of doing experiments as a way of testing hypotheses and rejecting those that do not pass the tests. For example, the ancient Greeks had some rather definite ideas about the motion of objects, such as projectiles, in the Earth's gravity. Yet they tested none of these by experiment, so convinced were they that the power of logical deduction *alone* could be used to discover the hidden and mysterious laws of nature and that once logic had been applied to understanding a problem, no experiments were necessary. If theory and experiment were to disagree, they would argue, then there must be something wrong with the experiment! This dominance of analysis and faith was so pervasive that it was 2000 years before Galileo, using an inclined plane and a crude timer (equipment surely within the abilities of the early Greeks to construct), discovered the laws of motion, which were later organized and analyzed by Newton.

In the case of modern physics, none of the fundamental concepts is obvious from reason alone. Only by doing often difficult and necessarily precise experiments do we learn about these unexpected and fascinating effects. We therefore emphasize the experiments that have been done to study such modern physics topics as relativity and quantum physics. These experiments have been done to unprecedented levels of precision—of the order of one part in 10^6 or better—and it can certainly be concluded that modern physics has been tested far better in the twentieth century than classical physics was tested in all of the preceding centuries.

Nevertheless, there is a persistent and often perplexing problem associated with modern physics, one which stems directly from your previous acquaintance with the "scientific method." This concerns the use of the word "theory," as in "theory of relativity" or "quantum theory," or even "atomic theory" or "theory of evolution." There are two contrasting and conflicting definitions of the word "theory" in the dictionary:

1. A hypothesis or guess.
2. An organized body of facts or explanations.

The "scientific method" refers to the first kind of "theory," while when we speak of the "theory of relativity" we refer to the second kind. Yet there is often confusion between the two definitions, and therefore relativity and quantum physics are sometimes incorrectly regarded as merely hypotheses, on which evidence is still being gathered, in the hope of someday submitting that evidence to some sort of international (or intergalactic) tribunal, which in turn might elevate the "theory" into a "law." Thus the "theory of relativity" might someday become the "law of relativity," like the "law of gravity." *Nothing could be further from the truth!*

The theory of relativity and the quantum theory, like the atomic theory or the theory of evolution, are truly "organized bodies of facts and explanations" and *not* "hypotheses." There is no question of these "theories" becoming "laws"—the "facts" (experiments, observations) of relativity and quantum physics, like those of atomism or evolution, are accepted by virtually all scientists today; whether one calls them theories or laws is merely a question of semantics and has nothing to do with their scientific merits. Like all scientific principles, they will continue to develop and change as new discoveries are made; that is the essence of scientific progress, and it must be remembered that the search for ultimate truths or eternal laws is *not* a goal of science.

Still, there are two remaining questions that you may find particularly bothersome as you study modern physics. First is the "how" of these theories. The experimental evidence that forms the basis of modern physics is almost always *indirect*—no one has ever "seen" a quantum or a pi meson or even a nucleus, and no one has ever traveled at nearly the speed of light and "seen" the effects of relativity; similarly, no one has ever "seen" single atoms joining to form compounds or "seen" one species evolving into another. Yet the experimental evidence for all of these processes is so compelling that no one who approaches them in the spirit of free and open inquiry can doubt them.

The second vexing question concerns the "why" of these theories. Why does Nature behave according to Einstein's relativity, rather than according to Galileo's? Why do particles sometimes behave as waves, and waves sometimes as particles? Why do atoms join to form compounds? Why do higher forms of life evolve from lower forms? Although scientists can provide extremely precise answers to the "how" of these theories, they cannot provide the answers to the "why," not because their powers of observation or experimental abilities are limited, but rather because the questions are outside the realm of experimentation. These questions are of extreme importance, and as potential practitioners of pure or applied science you should be aware of them and spend some time thinking about them. If answers to these questions are to be found at all, they will be found not in the field of science, but in the fields of philosophy or theology. As you begin to study the facts of modern science, you should keep these additional questions in mind and perhaps seek your instructor's opinions concerning them. Although such speculations are an exciting intellectual endeavor in their own right, they will not be discussed in this text.

SUGGESTIONS FOR FURTHER READING

If you need to review some topics from classical physics, here are some introductory texts that provide the necessary background:

H. C. Ohanian, *Physics,* 2nd ed. (New York, Norton, 1989).

R. Resnick, D. Halliday, and K. S. Krane, *Physics,* 4th ed. (New York, Wiley, 1992).

R. A. Serway, *Physics for Scientists and Engineers,* 3rd ed. (Philadelphia, Saunders, 1990).

P. A. Tipler, *Physics for Scientists and Engineers,* 3rd ed. (New York, Worth, 1991).

H. D. Young, *University Physics,* 8th ed. (Reading, MA, Addison-Wesley, 1992).

Sometimes it helps your understanding to read about a subject from a slightly different perspective or written in a slightly different way. Here are some other modern physics books at about the same level as this text:

A. Beiser, *Concepts of Modern Physics,* 4th ed. (New York, McGraw-Hill, 1987).

H. C. Ohanian, *Modern Physics* (Englewood Cliffs, NJ, Prentice-Hall, 1987).

S. T. Thornton and A. Rex, *Modern Physics for Scientists and Engineers* (Philadelphia, Saunders, 1993).

P. A. Tipler, *Elementary Modern Physics* (New York, Worth, 1992).

Some more advanced (but still undergraduate level) texts:

J. J. Brehm and W. J. Mullin, *Introduction to the Structure of Matter* (New York, Wiley, 1989).

R. Eisberg and R. Resnick, *Quantum Physics of Atoms, Molecules, Solids, Nuclei, and Particles,* 2nd ed. (New York, Wiley, 1985).

R. B. Leighton, *Principles of Modern Physics* (New York, McGraw-Hill, 1959).

Some descriptive, historical, philosophical, and nonmathematical texts which give good background material and are great fun to read:

A. Baker, *Modern Physics and Anti-Physics* (Reading, Addison-Wesley, 1970).

F. Capra, *The Tao of Physics* (Berkeley, Shambhala Publications, 1975).

G. Gamow, *Thirty Years that Shook Physics* (New York, Doubleday, 1966).

R. March, *Physics for Poets* (New York, McGraw-Hill, 1978).

E. Segrè, *From X-Rays to Quarks: Modern Physicists and their Discoveries* (San Francisco, W. H. Freeman, 1980).

G. L. Trigg, *Landmark Experiments in Twentieth Century Physics* (New York, Crane, Russak, 1975).

F. A. Wolf, *Taking the Quantum Leap* (San Francisco, Harper & Row, 1981).

G. Zukav, *The Dancing Wu Li Masters, An Overview of the New Physics* (New York, Morrow, 1979).

Gamow, Segrè, and Trigg have contributed directly to the development of modern physics and their books are written from a perspective that only those who were part of that development can offer. The books by Capra and Zukav draw interesting parallels between modern physics (especially quantum theory and particle physics) and oriental philosophy.

QUESTIONS

1. Under what conditions can you apply the law of conservation of energy? Conservation of linear momentum? Conservation of angular momentum?

2. Which of the conserved quantities are scalars and which are vectors? Is there a difference in how we apply conservation laws for scalar and vector quantities?

3. What other conserved quantities (besides energy, linear momentum, and angular momentum) can you name?

4. What is the difference between potential and potential energy? Do they have different dimensions? Different units?

5. In Section 1.1 we defined the electric force between two charges and the magnetic field of a current. Use these quantities to define the electric field of a single charge and the magnetic force on a moving electric charge.

6. Other than from the ranges of wavelengths shown in Figure 1.4, can you think of a way to distinguish radio waves from infrared waves? Visible from infrared?

That is, could you design a radio that could be tuned to infrared waves? Could living beings "see" in the infrared region?

7. What is the difference between dimensions and units?

8. Can a quantity have units but no dimensions? Dimensions but no units?

9. In the past century a macroscopic standard for time (the rotation of the Earth) has been replaced with a microscopic standard (the frequency of light emitted by an atom). Similarly, a macroscopic standard for length (a metal bar of length one meter) has been replaced with a microscopic standard (based on the basic unit of time and the speed of light). Might it be possible to replace the current macroscopic standard of mass (the standard kilogram) with a microscopic standard? If so, how would we connect the macroscopic measurement with the microscopic standard?

10. Consider the equation $y = x^{-1}$ with $x = 0.98$. By strictly following the rule for significant figures, we obtain $y = 1.0$. What is wrong with expressing the result in this way? How can we modify the rules for significant figures to include calculations such as this one?

PROBLEMS

1. An atom of mass m moving in the x direction with speed v collides elastically with an atom of mass $3m$ at rest. After the collision the first atom moves in the y direction. Find the direction of motion of the second atom and the speeds of both atoms (in terms of v) after the collision.

2. An atom of mass m moves in the positive x direction with speed v. It collides with and sticks to an atom of mass $2m$ moving in the positive y direction with speed $2v/3$. Find the resultant speed and direction of motion of the combination, and find the kinetic energy lost in this inelastic collision.

3. An atom of beryllium ($m \cong 8.00$ u) splits into two atoms of helium ($m \cong 4.00$ u) with the release of 92.0 keV of energy. If the original beryllium atom is at rest, find the kinetic energy, speed, and momentum of the two helium atoms.

4. Express the following speeds as a fraction of the speed of light: (a) a typical automobile speed (100 km/h); (b) the speed of sound (330 m/s); (c) the escape velocity of a rocket from the Earth's surface (11 km/s); (d) the orbital speed of the Earth about the Sun (Earth-Sun distance = 1.5×10^8 km).

5. Show that Planck's constant h has dimensions of linear momentum $\times$ displacement.

6. Starting from Coulomb's law, show that $e^2/4\pi\varepsilon_0$ has dimensions of energy $\times$ distance.

7. (a) Starting from Newton's universal law of gravitation, show that Gm^2 has dimensions of energy $\times$ distance. (b) Evaluate Gm^2 in units of eV·nm using the proton mass. (c) Evaluate the ratio $Gm^2/(e^2/4\pi\varepsilon_0)$. Is this a pure number? What is its significance?

8. Use the Avogadro constant to find the mass in kilograms equivalent to one atomic mass unit (u).

9. In the Bohr model of the hydrogen atom, the electron moves in a circular orbit of radius $h^2\varepsilon_0/\pi m_e e^2$ with angular momentum $h/2\pi$. Find the speed of the electron as a fraction of the speed of light. Express your result in terms of the fine structure constant, Equation 1.16.

10. A gas cylinder contains argon atoms ($m = 40.0$ u). The temperature is increased from 293 K (20°C) to 373 K (100°C). (a) What is the change in the average kinetic energy per atom? (b) The container is resting on a table in the Earth's gravity. Find the change in the height of the container that produces the same change in the average energy per atom found in part (a).

11. The *Bohr magneton* μ_B is equal to $eh/4\pi m_e$. (a) Show that μ_B can be expressed in units of joules per tesla. (b) Show that μ_B has the same dimensions as a magnetic moment (current × area).

12. The orbital radius of an electron in an atom of hydrogen is $h^2\varepsilon_0/\pi m_e e^2$. (a) Show that this quantity has the dimension of length. (b) Compute its value to four significant figures. (*Hint*: Multiply numerator and denominator by common factors to allow use of some of the combinations of constants computed in this chapter.)

13. A fundamental unit frequently encountered is the Compton wavelength of the electron, $h/m_e c$. (a) Show that this quantity has the dimension of length. (b) Compute the value of $h/m_e c$ to four significant figures.

14. A helium nucleus consists of two protons and two neutrons, and has a mass of 4.001506 u. Find the mass difference between a helium nucleus and its constituents. Express your result in u and in MeV/c^2.

15. Suppose the beryllium atom of Problem 3 were not at rest, but instead moved in the positive x direction and had a kinetic energy of 40.0 keV. One of the helium atoms is found to be moving in the positive x direction. Find the direction of motion of the second helium, and find the velocity of each of the two helium atoms. Solve this problem in two different ways: (a) By direct application of conservation of momentum and energy. (b) By applying the results of Problem 3 to a frame of reference moving with the original beryllium atom and then switching to the reference frame in which the beryllium is moving.

16. Suppose the beryllium atom of Problem 3 moves in the positive x direction and has kinetic energy 60.0 keV. One helium atom is found to move at an angle of 30° with respect to the x axis. Find the direction of motion of the second helium atom and find the velocity of each helium atom. Work this problem in two ways as you did the previous problem. (*Hint*: Consider one helium to be emitted with velocity components v_x and v_y in the beryllium rest frame. What is the relationship between v_x and v_y? How do v_x and v_y change when we move in the x direction at speed v?)

17. It is sometimes convenient to represent the behavior of a particle on a graph showing its position on one axis and its linear momentum on another (Figure 1.5 is an example for a particle confined to move in one dimension). The coordinate space defined by this system is called *phase space*. Draw the phase space graphs representing: (a) a particle at rest; (b) a particle moving at constant positive speed; (c) a particle originally at rest at $x = 0$ moving with a constant positive acceleration; (d) a particle in simple harmonic motion about a point x_0 on the positive x axis; (e) a ball thrown upward to a height $x = h$.

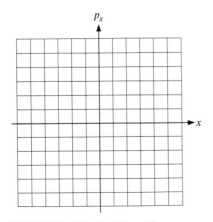

FIGURE 1.5   Problem 17.

THE SPECIAL THEORY OF RELATIVITY

Einstein's special theory of relativity and Planck's quantum theory burst forth on the physics scene almost simultaneously during the first decade of the twentieth century. Both theories caused profound changes in the way we view our universe at its most fundamental level.

In this chapter we study the special theory of relativity.* This theory has a completely undeserved reputation as being so exotic that few people can understand it. On the contrary, special relativity is simply a system of kinematics and dynamics, based on a set of postulates that is different from those of classical physics. The resulting formalism is not much more complicated than Newton's laws, but it does lead to several predictions that seem to go against our common sense. Even so, the special theory of relativity has been carefully and thoroughly tested by experiment and found to be correct in all its predictions.

We first review the classical relativity of Galileo and Newton, and then we show why Einstein proposed to replace it. We then discuss the mathematical aspects of special relativity, the predictions of the theory, and finally the experimental tests.

2.1 CLASSICAL RELATIVITY

A "theory of relativity" is in effect a way for observers in different frames of reference to compare the results of their observations. For example, consider an observer in a car parked by a highway near a large rock. To this observer, the rock is at rest. Another observer, who is moving along the highway in a car, sees the rock rush past as the car drives by. To this observer, the rock appears to be moving. A theory of relativity provides the conceptual framework and mathematical tools that enable the two observers to transform a statement such as "rock is at rest" in one frame of reference to the statement "rock is in motion" in another frame of reference. More generally, relativity gives a means for expressing the laws of physics in different frames of reference.

The mathematical basis for comparing the two descriptions is called a *transformation*. Figure 2.1 shows an abstract representation of the situation. Two observers O and O' are each at rest in their own frames of reference but move relative to one another with constant velocity **u**. (O and O' refer both to the observers and their reference frames or coordinate systems.) They observe the same *event*, which happens at a particular point in space and a particular time, such as a collision between two particles. According to O, the space and time coordinates of the event are x, y, z, t, while according to O' the coordinates of the *same event* are x', y', z', t'. The two observers use calibrated meter sticks and synchronized clocks, so any differences between the coordinates of the two events are due to their

---

* What is "special" about special relativity is that it applies to events described by observers in inertial reference frames. In Chapter 15 we discuss the *general* theory of relativity, which applies to all observers, including noninertial ones in accelerated reference frames.

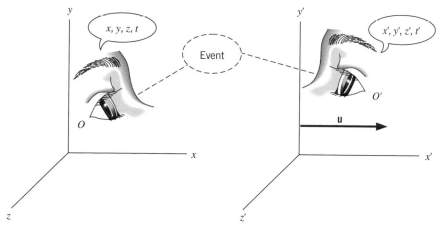

FIGURE 2.1   Two observers O and O' observe the same event. O' moves relative to O with a constant velocity **u**.

different frames of reference and not to the measuring process. We simplify the discussion by assuming that the relative velocity **u** always lies along the common xx' direction, as shown in Figure 2.1, and we let **u** represent the velocity of O' as measured by O (and thus O' would measure velocity $-$**u** for O).

In this discussion we make a particular choice of the kind of reference frames inhabited by O and O'. We assume that each observer has the capacity to test Newton's laws and finds them to hold in that frame of reference. For example, each observer finds that an object at rest or moving with a constant velocity remains in that state unless acted upon by an external force (Newton's first law, the law of inertia). Such frames of reference are called *inertial frames*. An observer in interstellar space floating in a nonrotating rocket with the engines off would be in an inertial frame of reference. An observer at rest on the surface of the Earth is *not* in an inertial frame, because the Earth is rotating about its axis and orbiting about the Sun; however, the accelerations associated with those motions are so small that we can usually regard our reference frame as approximately inertial. (The noninertial reference frame at the Earth's surface does produce important and often spectacular effects, such as the circulation of air around centers of high or low pressure.) An observer in an accelerating car, a rotating merry-go-round, or a descending roller coaster is *not* in an inertial frame of reference!

Inertial frames

We now derive the classical or *Galilean* transformation that relates the coordinates x, y, z, t to x', y', z', t'. We assume as a postulate of classical physics that $t = t'$, that is, time is the same for all observers. We also assume for simplicity that the coordinate systems are chosen so that their origins coincide at $t = 0$. Consider an object in O' at the coordinates x', y', z' (Figure 2.2). According to O, the y and z coordinates are the same as those in O'. Along the x direction, O would observe the object at $x = x' + ut$. We therefore have the *Galilean coordinate transformation*

Galilean transformation

$$x' = x - ut \qquad y' = y \qquad z' = z \qquad (2.1)$$

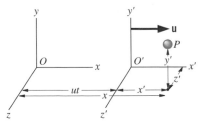

FIGURE 2.2 An object or event at point P is at coordinates x', y', z' with respect to O'. The x coordinate measured by O is $x = x' + ut$. The y and z coordinates in O are the same as those in O'.

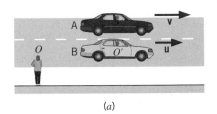

(a)

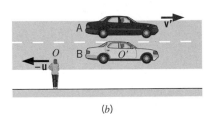

(b)

FIGURE 2.3 Example 2.1. (a) As observed by O at rest on the ground. (b) As observed by O' in car B.

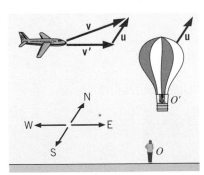

FIGURE 2.4 Example 2.2. As observed by O at rest on the ground, the balloon drifts north with the wind, while the plane flies north of east.

To find the velocities of the object as observed by O and O', we take the derivatives of these expressions with respect to t' on the left and with respect to t on the right (which we can do because we have assumed $t' = t$). This gives the *Galilean velocity transformation*

$$v'_x = v_x - u \qquad v'_y = v_y \qquad v'_z = v_z \qquad (2.2)$$

In a similar fashion, we can take the derivatives of Equation 2.2 with respect to time and obtain

$$a'_x = a_x \qquad a'_y = a_y \qquad a'_z = a_z \qquad (2.3)$$

Equation 2.3 shows again that Newton's laws hold for both observers. As long as u is constant ($du/dt = 0$), the observers measure identical accelerations and agree on the results of applying $\mathbf{F} = m\mathbf{a}$.

EXAMPLE 2.1

Two cars are traveling at constant speed along a road in the same direction. Car A moves at 60 km/h and car B moves at 40 km/h, each measured relative to an observer on the ground (Figure 2.3a). What is the speed of car A relative to car B?

SOLUTION

Let O be the observer on the ground, who observes car A to move at $v_x = 60$ km/h. Assume O' to be moving with car B at $u = 40$ km/h. Then

$$v'_x = v_x - u = 60 \, \text{km/h} - 40 \, \text{km/h} = 20 \, \text{km/h}$$

Figure 2.3b shows the situation as observed by O'.

EXAMPLE 2.2

An airplane is flying due east relative to still air at a speed of 320 km/h. There is a 65 km/h wind blowing toward the north, as measured by an observer on the ground. What is the velocity of the plane measured by the ground observer?

SOLUTION

Let O be the observer on the ground, and let O' be an observer who is moving with the wind, for example a balloonist (Figure 2.4). Then $u = 65$ km/h, and the xx' direction must be to the north. In this case we know the velocity with respect to O'; taking the y direction to the east, we have $v'_x = 0$ and $v'_y = 320$ km/h. Using Equation 2.2 we obtain

$$v_x = v'_x + u = 0 + 65 \, \text{km/h} = 65 \, \text{km/h}$$

$$v_y = v'_y = 320 \, \text{km/h}$$

Relative to the ground, the plane flies in a direction determined by $\phi = \tan^{-1}$ (65 km/h)/(320 km/h) = 11.5°, or 11.5° north of east.

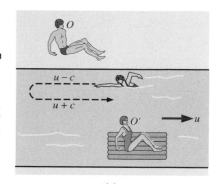

(a)

EXAMPLE 2.3

A swimmer capable of swimming at a speed c in still water is swimming in a stream in which the current is u (which we assume to be less than c). Suppose the swimmer swims upstream a distance L and then returns downstream to the starting point. Find the time necessary to make the round trip, and compare it with the time to swim across the stream a distance L and return.

SOLUTION

Let the frame of reference of O be the ground and the frame of reference of O' be the water, moving at speed u (Figure 2.5a). The swimmer always moves at speed c relative to the water, and thus $v'_x = -c$ for the upstream swim. (Remember u always defines the *positive x* direction.) According to Equation 2.2, $v'_x = v_x - u$, so $v_x = v'_x + u = u - c$. (As expected, the velocity relative to the ground has magnitude smaller than c; it is also *negative*, since the swimmer is swimming in the negative x direction, so $|v_x| = c - u$.) Therefore, $t_{\text{up}} = L/(c - u)$. For the downstream swim, $v'_x = c$, so $v_x = u + c$, $t_{\text{down}} = L/(c + u)$, and the total time is

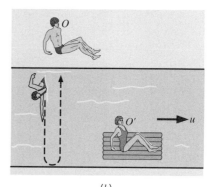

(b)

FIGURE 2.5 The motion of a swimmer as seen by observer O at rest on the bank of the stream. Observer O' moves with the stream at speed u.

$$t = \frac{L}{c + u} + \frac{L}{c - u} = \frac{L(c - u) + L(c + u)}{c^2 - u^2} = \frac{2Lc}{c^2 - u^2} = \frac{2L}{c} \frac{1}{1 - u^2/c^2} \quad (2.4)$$

In order to swim directly across the stream, the swimmer's efforts must be directed somewhat upstream in order to counter the effect of the current (Figure 2.5b). That is, in the frame of reference of O we would like to have $v_x = 0$, which requires $v'_x = -u$ according to Equation 2.2. Since the speed relative to the water is always c, $\sqrt{v'^2_x + v'^2_y} = c$; thus $v'_y = \sqrt{c^2 - v'^2_x} = \sqrt{c^2 - u^2}$, and the round-trip time is

$$t = 2t_{\text{across}} = \frac{2L}{\sqrt{c^2 - u^2}} = \frac{2L}{c} \frac{1}{\sqrt{1 - u^2/c^2}} \quad (2.5)$$

Notice the difference *in form* between this result and the result for the upstream-downstream swim, Equation 2.4.

2.2 THE MICHELSON-MORLEY EXPERIMENT

We have seen how Newton's laws remain valid with respect to a Galilean transformation from one inertial frame to another. It therefore is logical to ask whether other laws of physics are similarly invariant. Consider, for example, electromagnetism. The speed of light in vacuum is defined to have the value $c = 299{,}792{,}458$ m/s. Under the Galilean transformation, a light beam moving relative to observer O' in the x' direction at speed c would have a speed of $c + u$ relative to O. Direct measurements of this relative speed have become possible in recent years (as we discuss later in this chapter), but in the nineteenth century it was necessary to devise a more indirect measurement.

Suppose the swimmer in Example 2.3 is replaced by a light beam. Observer O' is in a frame of reference in which the speed of light is c. What is the speed of light as measured by observer O, who is in motion relative to O'? If the Galilean transformation is correct, we should expect to see a time difference between the upstream-downstream and cross-stream times, as in Example 2.3.

Physicists in the nineteenth century postulated just such a situation—a preferred frame of reference in which the speed of light has the precise value of c and other frames in relative motion in which the speed of light would differ, according to the Galilean transformation. The preferred frame, like that of observer O' in Example 2.3, is one that is at rest with respect to the medium in which light propagates at c (like the water of that example). What is the medium of propagation for light waves? It was inconceivable to physicists of the nineteenth century that a wave disturbance could propagate without a medium (consider mechanical waves such as sound or seismic waves, for example, which propagate due to mechanical forces in the medium). They postulated the existence of an invisible, massless medium, called the *ether,* which filled all space, was undetectable by any mechanical means, and existed solely for the propagation of light waves. It seemed reasonable then to obtain evidence for the ether by measuring the velocity of the Earth moving through the ether. This could be done in the geometry of Figure 2.5 by measuring the difference between the upstream-downstream and cross-stream times for a light wave. The calculation based on Galilean relativity would then give the relative velocity **u** between O (in the Earth's frame of reference) and the ether.

The first detailed and precise search for the preferred frame was performed in 1887 by the American physicist Albert A. Michelson and his associate E. W. Morley. Their apparatus consisted of a specially designed Michelson interferometer, illustrated in Figure 2.6. A monochromatic beam of light is split in two; the two beams travel different paths and are then recombined. Any phase difference between the combining beams causes bright and dark bands or "fringes" to appear, corresponding, respectively, to constructive and destructive interference, as shown in Figure 2.7.

There are two contributions to the phase difference between the beams. The first contribution comes from the path difference $AB - AC$; one of the beams may travel a longer distance. The second contribution, which would still be present even if the path lengths were equal, comes from the time difference between the upstream-downstream and cross-stream paths (as in Example 2.3) and indicates the motion of the Earth through the ether. Michelson and Morley used a clever method to isolate this second contribution—they rotated the entire apparatus by 90°! The rotation doesn't change the first contribution to the phase difference (because the

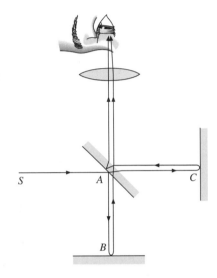

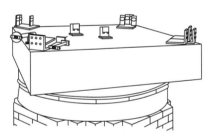

FIGURE 2.6 A schematic diagram of the Michelson interferometer. A beam of light from the source S is split in two by the half-silvered mirror at point A; one part is reflected by the mirror at B and the other is reflected at C. The beams are then recombined for observation of the interference. The bottom half of the figure shows a sketch of Michelson's apparatus. To improve sensitivity, mirrors were set up to make the beams travel each leg of the apparatus eight times, rather than just twice. To reduce vibrations from the surroundings, the interferometer was mounted on a stone slab about $1\frac{1}{2}$ m square floating in a pool of mercury.

lengths *AB* and *AC* don't change), but the second contribution changes sign, because what was an upstream-downstream path before the rotation becomes a cross-stream path after the rotation. As the apparatus is rotated through 90°, the fringes should change from bright to dark and back again as the phase difference changes. Each change from bright to dark represents a phase change of 180° (a half cycle), which corresponds to a time difference of a half period (about 10^{-15} s for visible light). Counting the number of fringe changes thus gives a measure of the time difference between the paths, which in turn gives the relative velocity *u*. (See Problem 1.)

When Michelson and Morley performed their experiment, there was no observable change in the fringe pattern—they deduced a shift of less than 0.01 fringe, corresponding to a speed of the Earth through the ether of at most 5 km/s. As a last resort, they reasoned that perhaps the orbital motion of the Earth just happened to cancel out the overall motion through the ether. If this were true, six months later (when the Earth would be moving in its orbit in the opposite direction) the cancellation should not occur. When they repeated the experiment six months later, they again obtained a null result. In no experiment were Michelson and Morley able to detect the motion of the Earth through the ether.

In summary, we have seen that there is a direct chain of reasoning that leads from Galileo's principle of inertia, through Newton's laws with their implicit assumptions about space and time, ending with the failure of the Michelson-Morley experiment to observe the motion of the Earth relative to the ether. Although several explanations were offered for the unobservability of the ether and the corresponding failure of the upstream-downstream and cross-stream velocities to add in the expected way, the most novel, revolutionary, and ultimately successful explanation is given by Einstein's special theory of relativity, which requires a serious readjustment of our traditional concepts of space and time, and therefore alters some of the very foundations of physics.

Albert A. Michelson (1852–1931, United States). He spent 50 years doing increasingly precise experiments with light, for which he became the first U. S. citizen to win the Nobel prize (1907).

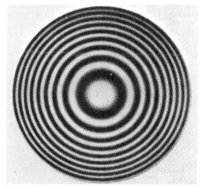

FIGURE 2.7 Interference fringes as observed with the Michelson interferometer of Figure 2.6. When the path length *ACA* changes by one-half wavelength relative to *ABA*, all light areas turn dark and all dark areas turn light.

2.3 EINSTEIN'S POSTULATES

The *special theory of relativity* is based on two postulates proposed by Albert Einstein in 1905:

The principle of relativity: *The laws of physics are the same in all inertial reference frames.*

The principle of the constancy of the speed of light: *The speed of light in free space has the same value c in all inertial reference frames.*

The first postulate declares that the laws of physics are absolute, universal, and the same for all inertial observers. Laws that hold for one inertial observer cannot be violated for *any* inertial observer.

The second postulate is more difficult to accept because it seems to go against our "common sense," which is based on the Galilean kinematics

Albert Einstein (1879–1955, Germany-United States). A gentle philosopher and pacifist, he was the intellectual guru to two generations of theoretical physicists and left his imprint on nearly every field of modern physics.

we observe in everyday experiences. Consider three observers *A*, *B*, and *C*. *B* is at rest, while *A* and *C* move away from *B* in opposite directions each at a speed of $c/4$. *B* fires a light beam in the direction of *A*. According to the Galilean transformation, if *B* measures a speed of *c* for the light beam, then *A* measures a speed of $c - c/4 = 3c/4$, while *C* measures a speed of $c + c/4 = 5c/4$. Einstein's second postulate, on the other hand, requires all three observers to measure the same speed of *c* for the light beam! This postulate immediately explains the failure of the Michelson-Morley experiment—the upstream-downstream and cross-stream speeds are identical (both are equal to *c*), so there is no phase difference between the two beams.

The two postulates also allow us to dispose of the ether hypothesis. The first postulate does not permit a preferred frame of reference (all inertial frames are equivalent), and the second postulate does not permit only a single frame of reference in which light moves at speed *c*, because light moves at speed *c* in *all* frames. The special theory of relativity eliminates the Galilean-Newtonian concept of absolute space and time, and considers both as being relative to the observer.

2.4 CONSEQUENCES OF EINSTEIN'S POSTULATES

Among their many consequences, Einstein's postulates require a new consideration of the fundamental nature of time and space. In this section we discuss how the postulates affect measurements of time and length intervals by observers in different frames of reference.

The Relativity of Time

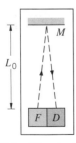

FIGURE 2.8   The clock ticks at intervals Δt_0 determined by the time for a light flash to travel the distance $2L_0$ from the flashing bulb *F* to the mirror *M* and back to the detector *D*. (The lateral distance between *F* and *D* is assumed to be negligible in comparison with L_0.)

To demonstrate the relativity of time, we use the timing device illustrated in Figure 2.8. It consists of a flashing light bulb *F* attached to a detector *D* and separated by a distance L_0 from a mirror *M*. A flash of light from the bulb is reflected by the mirror, and when the light returns to *D* the clock ticks and triggers another flash. The time interval between ticks is the distance $2L_0$ divided by the speed *c* (neglecting the lateral motion of the light beam between *F* and *D*):

$$\Delta t_0 = 2L_0/c \qquad (2.6)$$

This is the time interval that is measured when the clock is at rest with respect to the observer.

We consider two observers: *O* is at rest on the ground, and *O'* moves with speed *u*. Each observer carries a timing device. Figure 2.9 shows a sequence of events that *O* observes for the clock carried by *O'*. According

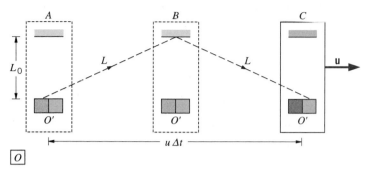

FIGURE 2.9 In the frame of reference of O, the clock carried by O' moves with speed u. The dashed line, of length $2L$, shows the path of the light beam according to O.

to O, the flash is emitted at A, reflected at B, and detected at C. In this interval Δt, O observes the clock to move forward a distance of $u\,\Delta t$ from the point at which the flash was emitted, and O concludes that the light beam travels a distance $2L$, where $L = \sqrt{L_0^2 + (u\,\Delta t/2)^2}$, as shown in Figure 2.9. Since O observes the light beam to travel at speed c, the time interval measured by O is

$$\Delta t = \frac{2L}{c} = \frac{2\sqrt{L_0^2 + (u\,\Delta t/2)^2}}{c} \qquad (2.7)$$

Substituting for L_0 from Equation 2.6 and solving Equation 2.7 for Δt, we obtain

$$\Delta t = \frac{\Delta t_0}{\sqrt{1 - u^2/c^2}} \qquad (2.8)$$

Equation 2.8 summarizes the effect known as *time dilation*. According to Equation 2.8, observer O measures a longer time interval than O' measures. This is a general result within special relativity, which we can summarize as follows. Consider an occurrence that has a duration Δt_0. An observer O' fixed with respect to that occurrence (i.e., its beginning and end take place at the same point in space, according to O') measures the interval Δt_0, which is known as the *proper time*. An observer O moving with respect to O' measures a longer time interval Δt for the same occurrence. The interval Δt is always longer than Δt_0, no matter what the magnitude or direction of $\mathbf{u}$.

Time dilation

This is a real effect that applies not only to clocks based on light beams but to time itself; all clocks run more slowly according to an observer in relative motion, biological clocks included. Even the growth, aging, and decay of living systems are slowed by the time dilation effect.

The time dilation effect has been verified experimentally with decaying elementary particles as well as with precise atomic clocks carried aboard aircraft. Some experimental tests are discussed in the last section of this chapter.

EXAMPLE 2.4

Muons are elementary particles with a (proper) lifetime of 2.2 μs. They are produced with very high speeds in the upper atmosphere when cosmic rays (high-energy particles from space) collide with air molecules. Take the height L_0 of the atmosphere to be 100 km in the reference frame of the Earth, and find the minimum speed that enables the muons to survive the journey to the surface of the Earth.

SOLUTION

The birth and decay of the muon can be considered as the "ticks" of a clock. In the frame of reference of the Earth (observer O) this clock is moving, and therefore its ticks are slowed by the time dilation effect. If the muon is moving at a speed that is close to c, the time necessary for it to travel from the top of the atmosphere to the surface of the Earth is

$$\Delta t = \frac{L_0}{c} = \frac{100\,\text{km}}{3.00 \times 10^8\,\text{m/s}} = 333\,\mu\text{s}$$

If the muon is to be observed at the surface of the Earth, it must live for at least 333 μs in the Earth's frame of reference. In the muon's frame of reference, the interval between its birth and decay is a proper time interval of 2.2 μs. The time intervals are related by Equation 2.8:

$$333\,\mu\text{s} = \frac{2.2\,\mu\text{s}}{\sqrt{1 - u^2/c^2}}$$

Solving, we find

$$u = 0.999978c$$

If it were not for the time dilation effect, muons would not survive to reach the Earth's surface. The observation of these muons is a direct verification of the time dilation effect of special relativity.

The Relativity of Length

For this discussion, the moving timing device of O' is turned sideways, so that the light travels parallel to the direction of motion of O'. Figure 2.10 shows the sequence of events that O observes for the moving clock. According to O, the length of the clock is L; as we shall see, this length is different from the length L_0 measured by O', relative to whom the clock is at rest.

The flash of light is emitted at A and reaches the mirror (position B) at time Δt_1 later. In this time interval, the light travels a distance $c\,\Delta t_1$, equal to the length L of the clock plus the additional distance $u\,\Delta t_1$ that the mirror moves forward in this interval. That is,

$$c\,\Delta t_1 = L + u\,\Delta t_1 \tag{2.9}$$

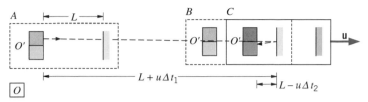

FIGURE 2.10 Here the clock carried by O' emits its light flash in the direction of motion.

The flash of light travels from the mirror to the detector in a time Δt_2 and covers a distance of $c\,\Delta t_2$, equal to the length L of the clock less the distance $u\,\Delta t_2$ that the clock moves forward in this interval:

$$c\,\Delta t_2 = L - u\,\Delta t_2 \tag{2.10}$$

Solving Equations 2.9 and 2.10 for Δt_1 and Δt_2, and adding to find the total time interval, we obtain

$$\Delta t = \Delta t_1 + \Delta t_2 = \frac{L}{c - u} + \frac{L}{c + u} = \frac{2L}{c}\frac{1}{1 - u^2/c^2} \tag{2.11}$$

From Equation 2.8,

$$\Delta t = \frac{\Delta t_0}{\sqrt{1 - u^2/c^2}} = \frac{2L_0}{c}\frac{1}{\sqrt{1 - u^2/c^2}} \tag{2.12}$$

Setting Equations 2.11 and 2.12 equal to one another and solving, we obtain

$$L = L_0\sqrt{1 - u^2/c^2} \tag{2.13}$$ *Length contraction*

Equation 2.13 summarizes the effect known as *length contraction*. Observer O', who is at rest with respect to the object, measures the *rest length* L_0 (also known as the *proper length*, in analogy with the proper time). All observers in motion relative to O' measure a shorter length, but only along the direction of motion; length measurements transverse to the direction of motion are unaffected (Figure 2.11).

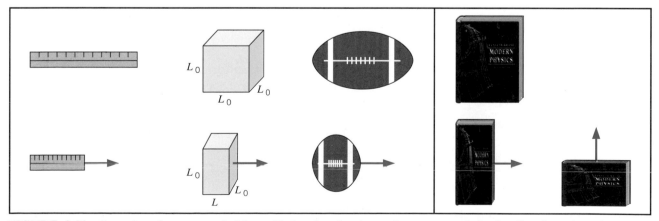

FIGURE 2.11 Some length-contracted objects. Notice that the shortening occurs only in the direction of motion.

For ordinary speeds ($u \ll c$), the effects of length contraction are too small to be observed. For example, a rocket of length 100 m traveling at the escape speed from Earth (11.2 km/s) would appear to an observer on Earth to contract only by about two atomic diameters!

Length contraction suggests that objects in motion are measured to have a shorter length than they do at rest. The objects do not actually shrink; there is merely a difference in the length measured by different observers. For example, to observers on Earth a high-speed rocket ship would appear to be contracted along its direction of motion (Figure 2.12a), but to an observer on the ship it is the passing Earth that appears to be contracted (Figure 2.12b).

These representations of length-contracted objects are somewhat idealized. The actual appearance of a rapidly moving object is determined by the time at which light leaves the various parts of the object and enters the eye or the camera. For more information on this subject, see the references at the end of this chapter.

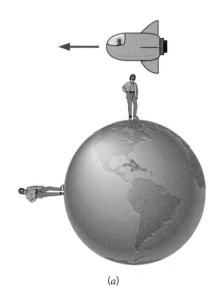

(a)

(b)

FIGURE 2.12 (a) The Earth views the passing contracted rocket. (b) From the rocket's frame of reference, the Earth appears contracted.

EXAMPLE 2.5

In the reference frame of the muon of Example 2.4, what is the apparent thickness of the Earth's atmosphere?

SOLUTION

In the reference frame of the muon, the Earth is rushing toward it at a speed of $u = 0.999978c$, as we found in Example 2.4. To an observer on the Earth, the height of the atmosphere is its rest length L_0 of 100 km. To the muon, the moving Earth has an atmosphere of height given by Equation 2.13:

$$L = L_0\sqrt{1 - u^2/c^2} = (100 \text{ km})\sqrt{1 - (0.999978)^2} = 0.66 \text{ km} = 660 \text{ m}$$

Note that what appears as a *time dilation* in one frame of reference can be regarded as a *length contraction* in another frame of reference.

EXAMPLE 2.6

An observer O is standing on a platform of length $D_0 = 65$ m on a space station. A rocket passes at a relative speed of $0.80c$ moving parallel to the edge of the platform. The observer O notes that the front and back of the rocket simultaneously line up with the ends of the platform at a particular instant (Figure 2.13a). (a) According to O, what is the time necessary for the rocket to pass a particular point on the platform? (b) What is the rest length L_0 of the rocket? (c) According to an observer O' on the rocket, what is the length D of the platform? (d) According to O', how long does it take for observer O to pass the entire length of the rocket? (e) According to O, the ends of the rocket simultaneously line up with the ends of the platform. Are these events simultaneous to O'?

SOLUTION

(a) According to O, the length L of the rocket matches the length D_0 of the platform. The time for the rocket to pass a particular point is measured by O to be

$$\Delta t_0 = \frac{L}{0.80c} = \frac{65 \text{ m}}{2.40 \times 10^8 \text{ m/s}} = 0.27 \; \mu s$$

This is a proper time interval, because O measures the interval between two events that occur at the same point in the frame of reference of O (the front of the rocket passes a point, and then the back of the rocket passes the same point).

(b) O measures the contracted length L of the rocket. We can find its rest length L_0 using Equation 2.13:

$$L_0 = \frac{L}{\sqrt{1 - u^2/c^2}} = \frac{65 \text{ m}}{\sqrt{1 - (0.80)^2}} = 108 \text{ m}$$

(c) According to O the platform is at rest, so 65 m is its rest length D_0. According to O', the contracted length of the platform is therefore

$$D = D_0\sqrt{1 - u^2/c^2} = (65 \text{ m})\sqrt{1 - (0.80)^2} = 39 \text{ m}$$

(d) For O to pass the entire length of the rocket, O' concludes that O must move a distance equal to its rest length, or 108 m. The time needed to do this is

$$\Delta t' = \frac{108 \text{ m}}{0.80c} = 0.45 \; \mu s$$

Note that this is *not* a proper time interval for O', who determines this time interval using one clock at the front of the rocket to measure the time at which O passes the front of the rocket, and another clock on the rear of the rocket to measure the time at which O passes the rear of the rocket. The two events therefore occur at different points in O' and so cannot be separated by a proper time in O'. The corresponding time interval measured by O for the same two events, which we calculated in part (a), *is* a proper time interval for O, because the two events *do* occur at the same point in O. The time intervals measured by O and O' should be related by the time dilation formula, as you should verify.

(e) According to O', the rocket has a rest length of $L_0 = 108$ m and the platform has a contracted length of $D = 39$ m. There is thus no way that O' could observe the two ends of both to align simultaneously. The sequence of events according to O' is illustrated in Figures 2.13b and 2.13c. The time interval $\Delta t'$ in O' between the two events that are simultaneous in O can be calculated by noting that, according to O', the time interval between the situations shown in Figures 2.13b and 2.13c must be that necessary for the platform to move a distance of 108 m − 39 m = 69 m, which takes a time

$$\Delta t' = \frac{69 \text{ m}}{0.80c} = 0.29 \; \mu s$$

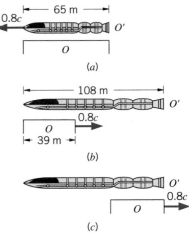

FIGURE 2.13 Example 2.6. (a) From the reference frame of O at rest on the platform, the passing rocket lines up simultaneously with the front and back of the platform. (b, c) From the reference frame O' in the rocket, the passing platform lines up first with the front of the rocket and later with the rear. Note the differing effects of length contraction in the two reference frames.

This result illustrates the relativity of simultaneity: two events that are simultaneous to O (the lining up of the two ends of the rocket with the two ends of the platform) *cannot* be simultaneous to O'.

Relativistic Velocity Addition

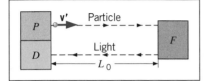

FIGURE 2.14   In this timing device, a particle is emitted by P at a speed v'. When the particle reaches F, it triggers the emission of a flash of light that travels to the detector D.

The timing device is now modified as shown in Figure 2.14. A source P emits particles that travel at speed v', according to an observer O' at rest with respect to the device. The flashing bulb F is triggered to flash when a particle reaches it. The flash of light makes the return trip to the detector D, and the clock ticks. The time interval Δt_0 between ticks measured by O' is composed of two parts: one for the particle to travel the distance L_0 at speed v' and another for the light to travel the same distance at speed c:

$$\Delta t_0 = L_0/v' + L_0/c \tag{2.14}$$

According to observer O, relative to whom O' moves at speed u, the sequence of events is similar to that shown in Figure 2.10. The emitted particle, which travels at speed v according to O, reaches F in a time interval Δt_1 after traveling the distance $v\,\Delta t_1$ equal to the (contracted) length L plus the additional distance $u\,\Delta t_1$ moved by the clock in that interval:

$$v\,\Delta t_1 = L + u\,\Delta t_1 \tag{2.15}$$

In the interval Δt_2, the light beam travels a distance $c\,\Delta t_2$ equal to the length L less the distance $u\,\Delta t_2$ moved by the clock in that interval:

$$c\,\Delta t_2 = L - u\,\Delta t_2 \tag{2.16}$$

We now solve Equations 2.15 and 2.16 for Δt_1 and Δt_2, add to find the total interval Δt between ticks according to O, use the time dilation formula, Equation 2.8, to relate this result to Δt_0 from Equation 2.14, and finally use the length contraction formula, Equation 2.13, to relate L to L_0. After doing the algebra, we find the result

Relativistic velocity addition law

$$v = \frac{v' + u}{1 + v'u/c^2} \tag{2.17}$$

Equation 2.17 is the *relativistic velocity addition law* for velocity components that are in the direction of u. Later in this chapter we use a different method to derive the corresponding results for motion in other directions.

We can also regard Equation 2.17 as a velocity transformation, enabling us to convert a velocity v' measured by O' to a velocity v measured by O. The corresponding classical law was given by Equation 2.2: $v = v' + u$. The difference between the classical and relativistic results is the denominator of Equation 2.17, which reduces to 1 in cases when the speeds are small compared with c. Example 2.7 shows how this factor prevents the measured speeds from exceeding c.

Equation 2.17 gives an important result when O' observes a light beam. For $v' = c$,

$$v = \frac{c + u}{1 + cu/c^2} = c \qquad (2.18)$$

That is, when $v' = c$, then $v = c$, *independent of the value of u.* All observers measure the same speed c for light, exactly as required by Einstein's second postulate.

EXAMPLE 2.7

A spaceship moving away from the Earth at a speed of $0.80c$ fires a missile parallel to its direction of motion (Figure 2.15). The missile moves at a speed of $0.60c$ relative to the ship. What is the speed of the missile as measured by an observer on the Earth?

SOLUTION

Here O' is on the ship and O is on Earth; O' moves with a speed of $u = 0.80c$ relative to O. The missile moves at speed $v' = 0.60c$ relative to O', and we seek its speed v relative to O. Using Equation 2.17, we obtain

$$v = \frac{v' + u}{1 + v'u/c^2} = \frac{0.60c + 0.80c}{1 + (0.60c)(0.80c)/c^2}$$

$$= \frac{1.40c}{1.48} = 0.95c$$

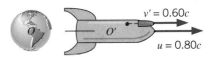

FIGURE 2.15 Example 2.7. A spaceship moves away from Earth at a speed of $0.80c$. An observer O' on the spaceship fires a missile and measures its speed to be $0.60c$ relative to the ship.

According to classical kinematics (the numerator of Equation 2.17), an observer on the Earth would see the missile moving at $0.60c + 0.80c = 1.40c$, thereby exceeding the maximum relative speed of c permitted by relativity. You can see how Equation 2.17 brings about this speed limit. Even if v' were $0.9999 \cdots c$ and u were $0.9999 \cdots c$, the relative speed v measured by O would remain less than c.

The Relativistic Doppler Effect

In the classical Doppler effect for sound waves, an observer moving relative to a source of waves (sound, for example) detects a different frequency than that emitted by the source. The frequency ν' heard by the observer O is related to the frequency ν emitted by the source S according to

$$\nu' = \nu \frac{v \pm v_O}{v + v_S} \qquad (2.19)$$

where v is the speed of the waves in the medium (such as still air, in the case of sound waves), v_S is the speed of the source *relative to the medium* and v_O is the speed of the observer *relative to the medium.* The upper signs in the numerator and denominator are chosen whenever S moves toward O or O moves toward S.

The classical Doppler shift for motion of the source differs from that for motion of the observer. For example, suppose the source emits sound waves at $\nu = 1000$ Hz. If the source moves at 30 m/s toward the observer who is at rest in the medium (which we take to be air, in which sound moves at $v = 340$ m/s), then $\nu' = 1097$ Hz, while if the source is at rest in the medium and the observer moves toward the source at 30 m/s, the frequency is 1088 Hz. Other possibilities in which the relative speed between S and O is 30 m/s, such as each moving toward the other at 15 m/s, give still different frequencies.

Here we have a situation in which it is not the relative speed of the source and observer that determines the Doppler shift—it is the speed of each with respect to the medium. This cannot occur for light waves, since there is no medium (no "ether") and no preferred reference frame by Einstein's first postulate. We therefore require a different approach to the Doppler effect for light waves, an approach that does not distinguish between source motion and observer motion, but involves only the relative motion between the source and the observer.

Consider a source of waves that is at rest in the reference frame of observer O. Observer O' moves relative to the source at speed u. We consider the situation from the frame of reference of O', as shown in Figure 2.16. Suppose O observes the source to emit N waves at frequency ν. According to O, it takes an interval $\Delta t_0 = N/\nu$ for these N waves to be emitted; this is a proper time interval in the frame of reference of O. The corresponding time interval to O' is $\Delta t'$, during which O moves a distance $u\,\Delta t'$. The wavelength according to O' is the total length interval occupied by these waves divided by the number of waves:

$$\lambda' = \frac{c\,\Delta t' + u\,\Delta t'}{N} = \frac{c\,\Delta t' + u\,\Delta t'}{\nu\,\Delta t_0} \tag{2.20}$$

The frequency according to O' is $\nu' = c/\lambda'$, so

$$\nu' = \nu\,\frac{\Delta t_0}{\Delta t'}\,\frac{1}{1 + u/c} \tag{2.21}$$

and using the time dilation formula, Equation 2.8, to relate $\Delta t'$ and Δt_0, we obtain

Relativistic Doppler shift

$$\nu' = \nu\,\frac{\sqrt{1 - u^2/c^2}}{1 + u/c} = \nu\,\sqrt{\frac{1 - u/c}{1 + u/c}} \tag{2.22}$$

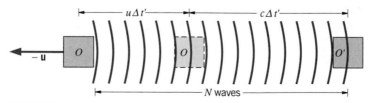

FIGURE 2.16   A source of waves, in the reference frame of O, moves at speed u away from observer O'. In the time $\Delta t'$ (according to O'), O moves a distance $u\,\Delta t'$ and emits N waves.

This is the formula for the *relativistic Doppler shift,* for the case in which the waves are observed in a direction parallel to **u**. Note that, unlike the classical formula, it does *not* distinguish between source motion and observer motion; the relativistic Doppler effect depends only on the relative speed u between the source and observer.

Equation 2.22 assumes that the source and observer are separating. If the source and observer are approaching one another, replace u by $-u$ in the formula.

EXAMPLE 2.8

A distant galaxy is moving away from the Earth at such high speed that the blue hydrogen line at a wavelength of 434 nm is recorded at 600 nm, in the red range of the spectrum. What is the speed of the galaxy relative to the Earth?

SOLUTION

Using Equation 2.22 with $\nu = c/\lambda$ and $\nu' = c/\lambda'$, we obtain

$$\frac{c}{\lambda'} = \frac{c}{\lambda} \sqrt{\frac{1 - u/c}{1 + u/c}}$$

$$\frac{c}{600 \text{ nm}} = \frac{c}{434 \text{ nm}} \sqrt{\frac{1 - u/c}{1 + u/c}}$$

Solving, we find

$$\frac{u}{c} - 0.31$$

Thus the galaxy is moving away from Earth at a speed of 9.4×10^7 m/s. Evidence obtained in this way indicates that nearly all the galaxies we observe are moving away from us. This suggests that the universe is expanding, and is usually taken to provide evidence in favor of the "Big Bang" theory of cosmology (see Chapter 16).

2.5 THE LORENTZ TRANSFORMATION

We have seen that the Galilean transformation of coordinates, time, and velocity is not consistent with Einstein's postulates. Although the Galilean transformation agrees with our "common sense" experience at low speeds, it does not agree with experiment at high speeds. We therefore need a new set of transformation equations that replaces the Galilean set and that is capable of predicting such relativistic effects as time dilation, length contraction, velocity addition, and the Doppler shift.

As before, we seek a transformation that enables observers O and O' in relative motion to compare their measurements of the space and time coordinates of the same event. The transformation equations relate the measurements of O (x, y, z, t) to those of O' (x', y', z', t'). This new transformation must have several properties: It must be linear (depending only on the first power of the space and time coordinates), which follows from the homogeneity of space and time; it must be consistent with Einstein's postulates; and it must reduce to the Galilean transformation when the relative speed between O and O' is small. We again assume that the velocity of O' relative to O is in the positive xx' direction.

The equations of the Lorentz transformation*, derived from these assumptions, are

Lorentz coordinate transformation

$$x' = \frac{x - ut}{\sqrt{1 - u^2/c^2}} \qquad (2.23a)$$

$$y' = y \qquad (2.23b)$$

$$z' = z \qquad (2.23c)$$

$$t' = \frac{t - (u/c^2)x}{\sqrt{1 - u^2/c^2}} \qquad (2.23d)$$

If O' moves *toward* (rather than away from) O, replace u with $-u$ in these equations.

The first three equations reduce directly to the Galilean transformation for space coordinates, Equations 2.1, when $u \ll c$. The fourth equation, which links the time coordinates, reduces to $t' = t$, which is a fundamental postulate of the Galilean-Newtonian world.

We now use the Lorentz transformation equations to derive some of the predictions of special relativity. The problems at the end of the chapter guide you in some other derivations. The results derived here are identical with those we obtained previously using Einstein's postulates, which shows that the equations of the Lorentz transformation are consistent with the postulates of special relativity.

Length Contraction

A rod of length L_0 is at rest in the reference frame of observer O'. The rod extends along the x' axis from x'_1 to x'_2; that is, O' measures the proper length $L_0 = x'_2 - x'_1$. Observer O, relative to whom the rod is in motion, measures the ends of the rod to be at coordinates x_1 and x_2. For O to determine the length of the moving rod, O must make a *simultaneous*

---

* H. A. Lorentz (1853–1928) was a Dutch physicist who shared the 1902 Nobel prize for his work on the influence of magnetic fields on light. In an unsuccessful attempt to explain the failure of the Michelson-Morley experiment, Lorentz developed the transformation equations that are named for him in 1904, a year *before* Einstein published his special theory of relativity. For a derivation of the Lorentz transformation, see R. Resnick and D. Halliday, *Basic Concepts in Relativity* (New York, Macmillan, 1992).

determination of x_1 and x_2, and then the length is $L = x_2 - x_1$. Suppose the first event is O' setting off a flash bulb at one end of the rod at x'_1 and t'_1, which O observes at x_1 and t_1, and the second event is O' setting off a flash bulb at the other end at x'_2 and t'_2, which O observes at x_2 and t_2. The equations of the Lorentz transformation relate these coordinates, specifically,

$$x'_1 = \frac{x_1 - ut_1}{\sqrt{1 - u^2/c^2}} \qquad x'_2 = \frac{x_2 - ut_2}{\sqrt{1 - u^2/c^2}} \qquad (2.24)$$

Subtracting these equations, we obtain

$$x'_2 - x'_1 = \frac{x_2 - x_1}{\sqrt{1 - u^2/c^2}} - \frac{u(t_2 - t_1)}{\sqrt{1 - u^2/c^2}} \qquad (2.25)$$

O' must arrange to set off the flash bulbs so that the flashes appear to be simultaneous to O. (They will *not* be simultaneous to O', as we discuss later in this section.) This enables O to make a simultaneous determination of the coordinates of the endpoints of the rod. If O observes the flashes to be simultaneous, then $t_2 = t_1$, and Equation 2.25 reduces to

$$x'_2 - x'_1 = \frac{x_2 - x_1}{\sqrt{1 - u^2/c^2}} \qquad (2.26)$$

With $x'_2 - x'_1 = L_0$ and $x_2 - x_1 = L$, this becomes

$$L = L_0\sqrt{1 - u^2/c^2} \qquad (2.27)$$

which is identical with Equation 2.8, which we derived earlier using Einstein's postulates.

Velocity Transformation

If O observes a particle to travel with velocity $\mathbf{v}$ (components v_x, v_y, v_z), what velocity $\mathbf{v}'$ does O' observe for the particle? The relationship between the velocities measured by O and O' is given by the *Lorentz velocity transformation*:

$$v'_x = \frac{v_x - u}{1 - v_x u/c^2} \qquad (2.28a)$$

$$v'_y = \frac{v_y\sqrt{1 - u^2/c^2}}{1 - v_x u/c^2} \qquad (2.28b) \qquad \textit{Lorentz velocity transformation}$$

$$v'_z = \frac{v_z\sqrt{1 - u^2/c^2}}{1 - v_x u/c^2} \qquad (2.28c)$$

The equation for v'_x is identical with Equation 2.17, a result we derived previously based on Einstein's postulates. Note that, in the limit of low speeds ($u \ll c$), the Lorentz velocity transformation reduces to the Galilean velocity transformation, Equation 2.2. Note also that $v'_y \neq v_y$, even though $y' = y$. This occurs because of the way the Lorentz transformation handles the time coordinate.

We can derive these transformation equations for velocity from the Lorentz coordinate transformation. By way of example, we derive the velocity transformation for $v_y' = dy'/dt'$. Differentiating the coordinate transformation $y' = y$, we obtain $dy' = dy$. Similarly, differentiating the time coordinate transformation, we obtain

$$dt' = \frac{dt - (u/c^2)dx}{\sqrt{1 - u^2/c^2}}$$

So

$$v_y' = \frac{dy'}{dt'} = \frac{dy}{[dt - (u/c^2)dx]/\sqrt{1 - u^2/c^2}} = \sqrt{1 - u^2/c^2}\,\frac{dy}{dt - (u/c^2)dx}$$

$$= \sqrt{1 - u^2/c^2}\,\frac{dy/dt}{1 - (u/c^2)dx/dt} = \frac{v_y\sqrt{1 - u^2/c^2}}{1 - uv_x/c^2}$$

Similar methods can be used to obtain the transformation equations for v_x' and v_z'. These derivations are left as exercises (Problem 13).

Simultaneity and Clock Synchronization

For most of us, synchronizing our watches or clocks is not a difficult process; for example, we can set our watches by looking at the nearest clock. However, this method ignores the time that it takes the light from the face of the clock to travel to our eye so we can set our watches. If we are 1 m from the clock, our watch will be behind by about 3 ns (3×10^{-9} s). Although this short time lag may not make you late for your physics lecture, it can be a serious problem for the experimental physicist, for whom the measurement of time intervals smaller than 1 ns is routine. Let us therefore try to be more precise. We build a device similar to that shown in Figure 2.17. Two clocks are located at $x = 0$ and $x = L$. A flash lamp is located at $x = L/2$, and the clocks are set running when they receive the flash of light from the lamp. Since the light takes the same interval of time to reach the two clocks, the clocks start together precisely at a time $L/2c$ after the flash is emitted, and the clocks are exactly synchronized.

Now let us examine the same situation from the point of view of the moving observer O'. If you have been paying attention this far into this chapter, you are probably suspecting that O' does not agree that the clocks are synchronized. In the frame of reference of O, two events occur: the receipt of a light signal by clock 1 at $x_1 = 0$, $t_1 = L/2c$ and the receipt of a light signal by clock 2 at $x_2 = L$, $t_2 = L/2c$. Using Equation 2.23d, we find that O' observes clock 1 to receive its signal at

$$t_1' = \frac{t_1 - (u/c^2)x_1}{\sqrt{1 - u^2/c^2}} = \frac{L/2c}{\sqrt{1 - u^2/c^2}} \tag{2.29}$$

while clock 2 receives its signal at

$$t_2' = \frac{t_2 - (u/c^2)x_2}{\sqrt{1 - u^2/c^2}} = \frac{L/2c - (u/c^2)L}{\sqrt{1 - u^2/c^2}} \tag{2.30}$$

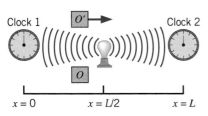

Clock 1 $O'\!\rightarrow$ Clock 2

O

$x = 0$ $x = L/2$ $x = L$

FIGURE 2.17 A flash of light, emitted from a point midway between the two clocks, starts the two clocks simultaneously according to O. Observer O' sees clock 2 start ahead of clock 1.

Thus t_2' is smaller than t_1' and clock 2 appears to receive its signal earlier than clock 1, so that the clocks start at times that differ by

$$\Delta t' = t_1' - t_2' = \frac{uL/c^2}{\sqrt{1 - u^2/c^2}} \tag{2.31}$$

according to O'. It is important to keep in mind that this is *not* a time dilation effect—time dilation comes from the *first* term of the Lorentz transformation (Equation 2.23d) for t', while the lack of synchronization arises from the *second* term. O' observes *both* clocks to run slow, due to time dilation; O' *also* observes clock 2 to be a bit ahead of clock 1.

We therefore reach the following conclusion: two events that are simultaneous in one reference frame are not simultaneous in another reference frame moving with respect to the first, unless the two events occur at the same point in space. (If $L = 0$, Equation 2.31 shows that the clocks are synchronized in all reference frames.) Clocks that appear to be synchronized in one frame of reference will not necessarily be synchronized in another frame of reference in relative motion.

EXAMPLE 2.9

Two rockets are leaving their space station along perpendicular paths, as measured by an observer on the space station. Rocket 1 moves at $0.60c$ and rocket 2 moves at $0.80c$, both measured relative to the space station. What is the velocity of rocket 2 as observed by rocket 1?

SOLUTION

Observer O is the space station, observer O' is rocket 1 (moving at $u = 0.60c$) and each observes rocket 2, moving (according to O) in a direction perpendicular to rocket 1. We take this to be the y direction of the reference frame of O. Thus O observes rocket 2 to have velocity components $v_x = 0$, $v_y = 0.80c$, as shown in Figure 2.18a.

We can find v_x' and v_y' using the Lorentz velocity transformation:

$$v_x' = \frac{0 - 0.60c}{1 - 0(0.60c)/c^2} = -0.60c$$

$$v_y' = \frac{0.80c\sqrt{1 - (0.60c)^2/c^2}}{1 - 0(0.60c)/c^2} = 0.64c$$

Thus, according to O', the situation looks like Figure 2.18b.

The speed of rocket 2 according to O' is $\sqrt{(0.60c)^2 + (0.64c)^2} = 0.88c$, less than c. According to the Galilean transformation, v_y' would be identical with v_y, and thus the speed would be $\sqrt{(0.60c)^2 + (0.80c)^2} = c$. Once again, the Lorentz transformation prevents relative speeds from reaching or exceeding the speed of light. If the two speeds were $0.999c$, the Lorentz transformation equations would give $v_x' = -0.999c$ and $v_y' = 0.0447c$, giving a relative speed of $0.999998c$. Notice how the Lorentz trans-

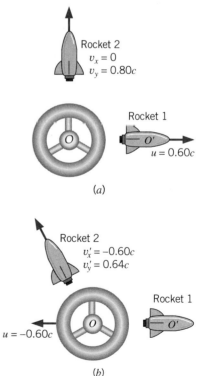

FIGURE 2.18 Example 2.9 (a) As viewed from the reference frame of O. (b) As viewed from the reference frame of O'.

formation causes the extreme shrinking of the *y* component from 0.999*c* to 0.0447*c*.

EXAMPLE 2.10

In Example 2.6, two events that were simultaneous to *O* (the lining up of the front and back of the rocket ship with the ends of the platform) were not simultaneous to *O'*. Find the time interval between these events according to *O'*.

SOLUTION

According to *O*, the two simultaneous events are separated by a distance of *L* = 65 m. For *u* = 0.80*c*, Equation 2.31 gives

$$\Delta t' = \frac{uL/c^2}{\sqrt{1 - u^2/c^2}} = \frac{(0.80)(65 \text{ m})/(3.00 \times 10^8 \text{ m/s})}{\sqrt{1 - (0.80)^2}} = 0.29 \ \mu\text{s}$$

which agrees with the result calculated in part (*e*) of Example 2.6.

2.6 THE TWIN PARADOX

We now turn briefly to what has become known as the twin paradox. Suppose there is a pair of twins on Earth. One, whom we shall call Casper, remains on Earth, while his twin sister Amelia sets off in a rocket ship on a trip to a distant planet. Casper, based on his understanding of special relativity, knows that his sister's clocks will run slow relative to his own and that therefore she should be younger than he when she returns, as our discussion of time dilation would suggest. However, recalling that discussion, we know that for two observers in relative motion, *each* thinks the *other's* clocks are running slow. We could therefore study this problem from the point of view of Amelia, according to whom Casper and the Earth (accompanied by the solar system and galaxy) make a round-trip journey away from her and back again. Under such circumstances, she will think it is her brother's clocks (which are now in motion relative to her own) that are running slow, and will therefore expect her brother to be younger than she when they meet again. While it is possible to disagree over whose clocks are running slow relative to his or her own, which is merely a problem of frames of reference, when Amelia returns to Earth (or when the Earth returns to Amelia), all observers must agree as to which twin has aged less rapidly. This is the paradox—each twin expects the other to be younger.

The resolution of this paradox lies in considering the asymmetric role of the two twins. The laws of special relativity apply only to inertial frames, those moving relative to one another at constant velocity. We may supply

Amelia's rockets with sufficient thrust so that they accelerate for a very short length of time, bringing the ship to a speed at which it can coast to the planet, and thus during her outward journey Amelia spends all but a negligible amount of time in a frame of reference moving at constant speed relative to Casper. However, in order to return to Earth, she must decelerate and reverse her motion. Although this also may be done in a very short time interval, Amelia's return journey occurs in a completely different inertial frame than her outward journey. It is Amelia's jump from one inertial frame to another that causes the asymmetry in the ages of the twins. Only Amelia has the necessity of jumping to a new inertial frame to return, and therefore *all observers will agree* that it is Amelia who is "really" in motion, and that it is her clocks that are "really" running slow; therefore she is indeed the younger twin on her return.

Let us make this discussion more quantitative with some numerical examples. We assume, as discussed above, that the acceleration and deceleration take negligible time intervals, so that all of Amelia's aging is done during the coasting. For simplicity, we assume the distant planet is at rest relative to the Earth; this does not change the problem, but it avoids the need to introduce yet another frame of reference. Suppose the planet to be 12 light-years distant from Earth, and suppose Amelia travels at a speed of $0.6c$. Then according to Casper it takes his sister 20 years (20 years $\times$ $0.6c = 12$ light-years) to reach the planet and 20 years to return, and therefore she is gone for a total of 40 years. (However, Casper doesn't know his sister has reached the planet until the light signal carrying news of her arrival reaches Earth. Since light takes 12 years to make the journey, it is 32 years after her departure when Casper sees his sister's arrival at the planet. Eight years later she returns to Earth.) From the frame of reference of Amelia aboard the rocket, the distance to the planet is contracted by a factor of $\sqrt{1 - (0.6)^2} = 0.8$, and is therefore $0.8 \times 12 = 9.6$ light-years. At a speed of $0.6c$, Amelia will measure 16 years for the trip to the planet, for a total round trip time of 32 years. Thus Casper ages 40 years while Amelia ages only 32 years and is indeed the younger on her return. We can confirm this analysis by having Casper send a light signal each year, on his birthday, to his sister. We know that the frequency of the signal as received by Amelia will be Doppler shifted. During the outward journey, she will receive signals at the rate of

$$(1 \text{ yr}^{-1}) \sqrt{\frac{1 - u/c}{1 + u/c}} = 0.5 \text{ yr}^{-1}$$

During the return journey, the Doppler-shifted rate will be

$$(1 \text{ yr}^{-1}) \sqrt{\frac{1 + u/c}{1 - u/c}} = 2 \text{ yr}^{-1}$$

Thus for the first 16 years, during Amelia's trip to the planet, she receives 8 signals, and during the return trip of 16 years, she receives 32 signals, for a total of 40. She receives 40 signals, indicating her brother has celebrated 40 birthdays during her 32-year journey.

It is left as an exercise (Problem 18) to consider the situation if it is Amelia who is sending the signals.

2.7 RELATIVISTIC DYNAMICS

We have seen how Einstein's postulates have led to a new "relative" interpretation of such previously absolute concepts as length and time, and that the classical concept of absolute velocity is not valid. It is reasonable then to ask how far this revolution is to go in changing our interpretation of physical concepts. We therefore discuss now the dynamical quantities energy and momentum, in order to examine these from the point of view of special relativity. Are our familiar relationships, such as $\mathbf{p} = m\mathbf{v}$ and $K = \frac{1}{2}mv^2$, still valid, or must we have new concepts of dynamical quantities also? Moreover, what of the fundamental conservation laws of classical physics, such as conservation of energy and conservation of linear momentum? These concepts are so important to classical physics that we are very reluctant to discard them. Finally, classical dynamics allows an object to be accelerated to speeds greater than the speed of light, if we apply a force for a sufficiently long time; our system of relativistic dynamics must prevent objects from reaching or exceeding the speed of light, no matter how great a force is applied.

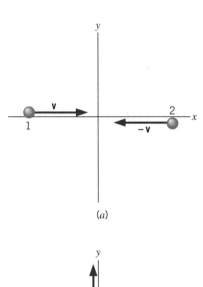

(a)

Consider the collision shown in Figure 2.19a, viewed from the reference frame of observer O. Two particles, each of mass m, move with equal and opposite velocities $\mathbf{v}$ and $-\mathbf{v}$ along the x axis. They collide at the origin, and the distance between their lines of approach has been adjusted so that after the collision the particles move along the y axis with equal and opposite final velocities, as shown in Figure 2.19b. We assume the collision to be perfectly elastic, so that no kinetic energy is lost. The final velocities must then be $\mathbf{v}$ and $-\mathbf{v}$.

Using the classical formula ($\mathbf{p} = m\mathbf{v}$), we find the components of the momentum of the two-particle system in the reference frame of O:

Initial: $p_{xi} = mv + m(-v) = 0$
            $p_{yi} = 0$

Final: $p_{xf} = 0$
            $p_{yf} = mv + m(-v) = 0$

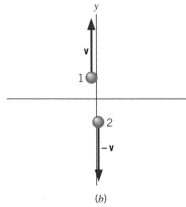

(b)

Thus $p_{xi} = p_{xf}$ and $p_{yi} = p_{yf}$; the initial (vector) momentum is equal to the final momentum, and momentum is conserved in the reference frame of O.

FIGURE 2.19 A collision between two particles of the same mass is shown in the reference frame of O (a) before the collision and (b) after the collision.

Let us now view the same collision from the reference frame of O', which moves relative to the reference frame of O with speed $u = -v$ (Figure 2.20a). Note that in the O' frame, particle 2 is at rest before the collision. We use the Lorentz velocity transformation, Equations 2.28a and 2.28b, to find the transformed x' and y' components of the initial and final velocities, as they would be observed by O'. These values, which you should calculate, are shown in Figures 2.20a and b.

We now use those velocities to find the components of the momentum of the system in the O' frame:

$$p'_{xi} = m \left(\frac{2v}{1 + v^2/c^2} \right) + m(0) = \frac{2mv}{1 + v^2/c^2}$$

$$p'_{yi} = 0$$

$$p'_{xf} = mv + mv = 2mv$$

$$p'_{yf} = mv\sqrt{1 - v^2/c^2} + m(-v\sqrt{1 - v^2/c^2}) = 0$$

We see that p'_{xi} is *not* equal to p'_{xf}, so O' concludes that momentum is *not* conserved.

It is clear from the above calculation that the law of conservation of linear momentum, which we have found useful in a variety of applications, does not satisfy Einstein's first postulate (the law must be the same in all inertial frames) if we calculate momentum as $\mathbf{p} = m\mathbf{v}$. Therefore, *if we are to retain the conservation of momentum as a general law consistent with Einstein's first postulate, we must find a new definition of momentum.* This new definition of momentum must have two properties: (1) It must yield a law of conservation of momentum that satisfies the principle of relativity; that is, if momentum is conserved according to an observer in one inertial frame, then it is conserved according to observers in *all* inertial frames. (2) At low speeds, the new definition must reduce to $\mathbf{p} = m\mathbf{v}$, which we know works perfectly well in the nonrelativistic case.

These requirements are satisfied by defining the relativistic momentum for a particle of mass m moving with velocity $\mathbf{v}$ as

$$\mathbf{p} = \frac{m\mathbf{v}}{\sqrt{1 - v^2/c^2}} \qquad (2.32) \qquad \textit{Relativistic momentum}$$

In terms of components, we can write Equation 2.32 as

$$p_x = \frac{mv_x}{\sqrt{1 - v^2/c^2}} \quad \text{and} \quad p_y = \frac{mv_y}{\sqrt{1 - v^2/c^2}} \qquad (2.33)$$

The velocity v that appears in the denominator of these expressions is *always* the velocity of the particle as measured in a particular inertial frame. It is *not* the velocity of an inertial frame. The velocity in the numerator can be any of the components of the velocity vector.

Let us see how this new definition restores conservation of momentum in the collision we considered. In the O frame, the velocities before and after are equal and opposite, and thus Equation 2.32 again gives zero for the initial and final momenta. In the O' frame, we can use the magnitudes of the velocities as given in Figure 2.20a and b to obtain, as you should verify,

$$p'_{xi} = p'_{xf} = \frac{2mv}{1 - v^2/c^2}$$

$$p'_{yi} = p'_{yf} = 0$$

$$(2.34)$$

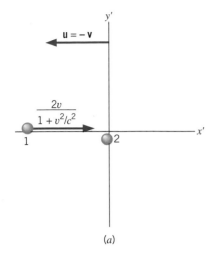

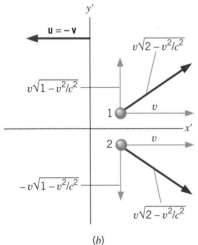

FIGURE 2.20 The same collision of Figure 2.19 is shown in the reference frame of O' (a) before the collision and (b) after the collision. This reference frame moves with velocity $-\mathbf{v}$ relative to O.

Thus the initial and final momenta are equal in the O' frame. Momentum is conserved in both the O and O' frames. In fact, the definition of momentum given in Equation 2.32 gives conservation of momentum in *all* inertial frames, as required by the principle of relativity.

Note also that, in the limit of low speeds, the denominator of Equation 2.32 is nearly equal to 1; at low speeds Equation 2.32 reduces to the familiar classical formula $\mathbf{p} = m\mathbf{v}$. Equation 2.32 thus also satisfies this necessary criterion of relativistic formulas.

EXAMPLE 2.11

What is the momentum of a proton moving at a speed of $v = 0.86c$?

SOLUTION

Using Equation 2.32, we obtain

$$
\begin{aligned}
p &= \frac{mv}{\sqrt{1 - v^2/c^2}} \\
&= \frac{(1.67 \times 10^{-27}\ \text{kg})(0.86)(3.00 \times 10^8\ \text{m/s})}{\sqrt{1 - (0.86)^2}} \\
&= 8.44 \times 10^{-19}\ \text{kg·m/s}
\end{aligned}
$$

The units of kg·m/s are generally not convenient in solving problems of this type. Instead, we manipulate Equation 2.32 to obtain

$$
pc = \frac{mcv}{\sqrt{1 - v^2/c^2}} = \frac{mc^2(v/c)}{\sqrt{1 - v^2/c^2}} = \frac{(938\ \text{MeV})(0.86)}{\sqrt{1 - (0.86)^2}}
$$

$$
= 1580\ \text{MeV}
$$

Here we have used the proton's *rest energy* mc^2, which is defined later in this section. The momentum is obtained from this result by dividing by the symbol c (not its numerical value), which gives

$$
p = 1580\ \text{MeV}/c
$$

The units of MeV/c for momentum are often used in relativistic calculations because, as we show later, the quantity pc often appears in these calculations. You should be able to convert MeV/c to kg·m/s and show that the two results obtained for p are equivalent.

In addition to momentum, special relativity also gives us a different approach to kinetic energy. Let us first indicate the difficulty by reconsidering the collision shown in Figure 2.19. If we use the classical expression $\frac{1}{2}mv^2$, the collision does not conserve kinetic energy in the O' frame. (We chose the final velocities in the O frame so that kinetic energy would be conserved.) Using the velocities shown in Figure 2.20a and b, you can show (see Problem 19) that, with $K = \frac{1}{2}mv^2$,

$$K_i' = \frac{2mv^2}{(1 + v^2/c^2)^2}$$

$$K_f' = mv^2(2 - v^2/c^2)$$

$$(2.35)$$

Thus K_i' is not equal to K_f', and the elastic collision apparently does not conserve kinetic energy in the O' frame. This situation violates the relativity postulate; the type of collision (elastic vs. inelastic) should depend on the properties of the colliding objects and not on the particular reference frame from which we happen to be viewing the collision. As was the case with momentum, we require a new definition of kinetic energy if we are to preserve the law of conservation of energy and the relativity postulate.

The classical expression for kinetic energy also violates the second relativity postulate by allowing speeds in excess of the speed of light. There is no limit (in either classical or relativistic dynamics) to the energy we can give to a particle. Yet, if we allow the kinetic energy to increase without limit, the classical expression $K = \frac{1}{2}mv^2$ implies that the velocity must correspondingly increase without limit, thereby violating the second postulate. We must therefore find a way to redefine kinetic energy, so that the kinetic energy of a particle can be increased without limit while its speed remains less than c.

The relativistic expression for the kinetic energy of a particle can be derived using essentially the same procedure used to derive the classical expression, starting with the particle form of the work-energy theorem (see Problem 21). The result of this calculation is

$$K = \frac{mc^2}{\sqrt{1 - v^2/c^2}} - mc^2 \qquad (2.36) \qquad \textit{Relativistic kinetic energy}$$

Equation 2.36 looks very different from the classical result $\frac{1}{2}mv^2$, but, as you should show (see Problem 27), Equation 2.36 reduces to the classical expression in the limit of low speeds ($v \ll c$). You can also see from the first term of Equation 2.36 that $K \to \infty$ as $v \to c$. Thus we can increase the kinetic energy of a particle without limit, and its speed will not exceed c.

We can also express Equation 2.36 as

$$K = E - E_0 \qquad (2.37)$$

where the *relativistic total energy* E is defined as

$$E = \frac{mc^2}{\sqrt{1 - v^2/c^2}} \qquad (2.38) \qquad \textit{Relativistic total energy}$$

and the *rest energy* E_0 is defined as

$$E_0 = mc^2 \qquad (2.39)$$

The rest energy is in effect the relativistic total energy of a particle measured in a frame of reference in which the particle is at rest.

The relativistic total energy is given by Equation 2.37 as

$$E = K + E_0 \qquad (2.40)$$

In interactions of particles at relativistic speeds, we can replace our previous principle of conservation of energy with one based on the total relativistic energy:

In an isolated system of particles, the relativistic total energy remains constant.

Using the relativistic form of kinetic energy given by Equation 2.36, we can show that kinetic energy is conserved in the O' frame of the collision of Figure 2.20 (see Problem 20). Because the rest energies of the initial and final particles are equal in this collision, conservation of relativistic total energy is equivalent to conservation of kinetic energy. In general, collisions of particles at high energies can result in the production of new particles, and thus the final rest energy may not be equal to the initial rest energy (see Example 2.17). Such collisions must be analyzed using conservation of total relativistic energy E; kinetic energy will *not* be conserved when the rest energy changes in a collision.

From Equation 2.40 we see that in an isolated system, for which the energy is constant ($\Delta E = 0$), we must have $\Delta K = -\Delta E_0$. That is, any change in the kinetic energy must be accompanied by a change in rest energy of the opposite sign. For example, in a reaction in which new particles are produced, the loss in kinetic energy of the reacting particles gives the increase in rest energy of the product particles. On the other hand, in a nuclear decay process such as alpha decay, the initial nucleus gives up some rest energy to account for the kinetic energy carried by the decay products.

Sometimes m in Equation 2.39 is called the *rest mass* m_0 and is distinguished from the "relativistic mass," which is defined as $m_0/\sqrt{1 - v^2/c^2}$. We choose not to use relativistic mass, because it can be a misleading concept. Whenever we refer to mass, we always mean rest mass.

Manipulation of Equations 2.32 and 2.38 gives a useful relationship among the total energy, momentum, and rest energy:

$$E = \sqrt{(pc)^2 + (mc^2)^2} \tag{2.41}$$

Figure 2.21 shows a useful mnemonic device for remembering this relationship, which has the form of the Pythagorean theorem for the sides of a right triangle.

When a particle travels at a speed close to the speed of light (say, $v > 0.99c$), which often occurs in high-energy particle accelerators, the particle's kinetic energy is much greater than its rest energy; that is, $K \gg E_0$. In this case, Equation 2.41 can be written, to a very good approximation,

$$E \cong pc \tag{2.42}$$

This is called the *extreme relativistic approximation* and is often useful for simplifying calculations.

For massless particles (such as photons or neutrinos), Equation 2.41 becomes exactly

$$E = pc \tag{2.43}$$

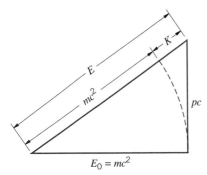

FIGURE 2.21   A useful mnemonic device for recalling the relationships between E_0, p, K, and E. Note that to put all variables in energy units, the quantity pc must be used.

All massless particles travel at the speed of light; otherwise, by Equations 2.36 and 2.38 their kinetic and total energies would be zero.

EXAMPLE 2.12

What are the kinetic and relativistic total energies of a proton ($E_0 = 938$ MeV) moving at a speed of $v = 0.86c$?

SOLUTION

In Example 2.11 we found the momentum of this particle to be $p = 1580$ MeV/c. The energy can be found from Equation 2.41:

$$E = \sqrt{(pc)^2 + (mc^2)^2} = \sqrt{(1580\,\text{MeV})^2 + (938\,\text{MeV})^2} = 1837\,\text{MeV}$$

The kinetic energy follows from Equation 2.37:

$$K = E - E_0 = 1837\,\text{MeV} - 938\,\text{MeV} = 899\,\text{MeV}$$

We could also have solved this problem by finding the kinetic energy directly from Equation 2.36.

EXAMPLE 2.13

Find the velocity and momentum of an electron ($E_0 = 0.511$ MeV) with a kinetic energy of 10.0 MeV.

SOLUTION

The total energy is $E = K + E_0 = 10.0$ MeV $+ 0.511$ MeV $= 10.51$ MeV. We can then find the momentum from Equation 2.41:

$$p = \frac{1}{c}\sqrt{E^2 - (mc^2)^2} = \frac{1}{c}\sqrt{(10.51\,\text{MeV})^2 - (0.511\,\text{MeV})^2} = 10.5\,\text{MeV}/c$$

Note that in this problem we could have used the extreme relativistic approximation, $p \cong E/c$, from Equation 2.42. The error we would make in this case would be only 0.1 percent.

The velocity can be found by solving Equation 2.38 for v.

$$\frac{v}{c} = \sqrt{1 - \left(\frac{mc^2}{E}\right)^2} = \sqrt{1 - \left(\frac{0.511\,\text{MeV}}{10.51\,\text{MeV}}\right)^2} = 0.9988 \qquad (2.44)$$

EXAMPLE 2.14

In the Stanford Linear Collider electrons are accelerated to a kinetic energy of 50 GeV. Find the speed of such an electron as (*a*) a fraction of *c*, and (*b*) a difference from *c*. The rest energy of the electron is 0.511 MeV $= 0.511 \times 10^{-3}$ GeV.

SOLUTION

(*a*) First we solve Equation 2.36 for v, cbtaining

$$v = c \sqrt{1 - \frac{1}{(1 + K/mc^2)^2}} \qquad (2.45)$$

and thus

$$v = c \sqrt{1 - \frac{1}{(1 + 50 \text{ GeV}/0.511 \times 10^{-3} \text{ GeV})^2}}$$

$$= 0.999\,999\,999\,948c$$

Calculators cannot be trusted to 12 significant digits. Here is a way to avoid this difficulty. We can write Equation 2.45 as $v = c(1 + x)^{1/2}$, where $x = -1/(1 + K/mc^2)^2$. Because $K \gg mc^2$, we have $x \ll 1$, and we can use the binomial expansion to write $v \approx c(1 + \frac{1}{2}x)$, or

$$v \approx c \left[1 - \frac{1}{2(1 + K/mc^2)^2} \right]$$

which gives

$$v = c(1 - 5.2 \times 10^{-11})$$

This leads to the value of v given above.

(*b*) From the above result, we have

$$c - v = 5.2 \times 10^{-11}c = 0.016 \text{ m/s} = 1.6 \text{ cm/s}$$

EXAMPLE 2.15

At a distance equal to the radius of the Earth's orbit (1.5×10^{11} m), the Sun's radiation has an intensity of about 1.4×10^3 W/m². Find the rate at which the mass of the Sun is decreasing.

SOLUTION

If we assume that the Sun's radiation is distributed uniformly over the surface area $4\pi r^2$ of a sphere of radius 1.5×10^{11} m, then the total radiative power emitted by the Sun is

$$4\pi(1.5 \times 10^{11} \text{ m})^2(1.4 \times 10^3 \text{ W/m}^2) = 4.0 \times 10^{26} \text{ W} = 4.0 \times 10^{26} \text{ J/s}$$

By conservation of energy, we know that the energy lost by the Sun through radiation must be accounted for by a corresponding loss in its rest energy. The change in mass Δm corresponding to a change in rest energy ΔE_0 of 4.0×10^{26} J each second is

$$\Delta m = \frac{\Delta E_0}{c^2} = \frac{4.0 \times 10^{26} \text{ J}}{9.0 \times 10^{16} \text{ m}^2/\text{s}^2} = 4.4 \times 10^9 \text{ kg}$$

The Sun loses mass at a rate of about 4 billion kilograms per second! If this rate were to remain constant, the Sun (with a present mass of 2×10^{30} kg) will shine "only" for another 10^{13} years.

EXAMPLE 2.16

A certain accelerator produces a beam of neutral K mesons or kaons ($m_K c^2 = 498$ MeV) with kinetic energy 325 MeV. Consider a kaon that decays in flight into two pi mesons or pions ($m_\pi c^2 = 140$ MeV). Find the kinetic energy of each pion in the special case in which the pions travel parallel or antiparallel to the direction of the kaon beam.

SOLUTION

The energy of the particles that remain after the decay can be found by applying principles of conservation of total relativistic energy and momentum. The initial relativistic total energy is, from Equation 2.40,

$$E_K = K + m_K c^2 = 325 \text{ MeV} + 498 \text{ MeV} = 823 \text{ MeV}$$

The initial momentum can be found from Equation 2.41:

$$p_K c = \sqrt{E_K^2 - (m_K c^2)^2} = \sqrt{(823 \text{ MeV})^2 - (498 \text{ MeV})^2} = 655 \text{ MeV}$$

The total energy of the final system consisting of the two pions is

$$
\begin{aligned}
E &= E_1 + E_2 \\
&= \sqrt{(p_1 c)^2 + (m_\pi c^2)^2} + \sqrt{(p_2 c)^2 + (m_\pi c^2)^2} = 823 \text{ MeV}
\end{aligned}
\tag{2.46}
$$

which, by conservation of total relativistic energy, we have equated to the initial total energy of 823 MeV. Thus we have one equation in the two unknowns p_1 and p_2, the momenta of the two pions.

To find a second equation in the two unknowns we apply conservation of momentum. The final momentum of the two-pion system along the beam direction is $p_1 + p_2$, and setting this equal to the initial momentum p_K gives

$$p_1 c + p_2 c = p_K c = 655 \text{ MeV} \tag{2.47}$$

We now have two equations in the two unknowns p_1 and p_2. Solving Equation 2.47 for $p_2 c$ and substituting this result into Equation 2.46, we obtain (after some algebraic manipulation) a quadratic equation for $p_1 c$, which can be solved by standard algebraic techniques to give

$$p_1 c = 668 \text{ MeV} \quad \text{or} \quad -13 \text{ MeV}$$

Since the labels 1 and 2 of the two pions are arbitrary, the solution gives one pion traveling parallel to the beam with momentum $p_1 = 668$ MeV/c, while the other pion travels in the opposite direction with momentum $p_2 = -13$ MeV/c. The corresponding kinetic energies are found using Equations 2.37 and 2.41, which give

$$K = \sqrt{(pc)^2 + (m_\pi c^2)^2} - m_\pi c^2$$

$$K_1 = \sqrt{(668 \text{ MeV})^2 + (140 \text{ MeV})^2} - 140 \text{ MeV} = 543 \text{ MeV}$$

$$K_2 = \sqrt{(-13 \text{ MeV})^2 + (140 \text{ MeV})^2} - 140 \text{ MeV} = 0.6 \text{ MeV}$$

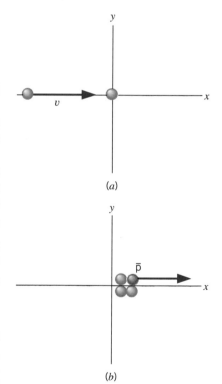

FIGURE 2.22 Example 2.17. (*a*) A proton moving with velocity v collides with another proton at rest. (*b*) The reaction produces three protons and an antiproton, which move together as a unit.

This problem can also be solved in a different way by making a Lorentz transformation to a reference frame in which the kaons are at rest (see Problem 37). The next Example demonstrates another application of this technique.

EXAMPLE 2.17

The discovery of the antiproton $\overline{p}$ (a particle with the same rest energy as a proton, 938 MeV, but with the opposite electric charge) took place in 1956 through the following reaction:

$$p + p \rightarrow p + p + p + \overline{p}$$

in which accelerated protons were incident on a target of protons at rest in the laboratory. The minimum incident kinetic energy needed to produce the reaction is called the *threshold* kinetic energy, for which the final particles move together as if they were a single unit (Figure 2.22). Find the threshold kinetic energy to produce antiprotons in this reaction.

SOLUTION

This problem is conceptually the reverse case of the previous Example. Here particles are coming together to form a composite. We will demonstrate an alternate method by solving in the center-of-momentum reference frame, in which the two protons come together with equal and opposite momenta to form a new particle at rest (Figure 2.23).

The final total energy in the center-of-momentum frame (observer O') is the rest energy of the products, which are produced at rest in this frame, so

$$E'_f = 4m_p c^2$$

The initial energy is the sum of the total energies of the two original reacting protons:

$$E'_i = E'_1 + E'_2$$

Conservation of energy requires $E'_i = E'_f$, and since the energies E'_1 and E'_2 are equal in the O' frame, we have

$$E'_1 = E'_2 = \tfrac{1}{2}E'_f = 2m_p c^2$$

The corresponding magnitude of the velocity of either reacting proton in the O' frame is found by solving Equation 2.38 for v/c, which gives

$$\frac{v'_1}{c} = \sqrt{1 - \left(\frac{m_p c^2}{E'_1}\right)^2} = \sqrt{1 - \left(\frac{1}{2}\right)^2} = \sqrt{\frac{3}{4}}$$

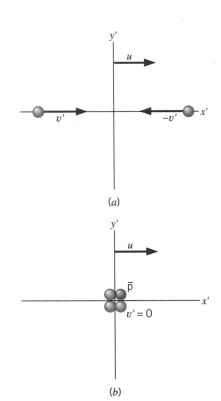

FIGURE 2.23 The reaction of Figure 2.22 viewed from the center-of-momentum reference frame. (a) Two protons moving with equal and opposite velocities collide. (b) The reaction products are formed at rest.

We now make a Lorentz transformation back to the laboratory using this as the transformation speed, which brings one of the protons to rest and gives the other a velocity v. The transformation equations from the O' frame to the O frame are the inverses of those given in Equation 2.28. Dropping the x subscript from Equation 2.28a and solving for v, we obtain (using $v' = c\sqrt{3/4}$ and $u = c\sqrt{3/4}$)

$$v = \frac{v' + u}{1 + uv'/c^2} = \frac{2c\sqrt{3/4}}{1 + (\sqrt{3/4})^2} = \frac{4\sqrt{3}}{7}c$$

This is the speed of the incident proton in the laboratory frame. Its total energy can be found from Equation 2.38:

$$E = \frac{mc^2}{\sqrt{1 - v^2/c^2}} = \frac{m_p c^2}{\sqrt{1 - (4\sqrt{3}/7)^2}} = 7m_p c^2$$

and the threshold kinetic energy is

$$K = E - m_p c^2 = 6m_p c^2 = 6(938 \text{ MeV}) = 5628 \text{ MeV} = 5.628 \text{ GeV}$$

The Bevatron accelerator at the Lawrence Berkeley Laboratory was designed with this experiment in mind, so that it could produce a beam of protons whose energy exceeded 5.6 GeV. The discovery of the antiproton in this reaction was honored with the award of the 1959 Nobel prize to the experimenters, Emilio Segrè and Owen Chamberlain.

2.8 EXPERIMENTAL TESTS OF SPECIAL RELATIVITY

Because special relativity provided such a radical departure from the notions of space and time in classical physics, it is important to perform detailed experimental tests that can clearly distinguish between the predictions of special relativity and those of classical physics. Many tests of increasing precision have been done since the theory was originally presented, and in every case the predictions of special relativity are upheld. Here we discuss a few of these tests.

Universality of the Speed of Light

The second relativity postulate asserts that the speed of light has the same value c for all observers. This leads to several types of experimental tests, of which we discuss two: (1) Does the speed of light change with the direction of travel? (2) Does the speed of light change with relative motion between source and observer?

The Michelson-Morley experiment provides a test of the first type. This experiment compared the upstream-downstream and cross-stream speeds of light and concluded that they were equal within the experimental error. Equivalently, we may say that the experiment showed that there is no preferred reference frame (no ether) relative to which the speed of light must be measured. If there is an ether, the speed of the Earth through the ether is less than 5 km/s, which is much smaller than the Earth's orbital speed about the Sun, 30 km/s. We can express their result as a difference Δc between the upstream-downstream and cross-stream speeds; the experiment showed that $\Delta c/c < 3 \times 10^{-10}$.

To reconcile the result of the Michelson-Morley experiment with classical physics, Lorentz proposed the "ether drag" hypothesis, according to which the motion of the Earth through the ether caused an electromagnetic drag that contracted the arm of the interferometer in the direction of motion. This contraction was just enough to compensate for the difference in the upstream-downstream and cross-stream times predicted by the Galilean transformation. This hypothesis succeeds only when the two arms of the interferometer are of the same length. To test this hypothesis, a similar experiment was done in 1932 by Kennedy and Thorndike; in their experiment, the lengths of the interferometer arms differed by about 16 cm, the maximum distance over which light sources available at that time could remain coherent. Their result was $\Delta c/c < 3 \times 10^{-8}$, which excludes the Lorentz contraction hypothesis as an explanation for the Michelson-Morley experiment.

In recent years, these fundamental experiments have been repeated with considerably improved precision using lasers as light sources. Experimenters working at the Joint Institute for Laboratory Astrophysics in Boulder, Colorado built an apparatus that consisted of two He-Ne lasers on a rotating granite platform. By electronically stabilizing the lasers, they improved the sensitivity of their apparatus by several orders of magnitude. Again expressing the result as a difference between the speeds along the two arms of the apparatus, this experiment corresponds to $\Delta c/c < 8 \times 10^{-15}$, an improvement of about 5 orders of magnitude over the original Michelson-Morley experiment. In a similar repetition of the Kennedy-Thorndike experiment using He-Ne lasers, they obtained $\Delta c/c < 1 \times 10^{-10}$, an improvement over the original experiment by a factor of 300. [See A. Brillet and J. L. Hall, *Physical Review Letters* **42**, 549 (1979); D. Hils and J. L. Hall, *Physical Review Letters* **64**, 1697 (1990).]

Now let us turn to the second type of test of the second postulate: Does the speed of light change with relative motion between the source and the observer? Special relativity asserts that all observers measure the same value of c. To test this prediction, it is necessary to measure the speed of a light beam emitted by a source in motion. Suppose we observe this beam along the direction of motion of the moving source, which might be moving toward us or away from us. In the rest frame of the source, the emitted light travels at speed c. We can express the speed of light in our reference frame as $c' = c + \Delta c$, where Δc is zero according to special relativity ($c' = c$) or is $\pm u$ according to classical physics ($c' = c \pm u$ in the Galilean transformation, depending on whether the motion is toward or away from the observer).

One experiment of this type is to study the X rays emitted by a binary pulsar, a rapidly pulsating source of X rays in orbit about another star, which would eclipse the pulsar as it rotated in its orbit. If the speed of light (in this case, X rays) were to change as the pulsar moved first toward and later away from the Earth in its orbit, the beginning and end of the eclipse would not be equally spaced in time from the midpoint of the eclipse. No such effect is observed, and from these observations it is concluded that $\Delta c/c < 2 \times 10^{-12}$, in agreement with predictions of special relativity. These experiments were done at $u/c = 10^{-3}$. [See K. Brecher, *Physical Review Letters* **39**, 1051 (1977).]

In another experiment of this type, the decay of pi mesons (pions) into gamma rays (another form of electromagnetic waves traveling at c) was observed. When pions (produced in laboratories with large accelerators) emit these gamma rays, they are traveling at speeds close to the speed of light, relative to the laboratory. Thus if Galilean relativity were valid, we should expect to find gamma rays emitted in the direction of motion of the decaying pions traveling at a speed c' in the laboratory of nearly $2c$, rather than always with c as predicted by special relativity. The observed laboratory speed of these gamma rays in one experiment was $(2.9977 \pm 0.0004) \times 10^8$ m/s when the decaying pions were moving at $u/c = 0.99975$. These results give $\Delta c/c < 2 \times 10^{-4}$, and $c' = c$ as expected from special relativity. This experiment shows directly that an object moving at a speed of nearly c relative to the laboratory emits "light" which travels at a speed of c relative to both the object *and* the laboratory, giving direct evidence for Einstein's second postulate. [See T. Alvager et al., *Physics Letters* **12**, 260 (1964).]

Time Dilation

We have already discussed the time dilation effect on the decay of muons produced by cosmic rays. Muon decay can also be studied in the laboratory. Muons can be produced following collisions in high-energy accelerators, and the decay of the muons can be followed by observing their decay products (ordinary electrons). These muons can either be trapped and decay at rest, or they can be placed in a beam and decay in flight. When muons are observed at rest, their decay lifetime is 2.198 μs. (As we discuss in Chapter 12, decays generally follow an exponential law. The lifetime is the time after which a fraction $1/e = 0.368$ of the original muons remain.) This is the *proper lifetime*, measured in a frame of reference in which the muon is at rest. In one particular experiment, muons were trapped in a ring and circulated at a momentum of $p = 3094$ MeV/c. The decays in flight occurred with a lifetime of 64.37 μs (measured in the laboratory frame of reference). For muons of this momentum, Equation 2.8 gives a dilated lifetime of (see Problem 38) 64.38 μs, which is in excellent agreement with the measured value and confirms the time dilation effect. [See J. Bailey et al., *Nature* **268**, 301 (1977).]

Another similar experiment was done with pions. The proper lifetime, measured for pions at rest, is known to be 26.0 ns. In one experiment, pions were observed in flight at $u/c = 0.913$, and their lifetime was measured to be 63.7 ns. (Pions decay to muons, so we can follow the exponential radioactive decay of the pions by observing the muons emitted as a result of the decay.) For pions moving at this speed, the expected dilated lifetime is in exact agreement with the measured value, once again confirming the time dilation effect. [See D. S. Ayres et al., *Physical Review D* **3**, 1051 (1971).]

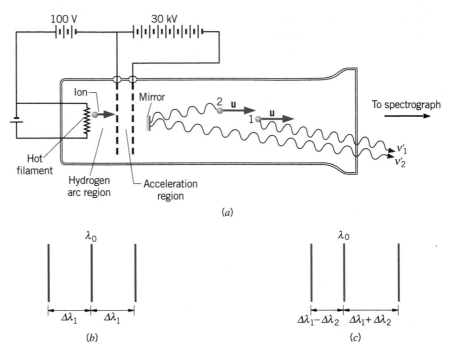

FIGURE 2.24 (*a*) Apparatus used in the Ives-Stilwell experiment. (*b*) Line spectrum expected from classical Doppler effect. (*c*) Line spectrum expected from relativistic Doppler effect.

The Doppler Effect

Confirmation of the relativistic Doppler effect first came from experiments done in 1938 by Ives and Stilwell. They sent a beam of hydrogen atoms, generated in a gas discharge, down a tube at a speed *u*, as shown in Figure 2.24. They could simultaneously observe light emitted by the atoms in a direction parallel to **u** (atom 1) and opposite to **u** (atom 2, reflected from the mirror). They used a spectrograph to photograph the characteristic spectral lines from these atoms and also, on the same photographic plate, from atoms at rest. If the classical Doppler formula were valid, the wavelengths of the lines from atoms 1 and 2 would be placed at symmetric intervals $\Delta\lambda_1 = \pm\lambda_0(u/c)$ on either side of the line from the atoms at rest (wavelength λ_0). The relativistic Doppler formula, on the other hand, gives a small additional asymmetric shift $\Delta\lambda_2 = +\frac{1}{2}\lambda_0(u/c)^2$ (computed for $u \ll c$, so that higher-order terms in u/c can be neglected). Figure 2.25 shows the results of Ives and Stilwell for one of the hydrogen lines (the blue line of the Balmer series at $\lambda_0 = 486$ nm). The agreement between the observed values and those predicted by the relativistic formula is impressive.

Recent experiments with lasers have verified the relativistic formula at greater accuracy. These experiments are based on the absorption of laser light by an atom; when the radiation is absorbed, the atom changes from its lowest-energy state (the ground state) to one of its excited states. The experiment consists essentially of comparing the laser wavelength needed

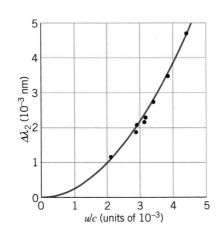

FIGURE 2.25 Results of the Ives-Stilwell experiment. According to classical theory, $\Delta\lambda_2 = 0$, while according to special relativity, $\Delta\lambda_2$ depends on $(u/c)^2$.

to excitc atoms at rest with that needed for atoms in motion. One experiment used a beam of hydrogen atoms with kinetic energy 800 MeV (corresponding to $u/c = 0.84$) produced in a high-energy proton accelerator. An ultraviolet laser was used to excite the atoms. This experiment verified the relativistic Doppler effect to an accuracy of about 3×10^{-4}. [See D. W. MacArthur et al., *Physical Review Letters* **56**, 282 (1986).] In another experiment, a beam of neon atoms moving with a speed of $u = 0.0036c$ was irradiated with light from a tunable dye laser. This experiment verified the relativistic Doppler shift to a precision of 2×10^{-6}. [See R. W. McGowan et al., *Physical Review Letters* **70**, 251 (1993).]

Relativistic Momentum and Energy

Nearly every time the nuclear or particle physicist enters the laboratory, a direct or indirect test of the momentum and energy relationships of special relativity is made. Principles of special relativity must be incorporated in the design of the high-energy accelerators used by nuclear and particle physicists, so even the existence of these successful building projects gives testimony to the validity of the formulas of special relativity.

The earliest direct confirmation of these relationships came just a few years after Einstein's 1905 paper. Simultaneous measurements were made of the momentum and velocity of high-energy electrons emitted in certain radioactive decay processes (nuclear beta decay, which is discussed in Chapter 12). Figure 2.26 shows the results of several different investigations plotted as *p/mv*, which should have the value 1 according to classical physics. The results agree with the relativistic formula and disagree with the classical one. Note that the relativistic and classical formulas give the same results at low speeds, and in fact the two cannot be distinguished for speeds below $0.1c$, which accounts for our failure to observe these effects in experiments with ordinary laboratory objects.

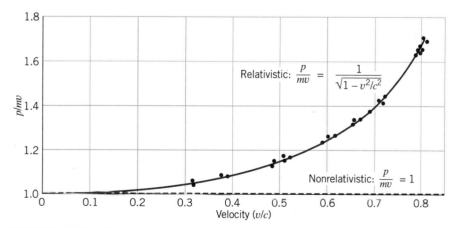

FIGURE 2.26 The ratio *p/mv* is plotted for electrons of various speeds. The data agree with the relativistic result and not at all with the nonrelativistic result ($p/mv = 1$).

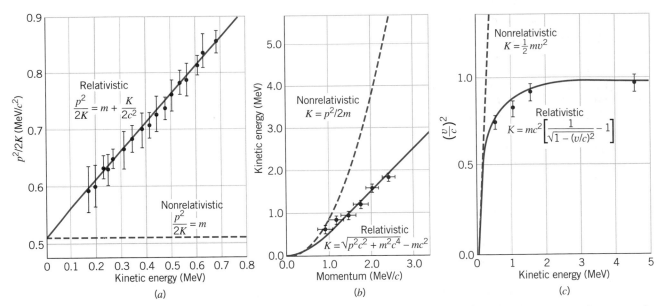

FIGURE 2.27   Confirmation of relativisitic kinetic energy relationships. In (*a*) and (*b*) the momentum and energy of radioactive decay electrons were measured simultaneously. In these two independent experiments, the data were plotted in different ways, but the results are clearly in good agreement with the relativistic relationships and in poor agreement with the classical, nonrelativistic relationships. In (*c*) electrons were accelerated to a fixed energy through a large electric field (up to 4.5 million volts, as shown) and the velocities of the electrons were determined by measuring the flight time over 8.4 m. Notice that at small kinetic energies $K \ll mc^2$), the relativistic and nonrelativistic relationships become identical. [*Sources:* (*a*) K. N. Geller and R. Kollarits, *Am. J. Phys.* **40**, 1125 (1972); (*b*) S. Parker, *Am. J. Phys.* **40**, 241 (1972); (*c*) W. Bertozzi, *Am. J. Phys.* **32**, 551 (1964).]

Other more recent experiments, in which the kinetic energies of fast electrons were measured, are shown in Figure 2.27. Once again, the data at high speeds agree with special relativity and disagree with the classical equations.

Twin Paradox

Although we cannot perform the experiment to test the twin paradox as we have described it, we can do an equivalent experiment. We take two clocks in our laboratory and synchronize them carefully. We then place one of the clocks in an airplane and fly it around the Earth. When we return the clock to the laboratory and compare the two clocks, we expect to find, if special relativity is correct, that the clock that has left the laboratory is the "younger" one—that is, it will tick away fewer seconds and appear to run behind its stationary twin. In this experiment, we use very precise clocks based on the atomic vibrations of cesium in order to measure the time differences between the clock readings, which amount to only about 10^{-7} s. This experiment is complicated by several factors, all of which can be computed rather precisely: the rotating Earth is *not* an inertial frame (there is a centripetal acceleration), clocks on the surface of the Earth are *already* moving because of the rotation of the Earth, and the *general* theory of relativity predicts that a change in the gravitational field strength, which our moving clock will experience as it changes altitude in its airplane flight, will also change the rate at which the clock runs. In this experiment, as in the others we have discussed, the results are entirely in agreement with the predictions of special relativity. [See J. C. Hafele and R. E. Keating, *Science* **177**, 166 (1972).]

In a similar experiment, a cesium atomic clock carried on the Space Shuttle was compared with an identical clock on the Earth. The comparison was made through a radio link between the Shuttle and the ground station. At an orbital height of about 328 km, the Shuttle moves at a speed of about 7712 m/s, or $2.5 \times 10^{-5}c$. A clock moving at this speed runs slower than an identical clock at rest by the time dilation factor. At low speed ($u \ll c$), we can approximate the difference between the clock rates as $\frac{1}{2}u^2/c^2$, or 3.30×10^{-10}. For every second the clock is in orbit, it loses 330 ps; equivalently, it loses about 1.8 μs per orbit. These time intervals can be measured with great precision, and the predicted asymmetric aging was verified to a precision of about 0.1 percent. [See E. Sappl, *Naturwissenschaften* **77**, 325 (1990).]

SUGGESTIONS FOR FURTHER READING

Special relativity has perhaps been the subject of more popular, nonmathematical books than any other area of science. Here are a few you can read for fun and to broaden your background:

L. Barnett, *The Universe and Dr. Einstein* (New York, Time Inc., 1962).

G. Gamow, *Mr. Tompkins in Paperback* (Cambridge, Cambridge University Press, 1967). Especially recommended; the master of popular science writing takes us on a fanciful journey to a world where *c* is so small that special relativistic effects are commonplace.

L. Marder, *Time and the Space Traveler* (Philadelphia, University of Pennsylvania Press, 1971).

B. Russell, *The ABC of Relativity* (New York, New American Library, 1958).

J. T. Schwartz, *Relativity in Illustrations* (New York, New York University Press, 1962).

Other introductions to relativity, more complete but not particularly more difficult than the present level, are the following:

A. P. French, *Special Relativity* (New York, Norton, 1968).

R. Resnick, *Introduction to Special Relativity* (New York, Wiley, 1968).

R. Resnick and D. Halliday, *Basic Concepts in Relativity* (New York, Macmillan, 1992).

For discussions of the appearance of objects traveling near the speed of light, see:

V. T. Weisskopf, "The Visual Appearance of Rapidly Moving Objects," *Physics Today,* September 1960.

I. Peterson, "Space-Time Odyssey," *Science News* **137**, 222 (April 14, 1990).

Some other useful works are:

L. B. Okun, "The Concept of Mass," *Physics Today,* June 1989, p. 31. This article explores the history of the "relativistic mass" concept and the connection between mass and rest energy.

C. Swartz, "Reference Frames and Relativity," *The Physics Teacher,* September 1989, p. 437. This article gives some mostly classical descriptions of inertial and noninertial reference frames.

R. Baierlein, "Teaching $E = mc^2$," *The Physics Teacher,* March 1991, p. 170. This article discusses some of the common misconceptions about mass and energy in special relativity.

Finally, a unique and delightful exploration of special relativity; elegant and witty, with all of the relativity paradoxes you could want, carefully explained and diagrammed, with many worked examples:

E. F. Taylor and J. A. Wheeler, *Spacetime Physics,* 2nd ed. (New York, Freeman, 1992).

QUESTIONS

1. Explain in your own words what is meant by the term "relativity." Are there different theories of relativity?

2. Suppose the two observers and the rock described in the first paragraph of Section 2.1 were isolated in interstellar space. Discuss the two observers' differing perceptions of the motion of the rock. Is there any experiment they can do to determine whether the rock is moving in any absolute sense?

3. Describe the situation of Figure 2.4 as it would appear from the reference frame of O'.

4. Does the Michelson-Morley experiment show that the ether does not exist or that it is merely unnecessary?

5. Suppose we made a pair of shears in which the cutting blades were many orders of magnitude longer than the handle. Let us in fact make it so long that, when we move the handles at angular velocity ω, a point on the tip of the blade has a tangential velocity $v = \omega r$ that is greater than c. Does this contradict special relativity? Justify your answer.

6. Light travels through water at a speed of about 2.25×10^8 m/s. Is it possible for a particle to travel through water at a speed v greater than 2.25×10^8 m/s?

7. Is it possible to have particles that travel at the speed of light? What does Equation 2.38 require of such particles?

8. How does relativity combine space and time coordinates into spacetime?

9. Einstein developed the relativity theory after trying unsuccessfully to imagine how a light beam would look to an observer traveling with the beam at speed c. Why is this so difficult to imagine?

10. Explain in your own words the terms *time dilation* and *length contraction.*

11. Does the Moon's disk appear to be a different size to a space traveler approaching it at $v = 0.99c$, compared with the view of a person at rest at the same location?

12. According to the time dilation effect, would the life expectancy of someone who lives at the equator be longer or shorter than someone who lives at the North Pole? By how much?

13. Criticize the following argument. "Here is a way to travel faster than light. Suppose a star is 10 light-years away. A radio signal sent from Earth would need 20 years to make the round trip to the star. If I were to travel to the star in my rocket at $v = 0.8c$, to me the distance to the star is contracted by $\sqrt{1 - (0.8)^2}$ to 6 light-years, and at that speed it would take me 6 light-years/$0.8c$ = 7.5 years to travel there. The round trip takes me only 15 years, and therefore I travel faster than light, which takes 20 years."

14. Is it possible to synchronize clocks that are in motion relative to each other? Try to design a method to do so. Which observers will believe the clocks to be synchronized?

15. Suppose event A causes event B. To one observer, event A comes before event B. Is it possible that in another frame of reference event B could come before event A? Discuss.

16. Is mass a conserved quantity in classical physics? In special relativity?

17. "In special relativity, mass and energy are equivalent." Discuss this statement and give examples.

18. Which is more massive, an object at low temperature or the same object at high temperature? A spring at its natural length or the same spring under compression? A container of gas at low pressure or at high pressure? A charged capacitor or an uncharged one?

19. Could a collision be elastic in one frame of reference and inelastic in another?

20. (a) What properties of nature would be different if there were a relativistic transformation law for electric charge? (b) What experiments could be done to prove that electric charge does *not* change with velocity?

PROBLEMS

1. A shift of one fringe in the Michelson-Morley experiment corresponds to a change in the round-trip travel time along one arm of the interferometer by one period of vibration of light (about 2×10^{-15} s) when the apparatus is rotated by 90°. Based on the results of Example 2.3, what velocity through the ether would be deduced from a shift of one fringe? (Take the length of the interferometer arm to be 11 m.)

2. The distance from New York to Los Angeles is about 5000 km and should take about 50 h in a car driving at 100 km/h. (a) How much shorter than 5000 km is the distance according to the car travelers? (b) How much less than 50 h do they age during the trip?

3. How fast must an object move before its length appears to be contracted to one-half its proper length?

4. An astronaut must journey to a distant planet, which is 200 light-years from Earth. What speed will be necessary if the astronaut wishes to age only 10 years during the round trip?

5. The proper lifetime of a certain particle is 100.0 ns. (a) How long does it live in the laboratory if it moves at $v = 0.960c$? (b) How far does it travel in the laboratory during that time? (c) What is the distance traveled in the laboratory according to an observer moving with the particle?

6. High-energy particles are observed in laboratories by photographing the tracks they leave in certain detectors; the length of the track depends on the speed of the particle and its lifetime. A particle moving at $0.995c$ leaves a track 1.25 mm long. What is the proper lifetime of the particle?

7. Carry out the missing steps in the derivation of Equation 2.17.

8. Two spaceships approach the Earth from opposite directions. According to an observer on the Earth, ship A is moving at a speed of $0.753c$ and ship B at a speed of $0.851c$. What is the speed of ship A as observed from ship B? Of ship B as observed from ship A?

9. Rocket A leaves a space station with a speed of $0.826c$. Later, rocket B leaves in the same direction with a speed of $0.635c$. What is the speed of rocket A as observed from rocket B?

10. One of the strongest emission lines observed from distant galaxies comes from hydrogen and has a wavelength of 122 nm (in the ultraviolet region). (a) How fast must a galaxy be moving away from us in order for that line to be observed in the visible region at 366 nm? (b) What would be the wavelength of the line if that galaxy were moving toward us at the same speed?

11. A physics professor claims in court that the reason he went through the red light ($\lambda = 650$ nm) was that, due to his motion, the red color was Doppler shifted to green ($\lambda = 550$ nm). How fast was he going?

12. A "cause" occurs at point 1 (x_1, t_1) and its "effect" occurs at point 2 (x_2, t_2). Use the Lorentz transformation to find $t_2' - t_1'$, and show that $t_2' - t_1' > 0$; that is, O' can never see the "effect" coming before its "cause."

13. Derive the Lorentz velocity transformations for v_x' and v_z'.

14. Observer O fires a light beam in the y direction ($v_y = c$). Use the Lorentz velocity transformation to find v_x' and v_y' and show that O' also measures the value c for the speed of light. Assume O' moves relative to O with velocity u in the x direction.

15. Observer O sees a red flash of light at the origin at $t = 0$ and a blue flash of light at $x = 3.26$ km at a time $t = 7.63$ μs. What are the distance and the time interval between the flashes according to observer O', who moves relative to O in the direction of increasing x with a speed of $0.625c$? Assume that the origins of the two coordinate systems line up at $t = t' = 0$.

16. A light bulb at point x in the frame of reference of O blinks on and off at intervals $\Delta t = t_2 - t_1$. Observer O', moving relative to O at speed u, measures the interval to be $\Delta t' = t_2' - t_1'$. Use the Lorentz transformation expressions to derive the time dilation expression relating Δt and $\Delta t'$.

17. Several spacecraft leave a space station at the same time. Relative to an observer on the station, A travels at $0.60c$ in the x direction, B at $0.50c$ in the y direction, C at $0.50c$ in the negative x direction, and D at $0.50c$ at 45° between

the y and negative x directions. Find the velocity components, directions, and speeds of B, C, and D as observed from A.

18. Suppose rocket traveler Amelia (Section 2.6) has a clock made on Earth. Every year on her birthday she sends a light signal to brother Casper on Earth. (a) At what rate does Casper receive the signals during Amelia's outward journey? (b) At what rate does he receive the signals during her return journey? (c) How many of Amelia's birthday signals does Casper receive during the journey that he measures to last 40 years?

19. Using the velocities shown in Figure 2.20a and b, obtain the kinetic energies given in Equation 2.35.

20. Using the relativistic formula for kinetic energy and the velocities given in Figure 2.20, show that kinetic energy is conserved according to observer O'.

21. The work-energy theorem relates the change in kinetic energy of a particle to the work done on it by an external force: $\Delta K = W = \int F\, dx$. Writing Newton's second law as $F = dp/dt$, show that $W = \int v\, dp$ and integrate by parts using the relativistic momentum to obtain Equation 2.36.

22. For what range of velocities of a particle of mass m can we use the classical expression for kinetic energy $\frac{1}{2}mv^2$ to within an accuracy of 1 percent?

23. For what range of velocities of a particle of mass m can we use the extreme relativistic approximation $E \cong pc$ to within an accuracy of 1 percent?

24. Use Equations 2.32 and 2.38 to derive Equation 2.41.

25. Suppose an observer O measures a particle of mass m moving in the x direction to have speed v, energy E, and momentum p. Observer O', moving at speed u in the x direction, measures v', E', and p' for the same object. (a) Use the Lorentz velocity transformation to find E' and p' in terms of m, u, and v. (b) Reduce $E'^2 - (p'c)^2$ to its simplest form and interpret the result.

26. Repeat Problem 25 for the mass moving in the y direction according to O. The velocity u of O' is still along the x direction.

27. Use the binomial expansion $(1 + x)^n = 1 + nx + [n(n - 1)/2!]x^2 + \cdots$ to show that Equation 2.36 for the relativistic kinetic energy reduces to the classical expression $\frac{1}{2}mv^2$ when $v \ll c$. This important result shows that our familiar expressions are correct at low speeds. By evaluating the first term in the expansion beyond $\frac{1}{2}mv^2$, find the speed necessary before the classical expression is off by 0.01 percent.

28. What is the change in mass when 1 g of copper is heated from 0 to 100°C? The specific heat of copper is 0.40 J/g·K.

29. Find the kinetic energy of an electron moving at a speed of (a) $v = 1.00 \times 10^{-4}c$; (b) $v = 1.00 \times 10^{-2}c$; (c) $v = 0.300c$; (d) $v = 0.999c$.

30. The 50-GeV accelerator described in Example 2.14 has a length of 3 km (about 2 miles). (a) How long would it take an electron to travel 3 km at the speed found in Example 2.14? (b) What is the length of the accelerator from the electron's frame of reference? (c) How long does it take, in the electron's frame of reference, to travel the length of the accelerator?

31. Electrons are accelerated to high speeds by a two-stage machine. The first stage accelerates the electrons from rest to $v = 0.99c$. The second stage accelerates the electrons from $0.99c$ to $0.999c$. (a) How much energy does the first

stage add to the electrons? (b) How much energy does the second stage add in increasing the velocity by only 0.9 percent?

32. An electron and a proton are each accelerated through a potential difference of 10.0 million volts. Find the momentum (in MeV/c) and the kinetic energy (in MeV) of each, and compare with the results of using the classical formulas.

33. In a nuclear reactor, each atom of uranium releases about 200 MeV when it fissions. What is the change in mass when 1 kg of uranium is fissioned?

34. An electron and a positron (an antielectron) make a head-on collision, each moving at $v = 0.99999c$. In the collision the electrons disappear and are replaced by two muons ($mc^2 = 105.7$ MeV) which move off in opposite directions. What is the kinetic energy of each of the muons?

35. It is desired to create a particle of mass 9700 MeV in a collision between a proton and an antiproton traveling at the same speed in opposite directions. What speed is necessary for this to occur?

36. A pion has a rest energy of 135 MeV. It decays into two gamma rays, bursts of electromagnetic radiation that travel at the speed of light. A pion moving through the laboratory at $v = 0.98c$ decays into two gamma rays of equal energies, making equal angles θ with the direction of motion. Find the angle θ and the energies of the two gamma rays. (*Hint*: Gamma rays are electromagnetic radiation with $E = pc$.)

37. Solve Example 2.16 by first making a Lorentz transformation to a reference frame in which the initial kaon is at rest. When a kaon at rest decays into two pions, they move in opposite directions with equal and opposite velocities, so they share the decay energy equally. Find the energies and velocities of the two pions in the kaon's rest frame. Then transform back to the lab frame to find their energies and velocities.

38. In the muon decay experiment discussed in Section 2.7 as a verification of time dilation, the muons move in the lab with a momentum of 3094 MeV/c. Find the dilated lifetime in the laboratory frame. (The proper lifetime is 2.198 μs.)

39. Derive the relativistic expression $p^2/2K = m + K/2c^2$, which is plotted in Figure 2.27a.

3

These solar cells turn the Sun's radiant energy directly into electricity through the photoelectric effect. Each cell produces a power of about 0.2 W, and the complete array of 120,000 cells provides enough power to pump water through an irrigation system.

THE PARTICLELIKE PROPERTIES OF ELECTROMAGNETIC RADIATION

We now turn to a discussion of *wave mechanics,* the second theory on which modern physics is based. One consequence of wave mechanics is the breakdown of the classical distinction between particles and waves. In this chapter we consider the three early experiments that provided evidence that light, which we have treated as a wave phenomenon, has properties that we normally associate with particles. Instead of spreading its energy smoothly over a wave front, the energy is delivered in concentrated bundles like particles; a discrete bundle (*quantum*) of electromagnetic energy is known as a photon.

Before we begin to discuss the experimental evidence that supports the existence of the photon and the particlelike properties of light, we first review some of the properties of electromagnetic waves.

3.1 REVIEW OF ELECTROMAGNETIC WAVES

An electromagnetic field is characterized by its electric field strength $\mathbf{E}$ and magnetic field intensity $\mathbf{B}$. For example, the radial electric field due to a point charge q at the origin is

$$\mathbf{E} = \frac{1}{4\pi\varepsilon_0}\frac{q}{r^2}\hat{\mathbf{r}} \tag{3.1}$$

where $\hat{\mathbf{r}}$ is a unit vector in the radial direction. The magnetic field at a distance r from a long, straight, current-carrying wire along the z axis is

$$\mathbf{B} = \frac{\mu_0 i}{2\pi r}\hat{\boldsymbol{\theta}} \tag{3.2}$$

where $\hat{\boldsymbol{\theta}}$ is a unit vector in the θ direction in cylindrical coordinates.

If the charges are accelerated, or if the current varies with time, an electromagnetic wave is produced, in which $\mathbf{E}$ and $\mathbf{B}$ vary not only with $\mathbf{r}$ but also with t. The mathematical expression that describes such a wave may have one of many different forms, depending on the properties of the source of the wave and of the medium through which the wave travels. One special form is the *plane wave,* in which the wave fronts are planes. (A point source, on the other hand, produces spherical waves, in which the wave fronts are spheres.) A plane electromagnetic wave traveling in the positive z direction is described by the expressions

Plane electromagnetic wave

$$\begin{aligned}\mathbf{E} &= \mathbf{E}_0 \sin\left(kz - \omega t + \phi\right) \\ \mathbf{B} &= \mathbf{B}_0 \sin\left(kz - \omega t + \phi\right)\end{aligned} \tag{3.3}$$

where the *wave number* k is found from the wavelength, $k = 2\pi/\lambda$, and the *angular frequency* ω is found from the frequency, $\omega = 2\pi\nu$. Since λ and ν are related by $c = \lambda\nu$, k and ω are also related by $c = \omega/k$. The angle ϕ represents an arbitrary phase angle.

The polarization of the wave is represented by the vector $\mathbf{E}_0$; the plane of polarization is determined by $\mathbf{E}_0$ and by the direction of propagation, the z axis in this case. Once we specify the direction of travel and the polarization $\mathbf{E}_0$, the direction of $\mathbf{B}_0$ is fixed by the requirements that $\mathbf{B}$ must be perpendicular to both $\mathbf{E}$ and the direction of travel, and that the vector product $\mathbf{E} \times \mathbf{B}$ point in the direction of travel. For example if $\mathbf{E}_0$ is in the x direction ($\mathbf{E}_0 = E_0\hat{\mathbf{i}}$, where $\hat{\mathbf{i}}$ is a unit vector in the x direction), then $\mathbf{B}_0$ must be in the y direction ($\mathbf{B}_0 = B_0\hat{\mathbf{j}}$). Moreover, the magnitude of $\mathbf{B}_0$ is determined by

$$B_0 = \frac{E_0}{c} \tag{3.4}$$

where c is the speed of light.

An electromagnetic wave transmits energy from one place to another; the energy flux is specified by the *Poynting vector* $\mathbf{S}$:

$$\mathbf{S} = \frac{1}{\mu_0} \mathbf{E} \times \mathbf{B} \tag{3.5}$$

For the plane wave, this reduces to

$$\mathbf{S} = \frac{1}{\mu_0} E_0 B_0 \sin^2 (kz - \omega t + \phi)\hat{\mathbf{k}} \tag{3.6}$$

where $\hat{\mathbf{k}}$ is a unit vector in the z direction. The Poynting vector has dimensions of energy per unit time per unit area—for example, J/s/m^2 or W/m^2. Figure 3.1 shows the orientation of the vectors $\mathbf{E}$, $\mathbf{B}$, and $\mathbf{S}$ for this special case.

Let us imagine the following experiment. We place a detector of electromagnetic radiation (a radio receiver or a human eye) at some point on the z axis, and we determine the electromagnetic power that the above plane

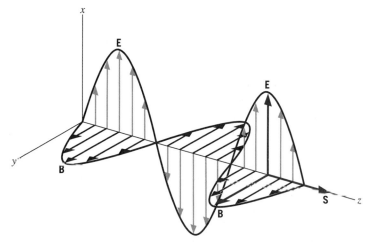

FIGURE 3.1 An electromagnetic wave traveling in the z direction. The electric field $\mathbf{E}$ lies in the xz plane and the magnetic field $\mathbf{B}$ lies in the yz plane.

wave delivers to the receiver. The receiver is oriented with its sensitive area A perpendicular to the z axis, so that the maximum signal is received; we correspondingly drop the vector representation of **S** and work only with its magnitude S. The power P entering the receiver is then

$$P = SA = \frac{1}{\mu_0} E_0 B_0 A \sin^2 (kz - \omega t + \phi) \tag{3.7}$$

which we can rewrite as

$$P = \frac{1}{\mu_0 c} E_0^2 A \sin^2 (kz - \omega t + \phi) \tag{3.8}$$

There are two important features of this expression that you should recognize:

Intensity of electromagnetic waves

1. The intensity (the average power per unit area) is proportional to E_0^2. This is a general property of waves: *the intensity is proportional to the square of the amplitude.* We will learn later that this same property also characterizes the waves that describe the behavior of material particles.

2. The intensity fluctuates with time, with the frequency $2\nu = 2(\omega/2\pi)$. Of course we don't usually observe this fluctuation; visible light, for example, has a frequency of about 10^{15} oscillations per second, and since our eye doesn't respond that quickly, we observe the time average of many (perhaps 10^{13}) cycles. If T is the observation time (perhaps 10^{-2} s in the case of the eye) then the average power is

$$P_{av} = \frac{1}{T} \int_0^T P \, dt \tag{3.9}$$

and using Equation 3.8 we obtain the intensity I:

$$I = \frac{P_{av}}{A} = \frac{1}{2\mu_0 c} E_0^2 \tag{3.10}$$

since the average value of $\sin^2 \theta$ is $\frac{1}{2}$.

The property that makes waves a unique physical phenomenon is the *principle of superposition,* which, for example, allows two waves to meet at a point, to cause a combined disturbance at the point that might be greater or less than the disturbance produced by either wave alone, and finally to emerge from the point of "collision" with all of the properties of each wave totally unchanged by the collision. To appreciate this important distinction between material objects and waves, imagine trying that trick with two automobiles!

This special property of waves leads to the phenomena of *interference* and *diffraction.* The simplest and best-known example of interference is *Young's double-slit experiment,* in which a monochromatic plane wave is incident on a barrier in which two narrow slits have been cut. (This experiment was first done with light waves, but in fact any wave will do as well, not only other electromagnetic waves, such as microwaves, but mechanical waves, such as water waves or sound waves. We assume that the experiment is being done with light waves.)

Young's double-slit experiment

Figure 3.2 illustrates this experimental arrangement. The plane wave is *diffracted* by each of the slits, so that the light passing through each slit covers a much larger area on the screen than the geometric shadow of the slit. This causes the light from the two slits to overlap on the screen, producing the interference. If we move away from the center of the screen just the right distance, we reach a point at which a wave crest passing through one slit arrives at exactly the same time as the previous wave crest that passed through the other slit. When this occurs, the intensity is a maximum, and a bright region appears on the screen. This is *constructive interference,* and it occurs continually at the point on the screen that is exactly one wavelength further from one slit than from the other. That is, if X_1 and X_2 are the distances from the point on the screen to the two slits, then a condition for maximum constructive interference is $|X_1 - X_2| = \lambda$. Constructive interference occurs when any wave crest from one slit arrives simultaneously with another from the other slit, whether it is the next, or the fourth, or the forty-seventh. The general condition for complete constructive interference is that the difference between X_1 and X_2 be an integral number of wavelengths:

$$|X_1 - X_2| = n\lambda \qquad n = 0, 1, 2, \ldots \qquad (3.11)$$ *Double-slit maxima*

It is also possible for the crest of the wave from one slit to arrive at a point on the screen simultaneously with the trough (valley) of the wave from the other slit. When this happens, the two waves cancel, giving a dark region

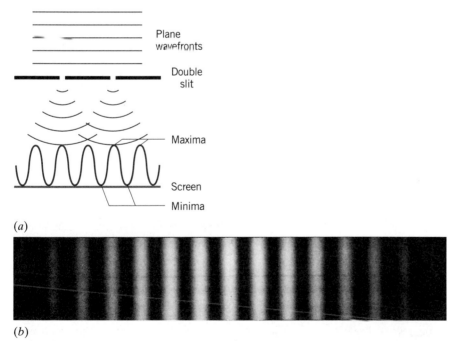

(a)

(b)

FIGURE 3.2 (a) Young's double-slit experiment. A plane wave front passes through both slits; the wave is diffracted at the slits, and interference occurs where the diffracted waves overlap on the screen. (b) The interference fringes observed on the screen.

on the screen. This is known as *destructive interference.* (The existence of destructive interference at intensity minima immediately shows that we must add the field vectors **E** of the waves from the two slits, and not their powers *P*, since *P* can never be negative.) Destructive interference occurs whenever the distances X_1 and X_2 are such that the phase of one wave differs from the other by one-half cycle, or by one and one-half cycles, two and one-half cycles, and so forth:

$$|X_1 - X_2| = \tfrac{1}{2}\lambda, \tfrac{3}{2}\lambda, \tfrac{5}{2}\lambda, \ldots$$

$$= (n + \tfrac{1}{2})\lambda \qquad n = 0, 1, 2, \ldots \tag{3.12}$$

Double-slit minima

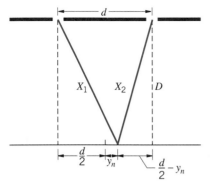

We can find the locations on the screen where the interference maxima occur in the following way. Let *d* be the separation of the slits, and let *D* be the distance from the slits to the screen. If y_n is the distance from the center of the screen to the *n*th maximum, then from the geometry of Figure 3.3 we find (assuming $X_1 > X_2$)

$$X_1^2 = D^2 + \left(\frac{d}{2} + y_n\right)^2$$

$$X_2^2 = D^2 + \left(\frac{d}{2} - y_n\right)^2 \tag{3.13}$$

FIGURE 3.3 The geometry of the double-slit experiment.

Subtracting, we obtain

$$X_1^2 - X_2^2 = 2y_n d \tag{3.14}$$

and so

$$y_n = \frac{(X_1 + X_2)(X_1 - X_2)}{2d} \tag{3.15}$$

In experiments with light, *D* is of order 1 m and y_n and *d* are typically at most 1 mm; thus $X_1 \cong D$ and $X_2 \cong D$, so $X_1 + X_2 \cong 2D$, and to a good approximation

$$y_n = (X_1 - X_2)\frac{D}{d} \tag{3.16}$$

Using Equation 3.11 for the values of $(X_1 - X_2)$ at the maxima, we find

$$y_n = n\frac{\lambda D}{d} \tag{3.17}$$

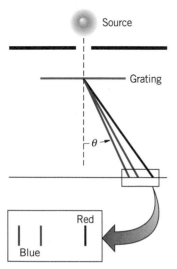

FIGURE 3.4 The use of a diffraction grating to analyze light into its constituent wavelengths.

Another device for observing the interference of light waves is the *diffraction grating,* which is a sort of *multiple-slit* device for producing the interference of light waves. The operation of this device is illustrated in Figure 3.4; interference maxima corresponding to different wavelengths appear at different angles θ, according to

$$d \sin \theta = n\lambda \tag{3.18}$$

where *d* is the slit spacing and *n* is the order number of the maximum ($n = 1, 2, 3, \ldots$). This relationship is derived in many introductory physics textbooks.

The advantage of the diffraction grating is its superior resolution—it enables us to get very good separation of wavelengths that are close to one another, and thus it is a very useful device for measuring wavelengths. Notice, however, that in order to get reasonable values of the angle θ, for example sin θ in the range of 0.3 to 0.5, we must have d of the order of a few times the wavelength. For visible light this is not particularly difficult, but for radiations of very short wavelength, mechanical construction of a grating is not possible. For example, for X rays with a wavelength of the order of 0.1 nm, we would need to construct a grating in which the slits were less than 1 nm apart, which is roughly the same as the spacing between the atoms of most materials. The solution to this problem has been known since the pioneering experiments of Laue and Bragg*—use the atoms themselves as a diffraction grating! A beam of X rays sees the regular spacings of the atoms in a crystal as a sort of three-dimensional diffraction grating. Consider the set of atoms shown in Figure 3.5, which represents a small portion of a two-dimensional slice of the crystal. The X rays are reflected from individual atoms in all directions, but in only one direction will the scattered "wavelets" constructively interfere to produce a reflected beam, and in this case we can regard the reflection as occurring from a plane drawn through the row of atoms. (This situation is identical with the reflection of light from a mirror—only in one direction will there be a beam of reflected light, and in that direction we can regard the reflection as occurring on a plane with the angle of incidence equal to the angle of reflection.) Suppose the rows of atoms are a distance d apart in the crystal. Then a portion of the beam is reflected from the front plane, and a portion is reflected from the second plane, and so forth. The wave fronts of the beam reflected from the second plane lag behind those reflected from the front plane, since the wave reflected from the second plane must travel an additional distance of 2d sin θ, where θ is the angle of incidence as *measured from the face of the crystal* (in optics we have previously always measured angles with respect to the *normal* to the surface). If this path difference is a whole number of wavelengths, the reflected beams interfere constructively and give an intensity maximum; thus the basic expression for the interference maxima in X-ray diffraction from a crystal is

X-ray diffraction

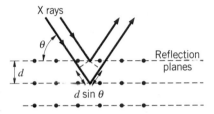

FIGURE 3.5 A beam of X rays reflected from a set of crystal planes of spacing d. The beam reflected from the second plane travels a distance 2d sin θ greater than the beam reflected from the first plane.

$$2d \sin \theta = n\lambda \qquad n = 1, 2, 3, \ldots \qquad (3.19)$$

Bragg's law

This result is known as *Bragg's law* for X-ray diffraction. Notice the factor of 2 that appears in Equation 3.19 but does *not* appear in the otherwise similar expression of Equation 3.18 for the ordinary diffraction grating.

EXAMPLE 3.1

A single crystal of table salt (NaCl) is irradiated with a beam of X rays of wavelength 0.250 nm, and the first Bragg reflection is observed at an angle of 26.3°. What is the atomic spacing of NaCl?

---

* Max von Laue (1879–1960, Germany) developed the method of X-ray diffraction for the study of crystal structures, for which he received the 1914 Nobel prize. Lawrence Bragg (1890–1971, English) developed the Bragg law for X-ray diffraction while he was a student at Cambridge University. He shared the 1915 Nobel prize with his father, William Bragg, for their research on the use of X rays to determine crystal structures.

SOLUTION

Solving Bragg's law for the spacing d, we have

$$d = \frac{n\lambda}{2 \sin \theta} = \frac{0.250 \text{ nm}}{2 \sin (26.3°)}$$

$$d = 0.282 \text{ nm}$$

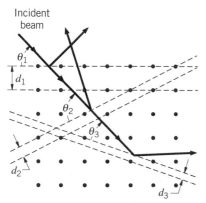

FIGURE 3.6 An incident beam of X rays can be reflected from many different crystal planes.

Our drawing of Figure 3.5 was very arbitrary—we had no basis for choosing which set of atoms to draw the reflecting planes through. Figure 3.6 shows a larger section of the crystal. As you can see, there are many possible reflecting planes, each with a different value of θ and d. (Of course, d_i, and θ_i are related and cannot be varied independently.) If we used a beam of X rays of a single wavelength, it might be difficult to find the proper angle and set of planes to observe the interference. However, if we use a beam of X rays of a continuous range of wavelengths, for each d_i and θ_i interference will occur for a certain wavelength λ_i and so there will be a pattern of interference maxima appearing at different angles of reflection as shown in Figure 3.6. The pattern of interference maxima depends not on the incident wavelengths (which form a continuous distribution) but rather on the spacing and the type of arrangement of the atoms in the crystal.

Figures 3.7 and 3.8 show sample patterns (called *Laue patterns*) that are obtained from X-ray scattering from two different crystals. The bright dots correspond to interference maxima for wavelengths from the range of incident wavelengths that happen to satisfy Equation 3.19. The three-dimensional pattern is more complicated than our two-dimensional drawings, but the individual dots have the same interpretation. Figure 3.9 shows the pattern obtained from a sample that consists of many tiny crystals, rather than one single crystal. (It looks like Figure 3.7 or 3.8 rotated rapidly about its center.) From such pictures it is also possible to deduce crystal structures and lattice spacing.

All of the examples we have discussed in this section depend on the wave properties of electromagnetic radiation. We now discuss some experiments with light and other electromagnetic radiation, which indicate that the wave interpretation is not a complete description of the properties of electromagnetic radiation.

3.2   THE PHOTOELECTRIC EFFECT

When a metal surface is illuminated with light, electrons can be emitted from the surface. This phenomenon, known as the *photoelectric effect,* was discovered by Heinrich Hertz in 1887 in the process of his research into electromagnetic radiation. The emitted electrons are called *photoelectrons.*

A sample experimental arrangement for observing the photoelectric effect is illustrated in Figure 3.10. Light falling on a metal surface (the

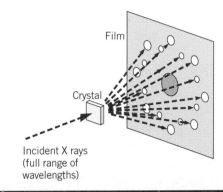

FIGURE 3.7 (Top) Apparatus for observing X-ray scattering by a crystal, An interference maximum (dot) appears on the film whenever a set of crystal planes happens to satisfy the Bragg condition for a particular wavelength. (Bottom) Laue pattern of NaCl crystal.

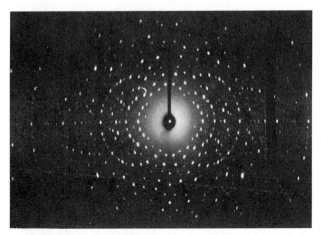

FIGURE 3.8 Laue pattern of a quartz crystal. The difference in crystal structure and spacing between quartz and NaCl makes this pattern look different from Figure 3.7.

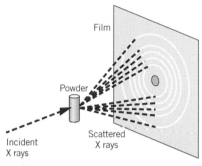

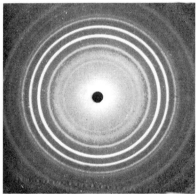

FIGURE 3.9 (Top) Apparatus for observing X-ray scattering from a powdered sample. Because the many crystals in a powder have all possible different orientations, each scattered ray of Figure 3.7 becomes a cone which forms a circle on the film. (Bottom) Diffraction pattern (known as *Debye-Scherrer* pattern) of a powder sample.

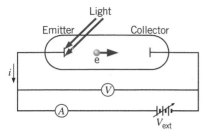

FIGURE 3.10 Apparatus for observing the photoelectric effect. The flow of electrons from the emitter to the collector is measured by the ammeter A as a current i in the external circuit. A variable voltage source V_{ext} establishes a potential difference between the emitter and collector, which is measured by the voltmeter V.

Work function

Predictions of wave theory

TABLE 3.1   SOME PHOTOELECTRIC WORK FUNCTIONS

Material	ϕ (eV)
Na	2.28
Al	4.08
Co	3.90
Cu	4.70
Zn	4.31
Ag	4.73
Pt	6.35
Pb	4.14

emitter) can release electrons, which travel to the collector. The experiment must be done in an evacuated tube, so that the electrons do not lose energy in collisions with molecules of the air. Among the properties that can be measured are the rate of electron emission and the maximum kinetic energy of the photoelectrons.* The rate of electron emission can be measured as an electric current i by an ammeter in the external circuit. The maximum kinetic energy of the electrons is determined by applying a negative potential to the collector that is just enough to repel the most energetic electrons, which then do not have enough energy to "climb" the potential hill. This potential, called the *stopping potential* V_s, is determined by increasing the magnitude of the voltage until the ammeter current drops to zero. At this point the maximum kinetic energy K_{max} is just equal to the energy eV_s lost by the electrons in "climbing" the hill:

$$K_{max} = eV_s \tag{3.20}$$

where e is the magnitude of the electric charge of the electron. Typical values of V_s are a few volts.†

In the classical picture, the surface of the metal is illuminated by an electromagnetic wave of intensity I. An electron absorbs energy from the wave until the binding energy of the electron to the metal is exceeded, at which point the electron is released. The minimum quantity of energy needed to remove an electron is called the *work function ϕ* of the material. Table 3.1 lists some values of the work function of different materials. You can see that the values are typically a few electron-volts.

What does the classical wave theory predict about the properties of the emitted photoelectrons?

1. *The maximum kinetic energy should be proportional to the intensity of the radiation.* As the brightness of the light source is increased, more energy is delivered to the surface (the electric field is greater) and the electrons should be released with greater kinetic energies. Equivalently, increasing the intensity of the light source increases the electric field **E** of the wave, which also increases the force $\mathbf{F} = -e\mathbf{E}$ on the electron and its kinetic energy when it eventually leaves the surface.

2. *The photoelectric effect should occur for light of any frequency or wavelength.* According to the wave theory, as long as the light is intense enough to release electrons, the photoelectric effect should occur no matter what the frequency or wavelength.

3. *The first electrons should be emitted in a time interval of the order of seconds after the radiation first strikes the surface.* In the wave theory, the energy of the wave is uniformly distributed over the wave front. If

---

* The electrons are emitted with a complex spectrum of energies, the study of which can reveal many details of the structure of the metal. Here we are concerned only with the *maximum* kinetic energy which, as is discussed later, depends on a simple property of the metal, the energy needed to remove the least tightly bound electron from its surface.
† The stopping potential is measured with the voltmeter in Figure 3.10. The potential difference V read by the voltmeter is not equal to that supplied by the variable source V_{ext}. Because the emitter and collector in the circuit are usually made of different metals, there is a *contact potential difference* V_{cpd} in the circuit, so that $V = V_{ext} + V_{cpd}$.

the electron absorbs energy directly from the wave, the amount of energy delivered to any electron is determined by how much radiant energy is incident on the surface area in which the electron is confined. Assuming this area is about the size of an atom, a rough calculation leads to an estimate that the time lag between turning on the light and observing the first photoelectrons should be of the order of seconds (see Example 3.2).

EXAMPLE 3.2

A laser beam with an intensity of 120 W/m² (roughly that of a small helium-neon laser) is incident on a surface of sodium. It takes a minimum energy of 2.3 eV to release an electron from sodium (the work function ϕ of sodium). Assuming the electron to be confined to an area of radius equal to that of a sodium atom (0.10 nm), how long will it take for the surface to absorb enough energy to release an electron?

SOLUTION

The average power P_{av} delivered by the wave of intensity I to an area A is IA. An atom on the surface displays a "target area" of $A = \pi r^2 = \pi(0.10 \times 10^{-9}\text{ m})^2 = 3.1 \times 10^{-20}\text{ m}^2$. If the entire electromagnetic power is delivered to the electron, energy is absorbed at the rate $\Delta E/\Delta t = P_{av}$. The time interval Δt necessary to absorb an energy $\Delta E = \phi$ can be expressed as

$$\Delta t = \frac{\Delta E}{P_{av}} = \frac{\phi}{IA}$$

$$= \frac{(2.3\text{ eV})(1.6 \times 10^{-19}\text{ J/eV})}{(120\text{ W/m}^2)(3.1 \times 10^{-20}\text{ m}^2)} = 0.10\text{ s}$$

In reality, electrons in metals are not always bound to individual atoms but instead can be free to roam throughout the metal. However, no matter what reasonable estimate we make for the area over which the energy is absorbed, the characteristic time for photoelectron emission is estimated to have a magnitude of the order of seconds, in a range easily accessible to measurement.

The experimental characteristics of the photoelectric effect were well known by the year 1902. How do the predictions of the classical theory compare with the experimental results?

Experimental results

1. *The maximum kinetic energy* (*determined from the stopping potential*) *is totally independent of the intensity of the light source.* Figure 3.11 shows a representation of the experimental results. Doubling the intensity of the source leaves the stopping potential unchanged, indicating no change in the kinetic energy of the electrons. This experimental result disagrees with the wave theory, which predicts that the maximum kinetic energy should depend on the intensity of the light.

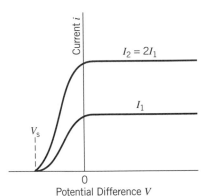

FIGURE 3.11 The photoelectric current i as a function of the potential difference V for two different values of the intensity of the light. When the intensity is doubled, the current is doubled (twice as many photoelectrons are emitted), but the stopping potential V_s remains the same.

Photons

Einstein's theory

2. *The photoelectric effect does not occur at all if the frequency of the light source is below a certain value.* This value, which is characteristic of the kind of metal surface used in the experiment, is called the *cutoff frequency* ν_c. Above ν_c, any light source, no matter how weak, will cause the emission of photoelectrons; below ν_c, no light source, no matter how strong, will cause the emission of photoelectrons. This experimental result also disagrees with the predictions of the wave theory.

3. *The first photoelectrons are emitted virtually instantaneously (within 10^{-9} s) after the light source is turned on.* The wave theory predicts a measurable time delay, so this result also disagrees with the wave theory.

These three experimental results all suggest the complete failure of the wave theory to account for the photoelectric effect.

A successful theory of the photoelectric effect was developed in 1905 by Albert Einstein. Five years earlier, in 1900, the German physicist Max Planck had developed a theory to explain the wavelength distribution of light emitted by hot, glowing objects (thermal radiation). (We discuss thermal radiation and Planck's theory in the next section.) Based on Planck's ideas, Einstein proposed that the energy of a light wave is not continuously distributed over the wavefront, but instead is concentrated in localized bundles, which later were called *photons*. The energy of a photon associated with an electromagnetic wave of frequency ν is

$$E = h\nu \tag{3.21}$$

where h is a proportionality constant known as *Planck's constant*. The photon energy can also be related to the wavelength of the electromagnetic wave by substituting $\nu = c/\lambda$, which gives

$$E = \frac{hc}{\lambda} \tag{3.22}$$

We often speak about photons as if they were particles, and as concentrated bundles of energy they have particlelike properties. Because photons travel with the electromagnetic wave at the speed of light, they must obey the relativistic relationship $p = E/c$. Combining this with Equation 3.22, we obtain

$$p = \frac{h}{\lambda} \tag{3.23}$$

Like other particles, photons carry linear momentum as well as energy.

Because a photon travels at the speed of light, it must have zero mass. Otherwise its energy and momentum would be infinite. Similarly, a photon's rest energy $E_0 = mc^2$ must also be zero. (However, under some circumstances, it is convenient to regard a photon of energy E as having a mass E/c^2 by virtue of its energy.)

In Einstein's interpretation, a photoelectron is released as a result of an encounter with a *single photon*. The entire energy of the photon is delivered instantaneously to a *single photoelectron*. If the photon energy $h\nu$ is greater than the work function ϕ of the material, the photoelectron will be released.

If the photon energy is smaller than the work function, the photoelectric effect will not occur. This explanation thus accounts for two of the failures of the wave theory: the existence of the cutoff frequency and the lack of any measurable time delay.

If the photon energy exceeds the work function, the excess energy appears as the kinetic energy of the electron:

$$K_{max} = h\nu - \phi \qquad (3.24)$$

The intensity of the light source does not appear in this expression! Doubling the intensity of the light means that twice as many photons strike the surface and twice as many photoelectrons are released, but they all have precisely the same maximum kinetic energy.

The maximum kinetic energy corresponds to the release of the least tightly bound electron. Some electrons may lose energy through interactions with other electrons in the material and emerge with smaller kinetic energies.

A photon that supplies an energy equal to ϕ, exactly the minimum amount needed to remove an electron, corresponds to light of frequency equal to the cutoff frequency ν_c. At this frequency, there is no excess energy for kinetic energy, so Equation 3.24 becomes $h\nu_c = \phi$, or

$$\nu_c = \frac{\phi}{h} \qquad (3.25)$$

The corresponding cutoff wavelength $\lambda_c = c/\nu_c$ is

$$\lambda_c = \frac{hc}{\phi} \qquad (3.26)$$

The photon theory appears to explain all of the observed features of the photoelectric effect. The most detailed test of the theory was done by Robert Millikan in 1915. Millikan measured the maximum kinetic energy (stopping potential) for different frequencies of the light and obtained a plot of Equation 3.24. A sample of his results is shown in Figure 3.12. From the slope of the line, Millikan obtained a value for Planck's constant

$$h = 6.57 \times 10^{-34} \text{ J} \cdot \text{s}$$

In part for his detailed experiments on the photoelectric effect, Millikan was awarded the 1923 Nobel prize in physics. Einstein was awarded the 1921 Nobel prize for his photon theory as applied to the photoelectric effect.

As we discuss in the next section, the wavelength distribution of thermal radiation also yields a value for Planck's constant, which is in good agreement with Millikan's value derived from the photoelectric effect. Planck's constant is one of the fundamental constants of nature; just as c is the characteristic constant of relativity, h is the characteristic constant of quantum mechanics. The value of Planck's constant has been measured to great precision in a variety of experiments. The presently accepted value is

$$h = 6.6260755 \times 10^{-34} \text{ J} \cdot \text{s}$$

Robert A. Millikan (1868–1953, United States). Perhaps the best experimentalist of his era, his work included the precise determination of Planck's constant using the photoelectric effect (for which he received the 1923 Nobel prize) and the measurement of the charge of the electron (using his famous "oil-drop" apparatus).

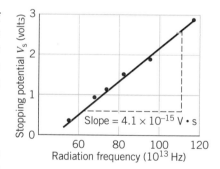

FIGURE 3.12 Millikan's results for the photoelectric effect in sodium. The slope of the line is h/e; the experimental determination of the slope gives a way of determining Planck's constant. The intercept should give the cutoff frequency; however, in Millikan's time the contact potentials of the electrodes were not known precisely and so the vertical scale is displaced by a few tenths of a volt. The slope not affected by this correction.

EXAMPLE 3.3

(*a*) What are the energy and momentum of a photon of red light of wavelength 650 nm? (*b*) What is the wavelength of a photon of energy 2.40 eV?

SOLUTION

(*a*) Using Equation 3.22 we obtain

$$E = \frac{hc}{\lambda} = \frac{(6.63 \times 10^{-34}\,\text{J·s})(3.00 \times 10^{8}\,\text{m/s})}{650 \times 10^{-9}\,\text{m}}$$

$$= 3.06 \times 10^{-19}\,\text{J}$$

Converting to electron-volts, we have

$$E = \frac{3.06 \times 10^{-19}\,\text{J}}{1.60 \times 10^{-19}\,\text{J/eV}} = 1.91\,\text{eV}$$

Problems of this sort are simplified if we express the combination *hc* in units of eV·nm, as we did in Chapter 1.

$$E = \frac{hc}{\lambda} = \frac{1240\,\text{eV·nm}}{650\,\text{nm}} = 1.91\,\text{eV}$$

The momentum is found in a similar way, using Equation 3.23

$$p = \frac{h}{\lambda} = \frac{1}{c}\frac{hc}{\lambda} = \frac{1}{c}\frac{1240\,\text{eV·nm}}{650\,\text{nm}} = 1.91\,\text{eV}/c$$

The momentum could also be found directly from the energy:

$$p = \frac{E}{c} = \frac{1.91\,\text{eV}}{c} = 1.91\,\text{eV}/c$$

(It may be helpful to review the discussion in Example 2.11 about these units of momentum.)

(*b*) Solving Equation 3.22 for λ, we find

$$\lambda = \frac{hc}{E} = \frac{1240\,\text{eV·nm}}{2.40\,\text{eV}} = 517\,\text{nm}$$

EXAMPLE 3.4

The work function for tungsten metal is 4.52 eV. (*a*) What is the cutoff wavelength λ_c for tungsten? (*b*) What is the maximum kinetic energy of the electrons when radiation of wavelength 198 nm is used? (*c*) What is the stopping potential in this case?

SOLUTION

(*a*) Equation 3.26 gives

$$\lambda_c = \frac{hc}{\phi} = \frac{1240\ \text{eV}\cdot\text{nm}}{4.52\ \text{eV}} = 274\ \text{nm}$$

in the ultraviolet region.

(*b*) At the shorter wavelength,

$$K_{max} = h\nu - \phi = \frac{hc}{\lambda} - \phi$$

$$= \frac{1240\ \text{eV}\cdot\text{nm}}{198\ \text{nm}} - 4.52\ \text{eV} = 1.74\ \text{eV}$$

(*c*) The stopping potential is the voltage corresponding to K_{max}:

$$V_s = \frac{K_{max}}{e} = \frac{1.74\ \text{eV}}{e} = 1.74\ \text{V}$$

3.3 BLACKBODY RADIATION

The first indication that all was not well with the classical wave picture of electromagnetic radiation (which worked so well to explain the nineteenth-century experiments of Young and of Hertz and which could be analyzed so precisely with the equations of Maxwell) followed from the failure of the wave theory to explain the observed spectrum of *thermal radiation*—that type of electromagnetic radiation emitted by all objects merely because of their temperature.

A typical experimental arrangement is shown in Figure 3.13. An object is maintained at a temperature T_1. The radiation emitted by the object is detected by an apparatus that is sensitive to the wavelength of the radiation. For example, a dispersive medium such as a prism can be used so that different wavlengths appear at different angles θ. By moving the radiation detector to different angles θ we can measure the intensity of the radiation at a specific wavelength. Since the detector is not a geometrical point (hardly an efficient detector!), but instead subtends a small range of angles $\Delta\theta$, what we really measure is the amount of radiation in some range $\Delta\theta$ at θ, or equivalently, in some range $\Delta\lambda$ at λ.

Let *I* represent the total intensity of electromagnetic radiation emitted at all wavelengths. In the interval $d\lambda$ at a particular wavelength λ (that is, between the limits λ and $\lambda + d\lambda$), the intensity dI is

$$dI = R(\lambda)\ d\lambda \qquad (3.27)$$

where $R(\lambda)$ is the *radiancy,* the intensity per unit wavelength interval. Our experiment measures dI and thus can determine $R(\lambda)$ at any desired

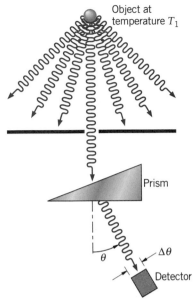

FIGURE 3.13 Measurement of the spectrum of thermal radiation. A device such as a prism is used to separate the wavelengths emitted by the object.

Stefan's law

Wien's displacement law

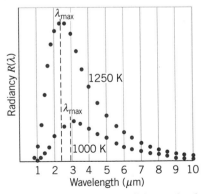

FIGURE 3.14 A possible result of the measurement of the radiancy over many different wavelengths. Each different temperature of the emitting body gives a different peak λ_{max}.

wavelength. Figure 3.14 shows a typical set of results when the object is at a temperature $T_1 = 1000$ K. If we now change the temperature of the object to a different value T_2, we obtain a different curve, as shown in Figure 3.14 for $T_2 = 1250$ K. If we repeat the measurement for many different temperatures, we obtain systematic results for the radiancy that reveal two important properties:

1. The total intensity radiated over all wavelengths, that is, the area under the curve of $R(\lambda)$, increases as the temperature is increased. This is not a surprising result: a glowing object does indeed radiate more energy as we increase its temperature. From careful measurement, we find that the total intensity increases as the fourth power of the temperature:

$$I = \int R(\lambda) \, d\lambda \propto T^4$$

or

$$I = \sigma T^4 \tag{3.28}$$

where we have introduced the proportionality constant σ. Equation 3.28 is called *Stefan's law* and the constant σ is called the *Stefan-Boltzmann constant*. Its value can be determined from experimental results such as those illustrated in Figure 3.14:

$$\sigma = 5.670399 \times 10^{-8} \text{ W/m}^2 \cdot \text{K}^4$$

2. The wavlength λ_{max} at which the radiancy $R(\lambda)$ reaches its maximum value decreases as the temperature is increased, in inverse proportion to the temperature:

$$\lambda_{max} \propto 1/T$$

From results such as those of Figure 3.14, we can determine the proportionality constant, so that

$$\lambda_{max} T = 2.898 \times 10^{-3} \text{ m} \cdot \text{K} \tag{3.29}$$

This result is known as *Wein's displacement law*; the term "displacement" refers to the way the peak is moved or displaced as the temperature is varied.

EXAMPLE 3.5

(a) At what wavelength does a room-temperature ($T = 20°C$) object emit the maximum thermal radiation? (b) To what temperature must we heat it until its peak thermal radiation is in the red region of the spectrum? (c) How many times as much thermal radiation does it emit at the higher temperature?

SOLUTION

(a) Using the absolute temperature, $T = 293$ K, Wien's displacement law gives

$$\lambda_{max} = \frac{2.898 \times 10^{-3} \text{ m} \cdot \text{K}}{293 \text{ K}} = 9.89 \ \mu\text{m}$$

(*b*) Taking the wavelength of red light to be $\lambda \cong 650$ nm, we again use Wien's displacement law to find T:

$$T = \frac{2.898 \times 10^{-3}\,\text{m} \cdot \text{K}}{650 \times 10^{-9}\,\text{m}} = 4460\,\text{K}$$

(*c*) Since the total intensity of radiation is proportional to T^4, the ratio of the total thermal emissions will be

$$\frac{T_2^4}{T_1^4} = \frac{(4460\,\text{K})^4}{(293\,\text{K})^4} = 5.37 \times 10^4$$

Be sure to notice the use of absolute (Kelvin) temperatures in this example.

It is now time to try to analyze and understand these results (the dependence of R on λ, Stefan's law, Wien's law) based on theories of thermodynamics and electromagnetism. We will not give the complete discussion here, but only a brief outline of the theory. At the end of this chapter you will find references to more detailed (and advanced) discussions of this topic.

Ordinary solid objects are seen because of the light that is reflected by them. At room temperature the thermal radiation is mostly in the *infrared* region of the spectrum ($\lambda_{\text{max}} \cong 10\ \mu$m), where our eyes are not sensitive. As we heat such objects, they begin to emit visible light; according to Equation 3.29, as T increases, λ_{max} decreases, and for relatively modest temperatures λ_{max} will move down to the visible region. For example, a piece of metal first begins to glow with a deep red color, and as the temperature is increased the color becomes more yellow.

Unfortunately, the amount of radiation emitted by a solid object depends not only on the temperature, but also on other factors, such as the properties of its surface (for example, its ability to reflect radiation incident upon it). To eliminate some of these difficulties we consider an object whose surface is completely black (*a blackbody*) which reflects none of the radiation that strikes it, and its surface properties can then not be seen. However, this generalization still does not simplify the problem sufficiently for us to be able to calculate the spectrum of emitted radiation, so we generalize still further to a special type of blackbody—a cavity, such as the inside of a metal box, with a small hole cut in one wall (Figure 3.15). *It is the hole, and not the box itself, that is the blackbody.* Radiation from outside that strikes the hole gets lost inside the box and has a negligible chance of reemerging from the hole; thus no reflections occur from the blackbody (the hole). Since the radiation that emerges from the hole is just a sample of the radiation inside the box, understanding the nature of the radiation inside the box allows us to understand the radiation that leaves through the hole.

Cavity radiation

The *classical* calculation of the radiant energy emitted at each wavelength now breaks down into several steps. Without proof, here are the essential parts of the derivation, which involves first computing the amount of radia-

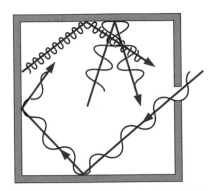

FIGURE 3.15 A cavity filled with electromagnetic radiation. The hole in the wall of the cavity represents an ideal blackbody.

tion (number of waves) at each wavelength, then the contribution of each wave to the total energy in the box, and finally the radiancy corresponding to that energy.

1. *The box is filled with electromagnetic standing waves.* If the walls of the box are metal, radiation is reflected back and forth with a node of the electric field at each wall (the electric field must vanish inside a conductor). This is the same condition that applies to other standing waves, like those on a stretched string or a column of air in an organ pipe.

2. *The number of standing waves with wavelengths between λ and $\lambda + d\lambda$ is*

$$N(\lambda)\, d\lambda = \frac{8\pi V}{\lambda^4}\, d\lambda \qquad (3.30)$$

where V is the volume of the box. For one-dimensional standing waves, as on a stretched string of length L, the allowed wavelength are $\lambda = 2L/n$, $(n = 1, 2, 3, \ldots)$. The number of possible standing waves with wavelengths between λ_1 and λ_2 is $n_2 - n_1 = 2L(1/\lambda_2 - 1/\lambda_1)$. In the small interval from λ to $\lambda + d\lambda$, the number of standing waves is $N(\lambda)\, d\lambda = |dn/d\lambda|\, d\lambda = (2L/\lambda^2)\, d\lambda$. Equation 3.30 can be obtained by extending this approach to three dimensions.

3. *Each individual wave contributes an energy kT to the radiation in the box.* This result follows from classical thermodynamics. The radiation in the box is in thermal equilibrium with the walls at temperature T. Radiation is reflected from the walls because it is absorbed and quickly reemitted by the atoms of the walls, which in the process oscillate at the frequency of the radiation. At temperature T, the mean thermal *kinetic* energy of an oscillating atom is $\frac{1}{2}kT$ (a result similar to one obtained for an ideal gas). For a simple harmonic oscillator, the mean *kinetic* energy is equal to the mean *potential* energy, so the mean *total* energy is kT.

4. *The radiancy is related to the energy density (energy per unit volume) $u(\lambda)$ at the wavelength λ according to*

$$R(\lambda) = \frac{c}{4}\, u(\lambda) \qquad (3.31)$$

This result follows from classical electromagnetism by calculating the amount of radiation passing through an element of surface area within the cavity.

Putting all these ingredients together, we first compute the energy density of radiation inside the cavity:

energy density = (number of standing waves per unit volume)

$\times$ (energy per standing wave)

or

$$u(\lambda) = \frac{8\pi}{\lambda^4}\, kT \qquad (3.32)$$

using Equation 3.30 for $N(\lambda)/V$, the number of standing waves per unit volume. Finally, Equation 3.31 gives us the radiancy:

$$R(\lambda) = \frac{8\pi}{\lambda^4} kT \frac{c}{4} \qquad (3.33)$$

Rayleigh-Jeans formula

(For more details of the derivation, see the references listed at the end of this chapter.)

This result is known as the *Rayleigh-Jeans formula*; based firmly on the classical theories of electromagnetism and thermodynamics, it represents our best attempt to apply classical physics to understanding the problem of blackbody radiation. In Figure 3.16 the radiancy calculated from the Rayleigh-Jeans formula is compared with our previous typical experimental results. The radiancy calculated with Equation 3.33 approaches the data at long wavelengths, but at short wavelengths, the classical theory (which predicts $R \to \infty$ as $\lambda \to 0$) fails miserably. The failure of the Rayleigh-Jeans formula at short wavelengths is known as the *ultraviolet catastrophe* and represents a serious problem for classical physics, because the theories of thermodynamics and electromagnetism on which the Rayleigh-Jeans formula is based have been carefully tested in many other circumstances and found to give extremely good agreement with experiment. It was apparent in the case of blackbody radiation that the classical theories would not work, and that a new kind of physical theory was needed.

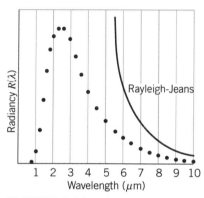

FIGURE 3.16 The failure of the classical Rayleigh-Jeans formula to fit the observed radiancy. At long wavelengths the theory approaches the data, but at short wavelengths the classical formula fails miserably.

The new physics that gave the correct interpretation of thermal radiation was proposed by the German physicist Max Planck in 1900. The ultraviolet catastrophe results because the Rayleigh-Jeans formula predicts too much radiancy at short wavelengths (or equivalently at high frequencies). What is needed is a way to make $R \to 0$ as $\lambda \to 0$, or as $\nu \to \infty$. Planck reasoned that the reflections at the walls of the cavity resulted from radiation being absorbed and then quickly reemitted by the atoms of the wall; during this time the atoms would oscillate at a frequency equal to the frequency of the radiation. Since the energy of an oscillating system depends on its frequency, Planck tried to find a way to reduce the number of high-frequency standing waves by reducing the number of high-frequency oscillators in the walls of the cavity. He did this by a bold assumption that formed the cornerstone of a new physical theory, *quantum physics*. Associated with this theory is a new version of mechanics, known as *wave mechanics*. We will discuss the methods of wave mechanics in Chapter 5; for now we will show how Planck's theory provided the correct interpretation of the emission spectrum of thermal radiation.

Planck suggested that an oscillating atom can absorb or reemit energy only in discrete bundles (called *quanta*). If the energy of the quanta were *proportional* to the frequency of the radiation, then as the frequencies became large, the energy would similarly become large; since no individual wave could contain more than kT of energy, no standing wave could exist whose energy quantum was larger than kT. This effectively limited the high-frequency (low-wavelength) radiant intensity, and solved the ultraviolet catastrophe.

Planck's theory

(As we will discover in our study of quantum physics, all calculated quantities are expressed in terms of *averages* or *probabilities*. Thus it would

Max Planck (1858–1947, Germany). His work on the spectral distribution of radiation, which led to the quantum theory, was honored with the 1918 Nobel prize. In his later years, he wrote extensively on religious and philosophical topics.

Planck's formula

be more correct to say that Planck's contribution was to find a way that the *probability* of finding a mode of oscillation with energy much greater than kT could be made negligibly small.)

In Planck's theory, each oscillator can emit or absorb energy only in quantities that are integer multiples of a certain basic quantity of energy ε,

$$E = n\varepsilon \qquad n = 1, 2, 3, \ldots \tag{3.34}$$

where n is the number of quanta. Furthermore, the energy of each of the quanta is determined by the frequency

$$\varepsilon = h\nu \tag{3.35}$$

where h is the constant of proportionality, now known as Planck's constant. Based on this assumption, Planck calculated the radiancy to be

$$R(\lambda) = \left(\frac{c}{4}\right)\left(\frac{8\pi}{\lambda^4}\right)\left[\left(\frac{hc}{\lambda}\right)\frac{1}{e^{hc/\lambda kT} - 1}\right] \tag{3.36}$$

(This result is derived in Section 10.6.) The agreement between experiment and Planck's formula is illustrated in Figure 3.17, and you can see how well Planck's formula fits the data.

Note that Planck's assumption about the discrete energy of the oscillations results in replacing the average energy kT, which appears in Equation 3.33, with the quantity in brackets in Equation 3.36. Since we know that the Rayleigh-Jeans law, Equation 3.33, works well at long wavelengths, we expect that in the limit of large λ, Equation 3.36 reduces to Equation 3.33. Expanding the exponential in the denominator gives this result directly.

In Problems 16 and 17 at the end of this chapter you are asked to demonstrate that Planck's formula can be used to deduce Stefan's law and Wien's displacement law. In fact, deducing Stefan's law from Planck's formula results in a relationship between the Stefan-Boltzmann constant and Planck's constant:

$$\sigma = \frac{2\pi^5 k^4}{15c^2 h^3} \tag{3.37}$$

By determining the value of the Stefan-Boltzmann constant from the radiancy data available in 1900, Planck was able to determine a value of the constant h:

$$h = 6.56 \times 10^{-34} \text{ J} \cdot \text{s}$$

which agrees very well with the value of h that Millikan deduced 15 years later based on the analysis of data from the photoelectric effect. The good agreement of these two values is remarkable, because they are derived from very different kinds of experiments—one involves the *emission* and the other the *absorption* of electromagnetic radiation. This suggests that the quantization property is not an accident arising from the analysis of one particular experiment, but is instead a property of the electromagnetic field itself. Along with many other scientists of his era, Planck was slow to accept this interpretation. However, later experimental evidence (including

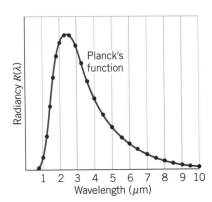

FIGURE 3.17   Planck's function fits the observed data perfectly.

the Compton effect) proved to be so compelling that it left no doubt about Einstein's photon theory and the particlelike structure of the electromagnetic field.

3.4 THE COMPTON EFFECT

Another way for radiation to interact with atoms is by means of the Compton effect, in which radiation scatters from loosely bound, nearly free electrons. Part of the energy of the radiation is given to the electron, which is released from the atom; the remainder of the energy is reradiated as electromagnetic radiation. According to the wave picture, the scattered radiation is less energetic than the incident radiation (the difference going into the kinetic energy of the electron) but has the same wavelength. As we will see, the photon concept leads to a very different prediction for the scattered radiation.

The scattering process is analyzed simply as an interaction (a "collision" in the classical sense of particles) between a single photon and an electron, which we assume to be at rest. Figure 3.18 shows the collision. Initially, the photon has energy E given by

$$E = h\nu = \frac{hc}{\lambda} \tag{3.38}$$

and linear momentum p,

$$p = \frac{E}{c} \tag{3.39}$$

The electron, at rest, has rest energy $m_e c^2$. After the scattering, the photon has energy E' and momentum p', and moves in a direction at an angle θ with respect to the direction of the incident photon. The electron has total energy E_e and momentum p_e and moves in a direction at an angle ϕ with respect to the initial photon. (To allow for the possibility of high-energy incident photons giving energetic scattered electrons, we use relativistic kinematics for the electron.) The usual conditions of conservation of energy and momentum are then applied:

$$E_{\text{initial}} = E_{\text{final}}$$
$$E + m_e c^2 = E' + E_e \tag{3.40a}$$
$$(p_x)_{\text{initial}} = (p_x)_{\text{final}}$$
$$p = p_e \cos \phi + p' \cos \theta \tag{3.40b}$$
$$(p_y)_{\text{initial}} = (p_y)_{\text{final}}$$
$$0 = p_e \sin \phi - p' \sin \theta \tag{3.40c}$$

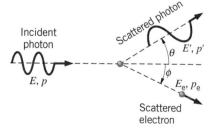

FIGURE 3.18 The geometry of Compton scattering.

We have three equations with four unknowns (θ, ϕ, E_e, E'; p_e and p' are not independent unknowns) that cannot be solved uniquely, but we can eliminate any two of the four unknowns by solving the equations simultaneously. If we choose to measure the energy and direction of the scattered photon, we eliminate E_e and ϕ. The angle ϕ is eliminated by combining the momentum equations:

$$p_e \cos \phi = p - p' \cos \theta$$

$$p_e \sin \phi = p' \sin \theta$$

Squaring and adding, we obtain

$$p_e^2 = p^2 - 2pp' \cos \theta + p'^2 \tag{3.41}$$

The relativistic relationship between energy and momentum is, according to Equation 2.41,

$$E_e^2 = c^2 p_e^2 + m_e^2 c^4$$

Substituting in this equation for E_e from Equation 3.40a and for p_e^2 from Equation 3.41, we obtain

$$(E + m_e c^2 - E')^2 = c^2(p^2 - 2pp' \cos \theta + p'^2) + m_e^2 c^4 \tag{3.42}$$

and after a bit of algebra, we find

Compton scattering formula

$$\frac{1}{E'} - \frac{1}{E} = \frac{1}{m_e c^2}(1 - \cos \theta) \tag{3.43}$$

This equation can also be written as

$$\lambda' - \lambda = \frac{h}{m_e c}(1 - \cos \theta) \tag{3.44}$$

where λ is the wavelength of the incident photon and λ' is the wavelength of the scattered photon. The quantity $h/m_e c$ is known as the *Compton wavelength of the electron* and has a value of 0.002426 nm; however, keep in mind that it is *not* a true wavelength but rather is a *change* of wavelength.

Equations 3.43 and 3.44 give the change in energy or wavelength of the photon, as a function of the *scattering angle* θ. Since the quantity on the right-hand side is never negative, E' is always less than E—the scattered photon has less energy than the original incident photon; the difference $E - E'$ is just the kinetic energy given to the electron, $E_e - m_e c^2$. Similarly, λ' is always greater than λ—the scattered photon has a longer wavelength than the incident photon; the change in wavelength ranges from 0 at $\theta = 0°$ to twice the Compton wavelength at $\theta = 180°$. Of course the descriptions in terms of energy and wavelength are equivalent, and the choice of which to use is merely a matter of convenience.

EXAMPLE 3.6

X rays of wavelength 0.2400 nm are Compton-scattered, and the scattered beam is observed at an angle of 60.0° relative to the incident beam. Find:

(*a*) the wavelength of the scattered X rays, (*b*) the energy of the scattered X-ray photons, (*c*) the kinetic energy of the scattered electrons, and (*d*) the direction of travel of the scattered electrons.

SOLUTION

(*a*) λ' can be found immediately from Equation 3.44:

$$\lambda' = \lambda + \frac{h}{m_e c}(1 - \cos\theta)$$

$$= 0.2400 \text{ nm} + (0.00243 \text{ nm})(1 - \cos 60°)$$

$$= 0.2412 \text{ nm}$$

(*b*) The energy E' can be found directly from λ':

$$E' = \frac{hc}{\lambda'} = \frac{1240 \text{ eV·nm}}{0.2412 \text{ nm}} = 5141 \text{ eV}$$

(*c*) From Equation 3.40*a* for conservation of energy, we have

$$E_e = (E - E') + m_e c^2 = K_e + m_e c^2$$
$$K_e = E - E'$$

$$(3.45)$$

The initial photon energy E is $hc/\lambda = 5167$ eV, and so

$$K_e = 5167 \text{ eV} - 5141 \text{ eV} = 26 \text{ eV}$$

(*d*) Solving Equations 3.40*b* and 3.40*c* for $p_e \cos\phi$ and $p_e \sin\phi$ as we did in deriving Equation 3.41, we divide (instead of adding and squaring):

$$\tan\phi = \frac{p' \sin\theta}{p - p' \cos\theta}$$

$$(3.46)$$

Multiplying top and bottom by c, and recalling that $E = pc$ and $E' = p'c$, we have

$$\tan\phi = \frac{E' \sin\theta}{E - E' \cos\theta} = \frac{(5141 \text{ eV})(\sin 60°)}{(5167 \text{ eV}) - (5141 \text{ eV})(\cos 60°)}$$

$$= 1.715$$

$$\phi = 59.7°$$

The first experimental demonstration of this type of scattering was done by Arthur Compton in 1923. A diagram of his experimental arrangement is shown in Figure 3.19. A beam of X rays is incident on a scattering target, for which Compton used carbon. (Although no scattering target contains actual "free" electrons, the outer or valence electrons in many materials are very weakly attached to the atom and behave like nearly free electrons. The kinetic energies of these electrons in the atom are so small compared with the energies of the incident X-ray photons that they can be regarded

Arthur H. Compton (1892–1962, United States). His work on X-ray scattering verified Einstein's photon theory and earned him the 1927 Nobel prize. He was a pioneer in research with X rays and cosmic rays. During World War II he directed a portion of the U.S. atomic bomb research.

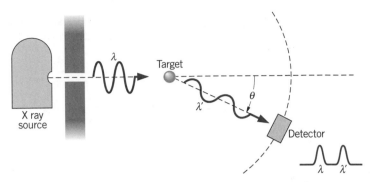

FIGURE 3.19 Schematic diagram of Compton scattering apparatus. The wavelength λ' of the scattered X rays is measured by the detector, which can be moved to different positions θ. The wavelength difference $\lambda' - \lambda$ varies with θ.

as nearly "free" electrons.) A movable detector measured the energy of the scattered X rays at various angles θ.

Compton's original results are illustrated in Figure 3.20. At each angle, two peaks appear, corresponding to scattered X-ray photons with two different energies or wavelengths. The wavelength of one peak does not change as the angle is varied; this peak corresponds to scattering from the tightly bound "inner" electrons of the atom. These electrons are so tightly attached to the atoms that the photon scatters and loses no energy. The wavelength of the other peak, however, varies strongly with angle; as can be seen from Figure 3.21, this variation is exactly as the Compton formula predicts.

Similar results can be obtained for the scattering of gamma rays, which are higher-energy (shorter wavelength) photons emitted in various radioactive decays. Compton also measured the variation in wavelength of scattered gamma rays, as illustrated in Figure 3.22. The change in wavelength deduced from the gamma-ray measurements is identical with the change in wavelength deduced from the X-ray measurements; the Compton formula 3.44 leads us to expect this, since the change in wavelength does not depend on the incident wavelength.

3.5   OTHER PHOTON PROCESSES

Although Compton scattering and the photoelectric effect provided the earliest experimental evidence in support of the quantization (particlelike behavior) of electromagnetic radiation, there are numerous other experiments that can also be interpreted correctly only if we assume the existence of photons as discrete quanta of electromagnetic radiation. In this section we discuss some of these processes, which cannot be understood if we consider only the wave nature of electromagnetic radiation.

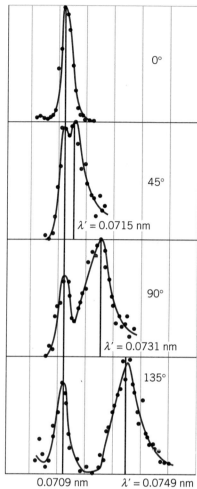

FIGURE 3.20 Compton's original results for X-ray scattering.

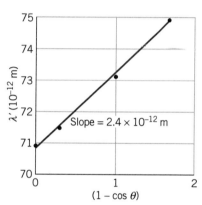

FIGURE 3.21 The scattered X-ray wavelengths λ', from Figure 3.20, for different scattering angles. The expected slope is 2.43×10^{-12} m, in agreement with the measured slope of Compton's data points.

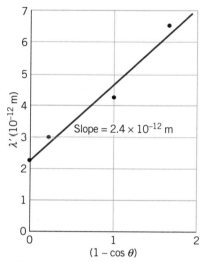

FIGURE 3.22 Compton's results for gamma-ray scattering. Even though the wavelengths are roughly two orders of magnitude smaller than the X-ray wavelengths, the slope is the same as in Figure 3.21, which the Compton formula, Equation 3.44, predicts.

Bremsstrahlung and X-ray Production

When an electric charge, such as an electron, is accelerated or decelerated, it radiates electromagnetic energy; in our present framework, we would say that it emits photons. Suppose we have a beam of electrons, which has achieved a kinetic energy $K = eV$ after having been accelerated through a potential V (Figure 3.23). When the electrons strike a target they are slowed down and eventually come to rest, because they make collisions with the atoms of the target material. In such a collision, momentum is transferred to the atom, the electron slows down, and photons are emitted. The recoil kinetic energy of the atom is small (because the atom is so massive) and can safely be neglected. If the electron has a kinetic energy K before the encounter and if it leaves after the collision with a smaller

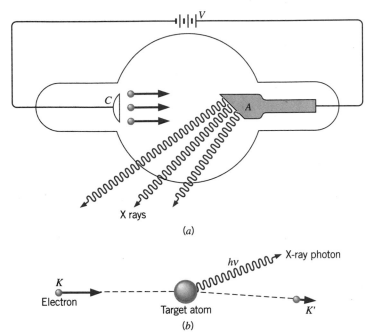

FIGURE 3.23   (*a*) Apparatus for producing bremsstrahlung. Electrons from a cathode *C* are accelerated to the anode *A* through the potential difference *V*. When an electron encounters a target atom of the anode, it can slow down, with the emission of an X-ray photon. (*b*) A schematic representation of the bremsstrahlung process.

kinetic energy K', then the photon energy is

$$h\nu = K - K' \tag{3.47}$$

The amount of energy lost, and therefore the energy and wavelength of the emitted photon, are not uniquely determined, since K is the only known energy in Equation 3.47. An electron usually will make many collisions, and therefore emit many different photons, before it is brought to rest; the photons then will range all the way from very small energies (large wavelengths) corresponding to small energy losses, up to a maximum energy of K, corresponding to an electron which loses all of its energy in a single encounter. The smallest emitted wavelength is therefore determined by the maximum possible energy loss,

$$h\nu = K$$

$$\frac{hc}{\lambda_{min}} = eV$$

Bremsstrahlung

$$\lambda_{min} = \frac{hc}{eV} \tag{3.48}$$

For typical accelerating voltages in the range of 10,000 V, λ_{min} is in the range of a few tenths of nm, which corresponds to the X-ray region of the spectrum. This *continuous* distribution of X rays (which is very different from the *discrete* X-ray energies that are emitted in atomic transitions; more about these in Chapter 8) is called *bremsstrahlung,* which is German for braking, or decelerating, radiation. Some sample bremstrahlung spectra are illustrated in Figure 3.24.

Symbolically we can write the bremsstrahlung process as

$$\text{electron} \rightarrow \text{electron} + \text{photon}$$

This is just the reverse process of the photoelectric effect, which is

$$\text{electron} + \text{photon} \rightarrow \text{electron}$$

However, *neither* process occurs for free electrons. In both cases there must be a heavy atom in the neighborhood to take care of the recoil momentum.

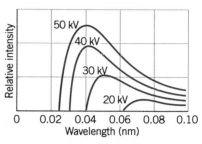

FIGURE 3.24 Some typical bremsstrahlung spectra. Each spectrum is labeled with the value of the accelerating voltage *V*.

Pair Production

Another process which can occur when photons encounter atoms is *pair production,* in which the photon loses all its energy and in the process two particles are created: an electron and a positron. (A positron is a particle that is identical in mass to the electron but which has a positive electric charge; more about *antiparticles* in Chapter 14.) Here we have an example of the creation of rest energy. The electron did not exist before the encounter of the photon with the atom (it was *not* an electron that was part of the atom). The photon energy is converted into the relativistic total energies E_+ and E_- of the positron and electron:

$$\begin{aligned} h\nu &= E_+ + E_- \\ &= (m_e c^2 + K_+) + (m_e c^2 + K_-) \end{aligned} \tag{3.49}$$

Pair production

Since K_+ and K_- are always positive, the photon must have an energy of at least $2m_e c^2 = 1.02$ MeV in order for this process to occur; such high-energy photons are in the region of *nuclear gamma rays.* Symbolically,

$$\text{photon} \rightarrow \text{electron} + \text{positron}$$

This process, like brehmsstrahlung, will not occur unless there is an atom nearby to supply the necessary recoil momentum. The reverse process,

$$\text{electron} + \text{positron} \rightarrow \text{photon}$$

also occurs; this process is known as *electron-positron annihilation* and can occur for free electrons and positrons as long as at least two photons are created. In this process the electron and positron disappear and are replaced by two photons. Conservation of energy requires that, if E_1 and E_2 are the photon energies,

Electron-positron annihilation

$$(m_e c^2 + K_+) + (m_e c^2 + K_-) = E_1 + E_2$$

Usually K_+ and K_- are negligibly small, and we can assume the positron and electron to be essentially at rest. Momentum conservation then requires the two photons to have equal energies of $m_e c^2$ (= 0.511 MeV) and to move in exactly opposite directions.

3.6 WHAT IS A PHOTON?

We have established in this chapter the firm experimental basis for the particlelike nature of light. We can describe photons by giving a few of their basic properties:

- like an electromagnetic wave, photons move with the speed of light;
- they have zero mass and rest energy;
- they carry energy and momentum, which are related to the frequency and wavelength of the electromagnetic wave by $E = h\nu$ and $p = h/\lambda$;
- they can be created or destroyed when radiation is emitted or absorbed;
- they can have particlelike collisions with other particles such as electrons.

Other properties emerge from more detailed studies, such as the role of photons in transmitting the electromagnetic force according to the modern field theory of electromagnetism, and the deflection of photons by a gravitational field (see Chapter 15).

From this list of properties, we begin to acquire a definite picture of the nature of the photon. This picture, however, is *very* different from the wave description of electromagnetic radiation. The energy of a wave is spread smoothly and continuously over the wavefronts; the energy of photons, in contrast, is confined to localized bundles. Clearly the wave and particle explanations of electromagnetic radiation are not consistent with each other.

In this chapter we have described some experiments that can be explained *only* through the photon picture. Other experiments, such as interference and diffraction, can be explained *only* through the wave picture. For example, the explanation of the double-slit interference experiment requires that the wavefront be divided so that some of its intensity can pass through each slit. A particle must choose to go through one slit or the other; only a wave can go through both.

If we regard the wave and particle pictures as valid but exclusive alternatives, we must assume that the light emitted by a source must travel *either* as waves *or* as particles. How does the source know what kind of light (particles or waves) to emit? Suppose we place a double-slit apparatus on one side of the source and a photoelectric cell on the other side. Light emitted toward the double slit behaves like a wave and light emitted toward the photocell behaves like particles. How did the source know in which direction to aim the waves and in which direction to aim the particles?

Perhaps nature has a sort of "secret code" in which the kind of experiment we are doing is signaled back to the source so that it knows whether to emit particles or waves. Let us repeat our dual experiment with light from a distant galaxy, light which has been traveling toward us for a time roughly equal to the age of the universe (15×10^9 years). Surely the kind of experiment we are doing could not be signaled back to the limits of the known universe in the time it takes us to remove the double-slit apparatus from the laboratory table and replace it with the photoelectric apparatus. Yet we find that the starlight can produce both the double-slit interference and also the photoelectric effect.

Figure 3.25 shows a recent experiment that was designed to test whether this dual nature is an intrinsic property of light or of our apparatus. A light beam from a laser goes through a beam splitter, which separates the beam into two components (A and B). The mirrors reflect the two component beams so that they can recombine to form an interference pattern. In beam A there is a switch that can deflect the beam into a detector. If the switch is off, beam A is not deflected and will combine with beam B to produce the interference pattern. If the switch is on, beam A is deflected and observed in the detector, indicating that the light traveled a definite path, as would be characteristic of a particle. To put this another way, if the switch is off, the light beam is observed as a wave; if it is on, the light beam is observed as particles.

If light behaves like *particles,* the beam splitter sends it along *either* path A or path B; either path can be randomly chosen for the particle, but each particle can travel only one path. If light behaves like a *wave,* on the other hand, the beam splitter sends it along *both* paths, dividing its intensity between the two. Perhaps the beam splitter can somehow sense whether the switch is open or closed, so that it knows whether we are doing a particle-type or a wave-type experiment. If this were true, then the beam splitter would "know" whether to send all of the intensity down one path (so that we would observe a particle) or to split the intensity between the paths (so that we would observe a wave). However, in this experiment, the laser intensity was set so low that only one photon at a time was in the apparatus. Furthermore, the experimenters used a very fast optical switch whose response time was shorter than the time it takes for a photon to travel through the apparatus to the switch. That is, the state of the switch could be changed *after* the light had already passed through the beam splitter, and so it was impossible for the beam splitter to "know" how the switch was set and thus whether a particle-type or a wave-type experiment was being done. This kind of experiment is called a "delayed choice" experiment, because the experimenter makes the choice of what kind of experiment to do after the light is already traveling on its way to the observation apparatus.

In this experiment, the investigators discovered that whenever they had the switch off, they observed the interference pattern characteristic of waves. When they had the switch on, they observed particles in the detector and no interference pattern. That is, whenever they did a wave-type experiment they observed waves, and whenever they did a particle-type experiment they observed particles. The wave and particle natures are both present simultaneously in the light, and this dual nature is clearly associated with the light and is not characteristic of the apparatus.

Another attempt to distinguish between the particle and wave natures of light was a variation of the double-slit experiment. In this experiment, light from a laser was scattered from two trapped mercury atoms, which in effect served as the two slits (Figure 3.26). If light behaves as a particle, it should be possible to determine through which "slit" it passes, or, in this case, from which atom it scatters. A wave, on the other hand, can be scattered by both atoms. From the polarization of the scattered light, it is possible to determine whether the scattering changes the state of only one

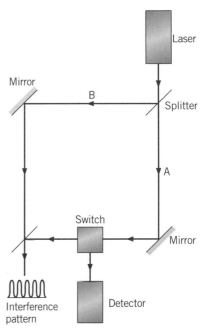

FIGURE 3.25 Apparatus for delayed choice experiment. Photons from the laser strike the beam splitter and can then travel paths A or B. The switch in path A can deflect the beam into a detector. If the switch is off, beam A recombines with beam B to form an interference pattern. [*Source*: A Shimony, "The Reality of the Quantum World," *Scientific American* **258**, 46 (January 1988).]

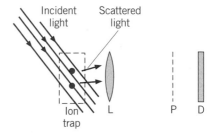

FIGURE 3.26 Schematic diagram of experiment to observe light scattering from two trapped atoms. The scattered light is collected by the lens L and focused through the polarizer P onto the detector D.

of the atoms (in which case the light behaves as a particle) or both atoms (in which case the light behaves as a wave).

Figure 3.27 shows the results of the experiment. When the light scatters from both atoms, the interference pattern is present; when it scatters from only one of the atoms, there is no interference. Since the same light is producing both outcomes, it can be concluded that the particle and wave natures are both present in the light. When we attempt to observe the particle nature of the light, the interference does not occur (we cannot simultaneously observe its wave nature).

Wave-particle duality

We are therefore trapped into an uncomfortable conclusion: Light is not *either* particles *or* waves; it is somehow *both* particles *and* waves, and only shows one or the other aspect, depending on the kind of experiment we are doing. A particle-type experiment shows the particle nature, while a wave-type experiment shows the wave nature. Our failure to classify light as *either* particle *or* wave is not so much a failure to understand the nature of light as it is a failure of our limited vocabulary (based on experiences with ordinary particles and waves) to describe a phenomenon that is more elegant and mysterious than either simple particles or waves.

The situation becomes even more difficult if we use our eye or a photographic film to observe this double-slit interference pattern. Our eye and the film both respond to individual photons. When a single photon is absorbed by a cell of the retina, an electrical impulse is produced that travels to the brain (of course, our vision is composed of many such impulses). When a single photon is absorbed by the film, a single grain of the photographic emulsion is darkened; a complete picture requires a large number of grains to be darkened.

Let us imagine for the moment that we could see individual grains of the film as they absorbed photons and darkened, and let us do the experiment with a light source that is so weak that there is a relatively long time interval between photons. We would see first one grain darken, then another, and so forth, until after a large number of photons we would see the interference pattern begin to emerge. Alternatively, the wave picture of the double-slit experiment suggests that we could find the net electric field of the wave that strikes the screen by superimposing the electric fields of the portions of the incident wave fronts that pass through the two slits; the intensity or power in that combined wave could then be found by a procedure similar to Equations 3.7 through 3.10, and we would expect that the resultant intensity should show maxima and minima just like the observed double-slit interference pattern.

In summary, the correct explanation of the origin and appearance of the interference pattern comes from the wave picture, and the correct interpretation of the evolution of the pattern on the film comes from the particle picture; the two explanations, which according to our limited vocabulary and common-sense experience cannot simultaneously be correct, must somehow be taken together to give a complete description of the properties of electromagnetic radiation.

This dilemma of wave-particle duality cannot be resolved with a simple explanation; physicists and philosophers have struggled with this problem ever since the quantum theory was introduced. The best we can do is to

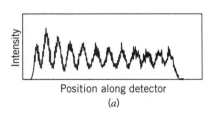

Position along detector

(a)

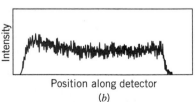

Position along detector

(b)

FIGURE 3.27   (a) Light scattered from both atoms shows interference. (b) Light scattered from only one atom shows no interference. [*Source*: U. Eichmann et al., *Physical Review Letters* **70,** 2359 (1993).]

say that neither the wave nor the particle picture is wholly correct all of the time, that both are needed for a complete description of physical phenomena, and that in fact the two are *complementary* to one another. In the case of the double-slit experiment, we might reason as follows: the interaction between a "source" of radiation and the electromagnetic field is quantized, and we can think of the atoms of the source as emitting individual photons. The interaction at the opposite end of the experiment, the photographic film, is also quantitized, and we have the similarly useful view of atoms absorbing individual photons. In between, the electromagnetic energy propagates smoothly and continuously as a wave and can show wave-type behavior (interference or diffraction). The effect of the double slit is to change the propagation of the wave (from a plane wave to the characteristic double-slit pattern, for example). Where the wave has large intensity, the film reveals the presence of many photons; where the wave has small intensity, few photons are observed. Recalling that the intensity of the wave is proportional to the square of its amplitude, we then have

$$\text{probability to observe photons} \propto |\text{electric field amplitude}|^2$$

It is this expression that provides the ultimate connection between the wave behavior and the particle behavior, and we will see in the next two chapters that a similar expression connects the wave and the particle aspects of those objects, such as electrons, which have been previously considered to behave as classical particles.

SUGGESTIONS FOR FURTHER READING

Electromagnetic radiation and interference are discussed in most introductory physics texts; see for example the list at the end of Chapter 1. Another good reference is H. D. Young, *Fundamentals of Waves, Optics, and Modern Physics* (New York, McGraw-Hill, 1976). A collection of photographs illustrating the phenomena of classical optics can be found in M. Cagnet, M. Francon, and J. C. Thrierr, *Atlas of Optical Phenomena* (Berlin, Springer-Verlag, 1962).

For more complete discussions of blackbody radiation, including derivations of the Rayleigh-Jeans law, see the following:

R. Eisberg and R. Resnick, *Quantum Physics of Atoms, Molecules, Solids, Nuclei, and Particles*, 2nd ed. (New York, Wiley, 1985).
F. K. Richtmeyer, E. H. Kennard, and J. N. Cooper, *Introduction to Modern Physics*, 6th ed. (New York, McGraw-Hill, 1969).

The properties of X rays, including diffraction and scattering, are discussed in the following:

L. Bragg, "X-Ray Crystallography," *Scientific American* **219**, 58 (July 1968).
G. L. Clark, *Applied X Rays* (New York, McGraw-Hill, 1940).
A. H. Compton and S. K. Allison, *X Rays in Theory and Experiment* (New York, Van Nostrand, 1935).

For more details of experiments discussed in this chapter and their importance in the development of quantum theory, see:

W. H. Cropper, *The Quantum Physicists* (New York, Oxford University Press, 1970), chapter 1.

A. Shimony, "The Reality of the Quantum World," *Scientific American* **258**, 46 (January 1988).

QUESTIONS

1. The diameter of an atomic nucleus is about 10×10^{-15} m. Suppose you wanted to study the diffraction of photons by nuclei. What energy of photons would you choose? Why?

2. How is the wave nature of light unable to account for the observed properties of the photoelectric effect?

3. In the photoelectric effect, why do some electrons have kinetic energies smaller than K_{max}?

4. Why doesn't the photoelectric effect work for free electrons?

5. What does the work function tell us about the properties of a metal? Of the metals listed in Table 3.1, which has the least tightly bound electrons? Which has the most tightly bound?

6. Electric current is charge flowing per unit time. If we increase the kinetic energy of the photoelectrons (by increasing the energy of the incident photons), shouldn't the current increase, because the charge flows more rapidly? Why doesn't it?

7. What might be the effects on a photoelectric effect experiment if we were to double the frequency of the incident light? If we were to double the wavelength? If we were to double the intensity?

8. In the photoelectric effect, how can a photon moving in one direction eject an electron moving in a different direction? What happens to conservation of momentum?

9. In Figure 3.11, why does the photoelectric current rise slowly to its saturation value instead of rapidly, when the potential difference is greater than V_s? What does this figure indicate about the experimental difficulties that might arise from trying to determine V_s in this way?

10. Suppose that the frequency of a certain light source is just above the cutoff frequency of the emitter, so that the photoelectric effect occurs. To an observer in relative motion, the frequency might be Doppler shifted to a lower value that is below the cutoff frequency. Would this moving observer conclude that the photoelectric effect does not occur? Explain.

11. Why do cavities that form in a wood fire seem to glow brighter than the burning wood itself? Is the temperature in such cavities hotter than the surface temperature of the exposed burning wood?

12. What are the fields of classical physics on which the classical theory of black-body radiation is based? Why don't we believe that the "ultraviolet catastrophe" suggests that something is wrong with one of those classical theories?

13. In what region of the electromagnetic spectrum do room-temperature objects radiate? What problems would we have if our eyes were sensitive in that region?

14. How does the total intensity of thermal radiation vary when the temperature of an object is doubled?

15. Compton-scattered photons of wavelength λ' are observed at 90°. In terms of λ', what is the scattered wavelength observed at 180°?

16. The Compton-scattering formula suggests that objects viewed from different angles should reflect light of different wavelengths. Why don't we observe a change in color of objects as we change the viewing angle?

17. You have a monoenergetic source of X rays of energy 84 keV, but for an experiment you need 70 keV X rays. How would you convert the X-ray energy from 84 to 70 keV?

18. Often we read about the problems of X-ray emission by TV sets. What is the origin of these X rays? Estimate their wavelengths.

19. The X-ray peaks of Figure 3.20 are not sharp but are spread over a range of wavelengths. What reasons might account for that spreading?

20. A beam of photons passes through a block of matter. What are the three ways discussed in this chapter that the photons can lose energy in interacting with the material?

21. Of the photon processes discussed in this chapter (photoelectric effect, thermal radiation, Compton scattering, bremsstrahlung, pair production, electron-positron annihilation), which conserve momentum? Energy? Mass? Number of photons? Number of electrons? Number of electrons minus number of positrons?

PROBLEMS

1. A double-slit experiment is performed with sodium light ($\lambda = 589.0$ nm). The slits are separated by 1.05 mm, and the screen is 2.357 m from the slits. Find the separation between adjacent maxima on the screen.

2. In Example 3.1, what angle of incidence will produce the second-order Bragg peak?

3. Monochromatic X rays are incident on a crystal in the geometry of Figure 3.5. The first-order Bragg peak is observed when the angle of incidence is 34.0°. The crystal spacing is known to be 0.347 nm. (a) What is the wavelength of the X rays? (b) Now consider a set of crystal planes that makes an angle of 45° with the surface of the crystal (as in Figure 3.6). For X rays of the same wavelength, find the angle of incidence measured from the surface of the crystal that produces the first-order Bragg peak. At what angle from the surface does the emerging beam appear in this case?

4. Find the momentum of (a) a 10.0-MeV gamma ray; (b) a 25-keV X ray; (c) a 1.0-μm infrared photon; (d) a 150-MHz radio-wave photon. Express the momentum in kg·m/s and eV/c.

5. Radio waves have a frequency of the order of 1 to 100 MHz. What is the range of energies of these photons? Our bodies are continuously bombarded by these photons. Why are they not dangerous to us?

6. (a) What is the wavelength of an X-ray photon of energy 10.0 keV? (b) What is the wavelength of a gamma-ray photon of energy 1.00 MeV? (c) What is the range of energies of photons of visible light with wavelengths 350 to 700 nm?

7. Assume that a 100-W light source delivers all its energy in the form of visible light, with average photon wavelength about 550 nm. How many photons per second strike a 20 cm × 30 cm sheet of paper 1 m from the light?

8. What is the cutoff wavelength for the photoelectric effect using an aluminum surface?

9. When sodium metal is illuminated with light of wavelength 4.20×10^2 nm, the stopping potential is found to be 0.65 V; when the wavelength is changed to 3.10×10^2 nm, the stopping potential is 1.69 V. Using *only these data* and the values of the speed of light and the electronic charge, find the work function of sodium and a value of Planck's constant.

10. A metal surface has a photoelectric cutoff wavelength of 325.6 nm. It is illuminated with light of wavelength 259.8 nm. What is the stopping potential?

11. When light of wavelength λ illuminates a copper surface, the stopping potential is V. In terms of V, what will be the stopping potential if the same wavelength is used to illuminate a sodium surface?

12. The cutoff wavelength for the photoelectric effect in a certain metal is 254 nm. (a) What is the work function for that metal? (b) Will the photoelectric effect be observed for $\lambda > 254$ nm or for $\lambda < 254$ nm?

13. A surface of zinc is illuminated and photoelectrons are observed. (a) What is the largest wavelength that will cause photoelectrons to be emitted? (b) What is the stopping potential when light of wavelength 220.0 nm is used?

14. A photon traveling in one direction strikes a surface and releases a photoelectron traveling in the opposite direction. By making any reasonable assumptions you may need, analyze this process from the standpoint of conservation of linear momentum.

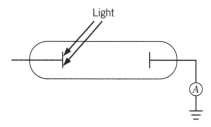

Light

FIGURE 3.28   Problem 15.

15. Suppose instead of the circuit of Figure 3.10, the photoelectric effect were observed with the apparatus shown in Figure 3.28. Assume the emitter to be a metal 1 cm by 1 cm in area and 0.1 mm thick, and assume it to be uniformly illuminated over its entire area by the light beam of intensity 4 W/m². Making any other reasonable assumptions you may need, estimate (a) the initial current just after the light is turned on, and (b) how long this current can flow.

16. By differentiating Equation 3.36, show that $R(\lambda)$ has its maximum as expected according to Wien's displacement law, Equation 3.29.

17. Integrate Equation 3.36 to obtain Equation 3.28. Use the definite integral $\int_0^\infty x^3 \, dx/(e^x - 1) = \pi^4/15$ to obtain Equation 3.37 relating the Stefan-Boltzmann constant to Planck's constant.

18. Use the numerical value of the Stefan-Boltzmann constant to find the numerical value of Planck's constant from Equation 3.37.

19. At what wavelength does the Sun emit its peak radiancy? The surface of the Sun has a temperature of about 6000 K. How does this compare with the peak sensitivity of the human eye?

20. The universe is filled with thermal radiation, which has a blackbody spectrum at an effective temperature of 2.7 K (see Chapter 16). What is the peak wavelength of this radiation? What is the energy (in eV) of quanta at the

peak wavelength? In what region of the electromagnetic spectrum is this peak wavelength?

21. A certain cavity has a temperature of 1150 K. (a) At what wavelength will the radiancy have its maximum value? (b) As a fraction of the maximum radiancy, what is the radiancy at twice the wavelength found in part (a)?

22. Estimate the power radiated by the human body.

23. A cavity is maintained at a temperature of 1650 K. At what rate does energy escape from the interior of the cavity through a hole in its wall of diameter 1.00 mm?

24. Show how Equation 3.43 follows from Equation 3.42.

25. Incident photons of energy 10.39 keV are Compton scattered, and the scattered beam is observed at 45.00° relative to the incident beam. (a) What is the energy of the scattered photons at that angle? (b) How much kinetic energy is given to the scattered electron?

26. X-ray photons of wavelength 0.02480 nm are incident on a target and the Compton-scattered photons are observed at 90.0°. (a) What is the wavelength of the scattered photons? (b) What is the momentum of the incident photons? Of the scattered photons? (c) What is the kinetic energy of the scattered electrons? (d) What is the momentum (magnitude and direction) of the scattered electrons?

27. In Compton scattering, calculate the maximum kinetic energy given to the scattered electron for a given photon energy.

28. High-energy gamma rays can reach a radiation detector by Compton scattering from the surroundings, as shown in Figure 3.29. This effect is known as backscattering. Show that, when $E \gg m_e c^2$, the backscattered photon has an energy of approximately 0.25 MeV, independent of the energy of the original photon, when the scattering angle is nearly 180°.

29. Gamma rays of energy 0.662 MeV are Compton scattered. (a) What is the energy of the scattered photon observed at a scattering angle of 60.0°? (b) What is the kinetic energy of the scattered electrons?

30. Prove that it is *not* possible to conserve both momentum and total relativistic energy in the following situation: A free electron moving at velocity **v** emits a photon and then moves at a slower velocity **v′**.

31. Suppose an atom of iron at rest emits an X-ray photon of energy 6.4 keV. Calculate the "recoil" momentum and kinetic energy of the atom. (*Hint*: Do you expect to need classical or relativistic kinetic energy for the atom? Is the kinetic energy likely to be much smaller than the atom's rest energy?)

32. What is the minimum X-ray wavelength produced in bremsstrahlung by electrons that have been accelerated through 2.50×10^4 V?

33. A photon of energy E strikes an electron at rest and undergoes pair production, producing a positive electron (positron) and an electron:

$$\text{photon} + e^- \rightarrow e^+ + e^- + e^-$$

The two electrons and the positron move off with identical momenta in the direction of the initial photon. Find the kinetic energy of the three final particles and find the energy E of the photon. (*Hint*: Conserve momentum and total relativistic energy.)

34. CUPS Exercise 3.1.

35. CUPS Exercise 3.2.

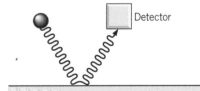

FIGURE 3.29 Problem 28.

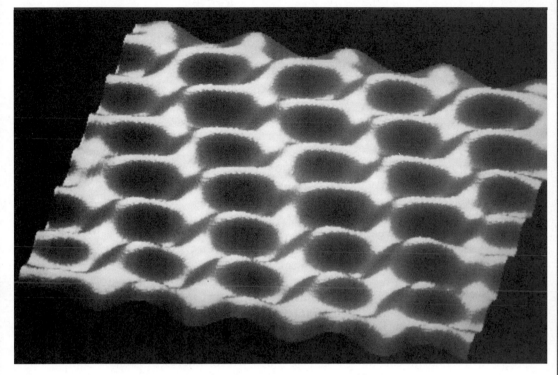

The wave nature of particles is essential to the operation of the scanning electron microscope, which uses electron waves to make images of objects whose features are too small to be seen with light waves. This photo shows the pattern of individual carbon atoms on the surface of a sample of graphite.

THE WAVELIKE PROPERTIES OF PARTICLES

The system of mechanics associated with quantum systems is sometimes called "wave mechanics." In this chapter we discuss the experimental evidence in support of the wave behavior of particles such as electrons. In classical physics, the laws governing the characteristics of waves and particles are fundamentally different. Projectiles obey particle-type laws, such as Newtonian mechanics. Waves undergo interference and diffraction, which cannot be explained by the Newtonian mechanics associated with particles. The energy carried by a particle (such as a projectile) is confined to a small region of space; a wave, on the other hand, distributes its energy throughout all space in wavefronts. In contrast to this clear distinction found in classical physics, the quantum theory requires that, in the microscopic domain, particles sometimes obey the rules that we have previously established for waves! We must discard some of our classically inspired notions of the distinction between particles and waves. Such classical considerations as grasping, confining, or locating a particle are not so easily applied to a wave. Imagine trying to grasp a sound wave or a water wave!

In discussing the logical and mathematical framework of quantum mechanics that resolves such dilemmas, we will refer to many axioms, analogies, and examples that have no counterparts in classical physics. The quantum theory must deal with the interactions of particles, but it must yield the wave nature of the resulting behavior. Furthermore, the wavelike behavior of atomic systems must disappear when we consider collections of atoms that make up ordinary objects, to which Newton's laws for particles can be successfully applied. This curious mix of wavelike and particlelike behavior is a fundamental part of the quantum theory. It may seem logically inconsistent, but *it works!* The full mathematical formulation enables us to compute the detailed properties of atoms and nuclei with incredible precision; when these computations are tested by experiment, they are found to be exceedingly accurate. Given its successes, we can excuse the theory for its apparent lack of a consistent framework.

As you study this chapter, notice the frequent references to such terms as the *probability* of the outcome of a measurement, the *average* of many repetitions of a measurement, and the *statistical* behavior of a system. These terms are a fundamental part of the quantum theory, and you cannot grasp the nature of the quantum theory until you feel comfortable with discarding such classical notions as fixed trajectories and certainty of outcome, while substituting the quantum mechanical notions of probability and statistically distributed outcomes.

4.1 DE BROGLIE'S HYPOTHESIS

Progress in physics can often be characterized by long periods of experimental and theoretical drudgery punctuated occasionally by flashes of insight that cause profound changes in the way we view the universe. Frequently the more profound the insight and the bolder the initial step, the simpler it seems in historical perspective, and the more likely we are to sit back and wonder "Why didn't I think of that?" Einstein's special theory of relativity is one example of such insight; the hypothesis of the Frenchman Louis de Broglie is another.

In the last chapter we discussed the double-slit experiment (which can be understood only if light behaves as a wave) and the photoelectric and Compton effects (which can be understood only if light behaves as a particle). Is this dual particle-wave nature a property only of light or of material objects as well? In a bold and daring hypothesis in his 1924 doctoral dissertation, de Broglie chose the latter alternative. Examining Equation 3.21, $E = h\nu$, and Equation 3.23, $p = h/\lambda$, we find some difficulty in applying the first in the case of particles, for we cannot be sure whether E should be the kinetic energy, total energy, or total relativistic energy (all, of course, are identical for light). No such difficulties arise from the second relationship. De Broglie suggested, lacking any experimental evidence in support of his hypothesis, that associated with any material particle moving with momentum p there is a wave of wavelength λ, related to p according to

$$\lambda = \frac{h}{p} \qquad (4.1) \qquad \textit{De Broglie wavelength}$$

The wavelength λ of a particle computed according to Equation 4.1 is called its *de Broglie wavelength.*

EXAMPLE 4.1

Compute the de Broglie wavelength of the following:

(*a*) A 1000-kg automobile traveling at 100 m/s (about 200 mi/h).
(*b*) A 10-g bullet traveling at 500 m/s.
(*c*) A smoke particle of mass 10^{-9} g moving at 1 cm/s.
(*d*) An electron with a kinetic energy of 1 eV.
(*e*) An electron with a kinetic energy of 100 MeV.

SOLUTION

(*a*) Using the classical relation between velocity and momentum,

$$\lambda = \frac{h}{p} = \frac{h}{mv} = \frac{6.6 \times 10^{-34}\,\text{J·s}}{(10^3\,\text{kg})(100\,\text{m/s})} = 6.6 \times 10^{-39}\,\text{m}$$

(*b*) As in part (*a*),

$$\lambda = \frac{h}{mv} = \frac{6.6 \times 10^{-34}\,\text{J·s}}{(10^{-2}\,\text{kg})(500\,\text{m/s})} = 1.3 \times 10^{-34}\,\text{m}$$

(*c*)

$$\lambda = \frac{h}{mv} = \frac{6.6 \times 10^{-34}\,\text{J·s}}{(10^{-12}\,\text{kg})(10^{-2}\,\text{m/s})} = 6.6 \times 10^{-20}\,\text{m}$$

Louis de Broglie (1892–1987, France). A member of an aristocratic family, his work contributed substantially to the early development of the quantum theory.

(*d*) The rest energy (mc^2) of an electron is 5.1×10^5 eV. Since the kinetic energy (1 eV) is much less than the rest energy, we can use nonrelativistic kinematics.

$$p = \sqrt{2mK} = \sqrt{(2)(9.1 \times 10^{-31}\,\text{kg})(1\,\text{eV})(1.6 \times 10^{-19}\,\text{J/eV})}$$
$$= 5.4 \times 10^{-25}\,\text{kg·m/s}$$

Then,

$$\lambda = \frac{h}{p} = \frac{6.6 \times 10^{-34}\,\text{J·s}}{5.4 \times 10^{-25}\,\text{kg·m/s}} = 1.2 \times 10^{-9}\,\text{m} = 1.2\,\text{nm}$$

We can also find this solution in the following way, using $hc = 1240$ eV·nm.

$$p = \sqrt{2mK} = \sqrt{\frac{2(mc^2)K}{c^2}} = \frac{1}{c}\sqrt{2(mc^2)K}$$
$$cp = \sqrt{2(5.1 \times 10^5\,\text{eV})(1\,\text{eV})} = 1.0 \times 10^3\,\text{eV}$$
$$\lambda = \frac{h}{p} = \frac{hc}{pc} = \frac{1240\,\text{eV·nm}}{1.0 \times 10^3\,\text{eV}} = 1.2\,\text{nm}$$

This method may seem artificial at first, but with practice it becomes quite useful, especially since energies are usually given in electron-volts in atomic and nuclear physics.

(*e*) In this case, the kinetic energy is much greater than the rest energy, and so we are in the extreme relativistic realm, where $K \approx E \approx pc$, as in Equation 2.42. The wavelength is

$$\lambda = \frac{hc}{pc} = \frac{1240\,\text{MeV·fm}}{100\,\text{MeV}} = 12\,\text{fm}$$

Note that the wavelengths computed in parts (*a*), (*b*), and (*c*) are far too small to be observed in the laboratory. Only in the last two cases, in which the wavelength is of the same order as atomic or nuclear sizes, do we have any chance of observing the wavelength. *Because of the smallness of h, only for particles of atomic or nuclear size will the wave behavior be observable.*

Two questions immediately follow. First, just what sort of wave is it which has this de Broglie wavelength? That is, what does the amplitude of the de Broglie wave measure? This is not an easy question to answer, so we defer a detailed answer until the last section of this chapter. For the present, we assume that, associated with the particle as it moves, there is a de Broglie wave of wavelength λ, which shows itself *when a wave-type experiment (such as diffraction) is performed on it.* The outcome of the wave-type experiment depends on this wavelength.

The second question then occurs: Why was this wavelength not directly observed before de Broglie's time? As parts (*a*), (*b*), and (*c*) of Example 3.1 showed, for ordinary objects the de Broglie wavelength is very small. Suppose we tried to demonstrate the wave nature of these objects through a double-slit type of experiment. Recall from Equation 3.17 that the spacing between adjacent fringes in a double-slit experiment is $\Delta y = \lambda D/d$. Putting in reasonable values for the slit separation d and slit-to-screen distance D, you will find that there is no achievable experimental configuration that can produce an observable separation of the fringes (see Problem 9). *There is no experiment that can be done to reveal the wave nature of macroscopic (laboratory-sized) objects.*

To show the wave nature of particles, we must use particles in the atomic realm. The indications of wave behavior come from interference and diffraction experiments. Interference experiments were discussed in Section 3.1; diffraction is discussed in most elementary physics texts and is illustrated for light waves in Figure 4.1.

The experimental difficulties of constructing double slits to do interference experiments with beams of particles such as electrons, neutrons, and atoms were not solved until long after the time of de Broglie's hypothesis. We discuss these experiments a bit later in this section. For now, we consider two experiments whose technological problems were solved soon after de Broglie's suggestion in 1924 of the wave nature of particles. These experiments involve *electron diffraction*, not through an artificially constructed single slit as in Figure 4.1 but instead through the atoms of a crystal. The outcomes of these experiments resemble those of the similar X-ray diffraction experiments illustrated in Section 3.1. The regular spacing of the atoms of the crystal produces the beautiful diffraction patterns illustrated for X rays in Figures 3.7–3.9.

To examine the wave nature of electrons we therefore adopt the following procedure. A beam of electrons is accelerated through a potential V, acquiring a nonrelativistic kinetic energy $K = eV$ and a momentum $p = \sqrt{2mK}$. Wave mechanics would describe the beam of electrons as a wave   *Electron diffraction*
of wavelength $\lambda = h/p$. The beam strikes a crystal in exactly the same way as the beam of X rays in Figure 3.7, and the scattered beam is photographed. Such a photograph is shown in Figure 4.2. The similarity between Figure 4.2 (electron diffraction) and Figures 3.7 and 3.8 (X-ray diffraction) strongly suggests that the electrons are behaving as waves.

FIGURE 4.1 (Top) Diffraction pattern for light waves incident on a single slit. (Bottom) Diffraction pattern for light waves incident on a fine wire.

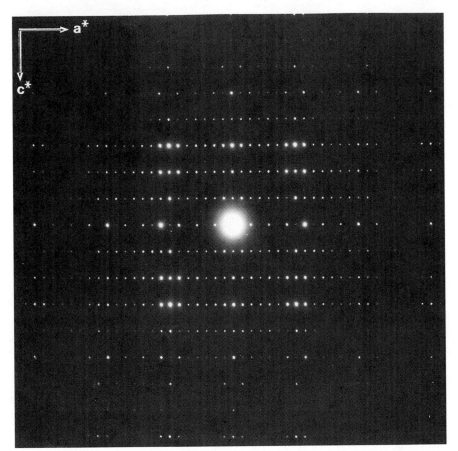

FIGURE 4.2 An electron diffraction pattern. Each bright dot is a region of constructive interference, as in the X-ray diffraction patterns of Figures 3.7 and 3.8. The target is a crystal of $Ti_2Nb_{10}O_{29}$.

A direct comparison of the "rings" produced in scattering by polycrystalline materials (see Figure 3.9) is shown in Figure 4.3. The effect of the comparison between electron scattering and X-ray scattering is striking, and there is again strong evidence for the similarity in the wave behavior of electrons and X rays. Experiments of the type illustrated in Figure 4.3 were first done in 1927 by G. P. Thomson, who shared the 1937 Nobel prize for this work. (Thomson's father, J. J. Thomson, received the 1906 Nobel prize for his discovery of the electron and measurement of its charge-to-mass ratio. It can thus be said that Thomson, the father, discovered the particle nature of the electron, while Thomson, the son, discovered its wave nature.)

The first experimental confirmation of the wave nature of electrons (and the quantitative confirmation of the de Broglie relationship $\lambda = h/p$) followed soon after de Broglie's original hypothesis. In 1926, at the Bell *Davisson-Germer experiment* Telephone Laboratories, Clinton Davisson and Lester Germer were investigating the reflection of electron beams from the surface of nickel crystals.

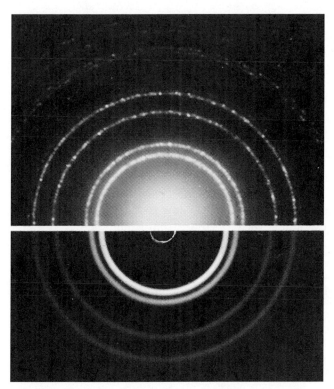

FIGURE 4.3 Comparison of X-ray diffraction and electron diffraction. The upper half of the figure shows the result of scattering of 0.071 nm X rays by an aluminum foil, and the lower half shows the result of scattering of 600 eV electrons by aluminum. (The wavelengths are different so the scales of the two halves have been adjusted.)

A schematic view of their apparatus is shown in Figure 4.4. A beam of electrons from a heated filament is accelerated through a potential difference V. Passing through a small aperture, the beam strikes a single crystal of nickel. Electrons are scattered in all directions by the atoms of the crystal, some of them striking a detector, which can be moved to any angle ϕ relative to the incident beam and which measures the intensity of the electron beam scattered at that angle.

Figure 4.5 shows the results of one of the experiments of Davisson and Germer. When the accelerating voltage is set at 54 V, there is an intense reflection of the beam at the angle $\phi = 50°$. It is difficult to account for this intense reflection through any effect other than the superposition of waves to give an intensity maximum. Let us see how these results give quantitative confirmation of the de Broglie wavelength.

If we assume that each of the atoms of the crystal can act as a scatterer, then the scattered *electron waves* can interfere, and we have a crystal diffraction grating for the electrons. Figure 4.6 shows a simplified representation of the nickel crystal used in the Davisson-Germer experiment. Be-

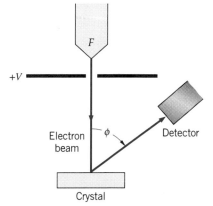

FIGURE 4.4 Apparatus used by Davisson and Germer to study electron diffraction. Electrons leave the filament F and are accelerated by the voltage V. The beam strikes a crystal and the scattered beam is detected at an angle ϕ relative to the incident beam. The detector can be moved in the range 0 to 90°.

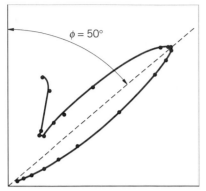

FIGURE 4.5   Results of Davisson and Germer. Each point on the plot represents the relative intensity when the detector in Figure 4.4 is located at the corresponding angle ϕ measured from the vertical axis. Constructive interference causes the intensity of the reflected beam to reach a maximum at $\phi = 50°$ for $V = 54$ V.

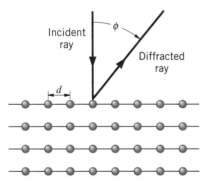

FIGURE 4.6   The crystal surface acts like a diffraction grating with spacing d.

Double-slit interference

cause the electrons were of low energy, they did not penetrate very far into the crystal, and it is sufficient to consider the diffraction to take place in the plane of atoms on the surface. The situation is entirely similar to using a reflection grating for light; the spacing d between the rows of atoms on the crystal is analogous to the spacing between the slits in the optical grating. The maxima for a diffraction grating occur at angles ϕ such that

$$d \sin \phi = n\lambda \tag{4.2}$$

where n (= 1, 2, 3, ...) is the order number of the maximum. From independent data, it is known that the spacing between the rows of atoms in a nickel crystal is $d = 0.215$ nm.

The peak at $\phi = 50°$ must be a first-order peak ($n = 1$), because no peaks were observed at smaller angles. If this is indeed an interference maximum, the corresponding wavelength is, from Equation 4.2,

$$\lambda = d \sin \phi = (0.215 \text{ nm})(\sin 50°) = 0.165 \text{ nm}$$

We can compare this value with that expected on the basis of the de Broglie theory. An electron accelerated through a potential difference of 54 V has a kinetic energy of 54 eV and therefore a momentum of

$$p = \sqrt{2mK} = \frac{1}{c}\sqrt{2mc^2K} = \frac{1}{c}(7430 \text{ eV})$$

The de Broglie wavelength is $\lambda = h/p = hc/pc$. Using $hc = 1240$ eV·nm,

$$\lambda = \frac{1240 \text{ eV·nm}}{7430 \text{ eV}} = 0.167 \text{ nm}$$

This is in excellent agreement with the value found from the diffraction maximum, and provides strong evidence in favor of the de Broglie theory. For this experimental work, Davisson shared the 1937 Nobel prize with G. P. Thomson.

The wave nature of paraticles is not exclusive to electrons; *any* particle with momentum p has de Broglie wavelength h/p. Neutrons are produced in nuclear reactors with kinetic energies corresponding to wavelengths of roughly 0.1 nm; these also should be suitable for diffraction by crystals. Figure 4.7 shows that diffraction of neutrons by a salt crystal produces the same characteristic patterns as the diffraction of electrons or X rays.

To study the nuclei of atoms, much smaller wavelengths are needed, of the order of 10^{-15} m. Figure 4.8 shows the diffraction pattern produced by the scattering of 1-GeV kinetic energy protons by oxygen nuclei. The maxima and minima of the diffraction pattern are similar to those of single-slit diffraction as shown in Figure 4.1. (The intensity at the minima does not fall to zero because nuclei do not have a sharp boundary. The determination of nuclear sizes from such diffraction patterns is discussed in Chapter 12.)

The definitive evidence for the wave nature of *light* was deduced from the double-slit experiment performed by Thomas Young in 1801. In principle, it should be possible to do double-slit experiments with *particles* and thereby directly observe their wavelike behavior. However, the technological difficulties of producing double slits for particles are formidable, and only in recent years have these problems been solved. The first double-slit experiment with electrons was done in 1961. A diagram of the apparatus is shown in Figure 4.9. The electrons from a hot filament were accelerated through 50 kV (corresponding to $\lambda = 5.4$ pm) and then passed through a double slit of separation 2.0 μm and width 0.5 μm. A photograph of the resulting intensity pattern is shown in Figure 4.10. The similarity with the double-slit pattern for light (Figure 3.2) is striking.

A similar experiment can be done for neutrons. In this case, the source of neutrons was a nuclear reactor, in which the emerging neutrons were slowed to a "thermal" energy distribution (average $K \approx kT \approx 0.025$ eV). Using crystal diffraction, specific wavelengths from this distribution could be selected through a process similar to what is shown in Figure 3.5—the neutrons (which have a continuous energy distribution) are incident on a crystal, and by choosing the scattering angle θ it is possible to choose a specific wavelength, as the Bragg scattering law, Equation 3.19, suggests (see Problem 12). The neutrons (of kinetic energy 0.00024 eV and de Broglie wavelength 1.85 nm) passed through a gap of diameter 148 μm in a material that absorbs virtually all of the neutrons incident on it (Figure 4.11). In the center of the gap was a boron wire (also highly absorptive for neutrons) of diameter 104 μm. The neutrons could pass on either side of

FIGURE 4.7 Diffraction of neutrons by a sodium chloride crystal.

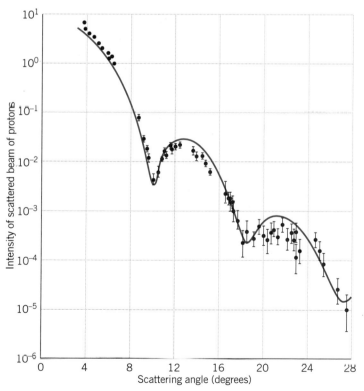

FIGURE 4.8 Diffraction of 1 GeV protons by oxygen nuclei. The pattern of maxima and minima is similar to that of single-slit diffraction of light waves.

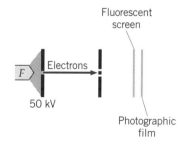

FIGURE 4.9 Double-slit apparatus for electrons. Electrons from the filament F are accelerated through 50 kV and pass through the double slit. They produce a visible pattern when they strike a fluorescent screen (like a TV screen), and the resulting pattern is photographed. A photograph is shown in Figure 4.10. [See C. Jönsson, *American Journal of Physics* **42**, 4 (1974).]

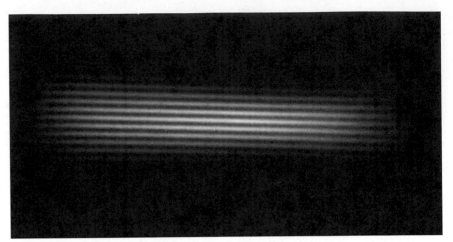

FIGURE 4.10   Double-slit interference pattern for electrons.

the wire through slits of width 22 μm. The intensity of neutrons that pass through this double slit was observed by sliding another slit across the beam and measuring the intensity of neutrons passing through this "scanning slit." Figure 4.12 shows the resulting pattern of intensity maxima and minima, which leaves no doubt that interference is occurring and that the neutrons have a corresponding wave nature. The wavelength can be deduced from the slit separation using Equation 3.17. Estimating the spacing Δy the adjacent maxima from Figure 4.12 to be about 75 μm, we obtain

$$\lambda = \frac{d \Delta y}{D} = \frac{(126\ \mu\text{m})(75\ \mu\text{m})}{5\ \text{m}} = 1.89\ \text{nm}$$

This result agrees very well with the de Broglie wavelength of 1.85 nm selected for the neutron beam.

It is also possible to do a similar experiment with atoms. In this case, a source of helium atoms formed a beam (of velocity corresponding to a kinetic energy of 0.020 eV) which passed through a double slit of separation 8 μm and width 1 μm. Again a scanning slit was used to measure the intensity of the beam passing through the double slit. Figure 4.13 shows

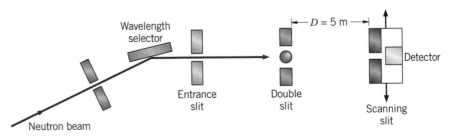

FIGURE 4.11   Double-slit apparatus for neutrons. Thermal neutrons from a reactor are incident on a crystal; scattering through a particular angle selects the energy of the neutrons. After passing through the double slit, the neutrons are counted by the scanning slit assembly, which moves laterally.

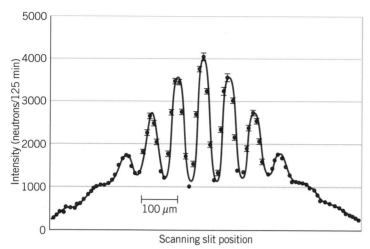

FIGURE 4.12 Intensity pattern observed for double-slit interference with neutrons. The spacing between the maxima is about 75 μm. [*Source*: R. Gähler and A. Zeilinger, *American Journal of Physics* **59**, 316 (1991).]

the resulting intensity pattern. Although the results are not as dramatic as those for electrons and neutrons, there is clear evidence of interference maxima and minima, and the separation of the maxima gives a wavelength that is consistent with the de Broglie wavelength (see Problem 7).

So far in this chapter we have discussed several different types of interference and diffraction experiments using different particles—electrons, protons, neutrons, and atoms. These experiments are not restricted to any particular type of particle nor to any particular type of observation. They are examples of a *general* phenomenon, the wave nature of particles, which was unobserved before 1920 because the necessary experiments had not yet been done. Today this wave nature is used as a basic tool by scientists. For example, neutron diffraction gives detailed information on the structure of solid crystals and of complex molecules (Figure 4.14). The electron microscope uses electron waves to illuminate and photograph objects; because the wavelength can be made thousands of times smaller than that of visible light, it is possible to resolve and observe small details that are not observable with visible light (see Color Plate 1).

When we do a double-slit experiment with particles such as electrons, it is tempting to try to determine through which slit the particle passes. For example, we could surround each slit with an electromagnetic loop that causes a meter to deflect whenever a charged particle or perhaps a particle with a magnetic moment passes through the loop (Figure 4.15). If we fire the particles through the slits at a slow enough rate, we can track each particle as it first passes through one slit or the other and then appears on the screen.

If we actually performed this experiment, we would no longer record an interference pattern on the screen. Instead, we would observe a pattern similar to that shown in Figure 4.15, with "hits" in front of each slit, but no interference fringes. No matter what sort of device we use to determine

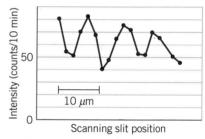

FIGURE 4.13 Intensity pattern observed for double-slit interference with helium atoms [*Source*: O. Carnal and J. Mlynek, *Physical Review Letters* **66**, 2689 (1991).]

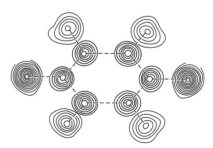

FIGURE 4.14 The atomic structure of solid benzene as deduced from neutron diffraction. The circles indicate contours of constant density. The black circles show the locations of the six carbon atoms that form the familiar benzene ring. The blue circles show the locations of the hydrogen atoms.

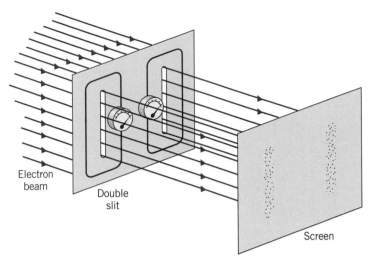

FIGURE 4.15   Apparatus to record passage of electrons through slits. Each slit is surrounded by a loop with a meter that signals the passage of an electron through the slit. No interference finges are seen on the screen.

through which slit the particle passes, the interference pattern will be destroyed. The classical *particle* must pass through one slit or the other; only a *wave* can reveal interference, which depends on parts of the wavefront passing through *both* slits and then recombining.

When we ask through which slit the particle passed, we are investigating only the *particle* aspects of its behavior, and we cannot observe its wave nature (the interference pattern). Conversely, when we study the wave nature, we cannot simultaneously observe the particle nature. The electron will behave as a particle *or* a wave, but we cannot observe *both* aspects of its behavior simultaneously.

Principle of complementarity   This is the basis for the *principle of complementarity*, which asserts that the complete description of a physical entity such as a photon or an electron cannot be made in terms of only particle properties or only wave properties, but that both aspects of its behavior must be considered. Moreover, the particle and wave natures cannot be observed simultaneously, and the aspect of the behavior that we observe depends on the kind of experiment we are doing: a particle-type experiment shows only particlelike behavior, and a wave-type experiment shows only wavelike behavior.

4.2   UNCERTAINTY RELATIONSHIPS FOR CLASSICAL WAVES

We explore in this section another important difference between classical particles and waves. Let us consider a wave of the form $y = y_1 \sin k_1 x$, as shown in Figure 4.16. This is a wave that repeats itself endlessly from $x =$

$-\infty$ to $x = +\infty$. If we ask the question "Where is the wave located?" we cannot provide an answer—it is everywhere. (Its wavelength, on the other hand, is precisely determined and is equal to $2\pi/k_1$.) If we are to use a wave to represent a particle, the wave must have one of the important attributes of a particle—it must be *localized,* or able to be confined to a relatively small (atom-sized or nucleus-sized, for example) region of space. Because of its infinite extent, the pure sine wave is of no use in localizing particles.

Now consider what happens when we add to our original wave another wave of slightly different wavelength (i.e., different k), so that $y = y_1 \sin k_1 x + y_2 \sin k_2 x$. The characteristic pattern, known in the case of sound waves as "beats," is produced as shown in Figure 4.17. The pattern still repeats endlessly from $x = -\infty$ to $x = +\infty$, but we know a bit more about the "position" of the wave—at certain values of x the medium is less likely to be "waving" than at others (or at least it is "waving" with a smaller amplitude). In Figure 4.17, we would observe vibration at the point $x = x_A$, but not at $x = x_B$. The state of our knowledge of the "position" of the wave has improved, but it is at the expense of our knowledge of its wavelength—we added together two different wavelengths and so the wavelength is no longer precisely determined.

If we continued to add waves of different wavelengths (different wave numbers k), with properly chosen amplitudes and phases, we could eventually reach a situation similar to that illustrated in Figure 4.18. This wave has virtually no amplitude outside a rather narrow region of space Δx (Δx is not precisely defined, but is a rough measure of the region over which the wave has reasonably large amplitude). In order to achieve this situation, we must add together a large number of waves of different wave numbers k. The wave thus represents a range of wave numbers (wavelengths) that we denote by Δk. When we had a single sine wave, Δk was zero (only one k) and Δx was infinite (the wave extended throughout all space). As we *increased* Δk (by adding more waves), we *decreased* Δx (the wave became more confined). We seem to have an inverse relationship between Δx and

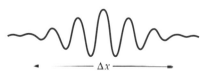

FIGURE 4.16 A pure sine wave, which extends from $-\infty$ to $+\infty$.

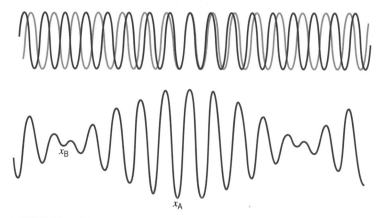

FIGURE 4.17 The superposition of two sine waves of nearly equal wavelengths to give beats. The two sine waves differ in wavelength by 10 percent but have the same amplitude.

FIGURE 4.18 The resultant of the addition of many sine waves (of different wavelengths and possibly different amplitudes).

Δk; as one decreases, the other increases. An approximate mathematical relationship between Δx and Δk is

$$\Delta x \, \Delta k \sim 1 \qquad (4.3)$$

where the wavy equal sign is taken to mean "of order of magnitude." Since we have not yet defined Δx and Δk precisely, they should be regarded as estimates, and Equation 4.3 is a rough indication of their relationship. Equation 4.3 asserts that the product of Δx, the spatial extent of the wave, and Δk, the range of wave numbers it contains, is of order of magnitude one.

For any type of wave, the position can be determined only at the expense of our knowledge of its wavelength. This statement, and its mathematical representation given in Equation 4.3, are the first of our "uncertainty relationships" for classical waves. The position and wavelength (wave number) are mutually "uncertain" to the degree given by Equation 4.3.

We can interpret this relationship in a slightly different way. Suppose we attempt to measure the wavelength of a classical wave, such as a water wave. We can do this by measuring the distance between two adjacent wave crests. Suppose the wave is a very short pulse with only one wave crest (Figure 4.19). Measuring λ is very difficult, and we are likely to make a large error, perhaps of the order of one wavelength. That is, when Δx, the spatial extent of the wave, is of the order of the wavelength ($\Delta x \sim \lambda$), then the uncertainty in the wavelength $\Delta\lambda$ is of the order of the wavelength ($\Delta\lambda \sim \lambda$). Combining these estimates, we obtain $\Delta x \, \Delta\lambda \sim \lambda^2$. Suppose now that the wave extends over N cycles, so that $\Delta x \sim N\lambda$. Now we can determine λ with greater precision. Counting the number of wavelengths in Δx, we still may make an error of the order of one wavelength (perhaps $\frac{1}{2}$ or $\frac{1}{3}$ or $\frac{1}{4}$, but still of order one) out of N, so now $\Delta\lambda \sim \lambda/N$, and once again $\Delta x \, \Delta\lambda \sim \lambda^2$. (A slightly different form of this relationship, with a smaller estimate of the uncertainties, is derived in Example 4.2.) This uncertainty relationship, which connects the "size" of a wave with the uncertainty in determining its wavelength, can be shown to be equivalent to Equation 4.3.

We can make similar arguments about the time dependence of the wave. Assume we have a wave that lasts for a certain time interval Δt. We want to determine the period T of this wave by counting the number of cycles in this time interval. If there are N cycles in the wave, then $\Delta t \sim NT$. In measuring the period, we may make an error of the order of one cycle in counting the number of cycles in the wave group, and so $\Delta T \sim T/N$.

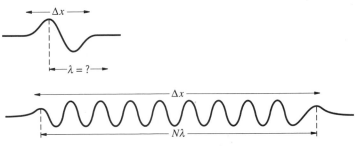

FIGURE 4.19 Two different groups of waves.

Combining these estimates, we obtain $\Delta T \, \Delta t \sim T^2$. Rewriting this estimate in terms of the angular frequency $\omega = 2\pi/T$, we obtain (again neglecting factors of order unity)

$$\Delta\omega \, \Delta t \sim 1 \qquad (4.4)$$

Frequency—time uncertainty relationship

This is the second of the uncertainty relationships for classical waves, and is similar to Equation 4.3 in that it gives a rough relationship between estimates of uncertainties.

EXAMPLE 4.2

In a measurement of the wavelength of water waves, 10 wave crests are counted in a distance of 200 cm. Estimate the minimum uncertainty in the wavelength that might be obtained from this experiment.

SOLUTION

In order to estimate $\Delta\lambda$, we need to relate it to Δk. We know k and λ are related by

$$k = \frac{2\pi}{\lambda}$$

and, after differentiating,

$$dk = \frac{-2\pi}{\lambda^2} d\lambda$$

Replacing the differentials with small intervals and taking magnitudes only, we find

$$\Delta k = \frac{2\pi}{\lambda^2} \Delta\lambda$$

(Think about how this is different from $\Delta k = 2\pi/\Delta\lambda$. Why is this latter equation incorrect? What happens when λ is known exactly?) From Equation 4.3, $\Delta x \, \Delta k \sim 1$, so

$$\Delta x \left(\frac{2\pi}{\lambda^2} \Delta\lambda \right) \sim 1$$

or

$$\Delta x \, \Delta\lambda \sim \frac{\lambda^2}{2\pi} \qquad (4.5)$$

Solving for $\Delta\lambda$ and inserting the numerical values, we obtain

$$\Delta\lambda \sim \frac{1}{\Delta x} \frac{\lambda^2}{2\pi} = \frac{1}{200 \text{ cm}} \frac{(20 \text{ cm})^2}{2\pi} \sim 0.3 \text{ cm}$$

EXAMPLE 4.3

An electronics salesman offers to sell you a frequency-measuring device. When hooked up to a sinusoidal signal, it automatically displays the frequency of the signal, and to account for frequency variations, the frequency is remeasured and the display updated once each second. The salesman claims the device to be accurate to 0.01 Hz. Is this claim valid?

SOLUTION

Based on Equation 4.4, we know that a measurement of frequency in a time $\Delta t = 1$ s must have an associated uncertainty of $\Delta\omega \sim 1$ rad/s. The salesman's quoted accuracy is $\Delta\omega = 2\pi(\Delta\nu) = 0.06$ rad/s, and thus we have reason to doubt the quoted specification.

4.3 HEISENBERG UNCERTAINTY RELATIONSHIPS

The uncertainty relationships discussed in the previous section apply to *all* waves, and we should therefore apply them to de Broglie waves. We can use the basic de Broglie relationship $p = h/\lambda$ along with the expression $k = 2\pi/\lambda$ to find $p = hk/2\pi$, which relates the momentum of a particle to the wave number of its de Broglie wave. The combination $h/2\pi$ occurs frequently in wave mechanics and is given the special symbol $\hbar$ ("*h*-bar")

$$\hbar = \frac{h}{2\pi} = 1.05 \times 10^{-34}\,\text{J·s}$$

$$= 6.58 \times 10^{-16}\,\text{eV·s}$$

In terms of $\hbar$,

$$p = \hbar k \tag{4.6}$$

Heisenberg uncertainty relationships

and so $\Delta k = \Delta p/\hbar$. From the uncertainty relationship (Equation 4.3) we then have

$$\Delta x\,\Delta p_x \sim \hbar \tag{4.7}$$

The x subscript has been added to the momentum to remind us that Equation 4.7 applies to motion in a given direction and relates the uncertainties in position and momentum in that direction only. Similar and independent relationships can be applied in the other directions as necessary; thus $\Delta y\,\Delta p_y \sim \hbar$ or $\Delta z\,\Delta p_z \sim \hbar$.

The energy-frequency relationship for photons is $E = h\nu$, which can also be written as $E = \hbar\omega$. As we did in the case of the momentum-wavelength relationship, we assume that this relationship can be carried over to particles for the purpose of calculating the uncertainty in energy, which involves small differentials only. We therefore take $\Delta E = \hbar \Delta \omega$, and by substituting for $\Delta \omega$ in Equation 4.4 we obtain

$$\Delta E \, \Delta t \sim \hbar \qquad (4.8)$$

Equations 4.7 and 4.8 are the *Heisenberg uncertainty relationships.* They are the mathematical representations of the *Heisenberg uncertainty principle,* which states:

It is not possible to make a simultaneous determination of the position and the momentum of a particle with unlimited precision,

and

It is not possible to make a simultaneous determination of the energy and the time coordinate of a particle with unlimited precision.

Werner Heisenberg (1901–1976, Germany). Best known for the uncertainty principle, he also developed a complete formulation of the quantum theory based on matrices.

These relationships give an estimate of the minimum uncertainty that can result from any experiment; measurement of the position and momentum of a particle will give a spread of values of widths Δx and Δp_x. We may, for other reasons, do much worse than Equations 4.7 or 4.8, but *we can do no better.* (Occasionally, you may see these relationships written with $\hbar/2$ or h, rather than $\hbar$, on the right-hand side, or else with $>$ rather than $\sim$ showing the equality. This difference is not very important, since Equations 4.7 and 4.8 give us only estimates. The actual uncertainties Δx and Δp_x depend on the distribution of different wave numbers (or wavelengths) we use to restrict the wave to the region Δx; the most compact distribution gives $\Delta x \, \Delta p_x = \hbar/2$, while for all other distributions $\Delta x \, \Delta p_x > \hbar/2$. We are therefore safe in using $\hbar$ as an estimate.)

These relationships have a profound impact on our view of nature. It is quite acceptable to say that there is an uncertainty in locating the position of a water wave. It is quite another matter to make the same statement about a de Broglie wave, since *there is an implied corresponding uncertainty in the position of the particle.* Equations 4.7 and 4.8 say that *nature imposes a limit on the accuracy with which we can do experiments;* no matter how well designed our measuring apparatus might be, we still can do no better than Equations 4.7 or 4.8. To determine the momentum accurately, we must measure over a long distance Δx; if we wish to confine a particle to a small region of space Δx, we lose our ability to measure its momentum accurately. To measure an energy with small uncertainty takes a long time Δt; if a particle lives only for a short time, its energy uncertainty will be large.

The following examples give applications of these uncertainty relationships.

EXAMPLE 4.4

An electron moves in the x direction with a speed of 3.6×10^6 m/s. We can measure its speed to a precision of 1%. (*a*) With what precision can we simultaneously measure its position? (*b*) What can we say about its motion in the *y* direction?

SOLUTION

(*a*) The electron's momentum is

$$p_x = mv_x = (9.11 \times 10^{-31} \text{ kg})(3.6 \times 10^6 \text{ m/s}) = 3.3 \times 10^{-24} \text{ kg·m/s}$$

The uncertainty Δp_x is 1% of this value, or 3.3×10^{-26} kg·m/s. The uncertainty in position is then

$$\Delta x \sim \frac{\hbar}{\Delta p_x} = \frac{1.05 \times 10^{-34} \text{ J·s}}{3.3 \times 10^{-26} \text{ kg·m/s}} = 3.2 \text{ nm}$$

which is roughly 10 atomic diameters.

(*b*) If the electron is moving in the *x* direction, then we know its speed in the *y* direction precisely; thus $\Delta p_y = 0$. The uncertainty relationship $\Delta y \, \Delta p_y \sim \hbar$ then requires that Δy be infinite. We can therefore know nothing at all about the electron's *y* coordinate.

EXAMPLE 4.5

Repeat the calculations of the previous Example in the case of a pitched baseball (*m* = 0.145 kg) moving at a speed of 95 mi/h (42.5 m/s). Again assume that its speed can be measured to a precision of 1%.

SOLUTION

(*a*) The baseball's momentum is

$$p_x = mv_x = (0.145 \text{ kg})(42.5 \text{ m/s}) = 6.16 \text{ kg·m/s}$$

The uncertainty in momentum is 6.16×10^{-2} kg·m/s, and the corresponding uncertainty in position is

$$\Delta x \sim \frac{\hbar}{\Delta p_x} = \frac{1.05 \times 10^{-34} \text{ J·s}}{6.16 \times 10^{-2} \text{ kg·m/s}} = 1.7 \times 10^{-33} \text{ m}$$

This uncertainty is 19 orders of magnitude smaller than the size of an atomic nucleus. The uncertainty principle cannot be blamed for the batter missing the pitch!

(*b*) It appears that the calculation of the uncertainty in the *y* coordinate (which we assume to be horizontal in the left-right direction according to the catcher) of the baseball will proceed along the same lines as that for the electron, which will lead to a conclusion that we can have no knowledge whatsoever about the horizontal coordinate of the baseball. Instead of passing over home plate, it might equally as well be found passing over first base or third base! Since we know this conclusion to be false, let us reconsider this calculation. Our error is in assuming that we can know that the baseball is moving *precisely* in the *x* direction. Suppose it has a tiny component of velocity in the *y* direction. This velocity component might result from air currents or our inability to throw the ball exactly in the x direction. Suppose this component to be 10^{-10} of the x component of the velocity; determining the direction of the baseball to this accuracy would be a remarkable achievement, but let's assume it can be done. A calculation similar to the previous one suggests that the uncertainty in the y coordinate is $\Delta y = 1.7 \times 10^{-25}$ m, again many times smaller than the diameter of a nucleus. There is no conceivable method that could be used to determine the y component of the baseball's velocity to a precision such that the uncertainty principle would affect our ability to follow its path. Once again we see that, because of the small magnitude of Planck's constant, quantum effects are not observable for ordinary objects.

Suppose we place a particle in a region of length L in which it can move back and forth between two walls. Between the walls, the particle moves freely (no forces act on it). The particle makes elastic collisions with the walls. Suppose the walls are initially located at $\pm\infty$. All we know about the position of the particle is that it is between the two walls ($\Delta x = \infty$). The uncertainty relationship then tells us that $\Delta p_x = 0$, that is, we can know the particle's momentum exactly. Let's assume that $p_x = 0$. We now move the walls toward one another until their separation is L. The particle is now confined to a smaller region of space ($\Delta x \sim L$), which means that the uncertainty in its momentum must grow ($\Delta p_x \sim \hbar/L$), and so an observation of the particle is likely to find it in motion.

Confining a particle

If we were to prepare a large number of systems and measure the momentum of the particle in each system, the average of our measurements would be zero. Since there is no preferred direction in this situation, the particle must be found moving in one direction as often as in the other, and the average of many measurements will be zero. Figure 4.20 shows a representative example of the outcome of a large number of measurements. The vertical scale shows how often a particular result is obtained. The values are symmetrically arranged about zero (the mean or average value), and the *width* of the distribution is measured by Δp_x.

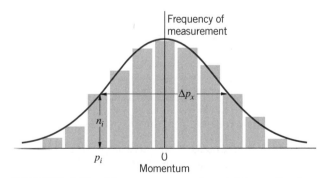

FIGURE 4.20 The momentum of a particle confined to a region Δx. If we repeat the measurement many times, a value p_i will be measured n_i times. The average momentum is zero, and the distribution has a width $\Delta p_x \sim \hbar/\Delta x$.

Figure 4.20 resembles a statistical distribution, and in fact the precise definition of Δp_x is similar to that of the standard deviation σ_A of a quantity A that has a mean or average value A_{av}:

$$\sigma_A = \sqrt{(A^2)_{av} - (A_{av})^2}$$

If there are N individual measurements of A, then $A_{av} = N^{-1}\Sigma A_i$ and $(A^2)_{av} = N^{-1}\Sigma A_i^2$.

We can therefore make a rigorous definition of the uncertainty in momentum as

$$\Delta p_x = \sqrt{(p_x^2)_{av} - (p_{x,av})^2} \tag{4.9}$$

In the case of the particle moving between the walls, $p_{x,av}$ is certainly zero. The remaining result,

$$\Delta p_x = \sqrt{(p_x^2)_{av}} \tag{4.10}$$

gives in effect a root-mean-square value of p_x. This can be taken to be a rough measure of the magnitude of p_x. Thus it is often said that Δp_x gives a measure of the magnitude of the momentum of the particle. As you can see from Figure 4.20, this is indeed true.

EXAMPLE 4.6

A beam of monoenergetic electrons (of momentum p_y) moving in the y direction passes through a slit of width a (Figure 4.21). Find the uncertainty in the x component of an electron's momentum after it passes through the slit, and compare the interpretation of the experiment according to the uncertainty principle with the conventional interpretation of single-slit diffraction.

SOLUTION

The beam of electrons is initially moving only in the y direction, so before passing through the slit it has no momentum component in the x direction ($p_x = 0$). Since we know p_x exactly ($\Delta p_x = 0$), Equation 4.7 gives $\Delta x = \infty$; that is, we have no knowledge at all about the location of the particles along the x direction.

Just after the beam passes through the slit, the uncertainty in its x coordinate has been reduced to the slit width, so now $\Delta x = a$. The corresponding uncertainty in its momentum is, by Equation 4.7,

$$\Delta p_x \sim \hbar/a$$

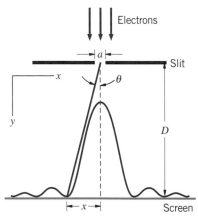

FIGURE 4.21 The intensity observed in single-slit diffraction. The slit has width a and the screen is at a distance D from the slit. The first diffraction minimum is a distance x from the center.

Measurements beyond the slit no longer show the particle moving precisely in the y direction; the momentum now has a small x component as well, with values distributed about zero as represented in Figure 4.20. In passing through the slit, a particle acquires on the average an x component of momentum of roughly $\hbar/a$, according to the uncertainty principle.

Let us now find the angle θ that specifies where a particle with this average p_x lands on the screen. For small angles, $\sin \theta \approx \tan \theta$ and so

$$\sin \theta \approx \tan \theta = \frac{p_x}{p_y} = \frac{\hbar/a}{p_y} = \frac{\lambda}{2\pi a}$$

using $\lambda = h/p_y$ for the de Broglie wavelength of the electrons. We can rewrite this expression as

$$a \sin \theta = \lambda/2\pi \tag{4.11}$$

which is similar to the expression $a \sin \theta = \lambda$ for the first minimum in diffraction by a slit of width a. The factor 2π in Equation 4.11 is somewhat arbitrary, depending on how we defined the uncertainties and the screen coordinates. Nevertheless, the calculation shows that the distribution of transverse momenta given by the uncertainty principle is roughly equivalent to the spreading of the beam into the central diffraction peak, and it illustrates again the close connection between wave behavior and uncertainty in particle location.

The diffraction (spreading) of a beam following passage through a slit is just the effect of the uncertainty principle on our attempt to specify the location of the particle. As we make the slit narrower, p_x increases and the beam spreads even more. In trying to obtain more precise knowledge of the location of the particle by making the slit narrower, we have lost knowledge of the direction of its travel. This trade-off between observations of position and momentum is the core of the Heisenberg uncertainty principle.

EXAMPLE 4.7

In nuclear beta decay, electrons are observed to be ejected from the atomic nucleus. Suppose we assume that electrons are somehow trapped within the nucleus, and that occasionally one escapes and is observed in the laboratory. Take the diameter of a typical nucleus to be 1.0×10^{-14} m, and use the uncertainty principle to estimate the range of kinetic energies that such an electron must have.

SOLUTION

If the electron were trapped in a region of width $\Delta x \sim 10^{-14}$ m, the corresponding uncertainty in its momentum would be

$$\Delta p_x = \frac{\hbar}{\Delta x} = \frac{1}{c}\frac{\hbar c}{\Delta x} = \frac{1}{c}\frac{197\,\text{MeV}\cdot\text{fm}}{10\,\text{fm}} = 19.7\,\text{MeV}/c$$

Note the use of $\hbar c = 197$ MeV·fm in this calculation. Since this momentum is clearly in the relativistic regime for electrons, we must use the relativistic formula to find the kinetic energy:

$$K = \sqrt{p^2c^2 + (mc^2)^2} - mc^2$$

and using Equation 4.10 to relate Δp_x to p_x^2, we obtain

$$K = 19\,\text{MeV}$$

That is, electrons confined to a nucleus-sized region of space would have a spread of kinetic energies with a typical value of around 19 MeV.

Electrons emitted from the nucleus in nuclear beta decay typically have kinetic energies of about 1 MeV, much smaller than the uncertainty principle requires for electrons confined inside the nucleus. This calculation thus suggests that beta-decay electrons of such low energies cannot be confined in a region of the size of the nucleus, and that another explanation must be found for the electrons observed in nuclear beta decay. (As we discuss in Chapter 12, these electrons did not pre-exist in the nucleus, which would violate the uncertainty principle, but are "manufactured" by the nucleus at the instant of the decay.)

EXAMPLE 4.8

(a) A charged pi meson has a rest energy of 140 MeV and a lifetime of 26 ns. Find the energy uncertainty of the pi meson, expressed in MeV and also as a fraction of its rest energy. (b) Repeat for the uncharged pi meson, with a rest energy of 135 MeV and a lifetime of 8.3×10^{-17} s. (c) Repeat for the rho meson, with a rest energy of 765 MeV and a lifetime of 4.4×10^{-24} s.

SOLUTION

(a) If the pi meson lives for 26 ns, we have only that much time in which to measure its rest energy, and Equation 4.8 tells us that *any* energy measurement done in a time Δt is uncertain by an amount of at least

$$\Delta E = \frac{\hbar}{\Delta t} = \frac{6.58 \times 10^{-16} \, \text{eV} \cdot \text{s}}{26 \times 10^{-9} \, \text{s}} = 2.5 \times 10^{-8} \, \text{eV} = 2.5 \times 10^{-14} \, \text{MeV}$$

$$\frac{\Delta E}{E} = \frac{2.5 \times 10^{-14} \, \text{MeV}}{140 \, \text{MeV}} = 1.8 \times 10^{-16}$$

(b) In a similar way,

$$\Delta E = \frac{\hbar}{\Delta t} = \frac{6.58 \times 10^{-16} \, \text{eV} \cdot \text{s}}{8.3 \times 10^{-17} \, \text{s}} = 7.9 \, \text{eV} = 7.9 \times 10^{-6} \, \text{MeV}$$

$$\frac{\Delta E}{E} = \frac{7.9 \times 10^{-6} \, \text{MeV}}{135 \, \text{MeV}} = 5.9 \times 10^{-8}$$

(c) For the rho meson,

$$\Delta E = \frac{\hbar}{\Delta t} = \frac{6.58 \times 10^{-16} \, \text{eV} \cdot \text{s}}{4.4 \times 10^{-24} \, \text{s}} = 1.5 \times 10^8 \, \text{eV} = 150 \, \text{MeV}$$

$$\frac{\Delta E}{E} = \frac{150 \, \text{MeV}}{765 \, \text{MeV}} = 0.20$$

In the first case, the uncertainty principle does not give a large enough effect to be measured—particle masses cannot be measured to a precision of 10^{-16} (about 10^{-6} is the best precision that we can obtain). In the second example, the uncertainty principle contributes at about the level of 10^{-7}, which approaches the limit of our measuring instruments and therefore might be observable in the laboratory. In the third example, we see that the uncertainty principle can contribute substantially to the precision of our knowledge of the rest energy of the rho meson; measurements of its rest energy will cluster about 765 MeV with a spread of 150 MeV, and no matter how precise an instrument we use to measure the rest energy, we can never reduce that spread.

The lifetime of a very short-lived particle such as the rho meson cannot be measured directly. In practice we reverse the procedure of the calculation of this example—we measure the rest energy, which gives a distribution of the form of Figure 4.20, and from the "width" ΔE of the distribution we deduce the lifetime using Equation 4.8. This procedure is discussed in Chapter 14.

EXAMPLE 4.9

Estimate the minimum velocity of a billiard ball ($m \sim 100$ g) confined to a billiard table of dimension 1 m.

SOLUTION

For $\Delta x \sim 1$ m, we have

$$\Delta p_x \sim \frac{\hbar}{\Delta x} = \frac{1.05 \times 10^{-34}\,\text{J·s}}{1\,\text{m}} = 1 \times 10^{-34}\,\text{kg·m/s}$$

so

$$\Delta v_x = \frac{\Delta p_x}{m} = \frac{1 \times 10^{-34}\,\text{kg·m/s}}{0.1\,\text{kg}} = 1 \times 10^{-33}\,\text{m/s}$$

Thus quantum effects suggest motion of the billiard ball with a speed the order of 1×10^{-33} m/s. At this speed, the ball would move a distance of $\frac{1}{100}$ of the diameter of an atomic nucleus in a time equal to the age of the universe! Once again, we see that quantum effects are not observable with macroscopic objects.

4.4 WAVE PACKETS

A pure sine wave is completely unlocalized—it extends from $-\infty$ to $+\infty$. A classical particle, on the other hand, is completely localized. Our quantum description mixes particles and waves. The particles are approximately, but not completely, localized. An electron, for example, is bound to a specific atom. We know its position to within an uncertainty of the order of the diameter of the atom (10^{-10} m), but we don't know exactly where it is within that atom. The method used in physics to describe such a situation is that of a *wave packet*. A wave packet can be considered to be the superposition of a large number of waves, which interfere constructively in the vicinity of the particle, giving the resultant wave a large amplitude, and interfere destructively far from the particle, so that the resultant wave has a small amplitude in regions where we don't expect to find the particle. The exact interpretation of these large and small amplitudes is discussed in the next section. For the present we will establish the mathematical description of the wave packet and discuss some of its properties.

An ideal wave packet would be one such as is pictured in Figure 4.18. Its amplitude is negligibly small, except for a region of space of dimension Δx. This corresponds to a particle that is localized in the region of dimension Δx. From our previous discussion, we know that such a situation results in a range of momenta Δp_x as specified by Equation 4.7; since each momentum corresponds to a unique de Broglie wavelength, a range of momenta Δp_x is equivalent to a range of wavelengths $\Delta\lambda$. Thus we expect that the mathematical description of the wave packet will be in terms of the addition (superposition) of a number of waves of varying wavelengths.

For simplicity, we consider sinusoidal waves of the form $y = A\cos kx$, where k is the wave number $2\pi/\lambda$. We also assume that all of the waves have the same amplitude. In general this will *not* be true, but it simplifies our calculations.

Let us begin with a wave of wave number k_1 and add to it a wave of nearly equal wave number $k_2 = k_1 + \Delta k$. The resulting wave form, shown in Figure 4.17, illustrates the phenomenon of "beats." The component waves start out in phase at $x = 0$, so that the resultant has its maximum amplitude there. Further along, the slight difference in wavelength causes the two waves to become out of phase, and the resultant has zero amplitude. A bit of trigonometric manipulation yields the result

$$y(x) = A\cos k_1 x + A\cos k_2 x$$

$$= 2A\cos\left(\frac{\Delta k}{2}x\right)\cos\left(\frac{k_1 + k_2}{2}x\right) \qquad (4.12)$$

The last term of Equation 4.12 gives the variation of the amplitude of the resultant within the envelope specified by the first cosine term.

Let us now consider these waves as traveling waves, whose mathematical description is obtained from Equation 4.12 by substituting $(kx - \omega t)$ for kx. The angular frequency of the oscillation is ω, and $v = \omega/k$ is the *phase velocity* of the wave—the speed with which a single component wave moves through the medium. The two waves are shown again in Figure 4.22 at $t = 0$ and at a later time t. In general, the phase velocities $v_1 = \omega_1/k_1$ and $v_2 = \omega_2/k_2$ may be different. Note that the envelope moves at a speed different from that of the individual waves. Again, we can derive an explicit expression for the resultant with a bit of trigonometric manipulation:

$$y(x, t) = A\cos(k_1 x - \omega_1 t) + A\cos(k_2 x - \omega_2 t)$$

$$= 2A\cos\left(\frac{\Delta k}{2}x - \frac{\Delta\omega}{2}t\right)\cos\left(\frac{k_1 + k_2}{2}x - \frac{\omega_1 + \omega_2}{2}t\right) \quad (4.13)$$

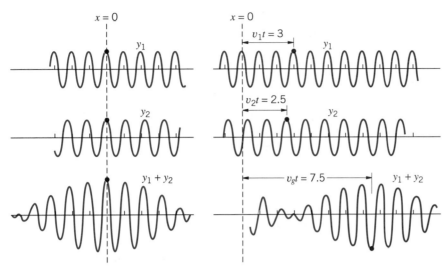

FIGURE 4.22 The group speed of a wave packet. At left is shown a "snapshot" at $t = 0$ of the waves y_1 and y_2 and their sum (y_1 has wavelength 1 unit and y_2 has wavelength $\frac{10}{9}$ unit.) Wave 1 moves with speed 3 units per second and wave 2 with speed 2.5 units per second. A snapshot at $t = 1$ s is shown at right. The two waves are not in phase until the point at 7.5 units; so the midpoint of the "beat" moves at a group speed of 7.5 units per second, in this case much greater than v_1 or v_2.

where $\Delta\omega = \omega_2 - \omega_1$. The envelope thus moves along with a speed $v = \Delta\omega/\Delta k$, while the wave within the envelope moves with speed $(\omega_1 + \omega_2)/(k_1 + k_2)$ which, if $\Delta\omega$ and Δk are small, does not differ greatly from v_1 or v_2.

The superposition of only two waves doesn't resemble the wave packet of Figure 4.18. We can make a better approximation at $t = 0$ by adding together more sine waves of different wave numbers k_i and possibly different amplitudes $A(k_i)$:

$$y(x) = \sum_{\substack{many \\ k_i}} A(k_i) \cos k_i x \tag{4.14}$$

If there are many different wave numbers and if they are very close together, the sum in Equation 4.14 can be replaced with an integral:

$$y(x) = \int A(k) \cos kx \, dk \tag{4.15}$$

where the integral is carried out over whatever range of wave numbers is permitted (possibly 0 to ∞).

For example, suppose we have a range of wave numbers from $k_0 - \Delta k/2$ to $k_0 + \Delta k/2$. If all of the waves have the same amplitude A, then from Equation 4.15 the form of the wave packet can be shown to be (see Problem 29 at the end of the chapter)

$$y(x) = \frac{2A}{x} \sin\left(\frac{\Delta k}{2} x\right) \cos k_0 x \tag{4.16}$$

The function $\cos k_0 x$ oscillates within the envelope $(2A/x) \sin (\Delta k\, x/2)$. This function is illustrated in Figure 4.23 and looks much more like the finite wave packet we are after. The wave has large amplitude only in a region Δx, but we have achieved this only by adding together many different wave numbers. (This is just as required by the uncertainty principle in the form of Equation 4.3—to make Δx small, Δk must be large.)

An even better approximation of the shape of the wave packet can be found by letting $A(k)$ vary; for example, the *Gaussian* $A(k) = e^{-(k-k_0)^2/2(\Delta k)^2}$ gives (see Problem 30)

$$y(x) \propto e^{-(\Delta k\, x)^2/2} \cos k_0 x \qquad (4.17)$$

Once again, there is an envelope that modulates the cosine and reduces its amplitude outside a region of width Δx, as shown in Figure 4.23. To restrict the wave to this small region, we have once again used a large range of wave numbers.

Figure 4.23 should be regarded as a snapshot of the wave packet at a specific time, such as $t = 0$. Similarly, Equations 4.16 and 4.17 represent the waves only at $t = 0$. To convert to traveling waves we must replace kx with $kx - \omega t$, as we did in Equation 4.13. In the case of the two waves used for Equation 4.13, we found that the envelope moved with speed $\Delta\omega/\Delta k$. We generalize this to the case in which there are many different wave numbers by defining the *group velocity* as follows:

Group velocity

$$v_{\text{group}} = \frac{d\omega}{dk} \qquad (4.18)$$

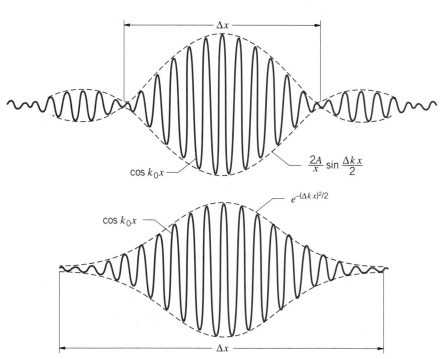

FIGURE 4.23 Example of two different wave packets. In each case there is a modulating function that reduces the amplitude of the cosine beyond the region Δx.

The envelope of the wave packet moves at the group velocity, while within the envelope each individual component wave moves with its *phase velocity*   *Phase velocity*

$$v_{\text{phase}} = \frac{\omega}{k} \qquad (4.19)$$

The phase velocity cannot be defined for the wave packet, but is meaningful only for a single component wave.

EXAMPLE 4.10

Certain ocean waves travel with a phase velocity $v_{\text{phase}} = \sqrt{g\lambda/2\pi}$, where g is the acceleration due to gravity. What is the group velocity of a "wave packet" of these waves? Express the result in terms of the phase velocity.

SOLUTION

The group velocity is found from Equation 4.18. Since $k = 2\pi/\lambda$, $v_{\text{phase}} = \sqrt{g/k}$. But with $v_{\text{phase}} = \omega/k$, we have $\omega/k = \sqrt{g/k}$, so $\omega = \sqrt{gk}$ and $d\omega = \sqrt{g}(\frac{1}{2}k^{-1/2})\, dk$. Therefore $d\omega/dk = \frac{1}{2}\sqrt{g/k}$, so $v_{\text{group}} = \frac{1}{2}v_{\text{phase}}$.

In summary, a particle that is localized in a certain region of space must be represented not by a single de Broglie wave of definite frequency and wavelength, but by a wave packet representing the superposition of a large number of waves. The envelope of the wave packet moves at the group velocity $d\omega/dk$.

Our discussion is not complete unless we can provide a physical interpretation of the group velocity. Suppose we have a localized particle, represented by a group of de Broglie waves. For each component wave, the energy of the particle is related to the frequency of the de Broglie wave by $E = h\nu = \hbar\omega$, and similarly the momentum of the particle is related to the wavelength of the de Broglie wave by $p = h/\lambda = \hbar k$. The group velocity $v_{\text{group}} = d\omega/dk$ of the de Broglie wave can then be expressed as

$$v_{\text{group}} = \frac{d\omega}{dk} = \left(\frac{d\omega}{dE}\right)\left(\frac{dE}{dp}\right)\left(\frac{dp}{dk}\right) = \left(\frac{1}{\hbar}\right)\left(\frac{dE}{dp}\right)(\hbar)$$

$$v_{\text{group}} = \frac{dE}{dp} \qquad (4.20)$$

In Equation 4.20, E represents the energy of the particle and p its momentum. For a classical particle having only kinetic energy $K = p^2/2m$, we can find dE/dp as

$$\frac{dE}{dp} = \frac{d}{dp}\left(\frac{p^2}{2m}\right) = \frac{p}{m} = v \qquad (4.21)$$

which is the velocity of the particle.

Combining Equations 4.20 and 4.21 we obtain an important result:

$$v_{\text{group}} = v_{\text{particle}} \qquad (4.22)$$

The velocity of a particle is equal to the group velocity of the corresponding wave packet. The wave packet and the particle move together—wherever the particle goes, its de Broglie wave packet moves along with it like a shadow. If we do a wave-type experiment on the particle, the de Broglie wave packet is always there to reveal the wave behavior of the particle. A particle can never escape its wave nature!

4.5 PROBABILITY AND RANDOMNESS

Any single measurement of the position or momentum of a particle can be made with as much precision as our experimental skill permits. How then does the wavelike behavior of a particle become observable? How does the uncertainty in position or momentum affect our experiment?

Suppose we prepare an atom by attaching an electron to a nucleus. (For this example we regard the nucleus as being fixed in space.) Some time after preparing our atom, we measure the position of the electron. We then repeat the procedure, preparing the atom *in an identical way,* and find that a remeasurement of the position of the electron yields a value different from that found in our first measurement. In fact, each time we repeat the measurement, we may obtain a different outcome. If we repeat the measurement a large number of times, we find ourselves led to a conclusion that runs counter to a basic notion of classical physics—*systems that are prepared in identical ways do not show identical subsequent behavior.* What hope do we then have of constructing a mathematical theory that has any usefulness at all in predicting the outcome of a measurement, if that outcome is completely random? The solution to this dilemma lies in the consideration of the *probability* of obtaining any given result from an experiment whose possible results are subject to the laws of statistics. We cannot predict the outcome of a single flip of a coin or roll of the dice, because any *single result* is as likely as any other *single result.* We can, however, predict the *distribution* of a large number of individual measurements. For example, on a single flip of a coin, we cannot predict whether the outcome will be "heads" or "tails"; the two are equally likely. If we make a large number of trials, we expect that approximately 50 percent will turn up "heads" and 50 percent will yield "tails"; even though we cannot predict the result of any single toss of the coin, we can predict reasonably well the result of a large number of tosses.

Our study of systems governed by the laws of quantum physics leads us to a similar situation. We cannot predict the outcome of any *single* measurement of the position of the electron in the atom we prepared, but if we do a large number of measurements, we ought to find a statistical distribution of results. If we cannot produce a mathematical theory that will predict the result of a single measurement, we can attempt to obtain a mathematical theory that predicts the statistical behavior of a system (or of a large number of identical systems). The quantum theory provides such a mathematical procedure, which enables us to calculate the average or probable outcome of measurements and the distribution of individual outcomes about the average. This is not such a disadvantage as it may seem, for in the realm of quantum physics, we seldom do measurements with, for example, a single atom. If we were studying the emission of light by a radiant system or the properties of a solid or the scattering of nuclear particles, we would be dealing with a large number of atoms, and so our concept of statistical averages is really quite useful.

In fact, such concepts are not as far removed from our daily lives as we might think. For example, what is meant when the TV weather forecaster "predicts" a 50 percent chance of rain tomorrow? Will it rain 50 percent of the time, or over 50 percent of the city? The proper interpretation of the forecast is that the existing set of atmospheric conditions will, in a large number of similar cases, result in rain in about half the cases. A surgeon who asserts that a patient has a 50 percent chance of surviving an operation means exactly the same thing—experience with a large number of similar cases suggests recovery in about half.

Quantum mechanics uses similar language. For example, if we say that the electron in a hydrogen atom has a 50 percent probability of circulating in a clockwise direction, we mean that in observing a large collection of similarly prepared atoms we find 50 percent to be circulating clockwise. Of course a *single measurement* shows *either* clockwise or counterclockwise circulation. (Similarly, it *either* rains or it doesn't; the patient *either* lives or dies.)

Of course, one could argue that the flip of a coin or the roll of the dice is not a random process, but that the apparently random nature of the outcome simply reflects our lack of knowledge of the state of the system. For example, if we knew exactly how the dice were thrown (magnitude and direction of initial velocity, initial orientation, rotational speed) and precisely what the laws are that govern their bouncing on the table, we should be able to predict exactly how they will land. (Similarly, if we knew a great deal more about atmospheric physics or physiology, we could predict with certainty whether or not it will rain tomorrow or an individual patient will survive.) When we instead analyze the outcomes in terms of probabilities, we are really admitting our inability to do the analysis exactly. There is a school of thought that asserts that the same situation exists in quantum physics. According to this interpretation, we could predict *exactly* the behavior of the electron in our atom if only we knew the nature of a set of so-called "hidden variables" that determine its motion. However, experimental evidence disagrees with this theory, and so we must conclude that the random behavior of a system governed by the laws of quantum physics is a fundamental aspect of nature and not a result of our limited knowledge of the properties of the sytstem.

4.6 THE PROBABILITY AMPLITUDE

One final problem remains to be discussed. What does the amplitude of the de Broglie wave represent? In any wave phenomenon, a physical quantity such as displacement or pressure varies with distance and time. What is the physical property that varies as the de Broglie wave propagates?

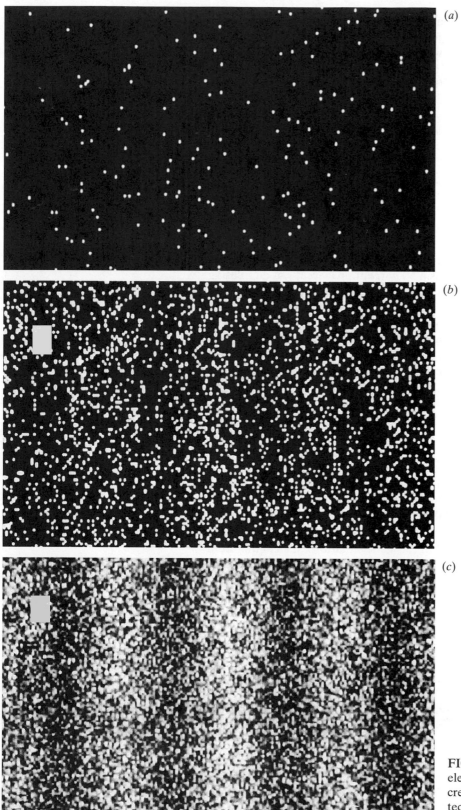

(a)

(b)

(c)

FIGURE 4.24 The buildup of an electron interference pattern as increasing numbers of electrons are detected: (a) 100 electrons; (b) 3000 electrons; (c) 70,000 electrons.

In a previous section, we discussed the representation of a localized particle by a wave packet. If a particle is confined to a region of space of dimension Δx, the wave packet that represents the particle has large amplitude only in a region of space of dimension Δx and has small amplitude elsewhere. That is, the amplitude is large where the particle is likely to be found and small where the particle is less likely to be found. *The probability of finding the particle at any point depends on the amplitude of its de Broglie wave at that point.* In analogy with classical physics, in which the intensity of any wave is proportional to the square of its amplitude, we have

$$\text{probability to observe particles} \propto |\text{de Broglie wave amplitude}|^2$$

Compare this with the similar relationship for photons discussed in Section 3.6:

$$\text{probability to observe photons} \propto |\text{electric field amplitude}|^2$$

Just as the electric field amplitude of an electromagnetic wave indicates regions of high and low probability for observing photons, the de Broglie wave performs the same function for particles. Figure 4.24 illustrates this effect, as individual electrons in a double-slit type of experiment eventually produce the characteristic interference fringes. The path of each electron is guided by its de Broglie wave toward the allowed regions of high probability. This statistical effect is not apparent for a small number of electrons, but it becomes quite apparent when a large number of electrons has been detected.

In the next chapter we discuss the mathematical framework for computing the wave amplitudes for a particle in various situations, and we also develop a more rigorous mathematical definition of the probability.

SUGGESTIONS FOR FURTHER READING

Many of the descriptive references listed in Chapter 1 are useful background reading for this chapter as well. For a delightful account of a world in which Planck's constant is so large that quantum effects are ordinary, see G. Gamow, *Mr. Tompkins in Paperback* (Cambridge, Cambridge University Press, 1967). Another nonmathematical discussion of quantum theory is B. Hoffmann, *The Strange Story of the Quantum* (New York, Dover, 1959). An imaginary dialogue, in which the protagonists of Galileo's dialogues are reunited to discuss quantum theory, measurement, and uncertainty, is in J. M. Jauch, *Are Quanta Real?* (Bloomington, Indiana University Press, 1973). The paradox of Schrödinger's cat is discussed in this last reference.

Other references, in which the philosophy of quantum theory is mixed with mathematics at about the same level as this text, are as follows:

R. P. Feynman, R. B. Leighton, and M. Sands, *The Feynman Lectures on Physics* (Reading, Addison-Wesley, 1965). Chapters 1 to 3 of Volume 3 are particularly good introductions to quantum waves and the philosophy of measurement.

R. Resnick and D. Halliday, *Basic Concepts in Relativity and Early Quantum Theory* (New York, Macmillan, 1992). Chapter 6 discusses the wave nature of particles and the uncertainty principle.

E. H. Wichmann, *Quantum Physics, Volume 4 of the Berkeley Physics Course* (New York, McGraw-Hill, 1971).

A more advanced work that is particularly careful about the philosophical background of quantum theory is:

D. Bohm, *Quantum Theory* (Englewood Cliffs, Prentice-Hall, 1951). See Chapters 5 and 6.

For a discussion of the uncertainty principle, see:

G. Gamow, "The Principle of Uncertainty," *Scientific American* **198**, 51 (January 1958).

A translation of Claus Jönsson's 1959 article on the electron double-slit experiment is given in *American Journal of Physics* **42**, 4 (1974). This short, clearly written paper is very readable and is highly recommended as an example of the careful experimental technique that is necessary in doing interference experiments to illustrate the wave nature of particles.

More recent experiments, in which an electron microscope has been used to demonstrate beautiful interference and diffraction effects with electrons, can be found in:

P. G. Merli, G. F. Missiroli, and G. Pozzi, *American Journal of Physics* **44**, 306 (1976).
G. Matteucci and G. Pozzi, *American Journal of Physics* **46**, 619 (1978).
A. Tonomura, J. Endo, T. Matsuda, T. Kawasaki, and H. Ezawa, *American Journal of Physics* **57**, 117 (1989).
G. Matteucci, *American Journal of Physics* **58**, 1143 (1990).

A summary of recent experiments demonstrating interference and diffraction effects with neutrons is:

R. Gähler and A. Zeilinger, *American Journal of Physics* **59**, 316 (1991).

QUESTIONS

1. When an electron moves with a certain de Broglie wavelength, does the electron shake back and forth or up and down at that wavelength?

2. Imagine a different world in which the laws of quantum physics still apply, but which has $h = 1$ J·s. What might be some of the difficulties of life in such a world? (See Gamow, *Mr. Tompkins in Paperback,* for a fanciful account of such a world.)

3. Suppose we try to measure an unknown frequency ν by listening for beats between ν and a known (and controllable) frequency ν'. (We assume ν' is known to arbitrarily small uncertainty.) The beat frequency is $|\nu' - \nu|$. If we hear no beats, then we conclude $\nu = \nu'$. (a) How long must we listen to hear "no" beats? (b) If we hear no beats in one second, how accurately have we determined ν? (c) If we hear no beats in 10 s, how accurately? In 100 s? (d) How is this experiment related to Equation 4.4?

4. What difficulties does the uncertainty principle cause in trying to pick up an electron with a pair of forceps?

5. Does the uncertainty principle apply to nature itself or only to the results of experiments? That is, is it the position and momentum that are *really* uncertain, or merely our knowledge of them? Does it make any difference?

6. The uncertainty principle states in effect that the more we try to confine an object, the faster we are likely to find it moving. Is this why you can't seem to keep money in your pocket or purse for long? Make a numerical estimate.

7. Consider a collection of gas molecules trapped in a container. As we move the walls of the container closer together (compressing the gas) the molecules move faster (the temperature increases). Does the gas behave this way because of the uncertainty principle? Justify your answer with some numerical estimates.

8. Many nuclei are unstable and undergo radioactive decay to other nuclei. The lifetimes for these decays are typically of the order of days to years. Do you expect that the uncertainty principle will cause a measurable effect in the precision to which we can measure the masses of atoms of these nuclei?

9. Just as the classical limit of relativity can be achieved by letting $c \to \infty$, the classical limit of quantum behavior is achieved by letting $h \to 0$. Consider the following in the $h \to 0$ limit and explain how they behave classically: the size of the energy quantum of an electromagnetic wave, the de Broglie wavelength of an electron, the Heisenberg uncertainty relationships.

10. Assume the electron beam in a television tube is accelerated through a potential difference of 25 kV and then passes through a deflecting capacitor of interior width 1 cm. Are diffraction effects important in this case? Justify your answer with a calculation.

11. The structure of crystals can be revealed by X-ray diffraction (Figures 3.7 and 3.8), electron diffraction (Figure 4.2), and neutron diffraction (Figure 4.7). In what ways do these experiments reveal similar structure? In what ways are they different?

12. Often it happens in physics that great discoveries are made inadvertently. What would have happened if Davisson and Germer had their accelerating voltage set below 32 V?

13. Suppose we cover one slit in the two-slit electron experiment with a very thin sheet of fluorescent material that emits a photon of light whenever an electron passes through. We then fire electrons one at a time at the double slit; whether or not we see a flash of light tells us which slit the electron went through. What effect does this have on the interference pattern? Why?

14. In another attempt to determine through which slit the electron passes, we suspend the double slit itself from a very fine spring balance and measure the "recoil" momentum of the slit as a result of the passage of the electron. Electrons that strike the screen near the center must cause recoils in opposite directions depending on which slit they pass through. Sketch such an apparatus and describe its effect on the interference pattern. (*Hint*: Consider the uncertainty principle $\Delta p \, \Delta x \sim \hbar$ as applied to the motion of the slits suspended from the spring. How precisely do we know the position of the slit?)

15. It is possible for v_{phase} to be greater than c? Can v_{group} be greater than c?

16. In a nondispersive medium, $v_{\text{group}} = v_{\text{phase}}$; this is another way of saying that all waves travel with the same phase velocity, no matter what their wavelengths. Is this true for (a) de Broglie waves? (b) Light waves in glass? (c) Light waves in vacuum? (d) Sound waves in air? What difficulties would be encountered in attempting communication (by speech or by radio signals for example) in a strongly dispersive medium?

PROBLEMS

1. Find the de Broglie wavelength of (a) a nitrogen molecule ($m = 28$ u) in air at room temperature. (b) A 5-MeV proton. (c) A 50-GeV electron. (d) An electron moving at $v = 10^6$ m/s.

2. The neutrons produced in a reactor are known as *thermal neutrons,* because their kinetic energies have been reduced (by collisions) until $K \cong \frac{3}{2}kT$ where T is room temperature. (a) What is the kinetic energy of such neutrons? (b) What is their de Broglie wavelength? Because this wavelength is of the same order as the lattice spacings of the atoms of a solid, neutron diffraction (like X-ray and electron diffraction) is a useful means of studying solid lattices.

3. To what voltages must we accelerate electrons (as in an electron microscope, for example) if we wish to resolve a virus of diameter 12 nm? An atom of diameter 0.12 nm? A proton of diameter 1.2 fm?

4. In an electron microscope we wish to study particles of diameter about 0.10 μm (about 1000 times the size of a single atom). (a) What should be the de Broglie wavelength of the electrons? (b) Through what voltage should the electrons be accelerated to have that de Broglie wavelength?

5. In order to study the atomic nucleus, we would like to observe the diffraction of particles whose de Broglie wavelength is about the same size as the nuclear diameter, about 14 fm for a heavy nucleus such as lead. What kinetic energy should we use if the diffracted particles are (a) electrons? (b) Neutrons? (c) Alpha particles ($m = 4$ u)?

6. A free electron bounces elastically back and forth in one dimension between two walls that are $L = 0.50$ nm apart. (a) Assuming that the electron is represented by a de Broglie standing wave with a node at each wall, show that the permitted de Broglie wavelengths are $\lambda = 2L/n$ ($n = 1, 2, 3, \ldots$). (b) Find the values of the kinetic energy of the electron for $n = 1, 2,$ and 3.

7. In the double-slit interference pattern for helium atoms (Figure 4.13), the beam of atoms had a kinetic energy of 0.020 eV. (a) What is the de Broglie wavelength of a helium atom with this kinetic energy? (b) Estimate the de Broglie wavelength of the atoms from the fringe spacing in Figure 4.13, and compare your estimate with the value obtained in part (a). The distance from the double slit to the scanning slit is 64 cm.

8. The atoms in a gas can be treated as classical particles if their de Broglie wavelength is much smaller than the average separation between the particles. Compare the average de Broglie wavelength and the average separation between the atoms in a container of (monatomic) helium gas at 1.00 atm pressure

and at room temperature (20°C). At what temperature or pressure would you expect quantum effects to become important?

9. Suppose we wish to do a double-slit experiment with a beam of the smoke particles of Example 4.1c. Assume we can construct a double slit whose separation is about the same size as the particles. Estimate the separation between the fringes if the double slit and the screen were on opposite coasts of the United States.

10. In the Davisson-Germer experiment using a Ni crystal, a second-order beam is observed at an angle of 55°. For what accelerating voltage does this occur?

11. A certain crystal is cut so that the rows of atoms on its surface are separated by a distance of 0.352 nm. A beam of electrons is accelerated through a potential difference of 175 V and is incident normally on the surface. If all possible diffraction orders could be observed, at what angles (relative to the incident beam) would the diffracted beams be found?

12. A beam of thermal neutrons (see Problem 2) emerges from a nuclear reactor and is incident on a crystal as shown in Figure 4.25. The beam is Bragg scattered, as in Figure 3.5, from a crystal whose scattering planes are separated by 0.247 nm. From the continuous energy spectrum of the beam we wish to select neutrons of energy 0.0105 eV. Find the Bragg scattering angle that results in a scattered beam of this energy. Will other energies also be present in the scattered beam?

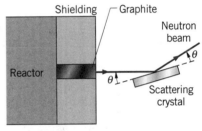

FIGURE 4.25 Problem 12.

13. Suppose a traveling wave has a speed v (where $v = \lambda\nu$). Instead of measuring waves over a distance Δx, we stay in one place and count the number of wave crests which pass in a time Δt. Show that Equation 4.5 is equivalent to Equation 4.4 for this case.

14. The speed of an electron is measured to within an uncertainty of 2.0×10^4 m/s. What is the size of the smallest region of space in which the electron can be confined?

15. An electron is confined to a region of space of the size of an atom (0.1 nm). (a) What is the uncertainty in the momentum of the electron? (b) What is the kinetic energy of an electron with a momentum equal to Δp? (c) Does this give a reasonable value for the kinetic energy of an electron in an atom?

16. The Σ^* particle has a rest energy of 1385 MeV and a lifetime of 2.0×10^{-23} s. What would be a typical range of outcomes of measurements of the Σ^* rest energy?

17. A pi meson (pion) and a proton can briefly join together to form a Δ particle. A measurement of the energy of the πp system (Figure 4.26) shows a peak at 1236 MeV, corresponding to the rest energy of the Δ particle, with an experimental spread of 120 MeV. What is the lifetime of the Δ?

18. A nucleus emits a gamma ray of energy 1.0 MeV from a state that has a lifetime of 1.2 ns. (a) What is the uncertainty in the energy of the gamma ray? (b) How does this compare with the experimental precision with which the best gamma-ray detectors can measure gamma-ray energies, which is of the order of several eV? Will this uncertainty be directly measurable?

19. A nucleus of helium with mass 5 u breaks up from rest into a nucleus of ordinary helium (mass = 4 u) plus a neutron (mass = 1 u). The rest energy liberated in the break-up is 0.89 MeV, which is shared (*not* equally) by the

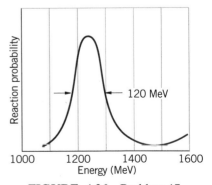

FIGURE 4.26 Problem 17.

products. (a) Using momentum conservation, find the kinetic energy of the neutron. (b) The lifetime of the original nucleus is 1.0×10^{-21} s. What range of values of the neutron kinetic energy might we measure in the laboratory as a result of the uncertainty relationship?

20. Alpha particles are emitted in nuclear decay processes with typical energies of 5 MeV. In analogy with Example 4.7, deduce whether the alpha particle can exist inside the nucleus.

21. In a metal, the conduction electrons are not attached to any one atom, but are relatively free to move throughout the entire metal. Consider a 1 cm × 1 cm × 1 cm piece of copper. (a) What is the uncertainty in any one component of the momentum of an electron confined to the metal? (b) What is the resulting estimate of the typical kinetic energy of an electron in the metal? (Assume $\Delta p = [(\Delta p_x)^2 + (\Delta p_y)^2 + (\Delta p_z)^2]^{1/2}$.) (c) Assuming the heat capacity of copper to be 24.5 J/mole·K, would the contribution of this motion to the internal energy of the copper be important at room temperature? What do you conclude from this? (See also Problem 23.)

22. A proton or a neutron can sometimes "violate" conservation of energy by emitting and then reabsorbing a pi meson, which has a mass of 135 MeV/c^2. This is possible as long as the pi meson is reabsorbed within a short enough time Δt consistent with the uncertainty principle. (a) Consider p → p + π. By what amount ΔE is energy conservation violated? (Ignore any kinetic energies.) (b) For how long a time Δt can the pi meson exist? (c) Assuming the pi meson to travel at very nearly the speed of light, how far from the proton can it go? (This procedure, as we discuss in Chapter 12, gives us an estimate of the *range* of the nuclear force, because we believe that protons and neutrons are held together in the nucleus by exchanging pi mesons.)

23. In a crystal, the atoms are a distance L apart; that is, each atom must be localized to within a distance of at most L. (a) What is the minimum uncertainty in the momentum of the atoms of a solid that are 0.20 nm apart? (b) What is the typical kinetic energy of such an atom of mass 65 u? (c) What would a collection of such atoms contribute to the internal energy of a typical solid, such as copper? Is this contribution important at room temperature? (See also Problem 21.)

24. An apparatus is used to prepare an atomic beam by heating a collection of atoms to a temperature T and allowing the beam to emerge through a hole of diameter d in one side of the oven. The beam then travels through a straight path of length L. Show that the diameter of the beam at the end of the path is larger than d by an amount of order $L\hbar/d\sqrt{3mkT}$ where m is the mass of an atom. Make a numerical estimate for typical values of $T = 1500$ K, $m = 7$ u (lithium atoms), $d = 3$ mm, $L = 2$ m.

25. Do the trigonometric manipulation necessary to obtain Equation 4.13.

26. Sound waves travel through air at a speed of 330 m/s. A whistle blast at a frequency of about 1.0 kHz lasts for 2.0 s. (a) Over what distance in space does the "wave train" representing the sound extend? (b) What is the wavelength of the sound? (c) Estimate the precision with which an observer could measure the wavelength. (d) Estimate the precision with which an observer could measure the frequency.

27. A stone tossed into a body of water creates a disturbance at the point of impact which lasts for 4.0 s. The wave speed is 25 cm/s. (a) Over what distance

on the surface of the water does the group of waves extend? (b) An observer counts 12 wave crests in the group. Estimate the precision with which the wavelength can be determined.

28. Show that the data used in Figure 4.21 are consistent with Equation 4.13, that is, use $\lambda_1 = 1$ and $\lambda_2 = 10/9$, $v_1 = 3$ and $v_2 = 2.5$ to show $v_{\text{group}} = 7.5$.

29. (a) Use a distribution of wave numbers of constant amplitude in a range Δk about k_0:

$$A(k) = A \qquad k_0 - \frac{\Delta k}{2} \leq k \leq k_0 + \frac{\Delta k}{2}$$

$$= 0 \qquad \text{otherwise}$$

and obtain Equation 4.16 from Equation 4.15. (b) Make a convenient choice of the width Δx of the wave packet, and show that $\Delta x \, \Delta k \sim 1$.

30. Use the distribution of wave numbers $A(k) = e^{-(k-k_0)^2/2(\Delta k)^2}$ for $k = -\infty$ to $+\infty$ to derive Equation 4.17. Ignore any multiplicative constant in $y(x)$.

31. (a) Show that the group velocity and phase velocity are related by:

$$v_{\text{group}} = v_{\text{phase}} - \lambda \frac{dv_{\text{phase}}}{d\lambda}$$

(b) When white light travels through glass, the phase velocity of each wavelength depends on the wavelength. (This is the origin of dispersion and the breaking up of white light into its component colors—different wavelengths travel at different speeds and have different indices of refraction.) How does v_{phase} depend on λ? Is $dv_{\text{phase}}/d\lambda$ positive or negative? Therefore, is $v_{\text{group}} > v_{\text{phase}}$ or $< v_{\text{phase}}$?

32. Certain surface waves in a fluid travel with phase velocity $\sqrt{b/\lambda}$, where b is a constant. Find the group velocity of a packet of surface waves, in terms of the phase velocity.

33. By a calculation similar to that of Equation 4.21, show that $dE/dp = v$ remains valid when E represents the relativistic kinetic energy of the particle.

5

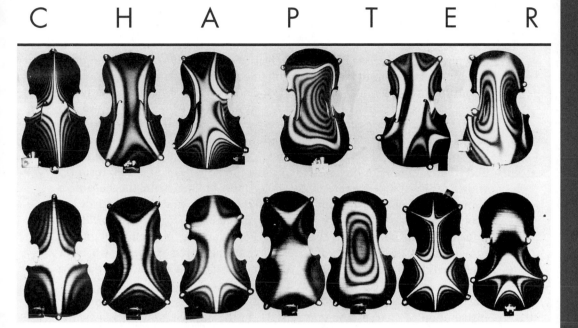

The mathematical techniques of quantum mechanics are formally similar to the techniques used to analyze classical waves. The resulting solutions of the quantum problems, such as that for a particle trapped in a two-dimensional region, often look similar to solutions of familiar classical wave problems, such as the vibrations of two-dimensional surfaces. The photograph shows the classical vibrations of violin plates, made visible using laser holograms.

THE SCHRÖDINGER EQUATION

The future behavior of a particle in a classical (nonrelativistic, nonquantum) situation may be predicted with absolute certainty using Newton's laws. If a known force $\mathbf{F}$ (which might be associated with a potential energy U) acts on a particle initially located at $\mathbf{r}_0$ and moving with velocity $\mathbf{v}_0$, we can do the mathematics necessary to solve Newton's second law, $\mathbf{F} = d\mathbf{p}/dt$ (a second-order, linear differential equation) and find the particle's location $\mathbf{r}(t)$ and velocity $\mathbf{v}(t)$ at all future times t. The mathematics may be difficult, and in fact it may not be possible to solve the equations in closed form (in which case an approximate solution can be obtained with the help of a computer). These are only mathematical difficulties, and the physics of the problem consists of writing down the original equation $\mathbf{F} = d\mathbf{p}/dt$ and interpreting its solutions $\mathbf{r}(t)$ and $\mathbf{v}(t)$. For example, a satellite or planet moving under the influence of a $1/r^2$ gravitational force can be shown, after the equations have been solved, to follow exactly an elliptical path. In a similar manner, we can use Maxwell's equations (a set of first-order differential equations) to find the electric and magnetic fields associated with any distribution of charges and currents. As in the case of Newton's laws, the physics of the problem consists of writing down the original equations and interpreting the solution.

In the case of nonrelativistic quantum physics, the basic equation to be solved is a second-order differential equation known as the *Schrödinger equation*. Like Newton's laws, the Schrödinger equation must be written down for a certain force acting on the particle (although we find it more convenient to work with the potential energy than with the force). Unlike Newton's laws, the Schrödinger equation does not give the trajectory of the particle; instead, its solution gives the *wave function* of the particle, which carries information about the particle's wavelike behavior. In this chapter we introduce the Schrödinger equation, obtain some of its solutions for certain potential energies, and learn how to interpret those solutions.

5.1   JUSTIFICATION OF THE SCHRÖDINGER EQUATION

Neither Newton's laws, nor Maxwell's equations, nor the Schrödinger equation can be derived from basic principles. Instead, they should be regarded as equations written to agree with previous experimental and theoretical results and to satisfy certain symmetry principles and conservation laws. In this section, we follow this procedure to develop the Schrödinger equation, which is the fundamental equation for calculating the wave behavior of nonrelativistic particles.

Let us put ourselves in the place of Erwin Schrödinger in 1926. Lacking direct experimental evidence to guide us in developing a fundamental equation for quantum behavior, we first list some properties that the equation must have.

1. It must conserve energy. While we are willing to sacrifice a great deal of the framework of classical physics, conservation of energy is one

principle that we expect to remain valid. We therefore take

$$K + U = E \tag{5.1}$$

where K, U, and E are respectively the kinetic, potential, and total energies. Because we are seeking an equation that describes nonrelativistic particles, we take $K = \frac{1}{2}mv^2 = p^2/2m$. Also, E represents the sum of kinetic and potential energies, not the relativistic total energy.

2. The equation, whatever its form, must be consistent with the de Broglie hypothesis—if we turn the mathematical crank for a free particle of momentum p, we must obtain a wave of wavelength λ equal to h/p. Using Equation 4.5, $p = \hbar k$, we obtain the kinetic energy associated with the free-particle de Broglie wave to be $K = \hbar^2 k^2/2m$.

3. The equation must be "well-behaved" in the mathematical sense. For example, the solution, which tells us something about the location or state of motion of the particle, must be *continuous*; we would be very surprised to find, for example, the particle suddenly disappearing at one point in space and reappearing at another. The solution must also be *single-valued*; there should be only one probability for the particle to be in a specific location at a specific time. The solution must also be *linear*, so that the de Broglie waves have the important *superposition* property we expect for waves.

Erwin Schrödinger (1887–1961, Austria). Although he disagreed with the probabilistic interpretation that was later given to his work, he developed the mathematical theory of wave mechanics that for the first time permitted the wave behavior of physical systems to be calculated.

Working backward, we begin with a solution to the equation for which we are searching—that of the free-particle de Broglie wave. From classical physics we know something about the mathematical form for a wave. For example, waves on a stretched string have the form $y(x,t) = A \sin (kx - \omega t)$, and electromagnetic waves have the similar form $\mathbf{E}(x,t) = \mathbf{E}_0 \sin (kx - \omega t)$ and $\mathbf{B}(x,t) = \mathbf{B}_0 \sin (kx - \omega t)$. We therefore postulate that the free-particle de Broglie wave, which we represent by $\Psi(x,t)$, has the basic form $\Psi(x,t) = A \sin (kx - \omega t)$, which represents a wave of amplitude A traveling in the positive x direction. The wave has wavelength $\lambda = 2\pi/k$ and frequency $\nu = \omega/2\pi$. To simplify the problem, we ignore the time dependence and deal only with a snapshot of the wave at a specific time, say $t = 0$. Defining $\psi(x)$ to be $\Psi(x,t = 0)$, we have

$$\psi(x) = A \sin kx \tag{5.2}$$

For now, we seek a time-independent form of the Schrödinger equation. (Later in this chapter we consider the time dependence.) The equation must involve the potential energy U. If U appears to the first power, then (to be dimensionally consistent and to conserve energy) the kinetic energy K must also appear to the first power. Using our previous result $K = \hbar^2 k^2/2m$, we can obtain a term in k^2 from Equation 5.2 by taking two derivatives with respect to x:

$$\frac{d^2\psi}{dx^2} = -k^2 A \sin kx = -k^2 \psi = -\frac{2m}{\hbar^2} K\psi = -\frac{2m}{\hbar^2}(E - U)\psi$$

or

$$-\frac{\hbar^2}{2m}\frac{d^2\psi}{dx^2} + U\psi = E\psi \tag{5.3}$$

One-dimensional, time-independent Schrödinger equation

Equation 5.3 is a second-order differential equation that we have constructed to have these properties: (1) it is consistent with conservation of energy; (2) it is linear and single-valued; (3) it gives a free-particle ($U = 0$) solution consistent with a single de Broglie wave. Other equations could be constructed with these properties, but only Equation 5.3 passes the stringent test of being consistent with experimental results in many physical situations. Equation 5.3 is the *time-independent Schrödinger equation* in one dimension. Although a complete specification of a wave must also include the time coordinate and real physical situations generally involve three dimensions, we can learn much about the mathematics and physics of quantum mechanics by studying the solutions of Equation 5.3. The inclusion of time dependence and the extension to three dimensions are discussed later in this chapter.

5.2   THE SCHRÖDINGER RECIPE

The techniques of solving Equation 5.3 are sufficiently similar, no matter what the form of the potential energy U (which is in general a function of x), that we may list a series of steps to be followed in obtaining the solutions. We assume that we know the potential energy $U(x)$, and we wish to obtain the wave function $\psi(x)$ and the energy E *for that potential energy*. This is a general example of a type of problem known as an *eigenvalue* problem; we find that it is possible to obtain solutions to the equation only for particular values of E, which are known as the *energy eigenvalues.*

1. Begin by writing Equation 5.3 with the appropriate $U(x)$. Note that if the potential energy changes discontinuously [$U(x)$ may be discontinuous; $\psi(x)$ may *not*], we may need to write different equations for different regions of space. Examples of this sort are given in Section 5.4.

2. Using general mathematical techniques suited to the form of the equation, find a mathematical function $\psi(x)$, which is a solution to the differential equation. Since there is no one specific technique for solving differential equations, we study examples to learn how to find solutions.

3. In general, several solutions may be found. By applying boundary conditions some of these may be eliminated and some arbitrary constants may be determined. It is the application of the boundary conditions that selects out the energy eigenvalues.

4. If you are seeking solutions for a potential that changes discontinuously, you must apply the continuity conditions on ψ (and usually on $d\psi/dx$) at the boundary between different regions.

5. Evaluate undetermined constants, for example, A in Equation 5.2. The method for doing this is discussed in the following section.

We consider now an example from classical physics that requires many of the same techniques of solution as do typical problems in quantum physics. The condition of continuity at the boundary between two regions is one that must frequently be applied in classical problems. By way of illustration, we study the following classical problem.

EXAMPLE 5.1

A mass m is dropped from rest at height H above a tank of water. On entering the water, it is subject to a buoyant force B greater than the weight of the object. (We neglect the viscous force exerted by the water.) Find the displacement and velocity of the object from the time it is released until it rises to the surface of the water.

SOLUTION

We choose a coordinate system in which y is positive upward and take $y = 0$ at the surface of the water. During the time the object is initially in free fall, it is subject only to the force of gravity. Then in region 1 (above the water) Newton's second law gives

$$-mg = m \frac{d^2 y_1}{dt^2}$$

which has the solutions

$$v_1(t) = v_0 - gt$$

$$y_1(t) = y_0 + v_0 t - \tfrac{1}{2} g t^2$$

where v_0 and y_0 are the initial velocity and height at $t = 0$.

When it enters the water (region 2) the force becomes $B - mg$, so Newton's second law becomes

$$B - mg = m \frac{d^2 y_2}{dt^2}$$

which has the solutions

$$v_2(t) = c_1 + \left(\frac{B}{m} - g \right) t$$

$$y_2(t) = c_2 + c_1 t + \frac{1}{2} \left(\frac{B}{m} - g \right) t^2$$

These solutions have four undetermined coefficients: y_0, v_0, c_1, and c_2. The first two constants are found by applying the *initial conditions*—at $t = 0$ (when the object is released) $y_0 = H$ and $v_0 = 0$, since it is dropped from rest. The solutions in region 1 are therefore

$$v_1(t) = -gt$$

$$y_1(t) = H - \tfrac{1}{2} g t^2$$

The next step is to apply the *boundary conditions* at the surface of the water to determine c_1 and c_2. Let t_1 be the time that the object enters the water. The boundary conditions require that v and y be continuous across the boundary between air and water:

$$y_1(t_1) = y_2(t_1)$$

and

$$v_1(t_1) = v_2(t_1)$$

The first condition states that the object does not disappear at one instant and reappear at a different point in space at the next instant. The second condition is equivalent to requiring that the speed changes smoothly at the water's surface. [If this were *not* so, then $v_1(t_1 - \Delta t) \neq v_2(t_1 + \Delta t)$ even as $\Delta t \to 0$, and the acceleration would be infinite.] To apply the boundary conditions, we must first find t_1, which is the time at which y_1 becomes zero.

$$y_1(t_1) = H - \tfrac{1}{2}gt_1^2 = 0$$

so

$$t_1 = \sqrt{\frac{2H}{g}}$$

We can then find the speed at which the object enters the water, $v_1(t_1)$:

$$v_1(t_1) = -gt_1 = -g\sqrt{\frac{2H}{g}} = -\sqrt{2gH}$$

The boundary conditions then give

$$y_2(t_1) = c_2 + c_1\sqrt{\frac{2H}{g}} + \frac{1}{2}\left(\frac{B}{m} - g\right)\left(\frac{2H}{g}\right) = 0$$

and

$$v_2(t_1) = c_1 + \left(\frac{B}{m} - g\right)\sqrt{\frac{2H}{g}} = -\sqrt{2gH}$$

These two equations may be solved simultaneously for c_1 and c_2, yielding $c_1 = -(B/m)\sqrt{2H/g}$ and $c_2 = H(1 + B/mg)$. The complete solutions in region 2 are thus

$$v_2(t) = -\frac{B}{m}\sqrt{\frac{2H}{g}} + \left(\frac{B}{m} - g\right)t$$

$$y_2(t) = H + \frac{HB}{mg} - \frac{B}{m}\sqrt{\frac{2H}{g}}t + \frac{1}{2}\left(\frac{B}{m} - g\right)t^2$$

The equations for v_1, y_1, v_2, and y_2 give the behavior of the object from $t = 0$ until it rises to the surface of the water.

We can *apply* these results to calculating other properties of the motion; for example, we can find the maximum depth reached by the object, which occurs when $v_2 = 0$. If we let t_2 be the time at which this occurs, then

$$v_2(t_2) = -\frac{B}{m}\sqrt{\frac{2H}{g}} + \left(\frac{B}{m} - g\right)t_2 = 0$$

$$t_2 = \frac{B}{B - mg}\sqrt{\frac{2H}{g}}$$

The depth D is just the value of y_2 at this time t_2:

$$D = y_2(t_2) = \left(H + \frac{HB}{mg}\right) - \frac{B}{m}\sqrt{\frac{2H}{g}}\, t_2 + \frac{1}{2}\left(\frac{B}{m} - g\right) t_2^2$$

$$= -\frac{mgH}{B - mg}$$

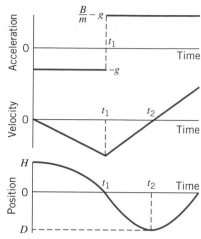

In summary, in this example we have used the equations of motion to find a solution, we have evaluated the undetermined constants in our solution by applying initial and boundary conditions, and we have used our resulting solution to calculate a feature of the future behavior of the object (in this case, the maximum depth D). *The same basic procedure can be used for problems in quantum physics.*

The behavior of the object is illustrated in Figure 5.1, which shows the acceleration, velocity, and position as functions of the time. Note that $v(t)$ and $y(t)$ are both continuous, as we have demanded by our application of the boundary conditions.

Suppose we replaced the water by a rigid surface from which the (likewise rigid) object rebounded elastically. Then in an idealized representation, we might picture this situation in Figure 5.2. In this case the object is subject to an infinite force during the instant it is in contact with the surface, so that its velocity changes discontinuously, but its position changes continuously (it still does not disappear and reappear somewhere else).

Our conclusions for the classical problem can be summarized as follows; the equivalent wording for applications to quantum mechanics is in brackets. When an object moves across the boundary between two regions in which it is subject to different $\left\{\begin{array}{l}\text{forces}\\\text{potential energies}\end{array}\right\}$, the basic behavior of the object is found by solving $\left\{\begin{array}{l}\text{Newton's second law}\\\text{the Schrödinger equation}\end{array}\right\}$. The $\left\{\begin{array}{l}\text{position}\\\text{wave function}\end{array}\right\}$ of the object is always continuous across the boundary, and the $\left\{\begin{array}{l}\text{velocity}\\\text{derivative } d\psi/dx\end{array}\right\}$ is also continuous as long as the $\left\{\begin{array}{l}\text{force}\\\text{change in potential energy}\end{array}\right\}$ remains finite.

Just as in the case of classical physics, each problem may require somewhat different techniques, and so it is difficult to establish a general procedure. The steps listed in this section should, however, suggest to you the general direction to take in seeking solutions. The best way to learn the techniques is by studying the examples given in this chapter. The recipe is incomplete at this point; we have discussed the mathematical technique for finding the solution $\psi(x)$, but we have not discussed the interpretation of the solution or its application to physical situations. These are discussed in the following sections.

FIGURE 5.1 The acceleration, velocity, and position of the mass described in Example 5.1. Notice that the acceleration is discontinuous, but the position and velocity are continuous.

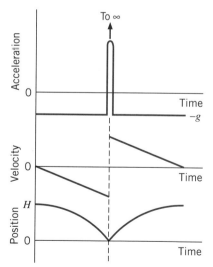

FIGURE 5.2 The acceleration, velocity, and position of a mass rebounding elastically from a hard surface. An infinite acceleration acts for an infinitesimally short time. The velocity is discontinuous (it changes instantly from $-v$ to $+v$) but the position is continuous.

5.3   PROBABILITIES AND NORMALIZATION

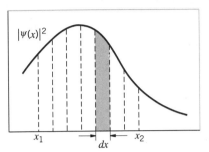

FIGURE 5.3   The total probability of finding the particle between any two coordinates x_1 and x_2 is determined by summing the probabilities in each interval dx.

The remaining steps in the Schrödinger recipe depend on the physical interpretation of the solution to the differential equation. Our original goal in solving the Schrödinger equation was to obtain the wave properties of the particle. These properties are described by the *wave function* $\psi(x)$. The function $\psi(x)$ represents a wave in the sense with which we are familiar—the wave may consist of a unique wavelength moving with a certain phase velocity or a mixture of wavelengths (a wave packet) moving with a certain group velocity. It is more difficult to give a physical interpretation to the amplitude of this wave. What does the amplitude of $\psi(x)$ represent, and what is the physical variable that is waving? It is certainly not a displacement, as in the case of a water wave or a wave on a stretched piano wire, nor is it a pressure wave, as in the case of sound. *It is a very different kind of wave, whose squared absolute amplitude gives the probability for finding the particle at a given location in space.*

If we define $P(x)$ as the *probability density* (probability per unit length, in one dimension), then according to the Schrödinger recipe

$$P(x)\, dx = |\psi(x)|^2\, dx \qquad (5.4)$$

Probability density

In Equation 5.4, $|\psi(x)|^2\, dx$ gives the probability to find the particle in the interval dx at x (that is, between x and $x + dx$).* Note that the absolute magnitude is needed to make the probability everywhere positive.

This interpretation of $|\psi(x)|^2$ helps us to understand the continuity condition of $\psi(x)$. We must not allow the probability to change discontinuously, but, like any well-behaved wave, the probability to locate the particle varies smoothly and continuously.

This interpretation of $\psi(x)$ now permits us to complete the Schrödinger recipe and to illustrate how to use the wave function to calculate quantities that we can measure in the laboratory. Steps 1 through 5 were given in the previous section; the recipe continues:

6. Suppose the interval between two points x_1 and x_2 is divided into a series of infinitesimal intervals of width dx (Figure 5.3). The total probability to find the particle between x_1 and x_2 is the sum of all the probabilities $P(x)\, dx$ in each interval dx. This sum can be expressed as an integral:

Probability of finding the particle between x_1 and $x_2 = \displaystyle\int_{x_1}^{x_2} P(x)\, dx$

Probability

$$= \int_{x_1}^{x_2} |\psi(x)|^2\, dx$$

$$(5.5)$$

---

* Sometimes for convenience we say (imprecisely) that $|\psi(x)|^2$ gives the probability to find the particle at the point x. However, a single point is a mathematical abstraction with no physical dimension. The probability of finding a particle *at a point* is zero, but there is a nonzero probability of finding the particle in an *interval*.

As a corollary to this rule, we require a 100 percent probability of finding the particle *somewhere* along the *x* axis, and thus

$$\int_{-\infty}^{+\infty} |\psi(x)|^2 \, dx - 1 \qquad (5.6) \qquad \textit{Normalization}$$

Equation 5.6 is known as the *normalization* condition, and shows us how to find the constant *A* discussed in step 5 of the recipe. Note that the constant *A* does not come out of the solution to the differential equation; in fact, as long as the Schrödinger equation is linear, if $\psi(x)$ is a solution, then *any* constant times $\psi(x)$ is also a solution. A wave function in which the arbitrary multiplicative constant is determined according to Equation 5.6 is said to be *normalized*; otherwise, it is *unnormalized*. Only a properly normalized wave function may be used for physically meaningful calculations. If the normalization has been done correctly, Equation 5.5 will always yield a probability that lies between 0 and 1.

7. Any solution to the Schrödinger equation for which $|\psi(x)|^2$ becomes infinite must be discarded—there can never be an infinite probability of finding the particle in any region. In practice, we "discard" a solution by setting its multiplicative constant equal to zero. For example, if the mathematical solution to the differential equation yields $\psi(x) = Ae^{kx} + Be^{-kx}$ for the *entire* region $x > 0$, then we must require $A = 0$ for the solution to be physically meaningful; otherwise $|\psi(x)|^2$ would become infinite as *x* goes to infinity. (However, if the solution is to be valid only in a small portion of the range of *x*, say $0 < x < L$, then we cannot set $A = 0$.) If this solution is to be valid in the *entire* region $x < 0$, then we must set $B = 0$.

8. Since we can no longer speak with certainty about the position of the particle, we can no longer guarantee the outcome of a single measurement of any physical quantity that depends on its position. However, if we can calculate the probability associated with any coordinate, we can find the *probable* outcome of any single measurement or (equivalently) the *average* outcome of a large number of measurements. For example, suppose we wish to find the average location of a particle by measuring its coordinate *x*. Performing a large number of measurements, we find the value x_1 a certain number of times n_1, x_2 a number of times n_2, etc., and in the usual way we can find the average value

$$x_{av} = \frac{n_1 x_1 + n_2 x_2 + \cdots}{n_1 + n_2 + \cdots} \qquad (5.7)$$

$$= \frac{\sum n_i x_i}{\sum n_i} \qquad (5.8)$$

The number of times n_i that we measure each x_i is proportional to the probability $P(x_i) \, dx$ to find the particle in the interval dx at x_i. Making this substitution and changing the sums to integrals, we have

$$x_{av} = \frac{\int_{-\infty}^{+\infty} P(x) \, x \, dx}{\int_{-\infty}^{+\infty} P(x) \, dx} \qquad (5.9)$$

and thus

Average or expectation value

$$x_{av} = \int_{-\infty}^{+\infty} |\psi(x)|^2 \, x \, dx \qquad (5.10)$$

where the last step can be made if the wave function is normalized, since the denominator of Equation 5.9 is then equal to one.

By analogy, the average value of any function of x can be found:

$$[f(x)]_{av} = \int_{-\infty}^{+\infty} P(x) f(x) \, dx = \int_{-\infty}^{+\infty} |\psi(x)|^2 f(x) \, dx \qquad (5.11)$$

Average values calculated according to Equation 5.10 or 5.11 are known as *expectation values.*

5.4 APPLICATIONS

The Free Particle

By a "free particle" we mean one that is moving with no forces acting on it in any region of space; that is, $F = 0$, and so $U(x) = $ constant for all x. We are free to choose the constant to be zero, since the potential energy is determined only to within an arbitrary constant.

We apply the recipe by writing Equation 5.3 with the appropriate potential energy ($U = 0$):

$$-\frac{\hbar^2}{2m} \frac{d^2\psi}{dx^2} = E\psi \qquad (5.12)$$

or

$$\frac{d^2\psi}{dx^2} = -k^2\psi \qquad (5.13)$$

where

$$k^2 = \frac{2mE}{\hbar^2} \qquad (5.14)$$

Equation 5.13 is a familiar one; written in this form, with k^2 always positive, its solution is

$$\psi(x) = A \sin kx + B \cos kx \qquad (5.15)$$

The allowed energy values can be found from Equation 5.14:

$$E = \frac{\hbar^2 k^2}{2m} \qquad (5.16)$$

Since our solution has placed no restrictions on k, the energy is permitted to have any value (in the language of quantum physics, we say that the energy is *not* quantized). We note that Equation 5.16 is the kinetic energy of a particle with momentum $p = \hbar k$ or, equivalently, $p = h/\lambda$; based on the discussion of Section 5.1, this is just what we would expect, since we have constructed the Schrödinger equation to yield the solution for the free particle corresponding to a single de Broglie wave.

Solving for A and B presents some difficulties because the normalization integral, Equation 5.6, cannot be evaluated from $-\infty$ to $+\infty$ for this wave function. (These difficulties would not occur if we made a linear super-position of many sine or cosine waves to form a wave packet, as we did in Section 4.4.) We therefore cannot determine probabilities from the wave function of Equation 5.15.

Particle in a Box (One Dimension)

In this case we have a particle moving freely in a one-dimensional "box" of length L; the particle is completely trapped within the "box." (Imagine a bead sliding without friction along a wire stretched between two rigid walls and making perfectly elastic collisions with the walls.) This potential energy may be expressed as:

$$U(x) = 0 \qquad 0 \le x \le L$$
$$\quad = \infty \qquad x < 0, x > L \tag{5.17}$$

The potential energy is shown in Figure 5.4 and is known as the *infinite potential energy well.* Of course, we are free to choose any constant value for U in the region $0 \le x \le L$; we choose it to be zero for convenience.

The recipe must now be applied separately to the regions inside and outside the box. We can analyze the outside region in either of two ways. If we examine Equation 5.3 for the region outside the box, we find that the only way to keep the equation from becoming meaningless when $U \to \infty$ is to require $\psi = 0$, so that $U\psi$ will not become infinite. Alternatively, we can go back to the original statement of the problem. If the walls of the box are perfectly rigid, the particle must always be in the box, and the probability for finding it elsewhere must be zero. To make the probability zero everywhere outside the box, we must make $\psi = 0$ outside the box. Thus we have

$$\psi(x) = 0 \qquad x < 0, x > L \tag{5.18}$$

The Schrödinger equation for $0 \le x \le L$, when $U(x) = 0$, is identical with Equation 5.12 and has the same solution:

$$\psi(x) = A \sin kx + B \cos kx \qquad (0 \le x \le L) \tag{5.19}$$

with

$$k^2 = \frac{2mE}{\hbar^2} \tag{5.20}$$

Our solution is not yet complete, for we have not evaluated A or B, nor have we found the allowed values of the energy E. To do this, we must apply the requirement that $\psi(x)$ must be continuous across any boundary. In this case, we require that our solutions for $x < 0$ and $x > 0$ match

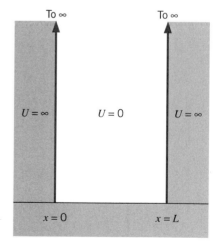

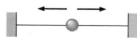

FIGURE 5.4 A particle moves freely in the one-dimensional region $0 \le x \le L$, but is excluded completely from $x < 0$ and $x > L$.

up at $x = 0$; similarly, the solutions for $x > L$ and $x < L$ must match at $x = L$.

Let us begin at $x = 0$. At $x < 0$, we have found that $\psi = 0$, and so we must set $\psi(x)$ of Equation 5.19 to zero at $x = 0$.

$$\psi(0) = A \sin 0 + B \cos 0 = 0$$

which gives

$$B = 0 \tag{5.21}$$

Since $\psi = 0$ for $x > L$, we must have $\psi(L) = 0$,

$$\psi(L) = A \sin kL + B \cos kL = 0 \tag{5.22}$$

Since we have already found $B = 0$, we must now have

$$A \sin kL = 0 \tag{5.23}$$

Either $A = 0$, in which case $\psi = 0$ *everywhere*, $\psi^2 = 0$ everywhere, and there is no particle (a meaningless solution) or else $\sin kL = 0$, which is true only when

$$kL = \pi, 2\pi, 3\pi, \ldots$$

or

$$kL = n\pi \qquad (n = 1, 2, 3, \ldots) \tag{5.24}$$

Since $k = 2\pi/\lambda$, we have $\lambda = 2L/n$; this is identical with the result obtained in introductory mechanics for the wavelengths of the standing waves in a string of length L fixed at both ends. *Thus the solution to the Schrödinger equation for a particle trapped in a linear region of length L is a series of standing de Broglie waves!* Not all wavelengths are permitted; only certain values, determined from Equation 5.24, may occur.

From Equation 5.20 we find that, since only certain values of k are permitted by Equation 5.24, only certain values of E may occur—*the energy is quantized!* Solving Equation 5.24 for k and substituting into Equation 5.20, we obtain

Energy values in infinite well

$$E_n = \frac{\hbar^2 k^2}{2m} = \frac{\hbar^2 \pi^2 n^2}{2mL^2} \qquad (n = 1, 2, 3, \ldots) \tag{5.25}$$

For convenience, let $E_0 = \hbar^2\pi^2/2mL^2$; this unit of energy is determined by the mass of the particle and the length of the box. Then $E_n = n^2 E_0$, and the only allowed energies for the particle are $E_0, 4E_0, 9E_0, 16E_0$, etc. All intermediate values, such as $3E_0$ or $6.2E_0$, are forbidden. Since the energy is purely kinetic in this case, our result means that only certain speeds are permitted for the particle. This is very different from the classical case, in which the bead (sliding without friction along the wire and colliding elastically with the walls) can be given any initial velocity and will move forever, back and forth, at the same speed. In the quantum case, this is not possible; only certain initial speeds can result in sustained states of motion; these special conditions are called "stationary states." (These states are "stationary" because, when the time dependence is included to make $\Psi(x,t)$, as in Section 5.6, $|\Psi(x,t)|^2$ is independent of time. Average values calculated according to Equation 5.11 likewise do not change with time. A particle initially in a pure stationary state remains in that state for all time.) The result of a measurement of the energy of a particle in a potential energy well *must* be one of the stationary state energies; no other result is possible.

Our solution for $\psi(x)$ is not yet complete, since we have not yet determined the constant A. To do this, we go back to the normalization condition given in Equation 5.6, $\int_{-\infty}^{+\infty} \psi^2 \, dx = 1$. Since $\psi = 0$ except for $0 \leq x \leq L$, the integral vanishes except inside that region, so that

$$\int_0^L A^2 \sin^2 \frac{n\pi x}{L} \, dx = 1 \qquad (5.26)$$

from which we find $A = \sqrt{2/L}$. The complete wave function for $0 \leq x \leq L$ is then

$$\psi_n(x) = \sqrt{\frac{2}{L}} \sin \frac{n\pi x}{L} \qquad (n = 1, 2, 3, \ldots) \qquad (5.27)$$

Wave functions in infinite well

In Figure 5.5, the allowed energy levels, wave functions, and probability densities ψ^2 are illustrated for the lowest several states. The lowest energy state, for which $n = 1$, is known as the *ground state,* and the states with higher energies ($n > 1$) are known as *excited states.*

Let us try to interpret the results of our calculation. Suppose we carefully place a particle with energy E_0 into our region (our "wire") and subsequently measure its position. After repeating this measurement a large number of times, we find a distribution of results similar to ψ^2 for the $n = 1$ case—the probability is greatest at $x = L/2$, tapers off as we move off center and falls to zero at the edges. (If we used a classical, nonquantum particle, we expect to find the same probability at all positions in the "box.") Suppose we repeat the measurement, except that now we prepare the system by giving the particle an energy of $4E_0$. We repeat our measurements of its position, and find a distribution of results in accordance with ψ^2 for $n = 2$; maxima in the probability at $x = L/4$ and $x = 3L/4$, and *zero* probability at $x = L/2$! The particle must travel so that it can be found occasionally at $L/4$ and at $3L/4$ without ever being found at $L/2$! Here we have a graphic illustration of the difference between classical and quantum physics. How can the particle get from $L/4$ to $3L/4$ without going through $L/2$? Our difficulty in answering this question comes from our desire to think in terms of particles, when quantum physics demands we think in terms of waves. The first overtone of a vibrating string of length L has a node in the center, and "information" travels from left to right and from right to left through the center, even though the midpoint does not move. When we speak of a position, we are referring to a *particle*; when we speak of motion from $L/4$ to $3L/4$, we are dealing with *waves.*

The calculation of probabilities and average values is illustrated by the following examples.

EXAMPLE 5.2

An electron is trapped in a one-dimensional region of length 1.0×10^{-10} m (a typical atomic diameter). (*a*) How much energy must be supplied to excite the electron from the ground state to the first excited state? (*b*) In the ground state, what is the probability of finding the electron in the region

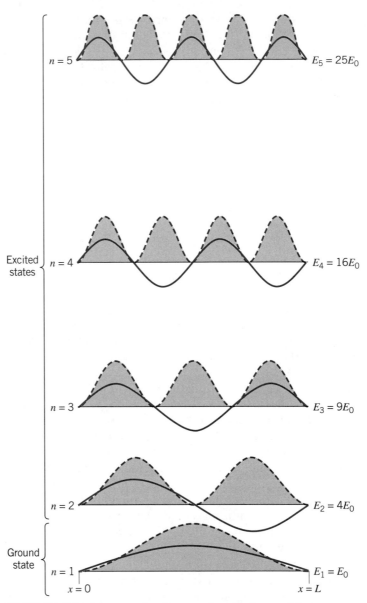

FIGURE 5.5   The permitted energy levels of a particle in a one-dimensional infinite well. The wave function for each level is shown by the solid curve, and the shaded region gives the probability density for each level.

from $x = 0.090 \times 10^{-10}$ m to 0.110×10^{-10} m? (*c*) In the first excited state, what is the probability of finding the electron between $x = 0$ and $x = 0.250 \times 10^{-10}$ m?

SOLUTION

(*a*) $E_1 = \dfrac{\hbar^2 \pi^2}{2mL^2} = \dfrac{(1.05 \times 10^{-34}\,\text{J·s})^2 (3.14)^2}{2(9.11 \times 10^{-31}\,\text{kg})(1.0 \times 10^{-10}\,\text{m})^2} = 6.0 \times 10^{-18}\,\text{J}$

$$= 37\,\text{eV}$$

With $E_2 = 4E_1$, the energy difference ΔE is

$$\Delta E = E_2 - E_1 = 4E_1 - E_1 = 3E_1$$
$$= 3(37 \text{ eV}) = 111 \text{ eV}$$

(*b*) From Equation 5.5,

$$\text{probability} = \int_{x_1}^{x_2} \psi^2 \, dx = \frac{2}{L} \int_{x_1}^{x_2} \sin^2 \frac{\pi x}{L} \, dx = \left(\frac{x}{L} - \frac{1}{2\pi} \sin \frac{2\pi x}{L} \right) \Big|_{x_1}^{x_2}$$

Evaluating this expression with $x_1 = 0.090 \times 10^{-10}$ m and $x_2 = 0.110 \times 10^{-10}$ m, we obtain

$$\text{probability} = 0.0038 = 0.38\%$$

(*c*)
$$\text{probability} = \int_{x_1}^{x_2} \left(\frac{2}{L} \right) \sin^2 \frac{2\pi x}{L} \, dx$$

$$= \left(\frac{x}{L} - \frac{1}{4\pi} \sin \frac{4\pi x}{L} \right) \Big|_{x_1}^{x_2}$$

$$= 0.25$$

(This result is of course what we would expect by inspection of the graph of ψ^2 for $n = 2$ in Figure 5.5. The interval from $x = 0$ to $x = L/4$ contains 25 percent of the total area under the ψ^2 curve.)

EXAMPLE 5.3

Show that the average value of x is $L/2$, independent of the quantum state.

SOLUTION

We use Equation 5.10; since $\psi = 0$ except for $0 \leq x \leq L$, we use 0 and L as the limits of integration

$$x_{av} = \frac{2}{L} \int_0^L \left(\sin^2 \frac{n\pi x}{L} \right) x \, dx$$

This can be integrated by parts or found in integral tables; the result is

$$x_{av} = \frac{L}{2}$$

Note that, as required, this result is independent of n. Thus a measurement of the average position of the particle yields no information about its quantum state.

Particle in a Box (Two Dimensions)

When we extend the previous situation to two and three dimensions, the principal features of the solution remain the same, but an important new

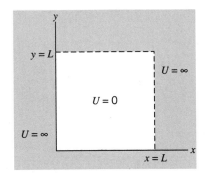

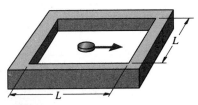

FIGURE 5.6 A particle moves freely in the two-dimensional region $0 \le x \le L$, $0 \le y \le L$.

feature is introduced. In this section we show how this occurs, since this new feature, known as *degeneracy,* is very important in our study of atomic physics.

To begin with, we need a Schrödinger equation that is valid in more than one dimension; our previous version, Equation 5.3, was a one-dimensional version. If the potential energy is a function of x and y, we expect that ψ also depends on both x and y, and the derivatives with respect to x must be replaced by derivatives with respect to x and y. In two dimensions, we then have*

$$-\frac{\hbar^2}{2m}\left(\frac{\partial^2\psi(x,y)}{\partial x^2} + \frac{\partial^2\psi(x,y)}{\partial y^2}\right) + U(x,y)\psi(x,y) = E\psi(x,y) \quad (5.28)$$

Our two-dimensional "box" can now be expressed as follows:

$$U(x,y) = 0 \qquad 0 \le x \le L; 0 \le y \le L$$
$$= \infty \qquad \text{otherwise} \qquad\qquad (5.29)$$

We picture a mass sliding without friction on a tabletop and colliding elastically with walls at $x = 0$, $x = L$, $y = 0$, and $y = L$, as in Figure 5.6. (For simplicity, we have made the box square; we could have made it rectangular by setting $U = 0$ when $0 \le x \le a$ and $0 \le y \le b$.)

Solving *partial* differential equations requires a technique more involved than we need to consider, so we will not give the details of the solution. We suspect that, as in the previous case, $\psi(x,y) = 0$ outside the box, in order to make the probability zero there. Inside the box, we consider solutions that are *separable*; that is, our function of x and y can be expressed as the product of one function that depends only on x and another that depends only on y:

$$\psi(x,y) = f(x)\, g(y) \qquad\qquad (5.30)$$

where the functions f and g are similar to Equation 5.15

$$f(x) = A \sin k_x x + B \cos k_x x$$
$$g(y) = C \sin k_y y + D \cos k_y y \qquad\qquad (5.31)$$

The wave number k of the previous problem has become the separate wave numbers k_x for $f(x)$ and k_y for $g(y)$. We show later how these are related. (See also Problem 18 at the end of this chapter.)

The continuity condition on $\psi(x,y)$ requires that the solutions inside and outside match at the boundary. Because $\psi = 0$ everywhere outside, the continuity condition then requires that $\psi = 0$ everywhere on the boundary. That is,

$$\psi(0,y) = 0 \quad \text{and} \quad \psi(L,y) = 0 \quad \text{for all } y$$
$$\psi(x,0) = 0 \quad \text{and} \quad \psi(x,L) = 0 \quad \text{for all } x$$

---

* The first two terms on the left side of this equation require *partial* derivatives; for well-behaved functions, these involve taking the derivative with respect to one variable while keeping the other constant. Thus if $f(x,y) = x^2 + xy + y^2$, $\partial f/\partial x = 2x + y$ and $\partial f/\partial y = 2y + x$.

In analogy with the previous problem, the condition at $x = 0$ gives $f(0) = 0$, which requires $B = 0$ in Equation 5.31. Similarly, the condition at $y = 0$ gives $g(0) = 0$, which requires $D = 0$. The condition $f(L) = 0$ requires that $\sin k_x L = 0$, and thus that $k_x L$ be an integer multiple of π; the condition $g(L) = 0$ similarly requires that $k_y L$ be an integer multiple of π. These two integers do not necessarily need to be the same, so we call them n_x and n_y. Making all these substitutions into Equation 5.30, we obtain

$$\psi(x,y) = A' \sin \frac{n_x \pi x}{L} \sin \frac{n_y \pi y}{L} \qquad (5.32)$$

where we have combined A and C into A'. The coefficient A' is once again found by the normalization condition, which in two dimensions becomes

$$\iint \psi^2 \, dx \, dy = 1 \qquad (5.33)$$

For our case this gives

$$\int_0^L dy \int_0^L A'^2 \sin^2 \frac{n_x \pi x}{L} \sin^2 \frac{n_y \pi y}{L} \, dx = 1 \qquad (5.34)$$

from which follows

$$A' = \frac{2}{L} \qquad (5.35)$$

(The solutions to this problem, which are standing de Broglie waves on a two-dimensional surface, are similar to the solutions of the classical problem of the vibrations of a stretched membrane such as a drumhead.)

Finally, we can plug our solution for $\psi(x,y)$ back into Equation 5.28 to find the energy:

$$E = \frac{\hbar^2 \pi^2}{2mL^2} (n_x^2 + n_y^2) \qquad (5.36)$$

Compare this result with Equation 5.25. Once again we let $E_0 = \hbar^2 \pi^2 / 2mL^2$ so that $E = E_0(n_x^2 + n_y^2)$. In Figure 5.7 the energies of the excited states are shown. You can see how different the energies are from those of the one-dimensional case shown in Figure 5.5.

Figure 5.8 shows the probability density ψ^2 for several different combinations of the *quantum numbers* n_x and n_y. The probability has maxima and minima, just like the probability in the one-dimensional problem. For example, if we gave the particle an energy of $8E_0$ and then made a large number of measurements of its position, we would expect to find it most often near the four points $(x,y) = (L/4,L/4)$, $(L/4,3L/4)$, $(3L/4,L/4)$ and $(3L/4,3L/4)$; we expect *never* to find it at $x = L/2$ or $y = L/2$. The *shape* of the probability density tells us something about the quantum numbers and therefore about the energy. Thus if we measured the probability density and found six maxima, as shown in Figure 5.8, we would deduce that the particle had an energy of $13E_0$ with $n_x = 2$ and $n_y = 3$, or else $n_x = 3$, $n_y = 2$.

Recently it has become possible to photograph the probability densities of electrons confined in a two-dimensional region. The tip of an electron

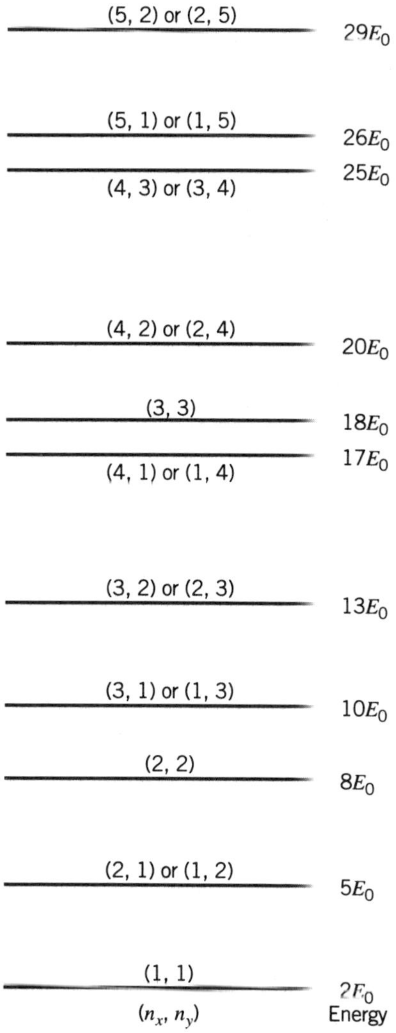

(5, 2) or (2, 5)	$29E_0$
(5, 1) or (1, 5)	$26E_0$
(4, 3) or (3, 4)	$25E_0$
(4, 2) or (2, 4)	$20E_0$
(3, 3)	$18E_0$
(4, 1) or (1, 4)	$17E_0$
(3, 2) or (2, 3)	$13E_0$
(3, 1) or (1, 3)	$10E_0$
(2, 2)	$8E_0$
(2, 1) or (1, 2)	$5E_0$
(1, 1)	$2E_0$
(n_x, n_y)	Energy

FIGURE 5.7 The lower permitted energy levels of the particle confined to the two-dimensional box.

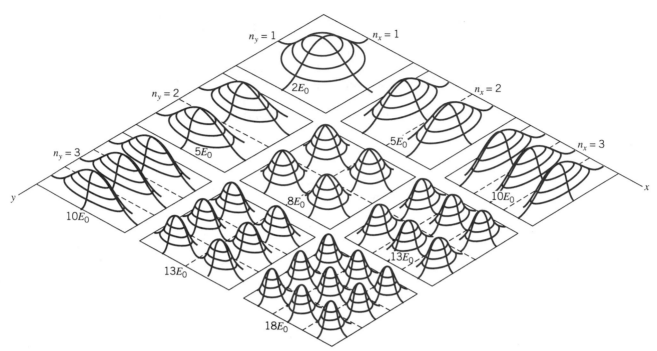

FIGURE 5.8   The probability density ψ^2 for some of the lower energy levels of the particle confined to the two-dimensional box.

microscope was used to place individual atoms on a metal surface to form the walls of the potential well. An electron on the surface was confined inside this well, and the probability density of the electron was observed also using the electron microscope. Color Plate 2 shows the results of this remarkable experiment. A ring of 48 iron atoms of radius 7.13 nm forms a "corral" that confines the electron. Inside the ring, the electron "waves" are clearly visible. The potential well is circular, rather than square, but otherwise the analysis follows the procedures described in this section; when the Schrödinger equation is solved in cylindrical polar coordinates with the potential energy for a circular well, the resulting probability density gives a close match with the observed one. These beautiful results are a dramatic confirmation of the wave functions obtained for the "particle in a box."

Degeneracy   Occasionally it happens that two different sets of quantum numbers n_x and n_y have exactly the same energy. This situation is known as *degeneracy,* and the energy levels are said to be *degenerate.* For example, the energy level at $E = 13E_0$ is degenerate, since both $n_x = 2, n_y = 3$ and $n_x = 3, n_y = 2$ have $E = 13E_0$. Since this degeneracy arises from interchanging n_x and n_y (which is the same as interchanging the x and y axes), the probability distributions in the two cases are not very different. However, consider the state with $E = 50E_0$, for which there are three sets of quantum numbers: $n_x = 7, n_y = 1$; $n_x = 1, n_y = 7$; and $n_x = 5, n_y = 5$. The first two result from the interchange of n_x and n_y and so have similar probability distributions, but the third represents a *very* different state of motion, as

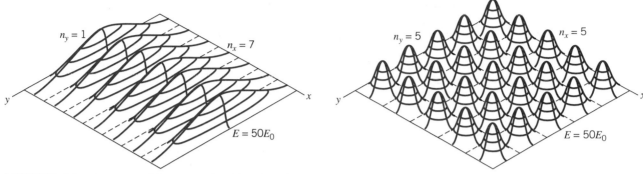

FIGURE 5.9 Two very different probability densities with exactly the same energy.

shown in Figure 5.9. The level at $E = 13E_0$ is said to be *two-fold* degenerate, while the level at $E = 50E_0$ is *three-fold* degenerate; we could also say that one level has a degeneracy of 2, while the other has a degeneracy of 3.

Degeneracy occurs in general whenever a system is labeled by two or more quantum numbers; as we have seen in the above calculation, different combinations of quantum numbers often can give the same value of the energy. The number of different quantum numbers required by a given physical problem turns out to be exactly equal to the number of dimensions in which the problem is being solved—one-dimensional problems need only one quantum number, two-dimensional problems need two, and so forth. When we get to three dimensions, as in some problems at the end of this chapter and especially in the hydrogen atom in Chapter 7, we find that the effects of degeneracy become more significant; in the case of atomic physics, the degeneracy is a major contributor to the structure and properties of atoms.

5.5 THE SIMPLE HARMONIC OSCILLATOR

Another situation that can be handled easily using the Schrödinger equation is the one-dimensional simple harmonic oscillator. The classical oscillator is an object of mass m attached to a spring of force constant k. The object is subject to a restoring force, $F = -kx$, where x is the displacement from its equilibrium position (which we take to be $x = 0$). Such an oscillator can be analyzed using Newton's laws to have a frequency $\omega_0 = \sqrt{k/m}$ and a period $T = 2\pi\sqrt{m/k}$. The oscillator has its maximum kinetic energy at $x = 0$; its kinetic energy vanishes at the *turning points* $x = \pm A_0$, where A_0 is the amplitude of the motion. At the turning points the oscillator comes to rest for an instant and then reverses its direction of motion. The motion is, of course, confined to the region $-A_0 \leq x \leq +A_0$.

Why analyze the motion of such a system using quantum mechanics? Although we never find in nature an example of a one-dimensional quantum oscillator, there are systems that behave approximately as one—a vibrating diatomic molecule, for example. In fact, any system in a potential energy minimum behaves approximately like a simple harmonic oscillator.

A force $F = -kx$ has the associated potential energy $U = \frac{1}{2}kx^2$, and so we have the Schrödinger equation:

$$-\frac{\hbar^2}{2m}\frac{d^2\psi}{dx^2} + \frac{1}{2}kx^2\psi = E\psi \tag{5.37}$$

(Since we are working in one dimension, U and ψ are functions only of x.) This differential equation is difficult to solve directly, and so we guess at its solutions. The solution must approach zero as $x \to \pm\infty$, and in the limit $x \to \pm\infty$ the solutions of Equation 5.37 behave like exponentials of $-x^2$. We therefore try $\psi(x) = Ae^{-ax^2}$, where A and a are constants that are determined by evaluating Equation 5.37 for this choice of $\psi(x)$. We begin by evaluating $d^2\psi/dx^2$.

$$\frac{d\psi}{dx} = -2ax(Ae^{-ax^2})$$

$$\frac{d^2\psi}{dx^2} = -2a(Ae^{-ax^2}) - 2ax(-2ax)Ae^{-ax^2}$$

and we plug $\psi(x)$ and $d^2\psi/dx^2$ into Equation 5.37 to see if there is a solution.

$$-\frac{\hbar^2}{2m}(-2aAe^{-ax^2} + 4a^2x^2Ae^{-ax^2}) + \frac{1}{2}kx^2(Ae^{-ax^2}) = EAe^{-ax^2} \tag{5.38}$$

Canceling the common factor Ae^{-ax^2} yields

$$\frac{\hbar^2 a}{m} - \frac{2a^2\hbar^2}{m}x^2 + \frac{1}{2}kx^2 = E \tag{5.39}$$

Equation 5.39 is *not* an equation to be solved for x, because we are looking for a solution which is valid for *any* x, not just for one specific value. In order for this to hold for *any* x, the coefficients of x^2 must cancel and the remaining constants must be equal. (That is, consider the equation $bx^2 = c$. This equation is of course valid for $x = \sqrt{c/b}$, but if we wish it to be true for *any* and *all* x, we must require that both $b = 0$ and $c = 0$.) Thus

$$-\frac{2a^2\hbar^2}{m} + \frac{1}{2}k = 0 \tag{5.40}$$

and

$$\frac{\hbar^2 a}{m} - E = 0 \tag{5.41}$$

which yield

$$a = \frac{\sqrt{km}}{2\hbar} \tag{5.42}$$

and

$$E = \tfrac{1}{2}\hbar\sqrt{k/m} \tag{5.43}$$

We can also write the energy in terms of the classical frequency $\omega_0 = \sqrt{k/m}$ as

$$E = \tfrac{1}{2}\hbar\omega_0 \tag{5.44}$$

The coefficient A must be found from the normalization condition (see Problem 20 at the end of the chapter).

The solution we have just found is illustrated in Figure 5.10. One feature of the solution is striking—there is a nonzero probability of finding the particle beyond the classical turning points $x = \pm A_0$. The total energy E is constant, and beyond $x = \pm A_0$, the potential energy is greater than E, so that the kinetic energy would become *negative.* This is an impossibility in the realm of classical physics, and so the classical *particle* can never be found at $|x| > A_0$. It is, however, possible for the quantum *wave* to penetrate into the classically forbidden region. We discuss this topic in the next section.

The solution we have found corresponds to the *ground state* of the oscillator. The mathematically difficult general solution is of the form $\psi_n(x) = Af_n(x)e^{-ax^2}$, where $f_n(x)$ is a polynomial in which the highest power of x is x^n. The corresponding energies are

$$E_n = (n + \tfrac{1}{2})\hbar\omega_0 \qquad (n = 0, 1, 2, \ldots) \tag{5.45}$$

Note that these levels, illustrated along with their probability densities in Figure 5.11, are *uniformly spaced,* in contrast to the one-dimensional particle in a box. All of the solutions have the property of penetration of

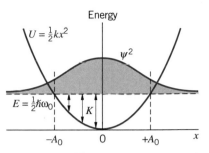

FIGURE 5.10 The ground state of the one-dimensional harmonic oscillator. The kinetic energy K is the difference between the total energy E and the potential energy $U = \tfrac{1}{2}kx^2$.

Harmonic oscillator energy levels

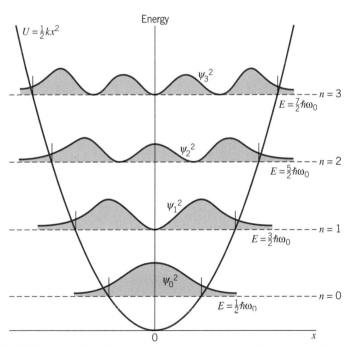

FIGURE 5.11 The lowest few energy levels and corresponding probability densities of the harmonic oscillator. The short vertical lines mark the classical turning points.

probability density into the forbidden region beyond the classical turning points. The probability density oscillates, somewhat like a sine wave, between the turning points, and decreases like e^{-2ax^2} to zero beyond the turning points.

A sequence of vibrational excited states similar to Figure 5.11 is commonly found in diatomic molecules such as HCl (see Chapter 9). The spacing between the states is typically 0.1–1 eV; the states are observed when photons (in the infrared region of the spectrum) are emitted or absorbed as the molecule jumps from one state to another. A similar sequence is observed in nuclei, where the spacing is 0.1–1 MeV and the radiations are in the gamma-ray region of the spectrum.

5.6   TIME DEPENDENCE

We have thus far not considered the time dependence of the Schrödinger equation or of its solutions. We will not consider the method of solution in detail, but will merely state the result: given a time-independent solution $\psi(x)$ of Equation 5.3 corresponding to the energy E, the *time-dependent wave function* $\Psi(x,t)$ is found according to*

$$\Psi(x,t) = \psi(x)e^{-i\omega t} \tag{5.46}$$

where the frequency ω is given by the de Broglie relationship

$$\omega = \frac{E}{\hbar} \tag{5.47}$$

As mentioned in Section 4.1, it is not clear whether the energy E in the de Broglie relationship should be the classical total energy or the relativistic total energy, since we get no clue from the corresponding relationship $E = h\nu$ for photons. In this chapter we have assumed the classical relationship $E = U + K$ and neglected the rest energy contribution to E. To be

---

* The imaginary number i is defined as $\sqrt{-1}$. A *complex number* can be represented as having a real part, which does not depend on i, and an imaginary part, which depends on i. We require no manipulations with complex numbers for this chapter, and need only the following relationship: the complex exponential $e^{i\theta}$ can be represented in terms of real trigonometric functions as

$$e^{i\theta} = \cos\theta + i\sin\theta$$

and

$$e^{-i\theta} = \cos\theta - i\sin\theta$$

Thus

$$\sin\theta = \frac{1}{2i}(e^{i\theta} - e^{-i\theta})$$

and

$$\cos\theta = \tfrac{1}{2}(e^{i\theta} + e^{-i\theta})$$

strictly correct, we should write $E - U + K + mc^2$ (but we still consider only cases where $v \ll c$ so that the classical $\frac{1}{2}mv^2$ is acceptable for K). The addition of the rest energy changes Equation 5.46 by introducing a factor $e^{-imc^2 t/\hbar}$, but since all of the measurable properties of $\Psi(x,t)$ depend on $\Psi\Psi^*$, the product of Ψ and its *complex conjugate* Ψ^* obtained by replacing i with $-i$, this additional factor has no observable consequences and can safely be ignored. As in Problem 33 of Chapter 4, the group velocity depends on dE/dp and the addition of constants such as mc^2 to E does not affect the motion of the wave packet.

In order to see how multiplying by $e^{-i\omega t}$ gives a wave, we consider how the wave function of the free particle, Equation 5.15, gives the time-dependent wave function $\Psi(x,t)$. This process is simplified if we first rewrite Equation 5.15 in terms of the complex exponentials e^{ikx} and e^{-ikx}, so that

$$\psi(x) = A'e^{ikx} + B'e^{-ikx} \tag{5.48}$$

The constants A' and B' can be found from the constants A and B. Now we have, for the time-dependent wave function,

$$\Psi(x,t) = (A'e^{ikx} + B'e^{-ikx})e^{-i\omega t}$$
$$= A'e^{i(kx-\omega t)} + B'e^{-i(kx+\omega t)} \tag{5.49}$$

The first term on the right represents a trigonometric function with phase $(kx - \omega t)$, and thus is a wave moving in the *positive x* direction; the second term corresponds to a wave moving in the *negative x* direction. The squared magnitudes of the coefficients give the intensities of the waves; thus the wave moving in the positive x direction has intensity $|A'|^2$ and the wave moving in the negative x direction has intensity $|B'|^2$.

Suppose we have a beam of monoenergetic particles moving in the positive x direction, which are represented by a wave function in the form of the *first* term of Equation 5.49. The probability for locating a particle is then given by $|A'|^2$. This is a constant, independent of the position x—a particle is equally likely to be found at any location along the x axis. (Of course, this is consistent with the uncertainty relationship. If the beam is monoenergetic, $\Delta p = 0$ and therefore $\Delta x = \infty$.)

If the wave function contains equal amplitudes of waves moving in both directions (i.e., if $|A'| = |B'|$), certain locations can be found at which the probability density $\Psi\Psi^*$ is equal to zero. *There can be points at which there is zero probability of finding the particle!* In analogy with classical physics, when we add two waves of equal amplitude traveling in opposite directions, we obtain a *standing wave*, in which there are certain points (known as "nodes") at which the combined wave amplitude vanishes at all times.

5.7 STEPS AND BARRIERS

In this general type of problem, we analyze what happens when a particle moving (again in one dimension) in a region of constant potential energy suddenly moves into a region of different, but also constant, potential energy. We will not discuss in detail the solutions to these problems, but the methods of solution of each are so similar that we can outline the steps to take in the solution. In this discussion, we let E be the (fixed) total energy of the particle and U_0 will be the value of the constant potential energy.

Potential Energy Step, $E > U_0$

Whenever E is greater than U_0, the solution to the Schrödinger equation is of the form

$$\psi(x) = A \sin kx + B \cos kx \tag{5.50}$$

where

$$k = \sqrt{\frac{2m}{\hbar^2}(E - U_0)} \tag{5.51}$$

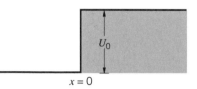

FIGURE 5.12 A step of height U_0.

A and B are constants to be found from the continuity and normalization conditions. For example, consider the potential energy step shown in Figure 5.12:

$$U(x) = 0 \qquad x < 0$$
$$= U_0 \qquad x \geq 0$$

If E is the total energy and is greater than U_0, then we can simply write down the solutions to the Schrödinger equation in the two regions:

Potential energy step, $E > U_0$

$$\psi_0(x) = A \sin k_0 x + B \cos k_0 x \qquad k_0 = \sqrt{\frac{2mE}{\hbar^2}} \qquad\qquad x < 0 \quad (5.52a)$$

$$\psi_1(x) = C \sin k_1 x + D \cos k_1 x \qquad k_1 = \sqrt{\frac{2m}{\hbar^2}(E - U_0)} \quad x > 0 \quad (5.52b)$$

Relationships among the four coefficients, A, B, C, and D, may be found by applying the condition that $\psi(x)$ and $\psi'(x) = d\psi/dx$ must be continuous at the boundary; thus $\psi_0(0) = \psi_1(0)$ and $\psi_0'(0) = \psi_1'(0)$. A typical solution might look like that sketched in Figure 5.13. Note that the application of

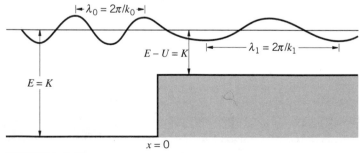

FIGURE 5.13 The wave function of a particle of energy E encountering a step of height U_0, for the case $E > U_0$. The de Broglie wavelength changes from λ_0 to λ_1 when the particle crosses the step, but ψ and $d\psi/dx$ are continuous at $x = 0$.

the continuity condition guarantees a smooth transition from one wave to the next at the boundary.

Once again, we may use the equation $e^{i\theta} = \cos\theta + i\sin\theta$ to transform these solutions from sines and cosines to complex exponentials:

$$\psi_0(x) = A'e^{ik_0x} + B'e^{-ik_0x} \qquad x < 0 \qquad (5.53a)$$

$$\psi_1(x) = C'e^{ik_1x} + D'e^{-ik_1x} \qquad x > 0 \qquad (5.53b)$$

When the time dependence has been added by multiplying each term by $e^{-i\omega t}$, we can then make the following identification of the component waves, recalling that $(kx - \omega t)$ is the phase of a wave moving in the positive x direction, while $(kx + \omega t)$ is the phase of a wave moving in the negative x direction, and assuming that *the squared magnitude of each coefficient gives the intensity of the corresponding component wave.* In the region $x < 0$, Equation 5.53a describes the superposition of a wave of intensity $|A'|^2$ moving in the positive x direction (from $-\infty$ to 0) and a wave of intensity $|B'|^2$ moving in the negative x direction. Suppose we had intended our solution to describe particles that are incident from the left on this step. Then $|A'|^2$ gives the intensity of the *incident* wave (more exactly, the de Broglie wave describing the incident beam of particles) and $|B'|^2$ gives the intensity of the *reflected* wave. The ratio $|B'|^2/|A'|^2$ tells us the reflected fraction of the incident wave intensity. In the region $x > 0$, the wave of intensity $|D'|^2$ moving in the negative x direction (from $x = +\infty$ to $x = 0$) cannot exist if we are firing particles from the negative x axis, so *in this particular experimental situation* we are justified in setting D' to zero. The intensity of the *transmitted* wave is then $|C'|^2$.

We can also analyze the solutions in terms of the kinetic energy in each region. Where the kinetic energy is greatest, the linear momentum p ($= \sqrt{2mK}$) is greatest, and the de Broglie wavelength λ ($= h/p$) is smallest. Thus in the region $x < 0$, the wavelength is smaller than in the region $x > 0$.

Potential Energy Step, $E < U_0$

Whenever E becomes less than U_0, a different solution results:

$$\psi(x) = Ae^{kx} + Be^{-kx} \qquad (5.54)$$

where

$$k = \sqrt{\frac{2m}{\hbar^2}(U_0 - E)} \qquad (5.55)$$

If the region in which this solution is to be valid extends to $+\infty$ or $-\infty$, we must keep ψ from becoming infinite by setting A or B to zero; if the region contains only finite x, this need not be done.

As an example, in the previous problem, if E were less than U_0, then the solution for ψ_0 (for $x < 0$) would still be given by Equation 5.52a or 5.53a, but the solution ψ_1 (for $x > 0$) becomes

$$\psi_1(x) = Ce^{k_1x} + De^{-k_1x} \qquad k_1 = \sqrt{\frac{2m}{\hbar^2}(U_0 - E)} \qquad (5.56)$$ *Potential energy step, $E < U_0$*

We set $C = 0$ to keep $\psi_1(x)$ from becoming infinite as $x \to \infty$, and we apply the boundary conditions on $\psi(x)$ and $\psi'(x)$ at $x = 0$. The resulting solution is shown in Figure 5.14.

This solution illustrates an important difference between classical and quantum mechanics. Classically, the particle can *never* be found in the region $x > 0$, since its total energy is not sufficient to overcome the potential energy step. However, quantum mechanics allows the wave function, and therefore the particle, to penetrate into the classically forbidden region. No experiment can ever *observe* the particle in the forbidden region (its kinetic energy would be negative there), but in certain experiments a particle can pass through a classically forbidden region and emerge into an allowed region where it *can* be observed. Examples of this effect are discussed later in this section.

Penetration into the forbidden region is associated with the wave nature of the particle, and we can show that the penetration distance is consistent with the uncertainty in defining the location or time coordinate of the particle. The probability density in the $x > 0$ region is $|\psi_1|^2$, which according to Equation 5.56 is proportional to $e^{-2k_1 x}$. If we define a representative penetration distance Δx to be the distance from $x = 0$ to the point at which the probability drops by $1/e$, then

$$e^{-2k_1 \Delta x} = e^{-1}$$

$$\Delta x = \frac{1}{2k_1} = \frac{1}{2} \frac{\hbar}{\sqrt{2m(U_0 - E)}} \tag{5.57}$$

If it were to enter the region with $x > 0$, the *particle* must gain an energy of at least $U_0 - E$ in order to get over the potential energy step; it must in addition gain some kinetic energy if it is to move in the region $x > 0$. Of course, it is a violation of conservation of energy for the particle to spontaneously gain *any* amount of energy, but according to the uncertainty relationship $\Delta E \, \Delta t \sim \hbar$ conservation of energy does not apply at times smaller than Δt except to within an amount $\Delta E \sim \hbar / \Delta t$. That is, if the particle "borrows" an amount of energy ΔE and "returns" the borrowed energy within a time $\Delta t \sim \hbar / \Delta E$, we observers will still believe energy is conserved. Suppose we borrow an energy sufficient to give the particle a

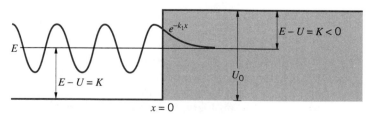

FIGURE 5.14 The wave function of a particle of energy E encountering a step of height U_0, for the case $E < U_0$. The wave function decreases exponentially in the classically forbidden region, where the classical kinetic energy would be negative. At $x = 0$, ψ and $d\psi/dx$ are continuous.

kinetic energy of K in the forbidden region. How far into the forbidden region does it penetrate?

The "borrowed" energy is $(U_0 - E) + K$; the term $(U_0 - E)$ gets the particle to the top of the step, and the extra K gives it its motion. We must return this energy within a time

$$\Delta t = \frac{\hbar}{U_0 - E + K} \tag{5.58}$$

The particle moves with speed $v = \sqrt{2K/m}$, and so the distance it can travel is

$$\Delta x = \tfrac{1}{2} v\, \Delta t$$

$$= \frac{1}{2} \sqrt{\frac{2K}{m}} \frac{\hbar}{U_0 - E + K} \tag{5.59}$$

(The factor of $\tfrac{1}{2}$ is present because in the time Δt the particle must penetrate the distance Δx and return.)

In the limit $K \to 0$, the penetration distance Δx goes to 0 according to Equation 5.59 because the particle has zero velocity; similarly, $\Delta x \to 0$ in the limit $K \to \infty$, because it moves for a vanishing time interval Δt. In between those limits, there must be a maximum value of Δx for some particular K. Differentiating Equation 5.59 with respect to K, we can find the maximum value

$$\Delta x_{\max} = \frac{1}{2} \frac{\hbar}{\sqrt{2m(U_0 - E)}} \tag{5.60}$$

This value of Δx is identical with Equation 5.57! This demonstrates that the penetration into the forbidden region given by the solution to the Schrödinger equation is entirely consistent with the uncertainty relationship. (The agreement between Equations 5.57 and 5.60 is really somewhat accidental, since the factor $1/e$ used to obtain Equation 5.57 was chosen arbitrarily. What we have really demonstrated is that the Schrödinger equation gives the same estimates of uncertainty as the Heisenberg relationships.)

Potential Energy Barrier

Consider now the potential energy barrier shown in Figure 5.15:

$$U(x) = 0 \qquad x < 0$$
$$= U_0 \qquad 0 \le x \le a$$
$$= 0 \qquad x > a$$

Particles with energy E less than U_0 are incident from the left. Our experience then leads us to expect solutions of the form shown in Figure 5.16—sinusoidal oscillation in the region $x < 0$ (an incident wave and a reflected wave), exponentials in the region $0 \le x \le a$, and sinusoidal oscillations in the region $x > a$ (the transmitted wave). The intensity of the transmitted

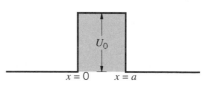

FIGURE 5.15 A barrier of height U_0 and width a.

wave can be found by proper application of the continuity conditions, which we do not discuss; the intensity is found to depend on the energy of the particle and on the height and thickness of the barrier. Classically, the particles should never appear at $x > a$, since they have not sufficient energy to overcome the barrier; this situation is an example of *barrier penetration,* sometimes called quantum mechanical *tunneling.* The particles can never be *observed* in the classically forbidden region $0 \le x \le a$, but can "tunnel" *through* that region and be observed at $x > a$.

Although the potential energy barrier of Figure 5.15 is quite schematic and hypothetical, there are many examples in nature of such tunneling. We consider four such examples.

1. *Alpha Decay.*   An atomic nucleus consists of protons and neutrons in a constant state of motion; occasionally these particles form themselves into an aggregate of two protons and two neutrons, called an alpha particle. In one form of radioactive decay, the nucleus can emit an alpha particle, which can be detected in the laboratory. However, in order to escape from the nucleus the alpha particle must penetrate a barrier of the form shown in Figure 5.17. The probability for the alpha particle to penetrate the barrier, and be detected in the laboratory, can be computed based on the energy of the alpha particle, and the height and thickness of the barrier. The decay probability can be measured in the laboratory, and it is found to be in excellent agreement with the value obtained from a quantum-mechanical calculation based on barrier penetration.

2. *Ammonia Inversion.*   Figure 5.18 is a representation of the ammonia molecule NH_3. If we were to try to move the nitrogen atom along the axis of the molecule, toward the plane of the hydrogen atoms, we find repulsion caused by the three hydrogen atoms, which produces a potential of the form shown in Figure 5.19. According to classical mechanics, unless we give the nitrogen atom sufficient energy, it should not be able to surmount the barrier and appear on the other side of the plane of hydrogens. According to quantum mechanics, the nitrogen can tunnel through the barrier and appear on the other side of the molecule. In fact, the N atom actually tunnels back and forth with a frequency in excess of 10^{10} oscillations per second.

3. *The Tunnel Diode.*   A tunnel diode is an electronic device that uses the phenomenon of tunneling. Schematically, the potential "seen" by an electron in a tunnel diode can be represented by Figure 5.20. The current that flows through such a device is produced by electrons tunneling through from one side to the other. The rate of tunneling, and therefore the current, can be regulated merely by changing the height of the barrier, which can be done with an applied voltage. This can be done rapidly, so that switching frequencies in excess of 10^9 Hz can be obtained. Ordinary semiconductor diodes depend on the diffusion of electrons across a junction, and therefore operate on much longer time scales (that is, at lower frequencies).

4. *The Scanning Tunneling Microscope.*   Images of individual atoms on the surface of materials (Color Plate 3) can be made with the scanning tunneling microscope. Electrons are trapped in a surface by a potential

*Barrier penetration (**tunneling**)*

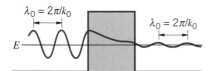

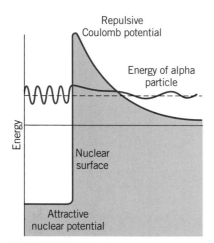

FIGURE 5.16   The wave function of a particle of energy $E < U_0$ encountering a barrier (the particle is incident from the left in the figure). The wavelength λ_0 is the same on both sides of the barrier, but the amplitude beyond the barrier is much less than the original amplitude. The particle can never be observed inside the barrier (where it would have negative kinetic energy) but it can be observed *beyond* the barrier.

FIGURE 5.17   An alpha particle penetrating the nuclear potential barrier. The probability to penetrate the barrier depends on its thickness and its height.

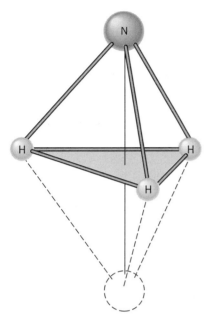

FIGURE 5.18 A schematic diagram of the ammonia molecule. The Coulomb repulsion of the three hydrogens establishes a barrier against the nitrogen atom moving to a symmetric position (shown in dashed lines) on the opposite side of the plane of hydrogens.

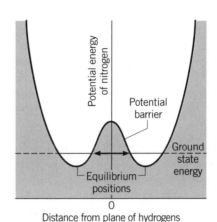

FIGURE 5.19 The potential energy seen by the nitrogen atom in an ammonia molecule. The nitrogen can penetrate the barrier and move from one equilibrium position to another.

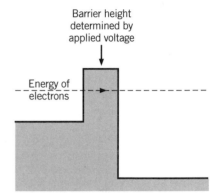

FIGURE 5.20 The potential energy barrier seen by an electron in a tunnel diode. The conductivity of the device is determined by the electron's probability to penetrate the barrier, which depends on the height of the barrier.

barrier (the work function of the material). When a needle-like probe is placed within about 1 nm of the surface (Figure 5.21), electrons can tunnel through the barrier between the surface and the probe and produce a current that can be recorded in an external circuit.

The current is very sensitive to the width of the barrier (the distance from the probe to the surface). In practice, a feedback mechanism keeps the current constant by moving the tip up and down. The motion of the tip gives a map of the surface that reveals details smaller than 0.01 nm, about $\frac{1}{100}$ the diameter of an atom!

For the development of the scanning tunneling microscope, Gerd Binnig and Heinrich Rohrer were awarded the 1986 Nobel prize in physics.

Before concluding this discussion, we return briefly to the question of particle-wave duality. The *particle* is never observed in the forbidden region; to find it there would violate energy conservation. Every particle that is incident from the left on the potential energy step of Figure 5.14 is reflected and ends up moving back in the negative x direction. Some are reflected at $x = 0$, while others may penetrate a distance Δx before turning around and reemerging. Every particle incident on the barrier of Figure 5.16 is either reflected or transmitted; the number of incident particles is equal to

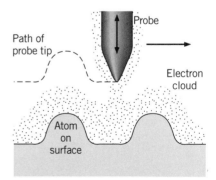

FIGURE 5.21 In a scanning tunneling microscope, a needle-like probe is scanned over a surface. The probe is moved vertically so that the distance between the probe and the surface remains constant as the probe scans laterally.

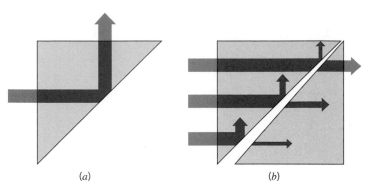

FIGURE 5.22 (*a*) Total internal reflection of light waves at a glass-air boundary. (*b*) Frustrated total internal reflection. The thicker the air gap, the smaller the probability to penetrate. Note that the light beam does not appear *in* the gap.

the number reflected back to $x < 0$ plus the number transmitted to $x > a$. None are "trapped" or ever seen in the forbidden region $0 < x < a$. How can the incident *particle* get from $x < 0$ to $x > a$? As a classical particle, *it can't!* However, the wave representing the particle *can* penetrate through the barrier, which allows the particle to be observed in the classically *allowed* region $x > a$.

This phenomenon of penetration of a forbidden region is a well-known property of classical waves. Quantum physics provides a new aspect to this observation by associating a particle with the wave, and thus allowing a particle to enter a classically forbidden region.

An example of the penetration effect for classical waves occurs for total internal reflection* of light waves. Figure 5.22*a* shows a light beam in glass incident on a boundary with air. The beam is totally reflected in the glass. However, if a second piece of glass is brought close to the first, as in Figure 5.22*b*, the beam can appear in the second piece of glass. This effect is called *frustrated total internal reflection*. The intensity of the beam in the second piece, represented by the widths of the arrows in Figure 5.22*b*, decreases rapidly as the thickness of the gap increases. A similar effect is observed for water waves, as shown in Figure 5.23.

We can explain these observations if we assume that a wave must penetrate a small distance, perhaps a few wavelengths, into the forbidden region to determine what is beyond the boundary, before returning to the allowed region. The wave does not propagate in the forbidden region; the laws of reflection and refraction forbid its presence there. If, however, a second allowed region is encountered within the penetration distance of a few wavelengths, the wave can propagate in this second allowed region. The

---

* Total internal reflection, which is discussed in many introductory physics texts, occurs when a light beam is incident on a boundary between two substances, such as glass and air, from the side with the higher index of refraction. If the angle of incidence inside the glass exceeds a certain critical value, the light beam is totally reflected back into the glass. In other cases, the beam is partially reflected and partially refracted across the boundary.

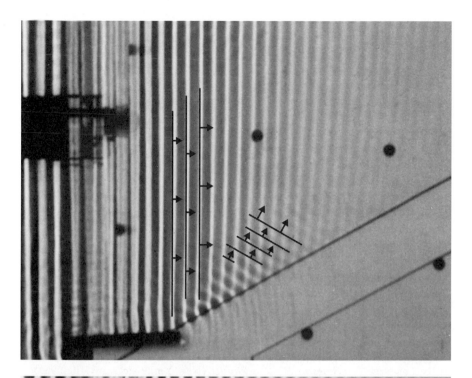

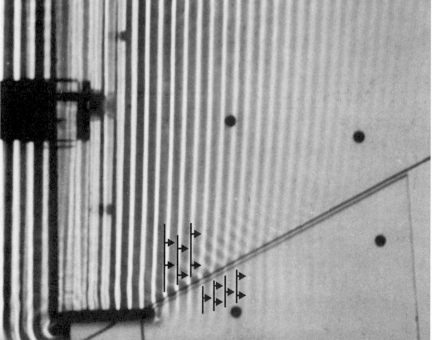

FIGURE 5.23 Frustrated total internal reflection for water waves. At the boundary the depth increases suddenly and the waves are totally reflected. When the gap is made narrow, the waves can penetrate and appear on the other side.

examples of tunneling discussed in this chapter can be regarded as applications of this classical wave effect to particles subject to quantum behavior.

SUGGESTIONS FOR FURTHER READING

An introduction to the formalism of quantum physics and its historical development is found in W. H. Cropper, *The Quantum Physicists and an Introduction to Their Physics* (New York, Oxford University Press, 1970).

Other discussions of the Schrödinger equation and examples of its use are given in the following books, in approximately increasing order of difficulty beginning at the level of this book:

N. Ashby and S. C. Miller, *Principles of Modern Physics* (San Francisco, Holden-Day, 1970).

P. A. Lindsay, *Introduction to Quantum Mechanics for Electrical Engineers* (New York, McGraw-Hill, 1967).

E. H. Wichmann, *Quantum Physics, Volume 4 of the Berkeley Physics Course* (New York, McGraw-Hill, 1971).

R. Eisberg and R. Resnick, *Quantum Physics of Atoms, Molecules, Solids, Nuclei, and Particles,* 2nd ed., (New York, Wiley, 1985).

R. B. Leighton, *Principles of Modern Physics* (New York, McGraw-Hill, 1959).

Illustrations showing the scattering of wave packets from barriers and the behavior of wave packets inside square-well and harmonic-oscillator potential energy wells may be found in:

S. Brandt and H. D. Dahmen, *The Picture Book of Quantum Mechanics* (New York, Wiley, 1985).

For a description of the scanning tunneling microscope, see:

G. Binnig and H. Rohrer, *Scientific American* **253,** 50 (August 1985).

QUESTIONS

1. Newton's laws can be solved to give the future behavior of a particle. In what sense does the Schrödinger equation also do this? In what sense does it not?

2. Why is it important for a wave function to be normalized? Is an unnormalized wave function a solution to the Schrödinger equation?

3. What is the physical meaning of $\int_{-\infty}^{+\infty} |\psi|^2 \, dx = 1$?

4. What are the dimensions of $\psi(x)$? Of $\psi(x, y)$?

5. None of the following are permitted as solutions of the Schrödinger equation. Give the reasons in each case.
   (a) $\psi(x) = A \cos kx \qquad x < 0$
   $\psi(x) = B \sin kx \qquad x > 0$

(b) $\psi(x) = \dfrac{Ae^{-kx}}{x}$ $-L \leq x \leq L$

(c) $\psi(x) = A \sin^{-1} kx$

(d) $\psi(x) = A \tan kx$ $x > 0$

6. What happens to the probability density in the infinite well when $n \to \infty$? Is this consistent with classical physics?

7. How would the solution to the infinite potential energy well be different if the well extended from $x = x_0$ to $x = x_0 + L$, where x_0 is a nonzero value of x? Would any of the measurable properties be different?

8. How would the solution to the one-dimensional infinite potential energy well be different if the potential energy were not zero for $0 \leq x \leq L$ but instead had a constant value U_0? What would be the energies of the excited states? What would be the wavelengths of the standing de Broglie waves? Sketch the behavior of the lowest two wave functions.

9. Assuming a pendulum to behave like a quantum oscillator, what are the energy differences between the quantum states of a pendulum of length 1 m? Are such differences observable?

10. For the potential energy barrier (Figure 5.15), is the wavelength for $x > a$ the same as the wavelength for $x < 0$? Is the amplitude the same?

11. Suppose particles were incident on the potential energy step from the *positive* x direction. Which of the four coefficients of Equation 5.53 would be set to zero? Why?

12. The energies of the excited states of the systems we have discussed in this chapter have been exact—there is no energy uncertainty. What does this suggest about the lifetime of particles in those excited states? Left on its own, will a particle ever make transitions from one state to another?

13. Explain how the behavior of a particle in a one-dimensional infinite well can be considered in terms of standing de Broglie waves.

14. How would you design an experiment to observe barrier penetration with sound waves? What range of thicknesses would you choose for the barrier?

15. If U_0 were negative in Figure 5.15, how would the wave functions look for $E > 0$? For $E < 0$?

16. Does Equation 5.20 imply that we know the momentum of the particle exactly? If so, what does the uncertainty principle indicate about our knowledge of its location? How can you reconcile this with our knowledge that the particle *must* be in the well?

17. Do sharp boundaries and steep jumps of potential energy occur in nature? If not, how would our analysis of potential energy steps and barriers be different?

PROBLEMS

1. A classical particle moves freely in the positive x direction with speed v_0. When it crosses the origin (at $t = 0$) it enters region 1 in which it feels a deceleration a_1. At time t_1 and position x_1 it leaves region 1 and enters region

2, where it feels a deceleration $a_2 = \frac{1}{2}a_1$. It leaves region 2 at time t_2 and coordinate x_2, when its velocity is $v_0/2$, and once again moves freely. Without setting up the equations of motion, make a sketch similar to Figure 5.1 showing the acceleration, velocity, and position from times less than zero to times greater than t_2. Think carefully about whether a, v, and x (and their slopes!) are continuous.

2. A wave has the form

$$y = -A \cos (2\pi x/\lambda + \pi/6)$$

when $x < 0$. For $x > 0$, the wavelength is $\lambda/2$. By applying continuity conditions at $x = 0$, find the amplitude (in terms of A) and phase of the wave in the region $x > 0$. Sketch the wave, showing both $x < 0$ and $x > 0$.

3. The ground-state energy of a particle in an infinite one-dimensional well is 4.4 eV. If the width of the well is doubled, what is the new ground-state energy?

4. A particle in an infinite well is in the ground state with an energy of 1.26 eV. How much energy must be added to the particle to reach the second excited state ($n = 3$)? The third excited state ($n = 4$)?

5. An electron is trapped in an infinitely deep one-dimensional well of width 0.251 nm. Initially the electron occupies the $n = 4$ state. (a) Suppose the electron jumps to the ground state with the accompanying emission of a photon. What is the energy of the photon? (b) Find the energies of other photons that might be emitted if the electron takes other paths between the $n = 4$ state and the ground state.

6. Show that Equation 5.26 gives the value $A = \sqrt{2/L}$.

7. A particle is trapped in an infinite one-dimensional well of width L. If the particle is in its ground state, evaluate the probability to find the particle (a) between $x = 0$ and $x = L/3$; (b) between $x = L/3$ and $x = 2L/3$; (c) between $x = 2L/3$ and $x = L$.

8. A particle is confined between rigid walls separated by a distance $L = 0.189$ nm. The particle is in the second excited state ($n = 3$). Evaluate the probability to find the particle in an interval of width 1.00 pm located at: (a) $x = 0.188$ nm; (b) $x = 0.031$ nm; (c) $x = 0.079$ nm. (*Hint*: No integrations are required for this problem; use Equation 5.4 directly.) What would be the corresponding results for a classical particle?

9. An electron is trapped in a one-dimensional well of width 0.132 nm. The electron is in the $n = 10$ state. (a) What is the energy of the electron? (b) What is the uncertainty in its momentum? (*Hint*: Use Equation 4.10.) (c) What is the uncertainty in its position? How do these results change as $n \rightarrow \infty$? Is this consistent with classical behavior?

10. What is the minimum energy of an electron trapped in a one-dimensional region the size of an atomic nucleus (1.0×10^{-14} m)? Compare this result with the equivalent value found in Example 4.7.

11. What is the minimum energy of a proton or neutron ($mc^2 \cong 940$ MeV) confined to a region of space of nuclear dimensions (1.0×10^{-14} m)?

12. Show that the average value of x^2 in the one-dimensional well is $(x^2)_{av} = L^2(\frac{1}{3} - 1/2n^2\pi^2)$.

13. Use the result of Problem 12 to show that, for the infinite one-dimensional well, defining $\Delta x = \sqrt{(x^2)_{av} - (x_{av})^2}$ gives

$$\Delta x = L \sqrt{\frac{1}{12} - \frac{1}{2\pi^2 n^2}}$$

14. (a) In the infinite one-dimensional well, what is p_{av}? (Use a symmetry argument.) (b) What is $(p^2)_{av}$? [*Hint*: What is $(p^2/2m)_{av}$?] (c) What is $\Delta p = \sqrt{(p^2)_{av} - (p_{av})^2}$? (d) Use the result of the previous problem to evaluate the smallest possible value of $\Delta x \, \Delta p$. How does this compare with the Heisenberg uncertainty relationship?

15. The problem of the particle in the *finite* potential energy well is similar to that of the infinite well, but the potential energy has the value U_0 for $x < 0$ and $x > L$. (a) Including six undetermined constants, write down the wave functions for the three regions $x < 0$, $0 < x < L$, and $x > L$, if $E < U_0$. (b) Two of the coefficients must be set to zero. Which ones? Why? (c) Keeping in mind the boundary conditions at $x = 0$ and $x = L$, sketch the three lowest-energy wave functions and probability densities. Do not attempt to apply the boundary conditions explicitly; just show the expected form of the wave function. Let the solutions for the infinite potential well guide you in deciding on the form of the wave functions inside the well.

16. What is the next level (above $E = 50E_0$) of the two-dimensional particle in a box in which the degeneracy is greater than 2?

17. A particle is confined to a two-dimensional box of length L and width $2L$. The energy values are $E = (\hbar^2 \pi^2 / 2mL^2)(n_x^2 + n_y^2/4)$. Find the two lowest degenerate levels.

18. Show by direct substitution that Equation 5.31 gives a solution to the two-dimensional Schrödinger equation, Equation 5.28. Find the relationship between k_x, k_y, and E.

19. A particle is confined to a three-dimensional region of space of dimensions L by L by L. The energy levels are $(\hbar^2 \pi^2 / 2mL^2)(n_x^2 + n_y^2 + n_z^2)$, where n_x, n_y, and n_z are integers ≥ 1. Sketch an energy level diagram, showing the energies, quantum numbers, and degeneracies for the lowest 10 energy levels.

20. Using the normalization condition, show that the constant A has the value $(m\omega_0/\hbar\pi)^{1/4}$ for the one-dimensional simple harmonic oscillator in its ground state.

21. (a) At the classical turning points $\pm A_0$ of the simple harmonic oscillator, $K = 0$ and so $E = U$. From this relationship, show that $A_0 = (\hbar\omega_0/k)^{1/2}$ for an oscillator in its ground state. (b) Find the turning points in the first and second excited states.

22. Use the ground-state wave function of the simple harmonic oscillator to find x_{av}, $(x^2)_{av}$, and Δx. Use the normalization constant $A = (m\omega_0/\hbar\pi)^{1/4}$.

23. (a) What value do you expect for p_{av} for the simple harmonic oscillator? Use a symmetry argument rather than a calculation. (b) Conservation of energy for the harmonic oscillator can be used to relate p^2 to x^2. Use this relation, along with the value of $(x^2)_{av}$ from Problem 22, to find $(p^2)_{av}$. (c) Evaluate Δp, using the results of (a) and (b).

24. From the results of Problems 22 and 23, evaluate $\Delta x \, \Delta p$ for the harmonic oscillator. Is the result consistent with the uncertainty relationship?

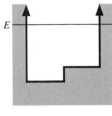

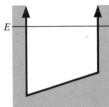

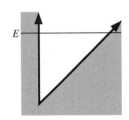

FIGURE 5.24 Problem 31.

25. The first excited state of the harmonic oscillator has a wave function of the form $\psi(x) = Axe^{-ax^2}$. Follow the method outlined in Section 5.5 to find a and the energy E. Find the constant A from the normalization condition.

26. Using the normalization constant A from Problem 20 and the value of a from Equation 5.42, evaluate the probability to find an oscillator in the ground state beyond the classical turning points $\pm A_0$. This problem cannot be solved in closed, analytic form. Develop an approximate, numerical method using a graph, calculator, or computer. Assume an electron bound to an atomic-sized region ($A_0 = 0.1$ nm) with an effective force constant of 1.0 eV/nm^2.

27. The ground state energy of an oscillating electron is 1.24 eV. How much energy must be added to the electron to move it to the second excited state? The fourth excited state?

28. Compare the probabilities for an oscillating particle in its ground state to be found in a small interval at the center of the well and at the classical turning points.

29. A two-dimensional harmonic oscillator has energy $E = \hbar\omega_0(n_x + n_y + 1)$, where n_x and n_y are integers beginning with zero. (a) Justify this result based on the energy of the one-dimensional oscillator. (b) Sketch an energy-level diagram similar to Figure 5.7, showing the values of E and the quantum numbers n_x and n_y. (c) Show that each level is degenerate, with degeneracy equal to $n_x + n_y + 1$.

30. Find the value of K at which Equation 5.59 has its maximum value, and show that Equation 5.60 is the maximum value of Δx.

31. Sketch a possible solution to the Schrödinger equation for each of the potential energies shown in Figure 5.24. In each case show several cycles of the wave function.

32. For a particle with energy $E < U_0$ incident on the potential energy step, use ψ_0 from Equation 5.52a and ψ_1 from Equation 5.56, and evaluate the constants B and D in terms of A by applying the boundary conditions at $x = 0$.

33. Using the wave functions of Equation 5.53 for the potential energy step, apply the boundary conditions of ψ and $d\psi/dx$ to find B' and C' in terms of A', for the potential step when particles are incident from the negative x direction. Evaluate the ratios $|B'|^2/|A'|^2$ and $|C'|^2/|A'|^2$ and interpret.

34. (a) Write down the wave functions for the three regions of the potential energy barrier (Figure 5.15) for $E < U_0$. You will need six coefficients in all. Use complex exponential notation. (b) Use the boundary conditions at $x = 0$ and at $x = a$ to find four relationships among the six coefficients. (Do not try to solve these relationships.) (c) Suppose particles are incident on the barrier from the left. Which coefficient should be set to zero? Why?

35. Repeat Problem 34 for the potential energy barrier when $E > U_0$, and sketch the solutions.

36. CUPS Exercise 7.1.

37. CUPS Exercise 7.2.

38. CUPS Exercise 7.9.

39. CUPS Exercise 7.11.

40. CUPS Exercise 7.12.

6

This model of the atom, based on the work of Rutherford and Bohr, shows the electrons circulating about the nucleus like planets circulating about the Sun. It can be a useful model for some purposes, but we shall see in Chapters 7 and 8 that it is not a correct picture of the structure of the atom.

THE RUTHERFORD-BOHR MODEL OF THE ATOM

Our goal in this chapter is to understand some of the details of atomic structure based on experimental studies of atoms. In particular, we discuss two types of experiments that are important in the development of our theory of atomic structure: the scattering of charged particles by atoms, which tells us about the distribution of electric charge in atoms, and the emission or absorption of radiation by atoms, which tells us about their excited states.

We use the information obtained from these experiments to develop an *atomic model,* which helps us understand and explain the properties of atoms. A model is usually an oversimplified picture of a more complex system, which provides some insight into its operation but may not be sufficiently detailed to explain *all* of its properties.

In this chapter, we discuss the experiments that led to the *Rutherford-Bohr model* (known simply the Bohr model), which is based on the familiar "planetary" structure in which the electrons orbit about the nucleus like planets about the Sun. Even though this model is not stricly valid from the standpoint of wave mechanics, it does help us understand many atomic properties, especially the excited states of the simplest atom, hydrogen. In Chapter 7, we show how wave mechanics changes our picture of the hydrogen atom, and in Chapter 8 we consider the structure of more complicated atoms.

6.1   BASIC PROPERTIES OF ATOMS

Before we begin to construct a model of the atom, let us summarize some of the basic properties of atoms.

1. *Atoms are very small,* about 0.1 nm (0.1×10^{-9} m) in radius. Thus any effort to "see" an atom using visible light ($\lambda \cong 500$ nm) is hopeless owing to diffraction effects. We can make a crude estimate of the *maximum* size of an atom in the following way. Consider a cube of elemental matter, for example iron. Iron has a density of about 8 g/cm^3 and a molar mass of 56 g. One mole of iron (56 g) contains Avogadro's number of atoms, about 6×10^{23}. Thus 6×10^{23} atoms occupy about 7 cm^3 and so 1 atom occupies about 10^{-23} cm^3. If we assume the atoms of a solid are packed together in the most efficient possible way, like hard spheres in contact, then the diameter of one atom is about $\sqrt[3]{10^{-23}\,\text{cm}^3} \cong 2 \times 10^{-8}$ cm = 0.2 nm.

2. *Atoms are stable*—they do not spontaneously break apart into smaller pieces or collapse; therefore the internal forces that hold the atom together must be in equilibrium. This immediately tells us that the forces that pull the parts of an atom together must be opposed in some way; otherwise atoms would collapse.

3. *Atoms contain negatively charged electrons, but are electrically neutral.* If we disturb an atom or collection of atoms with sufficient

force, electrons are emitted. We learn this fact from studying the Compton effect and the photoelectric effect. We also learned in Chapter 4 that even though electrons were emitted from the nuclei of atoms in certain radioactive decay processes, they don't "exist" in those nuclei but are manufactured there by some process. Electrons were excluded from the nucleus based on the uncertainty principle, which forbids emitted electrons of the energies observed in the laboratory from existing in the nucleus. The uncertainty principle places no such restriction on atoms (see Problem 1), so we assume the electron to be present in atoms.

We can also easily observe that bulk matter is electrically neutral, and we assume that this is likewise a property of the atoms. Experiments with beams of individual atoms support this assumption. From these experimental facts we deduce that an atom with Z negatively charged electrons must also contain a net positive charge of Ze.

4. *Atoms emit and absorb electromagnetic radiation.* This radiation may take many forms—visible light ($\lambda \sim 500$ nm), X rays ($\lambda \sim 1$ nm), ultraviolet rays ($\lambda \sim 10$ nm), infrared rays ($\lambda \sim 0.1$ μm), and so forth. In fact it is from observation of these emitted and absorbed radiations that we learn most of what we know about atoms. In a typical emission measurement, an electric current is passed through a tube containing a small sample of the gas phase of the element under study, and radiation is emitted when an excited atom returns to its ground state. The wavelengths of the many different emitted radiations can be measured with great precision, such as with a diffraction grating in the case of visible light. The absorption wavelengths can be measured by passing a beam of white light through a sample of the gas and noting which colors are removed from the white light by absorption in the gas. One particularly curious feature of the atomic radiations is that atoms don't always emit and absorb radiations at the same wavelengths—some wavelengths present in the *emission* experiment do not also appear in the *absorption* experiment. Any successful theory of atomic structure must be able to account for these emission and absorption wavelengths.

*6.2 THE THOMSON MODEL

An early model of the structure of the atom was proposed by J. J. Thomson, who was known for his previous identification of the electron and measurement of its charge-to-mass ratio e/m. The Thomson model incorporates many of the known properties of atoms: size, mass, number of electrons, and electrical neutrality. In this model, an atom contains Z electrons which are imbedded in a uniform sphere of positive charge (Figure 6.1). The total positive charge of the sphere is Ze, the mass of the sphere is essentially

* This is an optional section that may be skipped without loss of continuity.

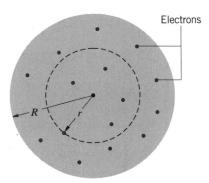

FIGURE 6.1 The Thomson model of the atom. Z tiny electrons are imbedded in a uniform sphere of positive charge Ze and radius R. An imaginary spherical surface of radius r contains a fraction r^3/R^3 of the charge.

the mass of the atom (the electrons don't contribute significantly to the total mass), and the radius R of the sphere is the radius of the atom. (This model is sometimes known as the "plum-pudding" model, since the electrons are distributed throughout the atom like raisins in a plum pudding.)

The force on an electron at a distance r from the center of a uniformly charged sphere of radius R can be computed using Gauss' law (see Problem 2):

$$F = \frac{Ze^2}{4\pi\varepsilon_0 R^3} r = kr \tag{6.1}$$

This linear restoring force permits the electrons to oscillate about their equilibrium positions just like a mass on a spring subject to the linear restoring force $F = kx$.

We therefore expect the electrons in the Thomson atom to oscillate about their equilibrium positions with a frequency $\nu = (2\pi)^{-1}\sqrt{k/m}$, where k is the constant defined by Equation 6.1. Since an oscillating electric charge radiates electromagnetic waves whose frequency is identical to the oscillation frequency, we might expect, based on the Thomson model, that the radiation emitted by atoms would show this characteristic frequency. This turns out *not* to be true (see Problem 3); the calculated frequencies do not correspond to the frequencies observed for radiation emitted by atoms.

The most serious failure of the Thomson model arises from the scattering of charged particles by atoms. Consider the passage of a single positively charged particle through the atom. The particle is deflected somewhat from its original trajectory owing to the electrical forces exerted on the particle by the atom. These forces are (1) a repulsive force due to the positive charge of the atom, and (2) an attractive force due to the negatively charged electrons. We assume that the mass of the deflected particle is both much greater than the mass of an electron and yet much less than the mass of the atom. In the encounter between the projectile and an electron, the forces exerted on each by the other are equal and opposite (by Newton's third law), and so the principal victim of the encounter is the much less massive electron; the effect on the projectile is negligible. (Imagine rolling a bowling ball through a field of Ping-Pong balls!) We thus need only consider the positively charged atom as a cause of the deflection of the particle. By the same argument, we neglect any possible motion of the more massive atom caused by the passage of the projectile. Our experiment, then, is the scattering of a positively charged projectile by the stationary positively charged massive part of the atom.

Figure 6.2 shows a representation of the deflection of the projectile, which is moving at speed v (we assume $v \ll c$ and use nonrelativistic mechanics, so $K = \frac{1}{2}mv^2$) along a line that would pass a distance b from the center of the atom if the projectile were not deflected. (The distance b is called the *impact parameter*.) The electrical repulsion causes a small deflection, and so the particle leaves the atom moving in a slightly different direction, at an angle θ with respect to the original direction.

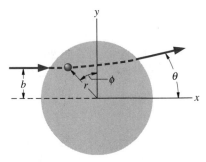

FIGURE 6.2 A positively charged alpha particle is deflected by an angle θ as it passes through a Thomson-model atom. The coordinates r and ϕ locate the alpha particle while it is inside the atom.

We can calculate the deflection angle θ by considering the *impulse* received by the projectile, which gives it some momentum in the y direction

$$\Delta p_y = \int F_y \, dt \qquad (6.2)$$

At an arbitrary point along the trajectory,

$$F_y = F \cos \phi$$

The projectile, which has a charge $q = ze$, experiences a force F given by analogy with Equation 6.1:

$$F = \frac{1}{4\pi\varepsilon_0} \frac{zZe^2}{R^3} r = zkr \qquad (6.3)$$

where k is the same constant defined by Equation 6.1. Since $\cos \phi \cong b/r$, we have

$$\Delta p_y \cong \int zkr \cdot \frac{b}{r} \cdot dt = zkb \int dt$$

$$= zkbT \qquad (6.4)$$

where T is the total time it takes for the projectile to travel through the atom, which is the total distance traveled in the atom divided by the average speed. Since the deflection is small, the path can be approximated as a straight line, as shown in Figure 6.3, and the average speed is very nearly equal to v, so

$$T \cong \frac{2\sqrt{R^2 - b^2}}{v} \qquad (6.5)$$

and Equation 6.4 then gives

$$\Delta p_y \cong \frac{2zkb}{v} \sqrt{R^2 - b^2} \qquad (6.6)$$

Assuming that the change in p_x is negligible, we have

$$\tan \theta = \frac{p_y}{p_x} \cong \frac{\Delta p_y}{p} \qquad (6.7)$$

and if θ is small, $\tan \theta \cong \theta$, so

$$\theta \cong \frac{\Delta p_y}{p} = \frac{2zkb}{mv^2} \sqrt{R^2 - b^2} \qquad (6.8)$$

Scattering angle for Thomson atom

The value of the scattering angle θ depends on the value of the impact parameter b. When $b = 0$, the projectile is undeflected ($\theta = 0$) because the net force on the projectile is zero. When $b = R$, the projectile is again undeflected, because we assume the force acts only when the projectile is within the atom. We can find a representative average value of θ corresponding to an average value of b, for which we choose the intermediate value $b = R/2$. In this case, Equation 6.8 gives

$$\theta_{\text{avg}} = \sqrt{\frac{3}{4}} \frac{zkR^2}{mv^2} \qquad (6.9)$$

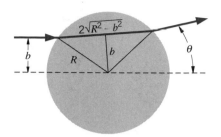

FIGURE 6.3 An approximate geometry of the deflection of an alpha particle by a Thomson atom. The scattering angle, whose maximum value is about 0.01°, has been greatly exaggerated.

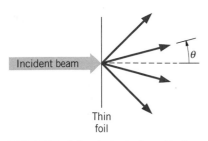

FIGURE 6.4 A typical scattering experiment. An incident beam is scattered by a thin foil; scattered particles are observed at all possible values of θ in the laboratory.

Equation 6.9 gives an estimate of the average scattering angle for a Thomson atom.

EXAMPLE 6.1

Using the Thomson model with $R = 0.1$ nm (a typical atomic radius), calculate the average deflection angle per collision when 5 MeV alpha particles ($z = 2$) are scattered from gold ($Z = 79$).

SOLUTION

The quantity zkR^2 can be computed to be

$$zkR^2 = 2\left(\frac{Ze^2}{4\pi\varepsilon_0 R^3}\right)R^2 = (2)(79)\frac{1.44\,\text{eV}\cdot\text{nm}}{0.1\,\text{nm}} \cong 2.3\,\text{keV}$$

(Recall that $e^2/4\pi\varepsilon_0 = 1.44$ eV·nm.) Also, $mv^2 = 2K = 10$ MeV, so

$$\theta_{\text{avg}} \cong \sqrt{\frac{3}{4}\frac{2.3\,\text{keV}}{10\,\text{MeV}}} \cong 2 \times 10^{-4}\,\text{rad} = 0.01°$$

Figure 6.4 shows the geometry for a typical scattering experiment. A beam of particles is incident on a thin foil. Instead of the scattering of a single projectile by a single atom, we observe the scattering of many projectiles by many atoms and we have no control over the impact parameter b for each scattering. Figure 6.5 shows an enlarged cross-section of the scattering foil. In passing through the foil, a projectile is scattered many times. For a typical foil thickness of 1 μm (10^{-6} m), the projectile is scattered by about 10^4 atoms, each of which deflects the projectile by an angle whose average value is θ_{avg}. The total scattering angle for any particular projectile is determined by statistical considerations, since some of the individual scatterings move the projectile toward larger scattering angles and some toward smaller angles, as shown in Figure 6.5. This is an example of a "random walk" problem—for N scatterings, the average net scattering angle θ is related to the average individual scattering angle by

$$\theta \simeq \sqrt{N}\,\theta_{\text{avg}} \tag{6.10}$$

For $N = 10^4$ and $\theta_{\text{avg}} = 0.01°$, we expect to see an average net scattering angle of about 1°, which is consistent with observations.

The Thomson model for scattering fails when we examine the probability for scattering at large angles. If each individual scattering deflects the projectile through an angle of around 0.01°, then to observe projectiles scattered through a total angle greater than 90°, we must have about 10^4 successive scatterings, *all* of which push the projectile toward larger angles. Since the probabilities of individual scatterings toward either larger or smaller angles are equal, the probability of having 10^4 successive scatterings toward larger angles, like the probability of finding 10^4 successive heads in tossing a coin, is about $(1/2)^{10,000} = 10^{-3000}$.

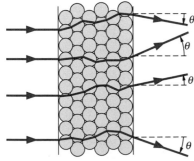

FIGURE 6.5 A microscopic representation of the scattering. Some individual scatterings tend to increase θ, while others tend to decrease θ.

An experiment of this sort was performed by Hans Geiger and Ernest Marsden in the laboratory of Ernest Rutherford in 1910. Their results showed that the probability of an alpha particle scattering at angles greater than 90° was about 10^{-4}. This remarkable discrepancy between the expected value (10^{-3000}) and the observed value (10^{-4}) was described by Rutherford in this way:

> *It was quite the most incredible event that ever happened to me in my life. It was as incredible as if you fired a 15-inch shell at a piece of tissue paper and it came back and hit you.*

The analysis of the results of such scattering experiments led Rutherford to propose that the mass and positive charge of the atom were not distributed uniformly over the volume of the atom, but instead were concentrated in an extremely small region, about 10^{-14} m in diameter, at the center of the atom. In the next section we will see how this proposal is consistent with the large-angle scattering results.

6.3 THE RUTHERFORD NUCLEAR ATOM

In analyzing the scattering of alpha particles, Rutherford concluded that the most likely way an alpha particle ($m = 4$ u) can be deflected through large angles is by a *single* collision with a more massive object. Rutherford therefore proposed that the charge and mass of the atom were concentrated at its center, in a region called the *nucleus*. Figure 6.6 illustrates the scattering geometry in this case. The projectile, of charge ze, experiences a repulsive force due to the positively charged nucleus:

$$F = \frac{(ze)(Ze)}{4\pi\varepsilon_0 r^2} \tag{6.11}$$

(We assume that the projectile is always outside the nucleus, so it feels the full nuclear charge Ze.) The atomic electrons, with their small mass, do not appreciably affect the path of the projectile and we neglect their effect on the scattering. We also assume that the nucleus is so much more massive than the projectile that it does not move during the scattering process; since no recoil motion is given to the nucleus, the initial and final kinetic energies K of the projectile are equal.

As Figure 6.6 shows, for each impact parameter b, there is a certain scattering angle θ, and we need the relationship between b and θ. The projectile can be shown[*] to follow a hyperbolic path; in polar coordinates

Ernest Rutherford (1871–1937, England). Founder of nuclear physics, he is known for his pioneering work on alpha-particle scattering and radioactive decays. His inspiring leadership influenced a generation of British nuclear and atomic scientists.

---

[*] See, for example, R. M. Eisberg and R. Resnick, *Quantum Physics of Atoms, Molecules, Solids, Nuclei, and Particles,* 2nd ed. (New York, Wiley, 1985).

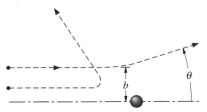

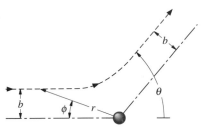

FIGURE 6.6   Scattering by a nuclear atom. The path of the scattered particle is a hyperbola. Smaller impact parameters give larger scattering angles.

FIGURE 6.7   The hyperbolic trajectory of a scattered particle.

r and ϕ, the equation of the hyperbola is

$$\frac{1}{r} = \frac{1}{b}\sin\phi + \frac{zZe^2}{8\pi\varepsilon_0 b^2 K}(\cos\phi - 1) \tag{6.12}$$

As shown in Figure 6.7, the initial position of the particle is $\phi = 0$, $r \to \infty$, and the final position is $\phi = \pi - \theta$, $r \to \infty$. Using the coordinates at the final position, Equation 6.12 reduces to

Rutherford scattering angle

$$b = \frac{zZe^2}{8\pi\varepsilon_0 K}\cot\tfrac{1}{2}\theta = \frac{zZ}{2K}\frac{e^2}{4\pi\varepsilon_0}\cot\tfrac{1}{2}\theta \tag{6.13}$$

(This result is written in this form so that $e^2/4\pi\varepsilon_0 = 1.44$ eV·nm can be easily inserted.) A projectile that approaches the nucleus with impact parameter b will be scattered at an angle θ; projectiles approaching with smaller values of b will be scattered through larger angles, as shown in Figure 6.6.

We divide our study of the scattering of charged projectiles by nuclei (which is commonly called *Rutherford scattering*) into three parts: (1) calculation of the fraction of projectiles scattered at angles greater than some value of θ, (2) the Rutherford scattering formula and its experimental verification, and (3) the closest approach of a projectile to the nucleus.

1. **The Fraction of Projectiles Scattered at Angles Greater than θ.**   From Figure 6.6 we see immediately that every projectile with impact parameters less than a given value of b will be scattered at angles greater than its corresponding θ. What is the chance of a projectile having an impact parameter less than a given value of b? Suppose the foil were one atom thick—a single layer of atoms packed tightly together, as in Figure 6.8. Each atom is represented by a circular disc, of area πR^2. If the foil contains N atoms, its total area is $N\pi R^2$. For scattering at angles greater than θ, the impact parameter must fall between zero and b—that is, the projectile must approach the atom within a circular disc of area πb^2. If the projectiles are spread uniformly over the area of the disc, then the fraction of projectiles that fall within that area is just $\pi b^2/\pi R^2$.

A real scattering foil may be thousands or tens of thousands of atoms thick. Let t be the thickness of the foil and A its area, and let ρ and M be the density and molar mass of the material of which the foil is made. The volume of the foil is then At, its mass is ρAt, the number of moles is

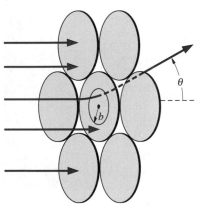

FIGURE 6.8   Scattering geometry for many atoms. For impact parameter b, the scattering angle is θ. If the particle enters the atom within the disc of area πb^2, its scattering angle will be larger than θ.

$\rho At/M$, and the number of atoms or nuclei per unit volume is

$$n = N_A \frac{\rho At}{M} \frac{1}{At} = \frac{N_A \rho}{M} \tag{6.14}$$

where N_A is Avogadro's number (the number of atoms per mole). As seen by an incident projectile, the number of nuclei per unit area is $nt = N_A \rho t/M$; that is, on the average, each nucleus contributes an area $(N_A \rho t/M)^{-1}$ to the field of view of the projectile. For scattering at angles greater than θ, it must once again be true that the projectile must fall within an area πb^2 of the center of an atom; the fraction scattered at angles greater than θ is just the fraction that approaches an atom within the area πb^2:

$$f_{<b} = f_{>\theta} = nt\pi b^2 \tag{6.15}$$

Fraction scattered at angles greater than **θ**

assuming that the incident particles are spread uniformly over the area of the foil.

EXAMPLE 6.2

A gold foil ($\rho = 19.3$ g/cm³, $M = 197$ g/mole) has a thickness of 2.0×10^{-4} cm. It is used to scatter alpha particles of kinetic energy 8.0 MeV. (*a*) What fraction of the alpha particles is scattered at angles greater than 90°? (*b*) What fraction of the alpha particles is scattered at angles between 90° and 45°?

SOLUTION

(*a*) For this case the number of nuclei per unit volume can be computed as

$$n = \frac{N_A \rho}{M} = \frac{(6.02 \times 10^{23} \text{ atoms/mole})(19.3 \text{ g/cm}^3)}{197 \text{ g/mole}}$$

$$= 5.9 \times 10^{22} \text{ atoms/cm}^3$$

$$= 5.9 \times 10^{28} \text{ atoms/m}^3$$

For scattering at 90°, the impact parameter b can be found from Equation 6.13:

$$b = \frac{(2)(79)}{2(8.0 \times 10^6 \text{ eV})} (1.44 \text{ eV} \cdot \text{nm}) \cot 45° = 1.4 \times 10^{-14} \text{ m}$$

so $\pi b^2 = 6.4 \times 10^{-28}$ m²/nucleus and we then have

$$f_{>90°} = (5.9 \times 10^{28} \text{ nuclei/m}^3)(2.0 \times 10^{-6} \text{ m})(6.4 \times 10^{-28} \text{ m}^2/\text{nucleus})$$

$$= 7.5 \times 10^{-5}$$

(*b*) Repeating the calculation for $\theta = 45°$, we find

$$b = \frac{(2)(79)}{2(8.0 \times 10^6 \text{ eV})} (1.44 \text{ eV} \cdot \text{nm}) \cot 22.5° = 3.4 \times 10^{-14} \text{ m}$$

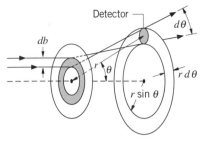

FIGURE 6.9 Particles entering the ring between b and $b + db$ are distributed uniformly along a ring of angular width $d\theta$. A detector is at a distance r from the scattering foil.

and $f_{>45°} = 4.4 \times 10^{-4}$. If a total fraction of 4.4×10^{-4} is scattered at angles greater than 45°, and of that, 7.5×10^{-5} is scattered at angles greater than 90°, the fraction scattered *between* 45° and 90° must be

$$4.4 \times 10^{-4} - 7.5 \times 10^{-5} = 3.6 \times 10^{-4}$$

---

2. **The Rutherford Scattering Formula and Its Experimental Verification.**   In order to find the probability that a projectile be scattered into a small angular range at θ (between θ and $\theta + d\theta$), we require that the impact parameter lie within a small range of values db at b (see Figure 6.9). The fraction, df, is then

$$df = nt(2\pi b \, db)$$

from Equation 6.15. Differentiating Equation 6.13 we find db in terms of $d\theta$:

$$db = \frac{zZ}{2K} \frac{e^2}{4\pi\varepsilon_0} (-\csc^2 \tfrac{1}{2}\theta)(\tfrac{1}{2} \, d\theta) \tag{6.16}$$

and so

$$|df| = \pi nt \left(\frac{zZ}{2K}\right)^2 \left(\frac{e^2}{4\pi\varepsilon_0}\right)^2 \csc^2 \tfrac{1}{2}\theta \cot \tfrac{1}{2}\theta \, d\theta \tag{6.17}$$

(This minus sign in Equation 6.16 is not important—it just tells us that θ increases as b decreases.) Suppose we place a detector for the scattered projectiles at the angle θ a distance r from the nucleus. The probability for a projectile to be scattered into the detector depends on df, which gives the probability for scattered particles to pass through the ring of radius $r \sin \theta$ and width $r \, d\theta$. The area of the ring is $dA = (2\pi r \sin \theta)r \, d\theta$. In order to calculate the rate at which projectiles are scattered *into the detector* we must know the probability *per unit area* for scattering into the ring. This is $|df|/dA$, which we call $N(\theta)$, and, after some manipulation, we find:

Rutherford scattering formula

$$N(\theta) = \frac{nt}{4r^2} \left(\frac{zZ}{2K}\right)^2 \left(\frac{e^2}{4\pi\varepsilon_0}\right)^2 \frac{1}{\sin^4 \tfrac{1}{2}\theta} \tag{6.18}$$

This is the *Rutherford scattering formula*.

In Rutherford's laboratory, Geiger and Marsden tested the predictions of this formula in a remarkable series of experiments requiring great care and experimental skill. They used *alpha particles* ($z = 2$) to scatter from nuclei in a variety of thin metal foils. In those days before electronic recording and processing equipment was available, Geiger and Marsden observed and recorded the alpha particles by counting the scintillations (flashes of light) produced when the alpha particles struck a zinc sulfide screen. A schematic view of their apparatus is shown in Figure 6.10. In all, four predictions of the Rutherford scattering formula were tested:

(*a*) $N(\theta) \propto t$. With a source of 8-MeV alpha particles from radioactive decay, Geiger and Marsden used scattering foils of varying thicknesses t while keeping the scattering angle θ fixed at about 25°. Their results are summarized in Figure 6.11, and the linear dependence of $N(\theta)$ on t is apparent. This is also evidence that, even at this moderate scattering angle, *single* scattering is much more important than *multiple* scattering. (In a random statistical theory of multiple scattering, the probability for scattering at a large angle would be proportional to the square root of the number of single scatterings, and we would expect $N(\theta) \propto t^{1/2}$. Figure 6.11 shows clearly that this is not true.)

This result emphasizes a significant difference between scattering by a Thomson model atom and a Rutherford nuclear atom: In the Thomson model, the projectile is scattered by *every* atom along its path as it passes through the foil (see Figure 6.5), while in the Rutherford nuclear model the nucleus is so tiny that the chance of even a single significant encounter is small and the chance of encountering more than one nucleus is negligible.

(*b*) $N(\theta) \propto Z^2$. In this experiment, Geiger and Marsden used a variety of different scattering materials, of approximately (*but not exactly*) the same thickness. This proportionality is therefore much more difficult to test than the previous one, since it involves the comparison of *different thicknesses* of *different materials*. However, as shown in Figure 6.12, the results are consistent with the proportionality of $N(\theta)$ to Z^2.

(*c*) $N(\theta) \propto K^{-2}$. In order to test this prediction of the Rutherford scattering formula, Geiger and Marsden kept the thickness of the scattering foil constant and varied the speed of the alpha particles. They accomplished this by slowing down the alpha particles emitted from the radioactive source, using thin sheets of mica. From independent measurements they knew the effect of different thicknesses of mica on the velocity of the alpha particles. The results of the experiment are shown in Figure 6.13; once again we see excellent agreement with the expected relationship.

(*d*) $N(\theta) \propto \sin^{-4} \frac{1}{2}\theta$. This dependence of N on θ is perhaps the most important and distinctive feature of the Rutherford scattering formula. It also produces the largest variation in N over the range accessible by experiment. In the previous tests, N varies by perhaps an order of magnitude; in this case N varies by about *five* orders of magnitude from the smaller to the larger angles. Geiger and Marsden used a gold foil and varied θ from 5 to 150°, to obtain the relationship between N and θ plotted in Figure 6.14. The agreement with the Rutherford formula is again very good.

Thus all predictions of the Rutherford scattering formula were confirmed by experiment, and the "nuclear atom" was verified.

3. **The Closest Approach of a Projectile to the Nucleus.** A positively charged projectile slows down as it approaches a nucleus, exchanging part of its initial kinetic energy for the electrostatic potential energy due to the nuclear repulsion. The closer the projectile gets to the nucleus, the more potential energy it gains, since

$$U = \frac{1}{4\pi\varepsilon_0} \frac{zZe^2}{r}$$

The maximum potential energy, and thus the minimum kinetic energy, occurs at the minimum value of r. We assume that $U = 0$ when the projectile is far from the nucleus, where it has total energy $E = K = \frac{1}{2}mv^2$. As the projectile approaches the nucleus, K decreases and U increases, but

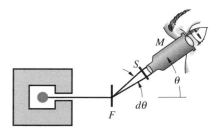

FIGURE 6.10 Schematic diagram of alpha-particle scattering experiment. A radioactive source of alpha particles is in a shield with a small hole. Alpha particles strike the foil F and are scattered into the angular range $d\theta$. Each time a scattered particle strikes the screen S a flash of light is emitted and observed with the movable microscope M.

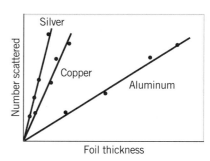

FIGURE 6.11 The dependence of scattering rate on foil thickness for three different scattering foils.

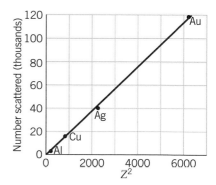

FIGURE 6.12 The dependence of scattering rate on the nuclear charge Z for foils of different materials. The data are plotted against Z^2.

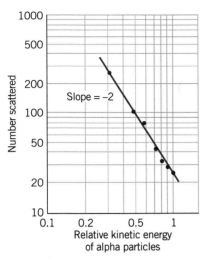

FIGURE 6.13 The dependence of scattering rate on the kinetic energy of the incident alpha particles for scattering by a single foil. Note the log-log scale; the slope of -2 shows that $\log N \propto -2 \log K$, or $N \propto K^{-2}$, as expected from the Rutherford formula.

Distance of closest approach

$U + K$ remains constant. At the distance $r_{\min}$, the speed is $v_{\min}$ and:

$$E = \frac{1}{2} m v_{\min}^2 + \frac{1}{4\pi\varepsilon_0} \frac{zZe^2}{r_{\min}} = \frac{1}{2} m v^2 \qquad (6.19)$$

(See Figure 6.15.)

Angular momentum is also conserved. Far from the nucleus, the angular momentum L is mvb, and at $r_{\min}$, the angular momentum is $m v_{\min} r_{\min}$, so:

$$mvb = m v_{\min} r_{\min}$$

or

$$v_{\min} = \frac{b}{r_{\min}} v \qquad (6.20)$$

Combining Equations 6.19 and 6.20, we find

$$\frac{1}{2} m v^2 = \frac{1}{2} m \left(\frac{b^2 v^2}{r_{\min}^2} \right) + \frac{1}{4\pi\varepsilon_0} \frac{zZe^2}{r_{\min}} \qquad (6.21)$$

This expression can be solved for the value of $r_{\min}$.

Notice that the kinetic energy of the projectile is not zero at $r_{\min}$, *unless* $b = 0$. (See Figure 6.15.) In this case, the projectile would lose all of its kinetic energy, and thus get closest to the nucleus. At this point its distance from the nucleus is d, *the distance of closest approach*. We find this distance by solving Equation 6.21 for $r_{\min}$ when $b = 0$, and obtain

$$d = \frac{1}{4\pi\varepsilon_0} \frac{zZe^2}{K} \qquad (6.22)$$

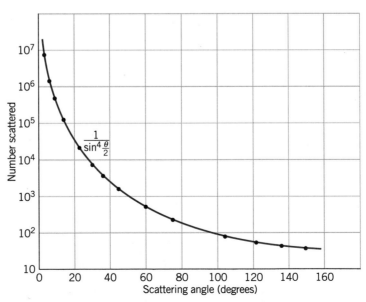

FIGURE 6.14 The dependence of scattering rate on the scattering angle θ, using a gold foil. The $\sin^{-4}(\theta/2)$ dependence is exactly as predicted by the Rutherford formula.

EXAMPLE 6.3

Find the distance of closest approach of an 8.0 MeV alpha particle incident on a gold foil.

SOLUTION

$$d = \frac{zZe^2}{4\pi\varepsilon_0}\frac{1}{K} = (2)(79)(1.44\text{ eV·nm})\frac{1}{8.0\times10^6\text{ eV}}$$

$$= 28\times10^{-6}\text{ nm}$$

$$= 2.8\times10^{-14}\text{ m}$$

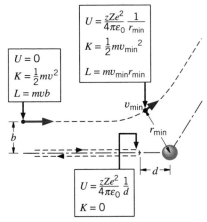

FIGURE 6.15 Closest approach of the projectile to the nucleus.

Although this distance is very small (much less than an atomic radius, for example) it is larger than the nuclear radius of gold (about 7×10^{-15} m). Thus the projectile is always *outside* of the nuclear charge distribution, and the Rutherford scattering law, which was derived assuming the projectile to remain outside the nucleus, correctly describes the scattering. If we increase the kinetic energy of the projectile, or decrease the electrostatic repulsion by using a target nucleus with low Z, this may not be the case. Under certain circumstances, the distance of closest approach can be less than the nuclear radius. When this happens, the projectile no longer feels the full nuclear charge, and the Rutherford scattering law no longer holds. In fact, as we discuss in Chapter 12, this gives us a convenient way of measuring the size of the nucleus.

6.4 LINE SPECTRA

The radiation from atoms can be classified into continuous spectra and discrete or line spectra. In a continuous spectrum, all wavelengths from some minimum, perhaps 0, to some maximum, perhaps approaching ∞, are emitted. The radiation from a hot glowing object is an example of this category. White light is a mixture of all of the different colors of visible light; an object that glows white hot is emitting light at all wavelengths of the visible spectrum. If, on the other hand, we force an electric discharge in a tube containing a small amount of the gas or vapor of a certain element, such as mercury, sodium, or neon, light is emitted at a few discrete wavelengths and not at any others. Examples of such "line" spectra are shown in Figure 6.16 and Color Plate 4. The strong 436 nm (blue) and 546 nm (green) lines in the mercury spectrum give mercury-vapor street lights their blue-green tint; the strong yellow line at 590 nm in the sodium spectrum (which is actually a *doublet*—two very closely spaced lines) gives sodium-vapor street lights a softer, yellowish color. The intense red lines of neon are responsible for the red color of "neon signs."

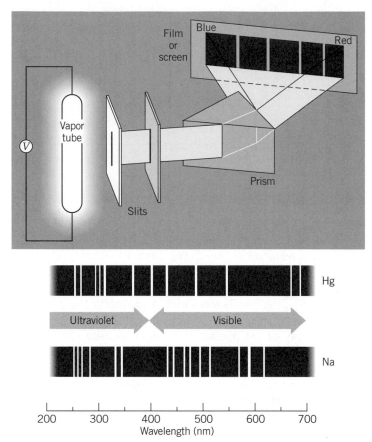

FIGURE 6.16 Apparatus for observing line spectra. Light is emitted when an electric discharge is created in a tube containing a vapor of an element. The light passes through a dispersive medium, such as a prism or a diffraction grating, which displays the individual component wavelengths at different positions. Sample line spectra are shown for mercury and sodium in the visible and near ultraviolet.

Another possible experiment is to pass a beam of white light, containing all wavelengths, through a sample of a gas. When we do so, we find that certain wavelengths have been absorbed from the light, and again a line spectrum results. These wavelengths correspond to many (*but not all*) of the wavelengths seen in the emission spectrum. Examples are shown in Figure 6.17.

In general, the interpretation of line spectra is very difficult in complex atoms, and so we will deal for now with the line spectra of the simplest atom, hydrogen. Regularities appear in both the emission and absorption spectra, as shown in Figure 6.18. Notice that, as with the mercury and sodium spectra, some lines present in the emission spectrum are missing from the absorption spectrum.

In our discussion of blackbody radiation, we found an example of the "reverse scientific method," in which, in the absence of a theory that explains the data, we try to find a function that fits the data, and then try

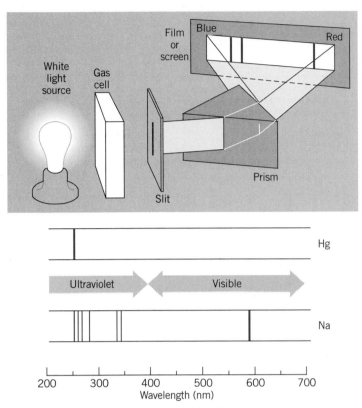

FIGURE 6.17 Apparatus for observing absorption spectra. A light source produces a continuous range of wavelengths, some of which are absorbed by a gaseous element. The light is dispersed, as in Figure 6.16. The result is a continuous "rainbow" spectrum, with dark lines at wavelengths where the light was absorbed by the gas.

to find a theory that explains the derived function. Johannes Balmer, a Swiss schoolteacher, noticed that the wavelengths of the group of emission lines of hydrogen in the visible region could be very accurately calculated from the formula

$$\lambda = 364.5 \frac{n^2}{n^2 - 4} \qquad (6.23) \qquad \textit{Balmer formula}$$

where λ is in units of nm and where n can take integer values beginning with 3. For example, for $n = 3$, $\lambda = 656.1$ nm. This formula is now known as the Balmer formula and the series of lines that it fits is called the Balmer series. The wavelength 364.5 nm, corresponding to $n \to \infty$, is called the *series limit*. It was soon discovered that all of the groupings of lines in the hydrogen spectrum could be fit with a similar formula of the form

$$\lambda = \lambda_{\text{limit}} \frac{n^2}{n^2 - n_0^2} \qquad (6.24)$$

where λ_{limit} is the wavelength of the appropriate series limit and where n takes integer values beginning with $n_0 + 1$. (For the Balmer series, $n_0 =$

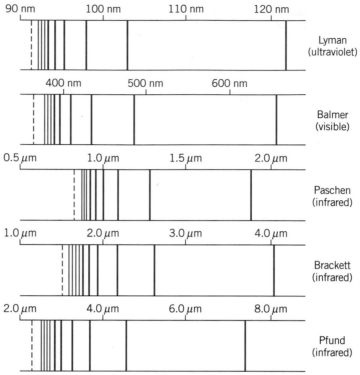

FIGURE 6.18   Emission and absorption spectral series of hydrogen. Note the regularities in the spacing of the spectral lines. The lines get closer together as the limit of each series (dashed line) is approached. Only the Lyman series appears in the absorption spectrum; all series are present in the emission spectrum.

2.) The other series are today known as Lyman ($n_0 = 1$), Paschen ($n_0 = 3$), Brackett ($n_0 = 4$), and Pfund ($n_0 = 5$).

Ritz combination principle

Another interesting property of the hydrogen wavelengths is summarized in the *Ritz combination principle*. If we convert the hydrogen emission wavelengths to frequencies, we find the curious property that certain pairs of frequencies added together give other frequencies which appear in the spectrum.

Any successful model of the hydrogen atom must be able to explain the occurrence of these interesting arithmetic regularities in the emission spectra.

EXAMPLE 6.4

The series limit of the Paschen series ($n_0 = 3$) is 820.1 nm. What are the three longest wavelengths of the Paschen series?

SOLUTION

From Equation 6.24,

$$\lambda = 820.1 \frac{n^2}{n^2 - 3^2} \qquad (n = 4, 5, 6, \ldots)$$

The three longest wavelengths are:

$$n = 4: \qquad \lambda = 820.1 \, \frac{4^2}{4^2 - 3^2} = 1875 \text{ nm}$$

$$n = 5: \qquad \lambda = 820.1 \, \frac{5^2}{5^2 - 3^2} = 1281 \text{ nm}$$

$$n = 6: \qquad \lambda = 820.1 \, \frac{6^2}{6^2 - 3^2} = 1094 \text{ nm}$$

These transitions are in the infrared region of the electromagnetic spectrum.

EXAMPLE 6.5

Show that the longest wavelength of the Balmer series and the longest *two* wavelengths of the Lyman series satisfy the Ritz combination principle. For the Lyman series, $\lambda_{\text{limit}} = 91.13$ nm.

SOLUTION

The longest wavelength of the Balmer series was found previously to be 656.1 nm, using Equation 6.23 with $n = 3$. Converting this to a frequency, we find $\nu = 4.57 \times 10^{14}$ Hz. Using Equation 6.24 for $n = 2$ and 3 with $n_0 = 1$, we find the longest two wavelengths of the Lyman series to be 121.5 nm and 102.5 nm, corresponding to frequencies of 24.67×10^{14} Hz and 29.24×10^{14} Hz. Adding the smallest frequency of the Lyman series to the smallest frequency of the Balmer series gives the next smallest Lyman frequency:

$$24.67 \times 10^{14} \text{ Hz} + 4.57 \times 10^{14} \text{ Hz} = 29.24 \times 10^{14} \text{ Hz}$$

demonstrating the Ritz combination principle.

Niels Bohr (1885–1962, Denmark). He developed a successful theory of the radiation spectrum of atomic hydrogen and also contributed the concepts of stationary states and complementarity to quantum mechanics. Later he developed a successful theory of nuclear fission. The institute of theoretical physics he founded in Copenhagen attracts scholars from throughout the world.

6.5 THE BOHR MODEL

Following Rutherford's proposal that the mass and positive charge are concentrated in a very small region at the center of the atom, the Danish physicist Niels Bohr in 1913 suggested that the atom was in fact like a miniature planetary system, with the electrons circulating about the nucleus like planets circulating about the Sun. The atom thus doesn't collapse under the influence of the electrostatic Coulomb attraction between nucleus and electrons for the same reason that the solar system doesn't collapse under the influence of the gravitational attraction between Sun and planets. In both cases, the attractive force provides the centripetal acceleration necessary to maintain the orbital motion.

We consider for simplicity the hydrogen atom, with a single electron circulating about a nucleus with a single positive charge, as in Figure 6.19.

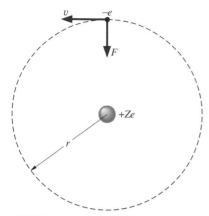

FIGURE 6.19 The Bohr model of the atom ($Z = 1$ for hydrogen).

The radius of the circular orbit is r, and the electron (of mass m) moves with constant tangential speed v. The attractive Coulomb force provides the centripetal acceleration v^2/r, so

$$F = \frac{1}{4\pi\varepsilon_0}\frac{q_1 q_2}{r^2} = \frac{1}{4\pi\varepsilon_0}\frac{e^2}{r^2} = \frac{mv^2}{r} \tag{6.25}$$

Manipulating this equation, we can find the kinetic energy of the electron (we are assuming the nucleus to remain at rest—more about this later)

$$K = \frac{1}{2}mv^2 = \frac{1}{8\pi\varepsilon_0}\frac{e^2}{r} \tag{6.26}$$

The potential energy of the electron-nucleus system is the Coulomb potential energy:

$$U = -\frac{1}{4\pi\varepsilon_0}\frac{e^2}{r} \tag{6.27}$$

The total energy $E = K + U$ is obtained by adding Equations 6.26 and 6.27:

$$E = -\frac{1}{8\pi\varepsilon_0}\frac{e^2}{r} \tag{6.28}$$

We have ignored one serious difficulty with this model thus far. Classical physics requires that an accelerated electric charge, such as our orbiting electron, must continuously radiate electromagnetic energy. As it radiates this energy, its total energy decreases, the electron spirals in toward the nucleus and the atom collapses. To overcome this difficulty, Bohr made a bold and daring hypothesis—he proposed that there are certain special states of motion, called *stationary states,* in which the electron may exist without radiating electromagnetic energy. In these states, according to Bohr, the angular momentum L of the electron takes values that are integer multiples of $\hbar$. In stationary states, the angular momentum of the electron may have magnitude $\hbar$, $2\hbar$, $3\hbar$, ..., but never such values as $2.5\hbar$ or $3.1\hbar$. This is called the *quantization of angular momentum.*

Quantization of angular momentum

In a circular orbit, the position vector **r** that locates the electron relative to the nucleus is always perpendicular to its linear momentum **p**. The angular momentum, which is defined as $\mathbf{L} = \mathbf{r} \times \mathbf{p}$, has magnitude $L = rp = mvr$ when **r** is perpendicular to **p**. Thus Bohr's postulate is

$$mvr = n\hbar \tag{6.29}$$

where n is an integer ($n = 1, 2, 3, \ldots$). We can use this expression with Equation 6.26 for the kinetic energy

$$\frac{1}{2}mv^2 = \frac{1}{2}m\left(\frac{n\hbar}{mr}\right)^2 = \frac{1}{8\pi\varepsilon_0}\frac{e^2}{r} \tag{6.30}$$

to find a series of allowed values of the radius r:

Hydrogen atom radii

$$r_n = \frac{4\pi\varepsilon_0\hbar^2}{me^2}n^2 = a_0 n^2 \tag{6.31}$$

where the *Bohr radius* a_0 is defined as

$$a_0 = \frac{4\pi\varepsilon_0\hbar^2}{me^2} = 0.0529 \text{ nm} \tag{6.32}$$

This important result is very different from what we expect from classical physics. A satellite may be placed into Earth orbit at any desired radius by boosting it to the appropriate altitude and then supplying the proper tangential speed. This is not true for an electron's orbit—only certain radii are allowed by the Bohr model. The radius of the electron's orbit may be a_0, $4a_0$, $9a_0$, $16a_0$, and so forth, but *never* $3a_0$ or $5.3a_0$.

Our expression for r may be combined with Equation 6.28 to give

$$E_n = -\frac{me^4}{32\pi^2\varepsilon_0^2\hbar^2}\frac{1}{n^2} \tag{6.33}$$

Hydrogen atom energy levels

We have added a subscript n to the energy to serve as an index to identify the energy levels. The constants may be evaluated, yielding

$$E_n = \frac{-13.6\,\text{eV}}{n^2} \tag{6.34}$$

The *energy levels* are indicated schematically in Figure 6.20. The electron's energy is *quantized*—only certain energy values are possible. In its lowest level, with $n = 1$, the electron has energy $E_1 = -13.6$ eV and orbits with a radius of $r_1 = 0.0529$ nm. This state is the *ground state*. The higher states ($n = 2$ with $E_2 = -3.4$ eV, $n = 3$ with $E_3 = -1.5$ eV, etc.) are the *excited states*.

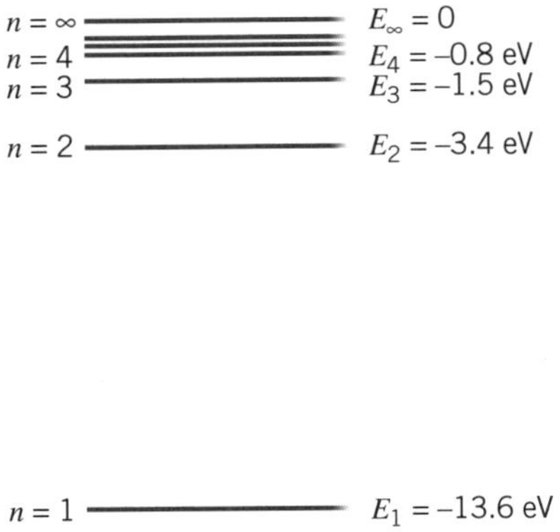

When the electron and nucleus are separated by an infinite distance, corresponding to $n = \infty$, we have $E = 0$. We might construct a hydrogen atom by beginning with the electron and nucleus separated by an infinite distance and then bring the electron closer to the nucleus until it is in orbit in a particular state n. Since this state has less energy than the $E = 0$ with which we began, we have released an amount of energy equal to $|E_n|$. Conversely, if we have an electron in a state n, we can "take the atom apart" by supplying an energy $|E_n|$. This energy is known as the *binding energy* of the state n. If we supply more energy than $|E_n|$ to the electron, the excess will appear as kinetic energy of the now free electron.

The *excitation energy* of an excited state n is the energy above the ground state, $E_n - E_1$. Thus the first excited state ($n = 2$) has excitation energy

$$\Delta E = E_2 - E_1 = -3.4\,\text{eV} - (-13.6\,\text{eV}) = 10.2\,\text{eV}$$

the second excited state has excitation energy 12.1 eV, and so forth.

We previously discussed the emission and absorption spectra of atomic hydrogen, and our discussion of the Bohr model is not complete without an understanding of the origin of these spectra. Bohr postulated that, even though the electron doesn't radiate when it remains in any particular stationary state, it can emit radiation when it moves to a lower energy level. In this lower level, the electron has less energy than in the original level, and the energy difference appears as a quantum of radiation whose energy $h\nu$ is equal to the energy difference between the levels. That is, if the electron jumps from $n = n_1$ to $n = n_2$, as in Figure 6.21, a photon appears with energy

$$h\nu = E_{n_1} - E_{n_2} \tag{6.35}$$

or, using Equation 6.33 for the energies,

$$\nu = \frac{me^4}{64\pi^3\varepsilon_0^2\hbar^3}\left(\frac{1}{n_2^2} - \frac{1}{n_1^2}\right) \tag{6.36}$$

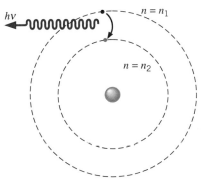

$n = \infty$ ———— $E_\infty = 0$
$n = 4$ ———— $E_4 = -0.8$ eV
$n = 3$ ———— $E_3 = -1.5$ eV

$n = 2$ ———— $E_2 = -3.4$ eV

$n = 1$ ———— $E_1 = -13.6$ eV

FIGURE 6.20 The energy levels of atomic hydrogen.

FIGURE 6.21 An electron jumps from the state n_1 to the state n_2 and emits a photon.

The wavelength of the emitted radiation is

$$\lambda = \frac{c}{\nu} = \frac{64\pi^2\varepsilon_0^2\hbar^3 c}{me^4}\left(\frac{n_1^2 n_2^2}{n_1^2 - n_2^2}\right) \tag{6.37}$$

The inverse of the numerical coefficient in Equation 6.37 is called the *Rydberg constant R_∞*

$$R_\infty = \frac{me^4}{64\pi^3\varepsilon_0^2\hbar^3 c} \tag{6.38}$$

The presently accepted numerical value is

$$R_\infty = 1.097 \times 10^7 \text{ m}^{-1}$$

EXAMPLE 6.6

Find the wavelengths of the transitions from $n_1 = 3$ to $n_2 = 2$ and from $n_1 = 4$ to $n_2 = 2$.

SOLUTION

Equation 6.37 gives

$$\lambda = \frac{1}{1.097 \times 10^7 \text{ m}^{-1}}\left(\frac{3^2 2^2}{3^2 - 2^2}\right) = 656.1 \text{ nm}$$

and

$$\lambda = \frac{1}{1.097 \times 10^7 \text{ m}^{-1}}\left(\frac{4^2 2^2}{4^2 - 2^2}\right) = 486.0 \text{ nm}$$

These wavelengths are remarkably close to the values of the two longest wavelengths of the Balmer series (Figure 6.18). In fact, Equation 6.37 gives

$$\lambda = (364.5 \text{ nm})\left(\frac{n_1^2}{n_1^2 - 4}\right)$$

for the wavelength of transitions from any state n_1 to $n_2 = 2$. This is identical with Equation 6.23 for the Balmer series. Thus we see that the radiations identified as the Balmer series correspond to transitions from higher levels to the $n = 2$ level. Similar identifications can be made for other series of radiations, as shown in Figure 6.22.

The Bohr formulas also explain the Ritz combination principle, according to which certain frequencies in the emission spectrum can be summed to give other frequencies. Let us consider a transition from a state n_3 to a state n_2, which is followed by a transition from n_2 to n_1. Equation 6.36 can be used for this case to give

$$\nu_{n_3 \to n_2} = cR_\infty\left(\frac{1}{n_3^2} - \frac{1}{n_2^2}\right)$$

$$\nu_{n_2 \to n_1} = cR_\infty\left(\frac{1}{n_2^2} - \frac{1}{n_1^2}\right)$$

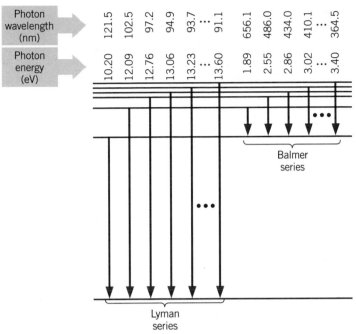

FIGURE 6.22 The transitions of the Lyman and Balmer series in hydrogen.

Thus

$$\nu_{n_3 \to n_2} + \nu_{n_2 \to n_1} = cR_\infty \left(\frac{1}{n_3^2} - \frac{1}{n_2^2} \right) + cR_\infty \left(\frac{1}{n_2^2} - \frac{1}{n_1^2} \right)$$

$$= cR_\infty \left(\frac{1}{n_3^2} - \frac{1}{n_1^2} \right)$$

which is equal to the frequency of the single photon emitted in a direct transition from n_3 to n_1, so

$$\nu_{n_3 \to n_2} + \nu_{n_2 \to n_1} = \nu_{n_3 \to n_1} \tag{6.39}$$

The Bohr model is thus entirely consistent with the Ritz combination principle. Since the frequency of an emitted photon is related to its energy by $E = h\nu$, the summing of frequencies is equivalent to the summing of energies. We may thus restate the Ritz combination principle in terms of energy: The energy of a photon emitted in a transition that skips or crosses over one or more states is equal to the step-by-step sum of the energies of the transitions connecting all of the individual states. (See Problem 24.)

The Bohr model also helps us understand why the atom doesn't absorb and emit radiation at all the same wavelengths. Isolated atoms are normally found only in the ground state; the excited states live for a very short time (less than 10^{-9} s) before decaying to the ground state. *The absorption spectrum therefore contains only transitions from the ground state.* From Figure 6.22, we see that only the radiations of the Lyman series can be found in the absorption spectrum of hydrogen. A hydrogen atom in its ground state can absorb radiation of 10.20 eV and reach the first excited state, or of 12.09 eV and reach the second excited state, and so forth. A hydrogen atom cannot absorb a photon of energy 1.89 eV (the first line of the Balmer series), because the atom is originally not in the $n = 2$ level. The Balmer series is therefore *not* found in the absorption spectrum.

The Bohr theory for hydrogen can be used for any atom with a single electron, even though the nuclear charge may be greater than 1. For example, we can calculate the energy levels of singly ionized helium (helium with one electron removed), doubly ionized lithium, etc. The nuclear electric charge enters the Bohr theory in only one place—in the expression for the electrostatic force between nucleus and electron, Equation 6.25. For a nucleus of charge Ze, the Coulomb force acting on the electron is

$$F = \frac{1}{4\pi\varepsilon_0} \frac{Ze^2}{r^2} \tag{6.40}$$

That is, where we had e^2 previously, we now have Ze^2. Making the same substitution in the final results, we can find the allowed radii:

Radii and energies of one-electron atoms

$$r_n = \frac{4\pi\varepsilon_0 \hbar^2}{Ze^2 m} n^2 = \frac{a_0 n^2}{Z} \tag{6.41}$$

and the energies become

$$E_n = -\frac{m(Ze^2)^2}{32\pi^2\varepsilon_0^2\hbar^2} \frac{1}{n^2} = -(13.6\,\text{eV})\frac{Z^2}{n^2} \tag{6.42}$$

The orbits in the higher-Z atoms are closer to the nucleus and have larger (negative) energies; that is, the electron is more tightly bound to the nucleus.

EXAMPLE 6.7

Calculate the two longest wavelengths of the Balmer series of triply ionized beryllium ($Z = 4$).

SOLUTION

Since the radiations of the Balmer series end with the $n = 2$ level, the two longest wavelengths are the radiations corresponding to $n = 3 \to n = 2$ and $n = 4 \to n = 2$. The energies of the radiations and their corresponding wavelengths are

$$E_3 - E_2 = -(13.6\,\text{eV})(4^2)\left(\frac{1}{9} - \frac{1}{4}\right) = 30.2\,\text{eV}$$

$$\lambda = \frac{hc}{E} = \frac{1240\,\text{eV}\cdot\text{nm}}{30.2\,\text{eV}} = 41.0\,\text{nm}$$

$$E_4 - E_2 = -(13.6 \text{ eV})(4^2)\left(\frac{1}{16} - \frac{1}{4}\right) = 40.8 \text{ eV}$$

$$\lambda = \frac{hc}{E} = \frac{1240 \text{ eV·nm}}{40.8 \text{ eV}} = 30.4 \text{ nm}$$

These radiations are in the ultraviolet region.

6.6 THE FRANCK-HERTZ EXPERIMENT

Let us imagine the following experiment, performed with the apparatus shown schematically in Figure 6.23. A filament heats the cathode, which then emits electrons. These electrons are accelerated toward the grid by the potential difference V, which we control. Electrons pass through the grid and reach the plate if V exceeds V_0, a small retarding voltage between the grid and the plate. The current of electrons reaching the plate is measured using the ammeter A.

Now suppose the tube is filled with atomic hydrogen gas at a low pressure. As the voltage is increased from zero, more and more electrons reach the plate, and the current rises accordingly. The electrons inside the tube may make collisions with atoms of hydrogen, *but lose no energy in these collisions*—the collisions are perfectly elastic. The only way the electron can give up energy in a collision is if the electron has enough energy to cause the hydrogen atom to make a transition to an excited state. Thus, when the energy of the electrons reaches and barely exceeds 10.2 eV (or when the voltage reaches 10.2 V), the electrons can make *inelastic* collisions, leaving 10.2 eV of energy with the atom (now in the $n = 2$ level), and the original electron moves off with very little energy. If it should pass through the grid, the electron might not have sufficient energy to overcome the

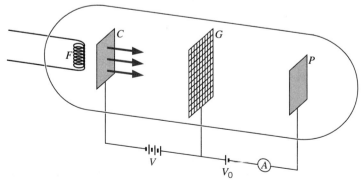

FIGURE 6.23 Franck-Hertz apparatus. Electrons leave the cathode C, are accelerated by the voltage V toward the grid G, and reach the plate P where they are recorded on the ammeter A.

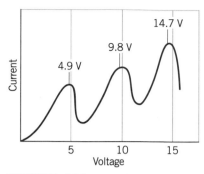

FIGURE 6.24 Result of Franck-Hertz experiment using mercury vapor. The current drops at voltages of 4.9 V, 9.8 V (= 2 × 4.9 V), 14.7 V (= 3 × 4.9 V).

small retarding potential and reach the plate. Thus when $V = 10.2$ V, a drop in the current is observed. As V is increased further, we begin to see the effects of multiple collisions. That is, when $V = 20.4$ V, an electron can make an inelastic collision, leaving the atom in the $n = 2$ state. The electron loses 10.2 eV of energy in this process, and so it moves off after the collision with a remaining 10.2 eV of energy, which is sufficient to excite a second hydrogen atom in an inelastic collision. Thus, if a drop in the current is observed at V, similar drops are observed at $2V$, $3V$, ...

This experiment should thus give rather direct evidence for the existence of atomic excited states. Unfortunately, it is not easy to do this experiment with hydrogen, because hydrogen occurs naturally in the molecular form H_2, rather than in atomic form. The molecules can absorb energy in a variety of ways, which would confuse the interpretation of the experiment. A similar experiment was done in 1914 by James Franck and Gustav Hertz, using a tube filled with mercury vapor. Their results are shown in Figure 6.24, which shows clear evidence for an excited state at 4.9 eV; whenever the voltage is a multiple of 4.9 V, a drop in the current appears. Coincidentally, the *emission* spectrum of mercury shows an intense ultraviolet line of wavelength 254 nm, which corresponds to an energy of 4.9 eV; this results from a transition between the same 4.9-eV excited state and the ground state. The Franck-Hertz experiment showed that an electron must have a certain minimum energy to make an inelastic collision with an atom; we now interpret that minimum energy as the energy of an excited state of the atom. Franck and Hertz were awarded the 1925 Nobel prize in physics for this work.

*6.7 THE CORRESPONDENCE PRINCIPLE

We have seen how Bohr's model permits calculations of transition wavelengths in atomic hydrogen that are in excellent agreement with the wavelengths observed in the emission and absorption spectra. However, in order to obtain this agreement, Bohr had to introduce two postulates that are radical departures from classical physics. In particular, an accelerated charged particle radiates electromagnetic energy according to classical physics, but an electron in Bohr's atomic model, accelerated as it moves in a circular orbit, does not radiate (unless it jumps to another orbit). Here we have a very different case than we did in our study of special relativity. You will recall, for example, that relativity gives us one expression for the kinetic energy, $K = E - E_0$, and classical physics gives us another, $K = \frac{1}{2}mv^2$; however, we showed that $E - E_0$ reduces to $\frac{1}{2}mv^2$ when $v \ll c$. Thus these two expressions are really not very different—one is merely a special case of the other. The dilemma associated with the accelerated electron is not simply a matter of atomic physics (as an example of quantum physics) being a special case of classical physics. Either the accelerated charge

* This is an optional section that may be skipped without loss of continuity.

radiates, or it doesn't! Bohr's solution to this serious dilemma was to propose the *correspondence principle*, which states that

Quantum theory must agree with classical theory in the limit in which classical theory is known to agree with experiment,

or equivalently

Quantum theory must agree with classical theory in the limit of large quantum numbers.

Let us see how we can apply this principle to the Bohr atom. According to classical physics, an electric charge moving in a circle radiates at a frequency equal to its frequency of rotation. For an atomic orbit, the period of revolution is the distance traveled in one orbit, $2\pi r$, divided by the orbital speed $v = \sqrt{2K/m}$, where K is the kinetic energy:

$$T = \frac{2\pi r}{\sqrt{2K/m}} = \frac{\sqrt{16\pi^3 \varepsilon_0 m r^3}}{e} \tag{6.43}$$

where we use Equation 6.26 for the kinetic energy. The frequency ν is the inverse of the period:

$$\nu = \frac{1}{T} = \frac{e}{\sqrt{16\pi^3 \varepsilon_0 m r^3}} \tag{6.44}$$

Using Equation 6.31 for the allowed orbits, we find

$$\nu_n = \frac{me^4}{32\pi^3 \varepsilon_0^2 \hbar^3} \frac{1}{n^3} \tag{6.45}$$

A "classical" electron moving in an orbit of radius r_n would radiate at this frequency ν_n.

If we made the radius of the Bohr atom so large that it went from a quantum-sized object (10^{-10} m) to a laboratory-sized object (10^{-3} m), the atom should behave classically. Since the radius increases with increasing n like n^2, this classical behavior should occur for n in the range 10^3–10^4. Let us then calculate the frequency of the radiation emitted by such an atom when the electron drops from the orbit n to the orbit $n - 1$. According to Equation 6.36, the frequency is

$$\nu = \frac{me^4}{64\pi^3 \varepsilon_0^2 \hbar^3} \left(\frac{1}{(n-1)^2} - \frac{1}{n^2} \right)$$

$$= \frac{me^4}{64\pi^3 \varepsilon_0^2 \hbar^3} \frac{2n-1}{n^2(n-1)^2} \tag{6.46}$$

If n is very large, then we can approximate $n - 1$ by n and $2n - 1$ by $2n$, which gives

$$\nu \cong \frac{me^4}{64\pi^3 \varepsilon_0^2 \hbar^3} \frac{2n}{n^4} = \frac{me^4}{32\pi^3 \varepsilon_0^2 \hbar^3} \frac{1}{n^3} \tag{6.47}$$

This is identical with Equation 6.45 for the "classical" frequency. The "classical" electron spirals slowly in toward the nucleus, radiating at the frequency given by Equation 6.45, while the "quantum" electron jumps from the orbit n to the orbit $n - 1$ and then to the orbit $n - 2$, and so forth, radiating at the frequency given by the identical Equation 6.47. (When the circular orbits are very large, this jumping from one circular orbit to the next smaller one looks very much like a spiral, as in Figure 6.25.)

In the region of large *n,* where classical and quantum physics overlap, the classical and quantum expressions for the radiation frequencies are identical. This is an example of an application of Bohr's correspondence principle. The applications of the correspondence principle go far beyond the Bohr atom, and this principle is important in understanding how we get from the domain in which the laws of classical physics are valid to the domain in which the laws of quantum physics are valid.

6.8   DEFICIENCIES OF THE BOHR MODEL

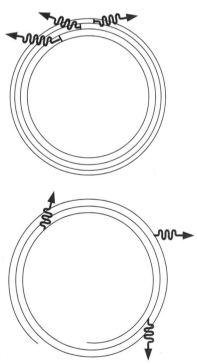

FIGURE 6.25 (Top) A large quantum atom. Photons are emitted in discrete transitions as the electron jumps to lower states. (Bottom) A classical atom. Photons are emitted continuously by the accelerated electron.

The Bohr model gives us a picture of how electrons move about the nucleus, and many of our attempts to explain the behavior of atoms refer to this picture, even though it is not strictly correct. It is remarkable (perhaps even fortunate or coincidental) that the successes of the model, with its new concepts of discrete energy levels and stationary states, occurred a full decade before de Broglie's work and the birth of wave mechanics.

In a way, the good agreement between our calculated values of the Balmer series wavelengths and the experimental values is fortuitous, because we made two errors. The first error is related to the finite mass of the nucleus and results from our neglect of the motion of the nucleus. The electron does not orbit about the nucleus, but instead the electron and the nucleus *both* orbit about their common center of mass. The kinetic energy should thus include an additional term due to the nuclear motion, and the effect of this inclusion is to reduce the Rydberg constant from the value R_∞ (so called because it would be correct if the nuclear mass were infinite) to the value $R = R_\infty/(1 + m/M)$, where M is the mass of the nucleus. This effect tends to *increase* the calculated wavelengths by a factor of about 1.00055. The second mistake we made was in converting the frequencies into wavelengths. The frequency of the emitted photon is directly related to the energy difference of the atomic levels; to find the wavelength, we use the expression $c = \lambda\nu$, which is correct *only in vacuum.* Since the experiment is usually performed in laboratories in air, the correct expression is $\lambda\nu = v_{air}$, where v_{air} is the speed of light in air. This effect tends to *decrease* the calculated wavelengths by a factor of about 1.00029. Thus our two mistakes offset one another to some extent.

In spite of its successes, the Bohr model is at best an incomplete model. It is useful only for atoms that contain one electron (hydrogen, singly ionized helium, doubly ionized lithium, etc.), but not for atoms with two or more electrons, since we have considered only the force between electron and nucleus, and not the force between the electrons themselves. Furthermore, if we look very carefully at the emission spectrum, we find that many lines are in fact not single lines, but very closely spaced combinations of two or more lines; the Bohr model is unable to account for these *doublets* of spectral lines. The model is also limited in its usefulness as a basis from which to calculate other properties of the atom; although we can accurately calculate the energies of the spectral lines, we cannot calculate their intensities. For example, how often will an electron in the $n = 3$ state jump directly to the $n = 1$ state, emitting the corresponding photon, and how often will it jump first to the $n = 2$ state and then to the $n = 1$ state, emitting two photons? A complete theory should provide a way to calculate this property.

A serious defect in the Bohr model is that it gives incorrect predictions for the angular momentum of the electron. For the ground state of hydrogen ($n = 1$), the Bohr theory gives $L = \hbar$, while experiment clearly shows $L = 0$.

A more serious deficiency of the model is that it completely violates the uncertainty relationship. (In Bohr's defense, remember that this was a decade before the introduction of wave mechanics, with its accompanying ideas of uncertainty.) The uncertainty relationship $\Delta x \, \Delta p_x \gtrsim \hbar$ is valid for any direction in space. If we choose the radial direction, then $\Delta r \, \Delta p_r \gtrsim \hbar$. For an electron moving in a circular orbit, we know the value of r exactly, and thus $\Delta r = 0$. If it is moving in a circle we also know p_r exactly (in fact it is exactly zero), and so $\Delta p_r = 0$. This simultaneous exact knowledge of r and p_r violates the uncertainty principle.

We do not wish, however, to discard the model completely. The Bohr model gives a mental picture of the structure of an atom that sometimes can be useful. There are many atomic properties, especially those associated with magnetism, which *can* be understood on the basis of Bohr orbits. When we treat the problem correctly in the next chapter, we find that the energy levels of hydrogen calculated by solving the Schrödinger equation are in fact identical with those of the Bohr model.

In this chapter we considered two problems of atomic physics—Rutherford scattering and the hydrogen emission spectrum. Both can be understood based on calculations done without reference to wave mechanics, even though we expect wave mechanics to be an important consideration on the atomic scale. In fact, these two examples are unique in that the "correct" wave-mechanical calculations of the Rutherford scattering formula and the hydrogen emission wavelengths give the same results as our classical calculation! This was not the case with many other phenomena we have studied, including blackbody radiation, Compton scattering, and the photoelectric effect. It is interesting to speculate on the development of physics in the era of Bohr and Rutherford if their classical calculations had not yielded correct results.

SUGGESTIONS FOR FURTHER READING

A discussion of many of the basic properties of atoms may be found in:

M. R. Wehr, J. A. Richards, and T. W. Adair, *Physics of the Atom* (Reading, Addison-Wesley, 1978).

For a historical perspective on the development of atomic theory, see:

H. A. Boorse and L. Motz, editors, *The World of the Atom* (New York, Basic Books, 1966).
G. K. T. Conn and H. D. Turner, *The Evolution of the Nuclear Atom* (London, Iliffe Books, 1965).
F. Friedman and L. Sartori, *The Classical Atom* (Reading, Addison-Wesley, 1965).

For more details on the history of the Thomson and Bohr models, see:

J. L. Heilbron, "J. J. Thomson and the Bohr Atom," *Physics Today,* April 1977, p. 23.
J. L. Heilbron, "Bohr's First Theories of the Atom," *Physics Today,* October 1985, p. 28.

For a popular summary of Rutherford's work, see:

E. N. da C. Andrade, "The Birth of the Nuclear Atom," *Scientific American* **195**, 93 (November 1956).

The early papers on the Rutherford model and its experimental confirmation illustrate the difficulty of the experiments and the care and abilities of the experimenters. They are easily readable and require no mathematics beyond the present level.

E. Rutherford, *Philosophical Magazine* **21**, 669 (1911).
H. Geiger, *Proceedings of the Royal Society of London* **A83**, 492 (1910).
H. Geiger and E. Marsden, *Philosophical Magazine* **25**, 604 (1913).

QUESTIONS

1. Does the Thomson model fail at large scattering angles or at small scattering angles? Why?

2. What principles of physics would be violated if we scattered a beam of alpha particles with a single impact parameter from a single target atom at rest?

3. Could we use the Rutherford scattering formula to analyze the scattering of:
   (a) Protons incident on iron?
   (b) Alpha particles incident on lithium ($Z = 3$)?
   (c) Silver nuclei incident on gold?
   (d) Hydrogen *atoms* incident on gold?
   (e) Electrons incident on gold?

4. What determines the angular range $d\theta$ in the alpha-particle scattering experiment (Figure 6.10)?

5. Why didn't Bohr use the concept of de Broglie waves in his theory?

6. In which Bohr orbit does the electron have the largest velocity? Are we justified in treating the electron nonrelativistically in that case?

7. How does an electron in hydrogen get from $r = 4a_0$ to $r = a_0$ without being anywhere in between?

8. How is the quantization of the energy in the hydrogen atom similar to the quantization of the systems discussed in Chapter 5? How is it different? Do the quantizations originate from similar causes?

9. In a Bohr atom, an electron jumps from state n_1, with angular momentum $n_1\hbar$, to state n_2, with angular momentum $n_2\hbar$. How can an isolated system change its angular momentum? (In classical physics, a change in angular momentum requires an external torque.) Can the photon carry away the difference in angular momentum? Estimate the maximum angular momentum, relative to the center of the atom, which the photon can have. Does this suggest another failure of the Bohr model?

10. The product $E_n r_n$ for the hydrogen atom is (1) independent of Planck's constant and (2) independent of the quantum number n. Does this observation have any significance? Is this a classical or a quantum effect?

11. (a) How does a Bohr atom violate the $\Delta x \, \Delta p$ uncertainty relationship? (b) How does a Bohr atom violate the $\Delta E \, \Delta t$ uncertainty relationship? (What is ΔE? What does this imply about Δt? What do you conclude about transitions between levels?)

12. List the assumptions made in deriving the Bohr theory. Which of these are a result of neglecting small quantities? Which of these violate basic principles of relativity or quantum physics?

13. List the assumptions made in deriving the Rutherford scattering formula. Which of these are a result of neglecting small quantities? Which of these violate basic principles of relativity or quantum physics?

14. In both the Rutherford theory and the Bohr theory, we used the classical expression for the kinetic energy. Estimate the velocity of an electron in the Bohr atom and of an alpha particle in a typical scattering experiment, and decide if the use of the classical formula is justified.

15. In both the Rutherford theory and the Bohr theory, we neglected any wave properties of the particles. Estimate the de Broglie wavelength of an electron in a Bohr atom and compare it with the size of the atom. Estimate the de Broglie wavelength of an alpha particle and compare it with the size of the nucleus. Is the wave behavior expected to be important in either case?

16. Why are the decreases in current in the Franck-Hertz experiment not sharp?

17. As indicated by the Franck-Hertz experiment, the first excited state of mercury is at an energy of 4.9 eV. Do you expect mercury to show absorption lines in the visible spectrum?

18. Is the correspondence principle a necessary part of quantum physics or is it merely an accidental agreement of two formulas? Where do we draw the line between the world of quantum physics and the world of classical, nonquantum physics?

PROBLEMS

1. Electrons in atoms are known to have energies in the range of a few eV. Show that the uncertainty principle allows electrons of this energy to be confined in a region the size of an atom (0.1 nm).

2. Consider an electron in Figure 6.1 subject to the electric field of the sphere of positive charge Ze in which it is embedded. (a) Using Gauss' Law, show that the electric field on the electron due to the positive charge is

$$E = \frac{1}{4\pi\varepsilon_0} \frac{Ze}{R^3} r$$

(b) For this electric field, show that the force on the electron is given by Equation 6.1.

3. (a) Compute the oscillation frequency of the electron and the expected absorption or emission wavelength in a Thomson-model hydrogen atom. Use $R = 0.053$ nm. Compare with the observed wavelength of the strongest emission and absorption line in hydrogen, 122 nm. (b) Repeat for sodium ($Z = 11$). Use $R = 0.18$ nm. Compare with the observed wavelength, 590 nm.

4. Use Equation 6.8 to find the maximum scattering angle for the Thomson atom and the impact parameter for which it occurs.

5. Consider the Thomson model for an atom with 2 electrons. Let the electrons be located along a diameter on opposite sides of the center of the sphere, each a distance x from the center. (a) Show that the configuration is stable if $x = R/2$. (b) Try to construct similar stable configurations for atoms with 3, 4, 5, and 6 electrons.

6. Alpha particles of kinetic energy 5.00 MeV are scattered at 90° by a gold foil. (a) What is the impact parameter? (b) What is the minimum distance between alpha particles and gold nucleus? (c) Find the kinetic and potential energies at that minimum distance.

7. How much kinetic energy must an alpha particle have before its distance of closest approach to a gold nucleus is equal to the nuclear radius (7.0×10^{-15} m)?

8. What is the distance of closest approach when alpha particles of kinetic energy 6.0 MeV are scattered by a thin copper foil?

9. Protons of energy 5.0 MeV are incident on a silver foil of thickness 4.0×10^{-6} m. What fraction of the incident protons is scattered at angles: (a) Greater than 90°? (b) Greater than 10°? (c) Between 5° and 10°? (d) Less than 5°?

10. Protons are incident on a copper foil 12 μm thick. (a) What should the proton kinetic energy be in order that the distance of closest approach equal the nuclear radius (5.0 fm)? (b) If the proton energy were 7.5 MeV, what is the impact parameter for scattering at 120°? (c) What is the minimum distance between proton and nucleus for this case? (d) What fraction of the protons is scattered beyond 120°?

11. Alpha particles of kinetic energy K are scattered either from a gold foil or a silver foil of identical thickness. What is the ratio of the number of particles scattered at angles greater than 90° by the gold foil to the same number for the silver foil?

12. The maximum kinetic energy given to the target nucleus will occur in a head-on collision with $b = 0$. (Why?) Evaluate the maximum kinetic energy given to the target nucleus when 8.0 MeV alpha particles are incident on a gold foil. Are we justified in neglecting this energy?

13. The maximum kinetic energy that an alpha particle can transmit to an *electron* occurs during a head-on collision. Compute the kinetic energy lost by an alpha

particle of kinetic energy 8.0 MeV in a head-on collision with an electron at rest. Are we justified in neglecting this energy in the Rutherford theory?

14. Alpha particles of energy 9.6 MeV are incident on a silver foil of thickness 7.0 μm. For a certain value of the impact parameter, the alpha particles lose exactly half their incident kinetic energy when they reach their minimum separation from the nucleus. Find the minimum separation, the impact parameter, and the scattering angle.

15. Alpha particles of kinetic energy 6.0 MeV are incident at a rate of 3.0×10^7 per second on a gold foil of thickness 3.0×10^{-6} m. A circular detector of diameter 1.0 cm is placed 12 cm from the foil at an angle of $30°$ with the direction of the incident alpha particles. At what rate does the detector measure scattered alpha particles?

16. In the $n = 3$ state of hydrogen, find the electron's velocity, kinetic energy, and potential energy.

17. Use the Bohr theory to find the series wavelength limits of the Lyman and Paschen series of hydrogen.

18. Show that the speed of an electron in the nth Bohr orbit of hydrogen is $\alpha c/n$, where α is the fine structure constant. What would be the speed in a hydrogenlike atom with a nuclear charge of Ze?

19. An electron is in the $n = 5$ state of hydrogen. To what states can the electron make transitions, and what are the energies of the emitted radiations?

20. A hydrogen atom is in the $n = 6$ state. (a) Counting all possible paths, how many different photon energies can be emitted if the atom ends up in the ground state? (b) Suppose only $\Delta n = 1$ transitions were allowed. How many different photon energies would be emitted? (c) How many different photon energies would occur in a Thomson-model hydrogen atom?

21. Continue Figure 6.22, showing the transitions of the Paschen series and computing their energies and wavelengths.

22. A collection of hydrogen atoms in the ground state is illuminated with ultraviolet light of wavelength 59.0 nm. Find the kinetic energy of the emitted electrons.

23. The *ionization energy* is the energy required to remove an electron from an atom. Find the ionization energy of: (a) The $n = 3$ level of hydrogen. (b) The $n = 2$ level of He$^+$ (singly ionized helium). (c) The $n = 4$ level of Li^{++} (doubly ionized lithium).

24. Use the Bohr formula to find the energy differences $E(n_1 \rightarrow n_2) = E_{n_1} - E_{n_2}$ and show that (a) $E(4 \rightarrow 2) = E(4 \rightarrow 3) + E(3 \rightarrow 2)$; (b) $E(4 \rightarrow 1) = E(4 \rightarrow 2) + E(2 \rightarrow 1)$. (c) Interpret these results based on the Ritz combination principle.

25. What is the difference in wavelength between the first line of the Balmer series in ordinary hydrogen ($M \cong 1.01$ u) and in "heavy" hydrogen ($M \cong 2.01$ u)?

26. Find the shortest and the longest wavelengths of the Lyman series of singly ionized helium.

27. Draw an energy-level diagram showing the lowest four levels of singly ionized helium. Show all possible transitions from the levels and label each transition with its wavelength.

28. An electron is in the $n = 8$ level of ionized helium. (a) Find the three longest wavelengths that are emitted when the electron makes a transition from the $n = 8$ level to a lower level. (b) Find the shortest wavelength that can be emitted. (c) Find the three longest wavelengths at which the electron in the $n = 8$ level will *absorb* a photon and move to a higher state, if we could somehow keep it in that level long enough to absorb. (d) Find the shortest wavelength that can be absorbed.

29. The lifetimes of the levels in a hydrogen atom are of the order of 10^{-8} s. Find the energy uncertainty of the first excited state and compare it with the energy of the state.

30. The *Handbook of Chemistry and Physics* lists the following emission wavelengths (in nm) for ionized helium:

23.73	30.38	121.5	251.1	468.6	1012.4
24.30	102.5	164.0	273.4	541.1	1162.6
25.63	108.5	238.5	320.3	656.0	1863.7

Using the same values of n_0 as the hydrogen spectrum, group these spectral lines into series, showing the index n for each line identified. Give the series limit of each series. In what region of the electromagnetic spectrum is each series located?

31. The *Handbook of Chemistry and Physics* lists the following emission wavelengths (in nm) for doubly ionized lithium: 11.39, 13.50, 54.00, 72.91. Identify these lines with the proper spectral series (as in hydrogen), giving the index n for each line and the series limit for each series.

32. When an atom emits a photon in a transition from a state of energy E_1 to a state of energy E_2, the photon energy is not precisely equal to $E_1 - E_2$. Conservation of momentum requires that the atom must recoil, and so some energy must go into recoil kinetic energy K_R. Show that $K_R \cong (E_1 - E_2)^2/2Mc^2$ where M is the mass of the atom. Evaluate this recoil energy for the $n = 2$ to $n = 1$ transition of hydrogen.

33. A long time ago, in a galaxy far, far away, electric charge had not yet been invented, and atoms were held together by gravitational forces. Compute the Bohr radius and the $n = 2$ to $n = 1$ transition energy in a gravitationally bound hydrogen atom.

34. (a) Find an expression for the Bohr radius a_0 in terms of the fine structure constant α (see Chapter 1), the rest energy of the electron, and the constant hc. (b) Do the same for the hydrogen ground-state energy E_1.

35. In a *muonic atom,* the electron is replaced by a negatively charged particle called the *muon.* The muon mass is 207 times the electron mass. (a) What is the shortest wavelength of the Lyman series in a muonic hydrogen atom? In what region of the electromagnetic spectrum does this belong? (b) How large is the correction for the finite nuclear mass in this case? (See the discussion at the beginning of Section 6.8.)

36. What is the radius of the first Bohr orbit of a muonic lead atom ($Z = 82$)? Compare with the nuclear radius of about 7 fm.

37. An alternative development of the Bohr theory begins by assuming that the stationary states are those for which the circumference of the orbit is an integral number of de Broglie wavelengths. (a) Show that this condition leads

to standing de Broglie waves around the orbit. (b) Show that this condition gives the angular momentum condition, Equation 6.29, used in the Bohr theory.

38. Show that the energy of the photon emitted when a hydrogen atom makes a transition from state n to state $n - 1$ is, when n is very large, $\Delta E \cong \alpha^2(mc^2/n^3)$ where α is the fine structure constant.

39. Suppose all of the excited levels of hydrogen had lifetimes of 10^{-8} s. As we go to higher and higher excited states, they get closer and closer together, and soon they are so close in energy that the energy uncertainty of each state becomes as large as the energy spacing between states, and we can no longer resolve individual states. Use the result of Problem 38 for the energy spacing and find the value of n for which this occurs. What is the radius of such an atom?

40. Compare the frequency of revolution of an electron with the frequency of the photons emitted in transitions from n to $n - 1$ for (a) $n = 10$; (b) $n = 100$; (c) $n = 1000$; (d) $n = 10,000$.

41. A hypothetical atom has only two excited states, at 4.0 and 7.0 eV, and has a ground-state ionization energy of 9.0 eV. If we used a vapor of such atoms for the Franck-Hertz experiment, for what voltages would we expect to see decreases in the current? List all voltages up to 20 V.

42. The first excited state of sodium decays to the ground state by emitting a photon of wavelength 590 nm. If sodium vapor is used for the Franck-Hertz experiment, at what voltage will the first current drop be recorded?

43. CUPS Exercise 2.8.

44. CUPS Exercise 2.9.

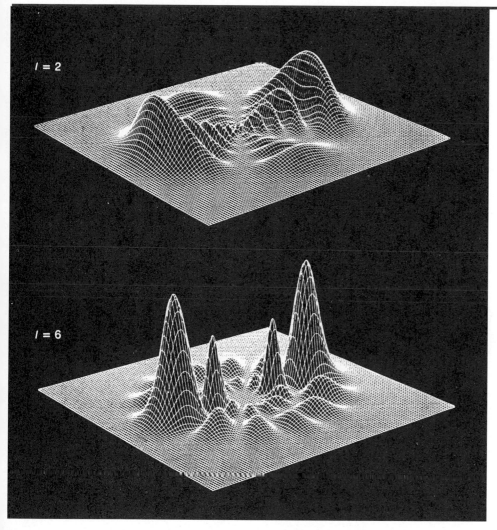

These computer-drawn graphs represent the probability to locate the electron in the n = 8 state of hydrogen for angular momentum quantum number l = 2 and l = 6; the vertical coordinate at any point gives the probability to find the electron in a small volume element at that point. The nucleus of the atom is at the center of each graph.

THE HYDROGEN ATOM IN WAVE MECHANICS

In this chapter we study the solutions of the Schrödinger equation for the hydrogen atom. We will see that these solutions, which lead to the same energy levels calculated in the Bohr model, differ from the Bohr model by allowing for the uncertainty in localizing the electron.

Other deficiencies of the Bohr model are not so easily eliminated by solving the Schrödinger equation. First, the so-called "fine structure" splitting of the spectral lines, which appears when we examine the lines very carefully, cannot be explained by our solutions; the proper explanation of this effect requires the introduction of a new property of the electron, the *intrinsic spin*. Second, the mathematical difficulties of solving the Schrödinger equation for atoms containing two or more electrons are formidable, so we restrict our discussion in this chapter to one-electron atoms, in order to see how wave mechanics enables us to understand some basic atomic properties. In the next chapter we discuss the structure of many-electron atoms.

7.1 THE SCHRÖDINGER EQUATION IN SPHERICAL COORDINATES

The Schrödinger equation in three dimensions has the following form:

$$-\frac{\hbar^2}{2m}\left(\frac{\partial^2\psi}{\partial x^2}+\frac{\partial^2\psi}{\partial y^2}+\frac{\partial^2\psi}{\partial z^2}\right) + U(x,y,z)\psi = E\psi \tag{7.1}$$

where ψ is a function of x, y, and z. The usual procedure for solving a partial differential equation of this type is to separate the variables. The potential energy for the force between the nucleus and electron is $U = -(1/4\pi\varepsilon_0)(e^2/r)$; since $r = \sqrt{x^2+y^2+z^2}$,

$$U(x,y,z) = -\frac{1}{4\pi\varepsilon_0}\frac{e^2}{\sqrt{x^2+y^2+z^2}} \tag{7.2}$$

The potential energy in this form does *not* lead to a separable equation, but if we work in the more convenient (at least for this calculation) spherical polar coordinates (r, θ, ϕ) instead of (x, y, z), we can separate the variables and find a set of solutions. The variables of spherical polar coordinates are illustrated in Figure 7.1. This simplification in the solution is at the expense of an increased complexity of the Schrödinger equation, which becomes:

$$-\frac{\hbar^2}{2m}\left[\frac{\partial^2\psi}{\partial r^2}+\frac{2}{r}\frac{\partial\psi}{\partial r}+\frac{1}{r^2\sin\theta}\frac{\partial}{\partial\theta}\left(\sin\theta\frac{\partial\psi}{\partial\theta}\right)+\frac{1}{r^2\sin^2\theta}\frac{\partial^2\psi}{\partial\phi^2}\right]$$
$$+ U(r,\theta,\phi)\psi = E\psi \tag{7.3}$$

where now $\psi = \psi(r, \theta, \phi)$. We consider only those solutions that are *separable* and can be factored as

$$\psi(r, \theta, \phi) = R(r)\Theta(\theta)\Phi(\phi) \tag{7.4}$$

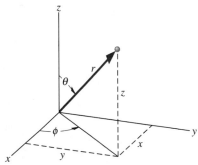

FIGURE 7.1 Spherical polar coordinates for the hydrogen atom. The proton is at the origin and the electron is at a radius *r,* in a direction determined by the polar angle θ and the azimuthal angle ϕ.

where the *radial function* $R(r)$, the *polar function* $\Theta(\theta)$, and the *azimuthal function* $\Phi(\phi)$ are each functions of a single variable. This procedure gives three differential equations, each of a single variable (r, θ, or ϕ).

7.2 THE HYDROGEN ATOM WAVE FUNCTIONS

When we solve a three-dimensional equation such as the Schrödinger equation, three parameters emerge in a natural way as indices or labels for the solutions, just as the single index n emerged from our solution of the one-dimensional infinite well in Section 5.4. These indices are the three *quantum numbers* that label the solutions. We do not discuss the mathematical details of obtaining the solutions, which can be found in the references at the end of this chapter.

Complete with quantum numbers, the separated solutions of Equation 7.4 can be written

$$\psi_{n,l,m_l}(r, \theta, \phi) = R_{n,l}(r)\Theta_{l,m_l}(\theta)\Phi_{m_l}(\phi) \tag{7.5}$$

The three indices (n, l, m_l) are the three quantum numbers that are necessary to describe the solutions. Wave functions corresponding to some values of the quantum numbers are shown in Table 7.1. The wave functions are written in terms of the Bohr radius a_0 defined in Equation 6.32.

The three quantum numbers and their allowed values are:

n	principal quantum number	$1, 2, 3, \ldots$
l	angular momentum quantum number	$0, 1, 2, \ldots, n \; 1$
m_l	magnetic quantum number	$0, \pm 1, \pm 2, \ldots, \pm l$

TABLE 7.1 SOME HYDROGEN ATOM WAVE FUNCTIONS

n	l	m_l	$R(r)$	$\Theta(\theta)$	$\Phi(\phi)$
1	0	0	$\dfrac{2}{a_0^{3/2}} e^{-r/a_0}$	$\dfrac{1}{\sqrt{2}}$	$\dfrac{1}{\sqrt{2\pi}}$
2	0	0	$\dfrac{1}{(2a_0)^{3/2}}\left(2 - \dfrac{r}{a_0}\right) e^{-r/2a_0}$	$\dfrac{1}{\sqrt{2}}$	$\dfrac{1}{\sqrt{2\pi}}$
2	1	0	$\dfrac{1}{\sqrt{3}(2a_0)^{3/2}}\dfrac{r}{a_0} e^{-r/2a_0}$	$\sqrt{\dfrac{3}{2}}\cos\theta$	$\dfrac{1}{\sqrt{2\pi}}$
2	1	± 1	$\dfrac{1}{\sqrt{3}(2a_0)^{3/2}}\dfrac{r}{a_0} e^{-r/2a_0}$	$\dfrac{\sqrt{3}}{2}\sin\theta$	$\dfrac{1}{\sqrt{2\pi}} e^{\pm i\phi}$

The principal quantum number n specifies the energy level, just as in the Bohr model. When we solve the Schrödinger equation, the quantized energy levels are obtained to be

$$E_n = -\frac{me^4}{32\pi^2\varepsilon_0^2\hbar^2}\frac{1}{n^2} \qquad (7.6)$$

which is identical to Equation 6.33. Note that the energy depends only on n and not on the other quantum numbers l or m_l. The permitted values of the angular momentum quantum number are limited by n, and those of the magnetic quantum number are limited by l.

For the ground state ($n = 1$), only $l = 0$ and $m_l = 0$ are allowed. The complete set of quantum numbers for the ground state is then $(n, l, m_l) = (1, 0, 0)$, and the wave function for this state is given in the top line of Table 7.1. The first excited state ($n = 2$) can have $l = 0$ or $l = 1$. For $l = 0$, only $m_l = 0$ is allowed. This state has quantum numbers $(2, 0, 0)$, and its wave function is given in the second line of Table 7.1. For $l = 1$, we can have $m_l = 0$ or ± 1. There are thus three possible sets of quantum numbers: $(2, 1, 0)$ and $(2, 1, \pm 1)$. The wave function for these states is given in the third and fourth lines of Table 7.1.

For the $n = 2$ level, there are four different possible sets of quantum numbers and correspondingly four different wave functions. All of these wave functions correspond to the same energy, so the $n = 2$ level is *degenerate*. (Degeneracy was introduced in Section 5.4.) If we listed the quantum numbers for the $n = 3$ level, we would find it to be degenerate with nine possible sets of quantum numbers. In general, the level with principal quantum number n has a degeneracy equal to n^2. Figure 7.2 illustrates the labelling of the first three levels.

If these different combinations of quantum numbers have exactly the same energy, what is the purpose of listing them separately? First, as we discuss in the last section of this chapter, the levels are not precisely degenerate, but are separated by a very small energy (about 10^{-5} eV). Second, in the study of the transitions between the levels, we find that the intensities of the individual transitions depend on the particular level from which the transition originates. Third, and perhaps most important, *each of these sets of quantum numbers corresponds to a very different wave function, and therefore represents a very different state of motion of the electron.* In under-

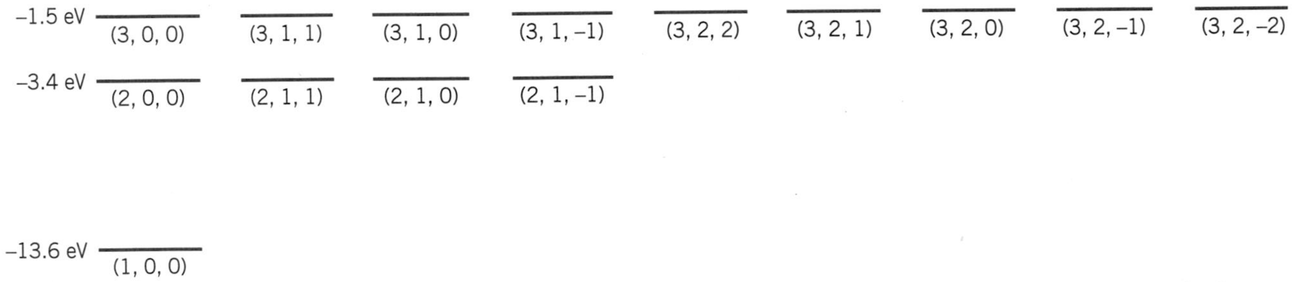

FIGURE 7.2   The lower energy levels of hydrogen, labeled with the quantum numbers (n, l, m_l). The first excited state is four-fold degenerate and the second excited state is nine-fold degenerate.

standing this last point, we must consider the geometrical interpretation of the quantum numbers, which is discussed in Section 7.4.

7.3 RADIAL PROBABILITY DENSITIES

As we learned in Chapter 5, the probability of finding the electron at a given location is determined by the square of the wave function. More specifically, $|\psi(r, \theta, \phi)|^2$ gives the *probability density* (probability per unit volume) of finding the electron at the location (r, θ, ϕ). To compute the actual probability of finding the electron, we multiply the probability per unit volume by the volume element dV located at (r, θ, ϕ). In spherical polar coordinates (see Figure 7.3) the volume element is

$$dV = r^2 \sin \theta \, dr \, d\theta \, d\phi \tag{7.7}$$

and therefore the probability is

$$|\psi_{n,l,m_l}(r, \theta, \phi)|^2 \, dV = |R_{n,l}(r)|^2 \, |\Theta_{l,m_l}(\theta)|^2 \, |\Phi_{m_l}(\phi)|^2 \, r^2 \sin \theta \, dr \, d\theta \, d\phi \tag{7.8}$$

Using this expression for the probability, we can calculate many features of the electron's spatial distribution. For example, we can find the radial probability $P(r) \, dr$ for locating the electron somewhere between r and $r + dr$ no matter what the values of θ and ϕ. To put it another way, we imagine a thin spherical shell of radius r and thickness dr, and ask what is the probability of the electron being within the volume of the shell. Since we are not interested in θ or ϕ, we integrate over all possible values of these variables:

$$P(r) \, dr = |R_{n,l}(r)|^2 \, r^2 \, dr \int_0^\pi |\Theta_{l,m_l}(\theta)|^2 \sin \theta \, d\theta \int_0^{2\pi} |\Phi_{m_l}(\phi)|^2 \, d\phi \tag{7.9}$$

The θ and ϕ integrals are each equal to unity, since each of the functions R, Θ, and Φ are individually *normalized*. Thus the *radial probability density* is

$$P(r) = r^2 |R_{n,l}(r)|^2 \tag{7.10}$$

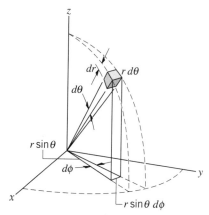

FIGURE 7.3   The volume element in spherical polar coordinates.

Radial probability density

Figure 7.4 shows this function for several of the lowest levels of hydrogen.

Notice in Figure 7.4 that the average radial coordinate appears to be about a_0 for the $n = 1$ wave function and about $5a_0$ for both of the $n = 2$ wave functions. Notice also that the average radius doesn't differ very much between the $l = 0$ and $l = 1$ wave functions of the $n = 2$ state. It appears from these graphs that the average radius depends mostly on n and not very much on l. (See Problems 16 and 17.) This trend continues for $n = 3$ and $n = 4$, as illustrated in Figure 7.5.

The principal quantum number n determines not only the energy level of the electron, it also determines to a great extent the average distance of the electron from the nucleus. As in the Bohr model, this average radius varies roughly as n^2, so that an $n = 2$ electron is on the average about 4 times farther from the nucleus than an $n = 1$ electron, an $n = 3$ electron is about 9 times farther from the nucleus than an $n = 1$ electron, and so forth.

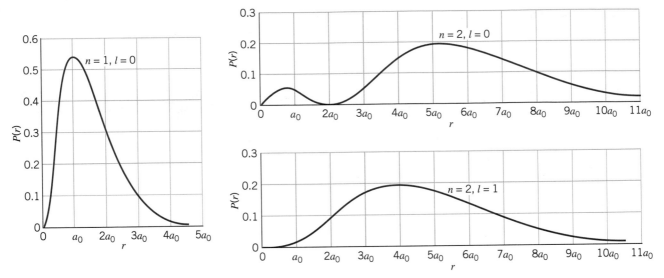

FIGURE 7.4   The radial probability density $P(r)$ for the three lowest states of hydrogen.

Another measure of the location of the electron is its most probable radius, determined from the location at which $P(r)$ has its maximum value. For each n, $P(r)$ for the state with $l = n - 1$ has only a single maximum, which occurs at the location of the Bohr orbit, $r = n^2 a_0$. The following Example illustrates this for the $n = 2$ state.

EXAMPLE 7.1

Prove that the most likely distance from the origin of an electron in the $n = 2, l = 1$ state is $4a_0$.

SOLUTION

In the $n = 2, l = 1$ level, the radial probability density is

$$P(r) = r^2 |R_{2,1}(r)|^2$$

$$= r^2 \frac{1}{24a_0^3} \frac{r^2}{a_0^2} e^{-r/a_0}$$

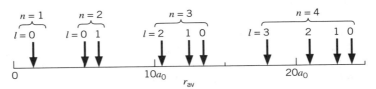

FIGURE 7.5   The average radius r_{av} for states of the hydrogen atom. The markers are labeled with the angular momentum quantum number l for each state.

We wish to find where this function has its maximum; in the usual fashion, we take the first derivative of $P(r)$ and set it equal to zero:

$$\frac{dP(r)}{dr} = \frac{1}{24a_0^5}\frac{d}{dr}(r^4 e^{-r/a_0})$$

$$= \frac{1}{24a_0^5}\left[4r^3 e^{-r/a_0} + r^4\left(-\frac{1}{a_0}\right)e^{-r/a_0}\right] = 0$$

$$\frac{1}{24a_0^5}e^{-r/a_0}\left[4r^3 - \frac{r^4}{a_0}\right] = 0$$

The only solution that yields a maximum is $r = 4a_0$.

EXAMPLE 7.2

An electron is in the $n = 1, l = 0$ state. What is the probability of finding the electron closer to the nucleus than the Bohr radius?

SOLUTION

We are again interested in the radial probability,

$$P(r)\,dr = r^2|R_{1,0}(r)|^2\,dr = r^2\frac{4}{a_0^3}e^{-2r/a_0}\,dr$$

The total probability of finding the electron between $r = 0$ and $r = a_0$ is

$$P = \int_0^{a_0} P(r)\,dr = \frac{4}{a_0^3}\int_0^{a_0} r^2 e^{-2r/a_0}\,dr$$

Letting $x = 2r/a_0$, we rewrite this as

$$P = \frac{1}{2}\int_0^2 x^2 e^{-x}\,dx$$

Evaluating the integral gives

$$P = 0.32$$

That is, 32 percent of the time the electron is closer than 1 Bohr radius to the nucleus.

EXAMPLE 7.3

For the $n = 2$ states ($l = 0$ and $l = 1$), compare the probabilities of the electron being found inside the Bohr radius.

SOLUTION

For the $n = 2, l = 0$ level, we have

$$P(r)\,dr = r^2|R_{2,0}(r)|^2\,dr = r^2\frac{1}{8a_0^3}\left(2 - \frac{r}{a_0}\right)^2 e^{-r/a_0}\,dr$$

The total probability of finding the electron between $r = 0$ and $r = a_0$ is

$$P = \int_0^{a_0} P(r)\, dr = \frac{1}{8a_0^3} \int_0^{a_0} \left(4r^2 - \frac{4r^3}{a_0} + \frac{r^4}{a_0^2} \right) e^{-r/a_0}\, dr$$

and again, letting $x = r/a_0$,

$$P = \frac{1}{8} \int_0^1 (4x^2 - 4x^3 + x^4)e^{-x}\, dx$$

Evaluating the integrals, we obtain

$$P = 0.034$$

For the $n = 2$, $l = 1$ level we have

$$P(r)\, dr = r^2 |R_{2,1}(r)|^2 = r^2 \frac{1}{24a_0^3} \frac{r^2}{a_0^2} e^{-r/a_0}\, dr$$

The total probability between $r = 0$ and $r = a_0$ is

$$P = \int_0^{a_0} P(r)\, dr = \frac{1}{24a_0^3} \int_0^{a_0} \frac{r^4}{a_0^2} e^{-r/a_0}\, dr$$

$$= \frac{1}{24} \int_0^1 x^4 e^{-x}\, dx$$

$$= 0.0037$$

In the $l = 1$ state, the probability of finding the electron inside a_0 is about 10 times smaller than in the $l = 0$ state. This is consistent with Figure 7.4, which shows a small peak in the radial probability density for $n = 2$, $l = 0$ at small r. There is clearly more area under the $P(r)$ curve between $r = 0$ and $r = a_0$ for $n = 2$, $l = 0$ than there is for $n = 2$, $l = 1$.

Combining this result with that of Figure 7.5, we see that, for the $n = 2$ state, the $l = 0$ electron spends more time close to the nucleus than the $l = 1$ electron *and it also spends more time farther away* (its average radius is larger). This is a general result that holds for any value of n—the smaller the l value, the larger is the probability to find the electron both close to the nucleus and far from the nucleus. How can the electron be both closer to the nucleus and farther away? The next section presents a geometrical interpretation for the angular momentum quantum number that makes this plausible.

7.4 ANGULAR MOMENTUM AND PROBABILITY DENSITIES

In this section, we consider the quantum numbers l and m_l and their physical interpretation. We also discuss the angular part of the electron's probability density.

In a classical orbit, such as a planet in the solar system, the total energy determines the average distance of the planet from the Sun. For a given total energy, many different orbits are possible, from the nearly circular orbit of the Earth to the highly elongated elliptical orbits of the comets. These orbits differ in their angular momentum L, which is largest for the circular orbit and smallest for the elongated ellipse. Figure 7.6 shows a variety of planetary orbits having the same total energy but different angular momentum.

The flattened ellipse, for which L is small, has the same property discussed at the end of the previous section—compared with the circular orbit of maximum L, the planet spends more time both close to the Sun and far from the Sun. This indicates a connection between the classical orbital angular momentum and the angular momentum quantum number l of the electron. However, as we shall see, there are important differences between the classical and quantum properties of angular momentum.

Classically, the angular momentum of a particle is represented by the vector $\mathbf{L} = \mathbf{r} \times \mathbf{p}$, where $\mathbf{r}$ is the position vector that locates the particle and $\mathbf{p}$ is its linear momentum. The direction of $\mathbf{L}$ is perpendicular to the plane of the orbit. Quantum theory gives a relationship between the length of the angular momentum vector and the angular momentum quantum number l:

$$|\mathbf{L}| = \sqrt{l(l+1)}\,\hbar \qquad (7.11)$$

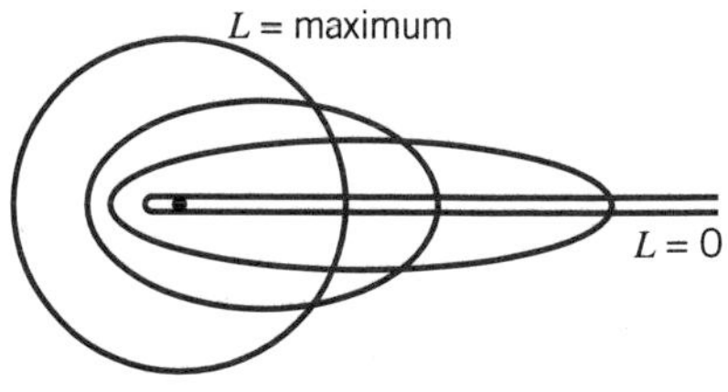

FIGURE 7.6 Planetary orbits of the same energy but different angular momentum L. As L decreases, the orbits become thinner and longer ellipses.

Length of L

EXAMPLE 7.4

Compute the length of the angular momentum vectors that represent the orbital motion of an electron in a state with $l = 1$ and in another state with $l = 2$.

SOLUTION

Equation 7.11 gives the relationship between the length of the vector and the associated quantum number l. For $l = 1$

$$|\mathbf{L}| = \sqrt{1(1+1)}\,\hbar = \sqrt{2}\,\hbar$$

and for $l = 2$

$$|\mathbf{L}| = \sqrt{2(2+1)}\,\hbar = \sqrt{6}\,\hbar$$

Note two important points here. First, the length of the vector $|\mathbf{L}|$ is always greater than $l\hbar$ since $\sqrt{l(l+1)}$ is always greater than l. The importance of this point is discussed later. Second, these values of $|\mathbf{L}|$, which we can interpret as the "magnitude" of the electron's angular momentum, are totally different from those found in the Bohr model. For example, an electron with $n = 3$ in the Bohr model has an angular momentum $|\mathbf{L}| = 3\hbar$ (see Section 6.5). In our quantum-mechanical vector model, an electron with $n = 3$ can have $l = 2$ (with $|\mathbf{L}| = \sqrt{6}\hbar$), or $l = 1$ (with $|\mathbf{L}| = \sqrt{2}\hbar$), or even $l = 0$ (with $|\mathbf{L}| = 0$).

Just like an ordinary classical vector, the vector **L** has components along any axis in space. Once again, the wave functions deduced using the Schrödinger equation give us the rules for computing the components of **L**. (We generally choose the z axis for special consideration, since it is an *z component of* **L** axis of reference in the spherical polar coordinate system.) The z component of **L**, which we denote by L_z, is restricted to the values

$$L_z = m_l \hbar \tag{7.12}$$

where m_l is the magnetic quantum number, which takes values $0, \pm 1, \pm 2, \ldots, \pm l$.

EXAMPLE 7.5

What are the possible z components of the vector **L** which represents the orbital angular momentum of a state with $l = 2$?

SOLUTION

The possible m_l values for $l = 2$ are $+2, +1, 0, -1, -2$, and so the **L** vector can have any of five possible z components: $L_z = 2\hbar, 1\hbar, 0, -1\hbar,$ or $-2\hbar$. The length of the vector **L**, as we found previously, is $\sqrt{6}\,\hbar$.

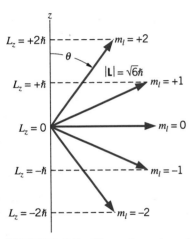

FIGURE 7.7 The orientations in space and z components of a vector with $l = 2$. There are five different possible orientations.

The components of the vector **L** for $l = 2$ are illustrated in Figure 7.7. Each different orientation in space of the vector **L** corresponds to a different m_l value. The polar angle θ that the vector **L** makes with the z axis can be found by referring to the figure. Since $L_z = |\mathbf{L}| \cos \theta$, we have

$$\cos \theta = \frac{L_z}{|\mathbf{L}|} = \frac{m_l \hbar}{\sqrt{l(l+1)}\,\hbar}$$

so

$$\cos \theta = \frac{m_l}{\sqrt{l(l+1)}} \tag{7.13}$$

Space quantization This behavior represents a curious aspect of quantum physics called *space quantization*, in which only certain orientations of angular momentum vectors are allowed. The number of these orientations is equal to $2l + 1$ (the number of different possible m_l values) and the magnitudes of their successive z components always differ by $\hbar$. As an example, suppose we could prepare a collection of hydrogen atoms in a state with $l = 1$. Choosing arbitrarily a z axis and using an appropriate experimental technique, we measure the z component of **L**. From such a measurement we find $L_z = \hbar, 0,$ or $-\hbar$. Choosing a completely different z axis, we repeat the measurement and once again we find $L_z = \hbar, 0,$ or $-\hbar$. This behavior is completely different from that of classical vectors. A classical vector of length 1.0 will have a z component of 1.0 if we choose our z axis along the vector, or -1.0 if we choose the opposite direction, or 0.5 if we choose our z axis at an angle of 60° to the vector, or 0.7 if we choose the z axis at 45° to the vector. A quantum mechanical vector representing $l = 1$ has z components that are restricted to $+\hbar$, 0, or $-\hbar$. The curious fact is that we observe this result no matter which direction we choose for the z axis!

You may perhaps be wondering why we have singled out the z axis for special attention. Aside from its convenience in polar coordinates, there is an important reason. According to quantum physics, we can have exact knowledge of only *one* of the three components of **L** (by convention, we choose the z component); the other components of **L** are completely uncertain. This follows from an additional form of the uncertainty principle,

$$\Delta L_z \, \Delta\phi \gtrsim \hbar \tag{7.14}$$

where ϕ is the azimuthal angle defined in Figure 7.1. If we know L_z exactly ($\Delta L_z = 0$), then we have no knowledge at all of the angle ϕ—all values are equally probable. This is equivalent to saying that we know nothing at all about L_x and L_y; whenever one component of **L** is determined, the other components are completely undetermined. Figure 7.8 gives a pictorial representation of the behavior of the **L** vector. We think of the vector as revolving, or *precessing,* about the z axis, so rapidly that we can never see the motion, all the while keeping L_z constant. In this interpretation, you can see how we can have no knowledge of the x and y components of **L**. You can also see why it *must* be true that $|\mathbf{L}| > l\hbar$. If it were possible to have $|\mathbf{L}| = l\hbar$, then when m_l had its maximum value ($m_l = +l$), we would have $L_z = m_l\hbar = l\hbar$. Since the length of the vector would be equal to its z component, it must lie along the z axis, so that $L_x = L_y = 0$. However, this simultaneous exact knowledge of all three components of **L** violates the form of the uncertainty principle as expressed in Equation 7.14, and therefore this situation is not permitted to occur.

Let us now turn to the angular part of the probability density, which is obtained from the squared magnitudes of the angular parts of the wave function:

$$P(\theta, \phi) = |\Theta(\theta)\Phi(\phi)|^2 \tag{7.15}$$

Angular probability density

Figure 7.9 shows the angular probablity densities for the $l = 0$ and $l = 1$ wave functions listed in Table 7.1.

Note that all of the probability densities are *cylindrically symmetric*—there is no dependence on the azimuthal angle ϕ. The $l = 0$ wave function

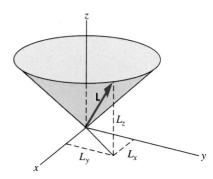

FIGURE 7.8 The vector **L** precesses rapidly about the z axis, so that L_z stays constant, but L_x and L_y are indeterminate.

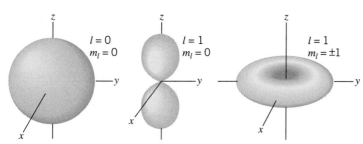

FIGURE 7.9 The angular dependence of the $l = 0$ and $l = 1$ probability densities.

is also *spherically symmetric*, that is, the probability density is independent of direction.

The $l = 1$ probability densities have two distinct shapes. For $m_l = 0$, the electron is found primarily in two regions of maximum probability along the positive and negative z axis, while for $m_l = \pm 1$, the electron is found primarily near the xy plane. For $m_l = 0$, the electron's angular momentum vector lies in the xy plane (Figure 7.7). Classically, the angular momentum vector is perpendicular to the orbital plane, so it should not be surprising that the electron spends most of its time in a direction perpendicular to the xy plane, that is, along the z axis. For $m_l = \pm 1$, the angular momentum vector has its maximum projection along the z axis; the electron, again orbiting perpendicular to $\mathbf{L}$, spends most of its time near the xy plane. These probability densities for locating the electron are consistent with the information given by the orientation of the angular momentum vector, and the cylindrical symmetry of the probability densities is consistent with the uncertainty in the knowledge of the orientation of $\mathbf{L}$ represented in Figure 7.8.

Putting together the radial and angular probability densities, we can obtain representations of the complete electron probability density $|\psi|^2$, as shown in Figure 7.10. We can regard these illustrations as representing the "smeared out" distribution of electronic charge in the atom, which results from the uncertainty in the electron's location. They also represent the statistical outcomes of a large number of measurements of the position of the electron in the atom. These spatial distributions have important consequences for the structure of atoms with many electrons, which is discussed in Chapter 8, and also for the joining of atoms into molecules, which is discussed in Chapter 9.

7.5 INTRINSIC SPIN

One way of observing space quantization is to place the atom in an externally applied magnetic field. From the interaction between the magnetic field and the *magnetic dipole moment* of the atom (which is related to the electron's orbital angular momentum), it is possible both to observe the separate components of $\mathbf{L}$ and also to determine l by counting the number of components (which, as we have seen, is equal to $2l + 1$). However, when this experiment is done, a surprising result emerges that indicates an unexpected property of the electron.

Figure 7.11 shows a classical magnetic moment, which results from a current loop or from the orbital motion of a charged object. (We assume, in analogy with the circulating electron, that the object carries a negative charge.) The magnetic moment $\boldsymbol{\mu}$ is a vector whose magnitude is equal to the product of the circulating current and the area enclosed by the orbital loop. The direction of $\boldsymbol{\mu}$ is perpendicular to the plane of the orbit, determined by the right-hand rule—with the fingers in the direction of the

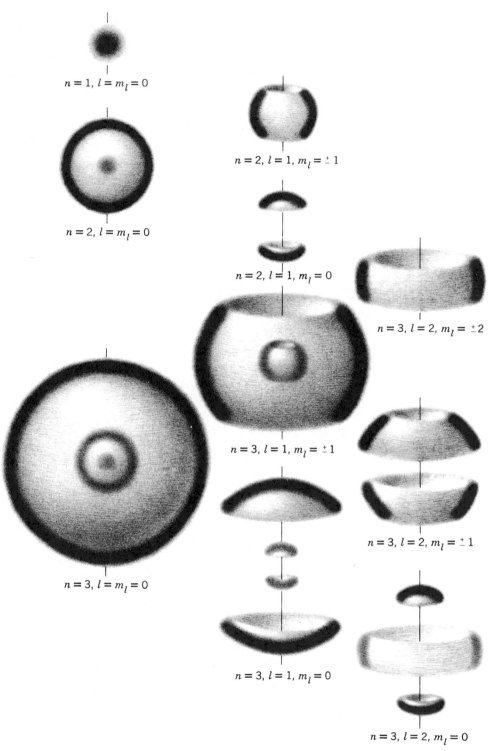

FIGURE 7.10   Representations of $|\psi|^2$ for different sets of quantum numbers. The z axis is the vertical axis. The intensity of each diagram at any point is proportional to the probability of locating an electron in a small volume element at that point.

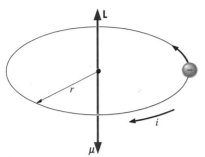

FIGURE 7.11   A circulating negative charge is represented as a current loop. Because the charge is negative, **L** and **μ** have opposite directions.

conventional (positive) current, the thumb indicates the direction of **μ**, as shown in Figure 7.11.

As we have seen, quantum mechanics forbids exact knowledge of the direction of **L** and therefore of **μ**. Figure 7.12 suggests the relationship between **L** and **μ** that is consistent with quantum mechanics. Only the z components of these vectors can be specified. Because the electron has a negative charge, **L** and **μ** have z components of opposite signs.

We can use the Bohr model with a circular orbit to obtain the relationship between **L** and **μ**, which turns out to be identical with the correct quantum mechanical result. We regard the circulating electron as a circular loop of current $i = dq/dt = q/T$, where q is the charge of the electron $(-e)$ and T is the time for one circuit around the loop. If the electron moves with speed $v = p/m$ around a loop of radius r, then $T = 2\pi r/v = 2\pi rm/p$. The magnetic moment is

$$\mu = iA = \frac{q}{2\pi rm/p}\,\pi r^2 = \frac{q}{2m}\,rp = \frac{q}{2m}\,|\mathbf{L}| \tag{7.16}$$

since $|\mathbf{L}| = rp$. Writing Equation 7.16 in terms of vectors and putting $-e$ for the electronic charge, we obtain

Orbital magnetic moment

$$\boldsymbol{\mu}_{\mathrm{L}} = -\frac{e}{2m}\mathbf{L} \tag{7.17}$$

The negative sign, which is present because the electron has a negative charge, indicates that the vectors **L** and **μ**$_{\mathrm{L}}$ point in opposite directions. The subscript **L** on **μ**$_{\mathrm{L}}$ reminds us that this magnetic moment arises from the *orbital* angular momentum **L** of the electron.

The z component of the magnetic moment is

$$\mu_{\mathrm{L},z} = -\frac{e}{2m}L_z = -\frac{e}{2m}m_l\hbar = -\frac{e\hbar}{2m}m_l = -m_l\mu_{\mathrm{B}} \tag{7.18}$$

The quantity $e\hbar/2m$ is defined to be the *Bohr magneton* μ_{B}:

Bohr magneton

$$\mu_{\mathrm{B}} = \frac{e\hbar}{2m} \tag{7.19}$$

The value of μ_{B} is

$$\mu_{\mathrm{B}} = 9.274 \times 10^{-24} \text{ J/T}$$

The Bohr magneton is a convenient unit for expressing atomic magnetic moments, which typically have values of the order of μ_{B}.

Before we consider further the behavior of **μ**$_{\mathrm{L}}$, we discuss the similar behavior of an *electric dipole,* which consists of two equal and opposite charges q separated by a distance r. The electric dipole moment **p** has magnitude qr and points from the negative charge to the positive charge. As shown in Figure 7.13, in a uniform *electric* field, the dipole experiences a torque that tends to rotate it into alignment with **E**. Suppose now that the field is not uniform—the field acting on the positive charge is *not* equal to the field that acts on the negative charge, as in Figure 7.14. There is still a net *torque* that tends to rotate the dipole, but there is also a net *force*

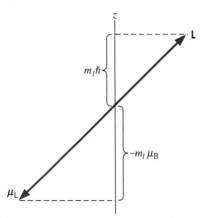

FIGURE 7.12   According to quantum mechanics, only the z components of **L** and **μ** can be specified.

that tends to move the dipole. Consider the two dipoles shown in Figure 7.15. Suppose the electric field near the bottom of the figure is greater in magnitude than the field near the top, and further suppose that the field points upward. Dipole *A*, with its dipole moment **p** inclined upward, experiences a net downward force, $\mathbf{F}_{net}$, since the (downward) force $\mathbf{F}_-$ on the negative charge is greater than the upward force $\mathbf{F}_+$ on the positive charge. On the other hand, dipole *B*, with its dipole moment **p** inclined downward, experiences a net upward force, since now $\mathbf{F}_+$ is greater than $\mathbf{F}_-$. We can state this result in another way that will be more applicable to our discussion of *magnetic* dipole moments. Let the field direction define the *z* axis. Then all dipoles with $p_z > 0$ (as dipole *A*) experience a net negative force and move in the negative *z* direction, and all dipoles with $p_z < 0$ (as dipole *B*) experience a net positive force and move in the positive *z* direction.

A magnetic dipole moment $\boldsymbol{\mu}$ behaves in an identical way. (In fact, if we imagine fictitious N and S poles, the behavior of a magnetic moment would be described by illustrations similar to Figures 7.13, 7.14, and 7.15.) A nonuniform *magnetic* field acting on the *magnetic* moments gives an unbalanced force that causes a displacement. Figure 7.16 illustrates the behavior of magnetic moments having different orientations in a nonuniform field. The two different orientations give net forces in opposite directions.

Imagine the following experiment, illustrated schematically in Figure 7.17. A beam of hydrogen atoms is prepared in the $n = 2$, $l = 1$ state. The beam consists of equal numbers of atoms in the $m_l = -1$, 0, and +1 states. (We assume we can do the experiment so quickly that the $n = 2$ state doesn't decay to the $n = 1$ state. In practice this may not be possible.) The beam passes through a region in which there is a nonuniform magnetic field. The atoms with $m_l = +1$ experience a net upward force and are deflected upward, while the atoms with $m_l = -1$ are deflected downward. The atoms with $m_l = 0$ are undeflected. After passing through the field,

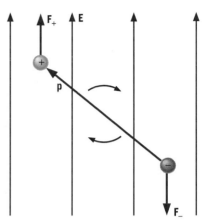

FIGURE 7.13 An electric dipole in a uniform electric field **E**. A force $\mathbf{F}_+$ on the positive charge and a force $\mathbf{F}_-$ on the negative charge produce a net torque on the dipole.

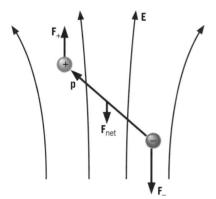

FIGURE 7.14 An electric dipole in a nonuniform field. The field decreases from the bottom to the top of the figure, so that the force $\mathbf{F}_-$ is greater than the force $\mathbf{F}_+$. There is a net downward force on the dipole.

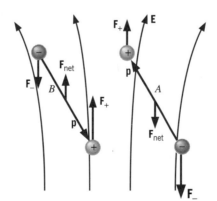

FIGURE 7.15 Two dipoles with oppositely directed moments in a nonuniform field. The dipoles move in opposite directions under the influence of the net force.

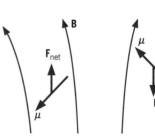

FIGURE 7.16 Two magnetic dipoles in a nonuniform magnetic field. Oppositely directed dipoles experience net forces in opposite directions.

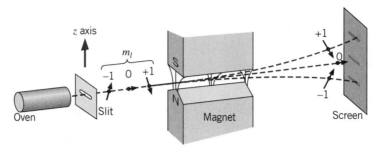

FIGURE 7.17   Schematic diagram of Stern-Gerlach experiment. A beam of atoms from an oven passes through a slit and then enters a region where there is a nonuniform magnetic field. Atoms with their magnetic dipole moments in opposite directions experience forces in opposite directions.

the beam strikes a screen where it makes a visible image. When the field is off, we expect to see one image of the slit in the center of the screen, since there is no deflection at all. When the field is on, we expect three images of the slit on the screen—one in the center (corresponding to $m_l = 0$), one above the center ($m_l = +1$), and one below the center ($m_l = -1$). If the atom were in the ground state ($l = 0$), we expect to see one image in the screen whether the field was off or on (recall that a $m_l = 0$ atom is not deflected). If we had prepared the beam in a state with $l = 2$, we would see five images with the field on. *The number of images that appears is just the number of different m_l values*, which is equal to $2l + 1$. Since l has possible values 0, 1, 2, 3, . . . , it follows that $2l + 1$ has the values 1, 3, 5, 7, . . . ; that is, we should always see an *odd number* of images on the screen. However, if we were actually to perform the experiment with hydrogen in the $l = 1$ state, we would find not three but *six* images on the screen! Even more confusing, when we do the experiment with hydrogen in the $l = 0$ state, we find not one but *two* images on the screen, one representing an upward deflection and one a downward deflection! In the $l = 0$ state, the vector **L** has length zero, and so we expect that there is *no magnetic moment* for the magnetic field to deflect. We observe this not to be true—even when $l = 0$, the atom still has a magnetic moment, in contradiction to Equation 7.16.

Stern-Gerlach experiment   The first experiment of this type was done by O. Stern and W. Gerlach in 1921. They used a beam of silver atoms; although the electronic structure of silver is more complicated than that of hydrogen (as we discuss in Chapter 8), the same basic principle applies—the silver must have $l = 0$, 1, 2, 3, . . . , and so an *odd number* of images is expected to appear on the screen. In fact, they observed the beam to split into *two* components, producing two images of the slits on the screen (see Figure 7.18).

The observation of separated images was the first conclusive evidence of *space quantization*; classical magnetic moments would have all possible orientations and would make a continuous smeared-out pattern on the screen, but the observation of a number of discrete images on the screen means that the atomic magnetic moments can take only certain discrete orientations in space. These correspond to the discrete orientations of the magnetic moment (or, equivalently, of the angular momentum).

However, the *number* of discrete images on the screen does not agree with our expectations that it be an odd number. We expect $2l + 1$ images, so for two images we should have $l = \frac{1}{2}$, which is not permitted by the Schrödinger equation. We can resolve this dilemma if there is another contribution to the angular momentum of the atom, the *intrinsic angular momentum* of the electron.

Associated with the motion of the Earth are two types of angular momentum—the *orbital angular momentum* of the motion about the Sun and the *intrinsic angular momentum* of the rotation of the Earth about its axis. Similarly, the electron has an *orbital angular momentum* **L**, which characterizes the motion of the electron about the nucleus, and an *intrinsic angular momentum* **S**, which behaves as if the electron were spinning about its axis. For this reason, **S** is usually called the intrinsic *spin*. (The concept of the electron as a tiny ball of charge spinning on its axis is a useful one, just like the Bohr model. Unfortunately it is not a correct concept. However, it frequently happens in the progress of science that the right idea is introduced for the wrong reason. After S. A. Goudsmit and G. E. Uhlenbeck introduced the concept of electron spin in 1925, P. A. M. Dirac showed that a proper *relativistic* quantum theory of the electron gives the electron spin directly as an additional quantum number.)

In order to explain the result of the Stern-Gerlach experiment, we must assign to the electron an intrinsic spin quantum number s of $\frac{1}{2}$. The intrinsic spin behaves much like the orbital angular momentum; there is the quantum number s (which we can regard as a label arising from the mathematics), the angular momentum vector **S**, an associated magnetic moment $\boldsymbol{\mu}_S$, a z component S_z, and a spin magnetic quantum number m_s. Figure 7.19 illustrates the vector properties of **S**, and Table 7.2 compares the properties of orbital and spin angular momentum.

The inclusion of spin gives a direct explanation for the Stern-Gerlach experiment. The outermost electron in a silver atom occupies a state with $l = 0$. (The other electrons do not contribute to the magnetic properties of the atom.) The magnetic behavior is therefore due entirely to the spin magnetic moment, which has only two possible orientations in the magnetic field corresponding to the two beams observed emerging from the magnet.

Every fundamental particle has a characteristic intrinsic spin and a corresponding spin magnetic moment. For example, the proton and neutron also have a spin quantum number of $\frac{1}{2}$. The photon has a spin quantum number of 1, while the pi meson (pion) has $s = 0$.

EXAMPLE 7.6

In a Stern-Gerlach type of experiment, the magnetic field varies with distance in the z direction according to $dB_z/dz = 1.4$ T/mm. The silver atoms travel a distance $x = 3.5$ cm through the magnet. The most probable speed of the atoms emerging from the oven is $v = 750$ m/s. Find the separation of the two beams as they leave the magnet. The mass of a silver atom is 1.8×10^{-25} kg, and its magnetic moment is about 1 Bohr magneton.

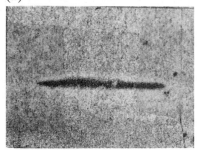

(a)

(b)

FIGURE 7.18 The results of the Stern-Gerlach experiment. (*a*) The image of the slit with the field turned off. (*b*) With the field on, two images of the slit appear on the screen. The scale at right represents 1 mm.

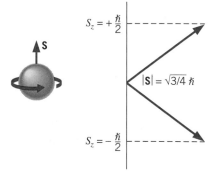

FIGURE 7.19 The spin angular momentum of an electron and the spatial orientation of the spin angular momentum vector.

TABLE 7.2   COMPARISON OF ORBITAL AND SPIN ANGULAR MOMENTUM

	Orbital	*Spin*
Quantum number	$l = 0, 1, 2, \ldots$	$s = \frac{1}{2}$
Length of vector	$|\mathbf{L}| = \sqrt{l(l+1)}\,\hbar$	$|\mathbf{S}| = \sqrt{s(s+1)}\,\hbar = \sqrt{3/4}\,\hbar$
z component	$L_z = m_l\hbar$	$S_z = m_s\hbar$
Magnetic quantum number	$m_l = 0, \pm 1, \pm 2, \ldots, \pm l$	$m_s = \pm\frac{1}{2}$
Magnetic moment	$\boldsymbol{\mu}_L = -(e/2m)\mathbf{L}$	$\boldsymbol{\mu}_S = -(e/m)\mathbf{S}$

SOLUTION

The potential energy of the magnetic moments in the magnetic field is

$$U = -\boldsymbol{\mu} \cdot \mathbf{B} = -\mu_z B_z$$

because the field along the central axis of the magnet has only a z component. The force on the atom can be found from the potential energy according to

$$F_z = -\frac{dU}{dz} = \mu_z \frac{dB_z}{dz}$$

The acceleration of a silver atom as it passes through the magnet is

$$a = \frac{F_z}{m} = \frac{\mu_z(dB_z/dz)}{m}$$

The vertical deflection Δz of either beam can be found from $\Delta z = \frac{1}{2}at^2$, where t, the time to traverse the magnet, equals x/v. Since each beam is deflected by this amount, the net separation d is $2\,\Delta z$, or

$$
\begin{aligned}
d &= \frac{\mu_z(dB_z/dz)x^2}{mv^2} \\
&= \frac{(9.27 \times 10^{-24}\,\text{J/T})\,(1.4 \times 10^3\,\text{T/m})\,(3.5 \times 10^{-2}\,\text{m})^2}{(1.8 \times 10^{-25}\,\text{kg})\,(750\,\text{m/s})^2} \\
&= 1.6 \times 10^{-4}\,\text{m} = 0.16\,\text{mm}
\end{aligned}
$$

This is consistent with the separation that can be read from the scale in Figure 7.18.

7.6 ENERGY LEVELS AND SPECTROSCOPIC NOTATION

We previously described all of the possible electronic states in hydrogen by three quantum numbers (n, l, m_l), but as we have seen, a fourth property of the electron, the intrinsic angular momentum or *spin*, requires the introduction of a fourth quantum number. We don't need to specify the spin s, since it is always $\frac{1}{2}$ (we regard it as a fundamental property of the electron, like its electric charge or its mass), but we do need to specify the value of the quantum number m_s ($+\frac{1}{2}$ or $-\frac{1}{2}$), which tells us about the z component of **S**. Thus the complete description of the state of an electron in an atom requires the four quantum numbers (n, l, m_l, m_s).

For example, the ground state of hydrogen was previously labeled as $(n, l, m_l) = (1, 0, 0)$. With the addition of m_s, this would become either $(1, 0, 0, +\frac{1}{2})$ or $(1, 0, 0, -\frac{1}{2})$. The degeneracy of the ground state is now 2. The first excited state would have eight possible labels: $(2, 0, 0, +\frac{1}{2})$, $(2, 0, 0, -\frac{1}{2})$, $(2, 1, 1, +\frac{1}{2})$, $(2, 1, 1, -\frac{1}{2})$, $(2, 1, 0, +\frac{1}{2})$, $(2, 1, 0, -\frac{1}{2})$, $(2, 1, -1, +\frac{1}{2})$, and $(2, 1, -1, -\frac{1}{2})$. Since there are now two possible labels for each previous single label (each n, l, m_l becomes n, l, m_l, $+\frac{1}{2}$ and n, l, m_l, $-\frac{1}{2}$), the degeneracy of each level is $2n^2$ instead of n^2.

When we place an atom in a magnetic field, it is necessary to distinguish between different m_l or m_s values. For most other applications, the values of m_l and m_s are of no importance, and it is cumbersome to write them each time we wish to refer to a certain level of an atom. We therefore use a different notation, known as *spectroscopic notation*, to label the levels. In this system we use letters to stand for the different l values: for $l = 0$, we use the letter s (do not confuse this with the quantum numbers s), for $l = 1$, we use the letter p, and so on. The complete notation is as follows:

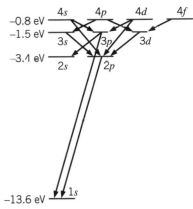

FIGURE 7.20 A partial energy level diagram of hydrogen, showing the spectroscopic notation of the levels and some of the transitions that satisfy the $\Delta l = \pm 1$ selection rule.

value of l	0	1	2	3	4	5	6
designation	s	p	d	f	g	h	i

Spectroscopic notation

(The first four letters stand for sharp, principal, diffuse, and fundamental, which were terms used to describe atomic spectra before atomic theory was developed.) In spectroscopic notation, the ground state of hydrogen is labeled $1s$, where the value $n = 1$ is specified before the s. Figure 7.20 illustrates the labeling of the hydrogen atom levels in this notation.

Also shown on Figure 7.20 are lines representing some different photons that can be emitted when the atom makes a transition from one state to a lower state. These lines indicate an additional feature of the level diagram, known as a *selection rule*. Not all transitions are allowed to occur. By solving the Schrödinger equation and using the solutions to compute *transition probabilities*, we find that the transitions most likely to occur are those that change l by one unit, and thus the selection rule is

$$\Delta l = \pm 1 \tag{7.20}$$

Selection rule

For example, the $3s$ level cannot emit a photon in a transition to the $2s$ level ($\Delta l = 0$), but rather must go to the $2p$ level ($\Delta l = 1$). There is no selection rule for n, so the $3p$ level can go to $2s$ or $1s$ (but not $2p$).

*7.7 THE ZEEMAN EFFECT

Let us consider for the moment a hypothetical (and less interesting) world in which the electron has no spin, and therefore no spin magnetic moment.

* This is an optional section that may be skipped without loss of continuity.

Suppose we prepared a hydrogen atom in a $2p$ ($l = 1$) level and placed it in an external uniform magnetic field **B** (supplied by a laboratory electromagnet, for example). The magnetic moment $\boldsymbol{\mu}_L$ associated with the orbital angular momentum then interacts with the field, and the energy associated with this interaction is

$$U = -\boldsymbol{\mu}_L \cdot \mathbf{B} \qquad (7.21)$$

That is, magnetic moments aligned in the direction of the field have less energy than those aligned oppositely to the field. Let us assume that the field is in the z direction. Using Equation 7.18 for the z component of the magnetic moment, we have

$$U = -\mu_{L,z} B = m_l \mu_B B \qquad (7.22)$$

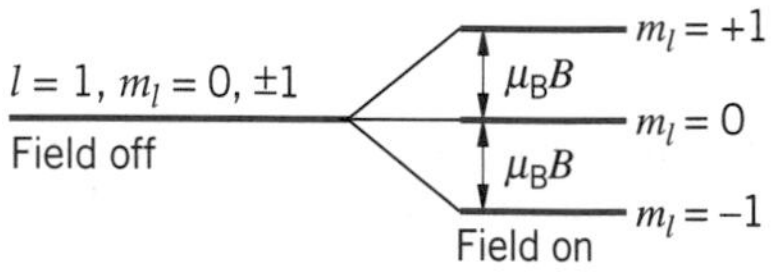

$l = 1$, $m_l = 0, \pm 1$
Field off
$\mu_B B$
$\mu_B B$
$m_l = +1$
$m_l = 0$
$m_l = -1$
Field on

FIGURE 7.21   The splitting of an $l = 1$ level in an external magnetic field. (The effects of the electron's spin angular momentum are ignored.) The energy in a magnetic field is different for different values of m_l.

in terms of the Bohr magneton μ_B defined in Equation 7.19. In the absence of a magnetic field, the $2p$ level has a certain energy E_0 (-3.4 eV). When the field is turned on, the energy becomes $E_0 + U = E_0 + m_l \mu_B B$; that is, there are now three different possible energies for the level, depending on the value of m_l. Figure 7.21 illustrates this situation.

Now suppose the atom emits a photon in a transition to the ground state ($1s$). In the absence of the magnetic field, a single photon is emitted with an energy of 10.2 eV and a corresponding wavelength of 122 nm. When the magnetic field is present, three photons can be emitted, with energies of 10.2 eV + $\mu_B B$, 10.2 eV, and 10.2 eV $- \mu_B B$. Let us examine how a small change in energy ΔE ($= \mu_B B$) affects the wavelength. Differentiating the expression $E = hc/\lambda$, we have

$$dE = \frac{-hc}{\lambda^2} d\lambda \qquad (7.23)$$

We replace the differentials with small differences, take absolute magnitudes, and solve for $\Delta \lambda$, which gives

$$\Delta \lambda = \frac{\lambda^2}{hc} \Delta E \qquad (7.24)$$

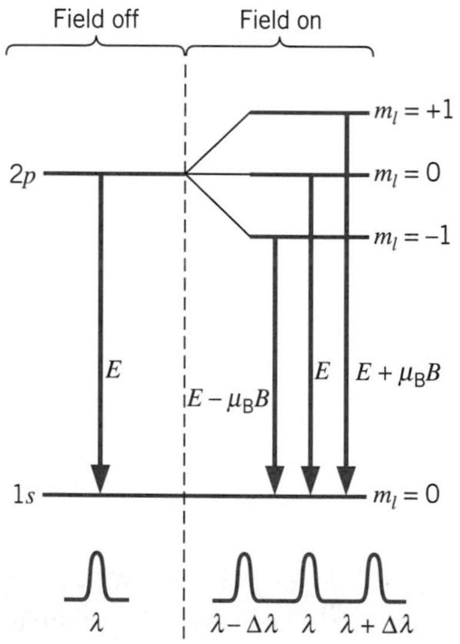

Figure 7.22 illustrates the three transitions, and shows an example of the result of a measurement of the emitted wavelengths.

In analyzing transitions between different m_l states, often we need to use a second *selection rule*: the only transitions that occur are those that change m_l by 0, +1, or −1:

$$\Delta m_l = 0, \pm 1 \qquad (7.25)$$

Changes in m_l of two or more are not permitted.

Field off   Field on

$2p$

$1s$

$m_l = +1$
$m_l = 0$
$m_l = -1$

E $E - \mu_B B$ E $E + \mu_B B$

$m_l = 0$

λ $\lambda - \Delta\lambda$ λ $\lambda + \Delta\lambda$

FIGURE 7.22   The normal Zeeman effect. When the field is turned on, the single wavelength λ becomes three separate wavelengths.

EXAMPLE 7.7

Compute the change in wavelength of the $2p - 1s$ photon when a hydrogen atom is placed in a magnetic field of 2.00 T.

SOLUTION

The energy of the photon is

$$E = -13.6 \, \text{eV} \left(\frac{1}{2^2} - \frac{1}{1^2} \right) = 10.2 \, \text{eV}$$

and its wavelength is

$$\lambda = \frac{hc}{E} = \frac{1240 \, \text{eV} \cdot \text{nm}}{10.2 \, \text{eV}} = 122 \, \text{nm}$$

The energy change ΔE of the levels is

$$\Delta E = \mu_B B = (9.27 \times 10^{-24} \, \text{J/T}) \, (2 \, \text{T})$$

$$= 18.5 \times 10^{-24} \, \text{J}$$

$$= 11.6 \times 10^{-5} \, \text{eV}$$

and so, from Equation 7.24,

$$\Delta \lambda = \frac{\lambda^2}{hc} \Delta E = \frac{(122 \, \text{nm})^2}{1240 \, \text{eV} \cdot \text{nm}} 11.6 \times 10^{-5} \, \text{eV}$$

$$= 0.00139 \, \text{nm}$$

Even for such a relatively large laboratory magnetic field as 2 T, the change in wavelength is very small, but easily measurable using an optical spectrometer.

The experiment we have just considered is an example of the *Zeeman effect*—the splitting of a spectral line with a single wavelength into lines with several different wavelengths when the emitting atoms are in an externally applied magnetic field. In the *normal Zeeman effect* a single spectral line splits into three components; this occurs only in atoms without spin. (All electrons of course have spin, unlike the hypothetical spinless electrons we considered; however, in certain atoms with several electrons, the spins can pair off and cancel, so that the atom behaves like a spinless one.) When spin *is* present, we must consider not only the effect of the orbital magnetic moment but also the spin magnetic moment. The resulting pattern of level splittings is more complicated, and spectral lines may split into more than three components. This case is known as the *anomalous Zeeman effect*, an example of which is shown in Figure 7.23.

Magnet

OFF

ON

FIGURE 7.23 The Zeeman effect in rhodium. The bottom spectrum shows the splitting of the spectral lines when the magnet is turned on.

*7.8   FINE STRUCTURE

In our discussion of the hydrogen spectrum in Chapter 6, it was mentioned that a careful inspection of the emission lines shows that many of them are in fact not single lines but very closely spaced combinations of two lines. In this section we examine the origin of that effect, known as *fine structure*.

In this calculation it is more convenient for us to examine the hydrogen atom from the electron's frame of reference, in which the proton *appears* to travel around the electron, just as the Sun *appears* to travel around the Earth. For convenience, we treat this problem in the context of the Bohr model to obtain an estimate of the effect.

Figure 7.24 shows the atom from the frames of reference of the proton and the electron. In the electron reference frame the motion of the proton in a circular orbit of radius r can be considered to be a current loop, which causes a magnetic field $\mathbf{B}$ at the electron. This magnetic field interacts with the spin magnetic moment of the electron, $\boldsymbol{\mu}_S = (-e/m)\mathbf{S}$. The interaction energy of the magnetic moment $\boldsymbol{\mu}_S$ in a magnetic field is

$$U = -\boldsymbol{\mu}_S \cdot \mathbf{B} \tag{7.26}$$

That is, when $\boldsymbol{\mu}_S$ and $\mathbf{B}$ are parallel, the energy ($U = -\mu_S B$) is lower than it is when $\boldsymbol{\mu}_S$ and $\mathbf{B}$ are antiparallel ($U = +\mu_S B$). We define the z direction to be the direction of $\mathbf{B}$; with $\boldsymbol{\mu}_S = (-e/m)\mathbf{S}$, we have

$$U = \frac{e}{m}\mathbf{S} \cdot \mathbf{B} = \frac{e}{m}S_z B \tag{7.27}$$

Since $S_z = \pm\frac{1}{2}\hbar$, the energy is

$$U = \pm\frac{e\hbar}{2m}B = \pm\mu_B B \tag{7.28}$$

The situation shown in Figure 7.24 has $S_z = +\frac{1}{2}\hbar$, and thus $U = +\mu_B B$. When $\mathbf{S}$ has the opposite orientation, $U = -\mu_B B$. The effect is to split

---

* This is an optional section that may be skipped without loss of continuity.

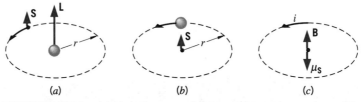

(a)   (b)   (c)

FIGURE 7.24   (a) An electron circulates about the proton with orbital angular momentum $\mathbf{L}$. The spin of the electron is parallel to $\mathbf{L}$. (b) From the point of view of the electron, the proton circulates as shown. (c) The apparently circulating proton is represented by the current i and causes a magnetic field $\mathbf{B}$ at the electron. The spin magnetic moment of the electron is opposite to its spin angular momentum.

each level into two, a higher state with **L** and **S** parallel and a lower state with **L** and **S** antiparallel, as shown in Figure 7.25. The energy difference between the states is $\Delta E = 2\mu_B B$.

At this point, the result looks rather similar to that of our previous discussion of the Zeeman effect, but it is important to note one significant difference: the magnetic field B in this case is *not* a field in the laboratory that can be turned on or off; it is, instead, a field produced by the apparent motion of the proton, that is *always* present.

We can use the Bohr model to make a rough estimate of the magnitude of this energy splitting. A circular loop of radius r carrying current i establishes at its center a magnetic field

$$B = \frac{\mu_0 i}{2r} \qquad (7.29)$$

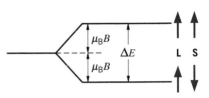

FIGURE 7.25 The fine-structure splitting in hydrogen. The state with **L** and **S** parallel is slightly higher in energy than the state with **L** and **S** antiparallel.

The current i is the charge carried around the loop ($+e$ in this case) divided by the time T for one orbit. The time for one orbit is the distance traveled ($2\pi r$) divided by the speed v.

$$B = \frac{\mu_0}{2r}\frac{e}{T} = \frac{\mu_0}{2r}\frac{ev}{2\pi r} \qquad (7.30)$$

The energy difference between the states is then

$$\Delta E = 2\mu_B B = \frac{\mu_0 ev}{2\pi r^2}\mu_B \qquad (7.31)$$

Since $|\mathbf{L}| = mvr = n\hbar$, (remember, we are using the Bohr model), the speed v is $n\hbar/mr$, so

$$\Delta E = \frac{\mu_0 e^2 \hbar^2 n}{4\pi m^2 r^3} \qquad (7.32)$$

Substituting Equation 6.31 for the Bohr radius, we obtain

$$\Delta E = \frac{\mu_0 e^2 \hbar^2 n}{4\pi m^2}\left(\frac{me^2}{4\pi\varepsilon_0\hbar^2}\frac{1}{n^2}\right)^3$$

$$= \frac{\mu_0 me^8}{256\pi^4\varepsilon_0^3\hbar^4}\frac{1}{n^5} \qquad (7.33)$$

We can rewrite this in a somewhat simpler form by recalling that $c^2 = 1/\varepsilon_0\mu_0$ and using the dimensionless constant defined in Equation 1.15

$$\alpha = \frac{e^2}{4\pi\varepsilon_0\hbar c} \qquad (7.34)$$

which gives

$$\Delta E = (mc^2)\alpha^4\frac{1}{n^5} \qquad (7.35)$$

Estimate of fine-structure splitting in hydrogen

α is known as the *fine structure constant* and is a dimensionless constant with a value very nearly equal to $\frac{1}{137}$. For the $n = 2$ state of hydrogen, we expect the state with **L** and **S** parallel to differ in energy from the state with **L** and **S** antiparallel by

$$\Delta E = (0.511 \text{ MeV}) \left(\frac{1}{137}\right)^4 \frac{1}{2^5} = 4.53 \times 10^{-5} \text{ eV}$$

We can compare this estimate with the experimental value, based on the observed splitting of the first line of the Lyman series, which gives 4.54×10^{-5} eV. We see that in spite of the assumptions we have made, our use of the Bohr model, and our failure to use the hydrogen wave functions to do this calculation, the agreement with the experimental value is remarkably good. (In fact, the agreement is so good as to be embarrassing, for we neglected to consider the important *relativistic* effect of the motion of the electron, which contributes to the fine structure about equally as the *spin-orbit* interaction discussed in this section. We really should regard this calculation as an *order-of-magnitude* estimate, which happens by chance to give a numerical result close to the observed value.)

SUGGESTIONS FOR FURTHER READING

A more detailed treatment of the hydrogen atom, especially of the fine structure, is in the following:

J. Norwood, *Twentieth Century Physics* (Englewood Cliffs, Prentice-Hall, 1976).

An excellent and detailed full-scale treatment of the solutions of the Schrödinger equation for the hydrogen atom is Chapter V of the following:

L. Pauling and E. B. Wilson, *Introduction to Quantum Mechanics* (New York, McGraw-Hill, 1935).

Representing the three-dimensional probability distributions on a two-dimensional paper is a great challenge for illustrators, and the interpretation of such illustrations is often a similar challenge for students. It might be helpful to look at some other representations:

N. Ashby and S. C. Miller, *Principles of Modern Physics* (San Francisco, Holden-Day, 1970).
R. B. Leighton, *Principles of Modern Physics* (New York, McGraw-Hill, 1959).
S. Brandt and H. D. Dahmen, *The Picture Book of Quantum Mechanics* (New York, Wiley, 1985).

A remarkable set of computer-drawn probability distributions can be found in the following:

D. Kleppner, M. G. Littman, and M. L. Zimmerman, "Highly Excited Atoms," *Scientific American* **244**, 130 (May 1981).

QUESTIONS

1. How does the quantum-mechanical interpretation of the hydrogen atom differ from the Bohr model?

2. How does a quantized angular momentum vector differ from a classical angular momentum vector?

3. What are the meanings of the quantum numbers n, l, m_l according to (a) the quantum-mechanical calculation; (b) the vector model; (c) the Bohr (orbital) model?

4. List the dynamical quantities that are constant for a specific choice of n and l. List the dynamical quantities that are *not* constant. Compare these lists with the Bohr model.

5. How does the orbital angular momentum differ between the Bohr model and the quantum-mechanical calculation?

6. What does it mean that **L** precesses about the z axis? Can we observe the precession?

7. In the Bohr model, we calculated the total energy from the potential energy and kinetic energy for each orbit. In the quantum-mechanical calculation, is the potential energy constant for any set of quantum numbers? Is the kinetic energy? Is the total energy?

8. What is meant by the term *space quantization*? Is space really quantized?

9. A deficiency of the Bohr model is the problem of angular momentum conservation in transitions between levels. Discuss this problem in relation to the quantum-mechanical angular momentum properties of the atom, especially the selection rule Equation 7.20. The photon can be considered to carry angular momentum $\hbar$.

10. The $2s$ electron has a greater probability to be close to the nucleus than the $2p$ electron (see Example 7.3) and also a greater probability to be farther away (see Problem 14). How is this possible?

11. The probability density $\psi^*\psi$ does not depend on ϕ for the wave functions listed in Table 7.1. What is the significance of this?

12. How would the wave functions of Table 7.1 change if the nuclear charge were Ze instead of e? (Recall how we made the same change in the Bohr model in Section 6.5.) What effect would this have on the radial probability densities $P(r)$?

13. Can a hydrogen atom in its ground state absorb a photon (of the proper energy) and end up in the $3d$ state?

14. Is it *correct* to think of the electron as a tiny ball of charge spinning on its axis? Is it *useful*? Is this situation similar to using the Bohr model to represent the electron's orbital motion?

15. The photon has a spin quantum number of 1, but its spin magnetic moment is zero. Explain.

16. What are the similarities and differences between Zeeman splitting and fine-structure splitting?

17. How would the calculated fine structure be different in an atom with a single electron and a nuclear charge of Ze?

18. Does the fine structure, as we have calculated it, have any effect on the $n = 1$ level?

19. How would (a) the Zeeman effect and (b) the fine structure be different in a muonic hydrogen atom? (See Problems 35 and 36 in Chapter 6.) The muon has the same spin as the electron, but is 207 times as massive.

20. Even though our calculation of the fine structure was based on a very simplified model, it does yield a result similar to the more correct calculation: the fine-structure splitting decreases as we go to higher excited states. Give at least two qualitative reasons for this.

PROBLEMS

1. List the 16 possible sets of quantum numbers of the $n = 4$ level of hydrogen without the inclusion of electron spin (as in Figure 7.2).

2. (a) What are the possible values of l for $n = 6$? (b) What are the possible values of m_l for $l = 6$? (c) What is the smallest possible value of n for which l can be 4? (d) What is the smallest possible l that can have a z component of $4\hbar$?

3. (a) Including the electron spin, what is the degeneracy of the $n = 5$ energy level of hydrogen? (b) By adding up the number of states for each value of l permitted for $n = 5$, show that the same degeneracy as part (a) is obtained.

4. For each l value, the number of possible states is $2(2l + 1)$. Show explicitly that the total number of states for each principal quantum number is

$$\sum_{l=0}^{n-1} 2(2l + 1) = 2n^2$$

This gives the degeneracy of each energy level.

5. Explain why each of the following sets of quantum numbers (n, l, m_l, m_s) is not permitted for hydrogen.
   (a) $(2, 2, -1, +\frac{1}{2})$
   (b) $(3, 1, +2, -\frac{1}{2})$
   (c) $(4, 1, +1, -\frac{3}{2})$
   (d) $(2, -1, +1, +\frac{1}{2})$

6. An electron is in the $n = 4$, $l = 3$ state of hydrogen. (a) What is the length of the electron's angular momentum vector? (b) How many different possible z components can the angular momentum vector have? List the possible z components. (c) What are the values of the angle that the **L** vector makes with the z axis? (d) Would your answers to (a), (b), or (c) change if the principal quantum number n were 5 instead of 4?

7. What angles does the **L** vector make with the z axis when $l = 2$?

8. Show that the $(1, 0, 0)$ and $(2, 0, 0)$ wave functions listed in Table 7.1 are properly normalized.

9. Show by direct substitution that the $n = 2$, $l = 0$, $m_l = 0$ and $n = 2$, $l = 1$, $m_l = 0$ wave functions of Table 7.1 are both solutions of Equation 7.3 corresponding to the energy of the first excited state of hydrogen.

10. Show by direct substitution that the wave function corresponding to $n = 1$, $l = 0$, $m_l = 0$ is a solution of Equation 7.3 corresponding to the ground-state energy of hydrogen.

11. Show that the radial probability density of the ls level has its maximum value at $r = a_0$.

12. Find the values of the radius where the $n = 2$, $l = 0$ radial probability density has its maximum values.

13. What is the probability of finding a $n = 2$, $l = 1$ electron between a_0 and $2a_0$?

14. Find the probabilities for the $n = 2$, $l = 0$ and $n = 2$, $l = 1$ states to be further than $5a_0$ from the nucleus. Which has the greater probability to be far from the nucleus?

15. For a hydrogen atom in the ground state, what is the probability to find the electron between $1.00a_0$ and $1.01a_0$? (*Hint:* It is not necessary to evaluate any integrals to solve this problem.)

16. The mean or average value of the radius r can be found according to $r_{av} = \int_0^\infty rP(r)\, dr$. Show that the mean value of r for the $1s$ state of hydrogen is $\frac{3}{2}a_0$. Why is this greater than the Bohr radius?

17. Find the value of r_{av} (see Problem 16) for the $2s$ and $2p$ levels.

18. The mean or average value of the potential energy of the electron in a hydrogen atom can be found from $U_{av} = \int_0^\infty U(r)\, P(r)\, dr$. Find U_{av} in the $1s$ state and compare with the potential energy computed with the Bohr model when $n = 1$.

19. (a) For the $1s$, $2s$, and $2p$ states of hydrogen, show that $(r^{-1})_{av} = 1/n^2a_0$. (*Hint:* Use $\int_0^\infty x^n e^{-ax}\, dx = n!/a^{n+1}$.) (b) This turns out to be a general result for any state of hydrogen. Based on this result, explain why the Bohr model gives such a good estimate for the fine-structure splitting as well as for other magnetic effects due to the circulating electron.

20. Suppose the source of atoms in a Stern-Gerlach experiment were an oven of temperature 1000 K. Assume the magnetic field gradient to be 10 T/m, and take the length of the magnetic field region and the field-free region between magnet and screen to be 1 m each. Make any other assumptions you may need and estimate the separation of the images obseved on the screen.

21. List the excited states (in spectroscopic notation) to which the $4p$ state can make downward transitions.

22. A hydrogen atom is in an excited $5g$ state, from which it makes a series of transitions, ending in the $1s$ state. Show, on a diagram similar to Figure 7.20, the sequence of transitions that can occur. (b) Repeat part (a) if the atom begins in the $5d$ state.

23. Consider the normal Zeeman effect applied to the $3d$ to $2p$ transition. (a) Sketch an energy-level diagram that shows the splitting of the $3d$ and $2p$ levels in an external magnetic field. Indicate all possible transitions from each m_l state of the $3d$ level to each m_l state of the $2p$ level. (b) Which transitions satisfy the $\Delta m_l = \pm 1$ or 0 selection rule? (c) Show that there are only three different transition energies emitted.

24. A collection of hydrogen atoms is placed in a magnetic field of 3.50 T. Ignoring the effects of electron spin, find the wavelengths of the three normal Zeeman components (a) of the 3*d* to 2*p* transition; (b) of the 3*s* to 2*p* transition.

25. Calculate the wavelengths of the components of the first line of the Lyman series, taking the fine structure of the 2*p* level into account.

26. Calculate the energies and wavelengths of the 3*d* to 2*p* transition, taking into account the fine structure of *both* levels. How many component wavelengths might there be in the transition?

27. CUPS Exercise 8.2.

28. CUPS Exercise 8.3.

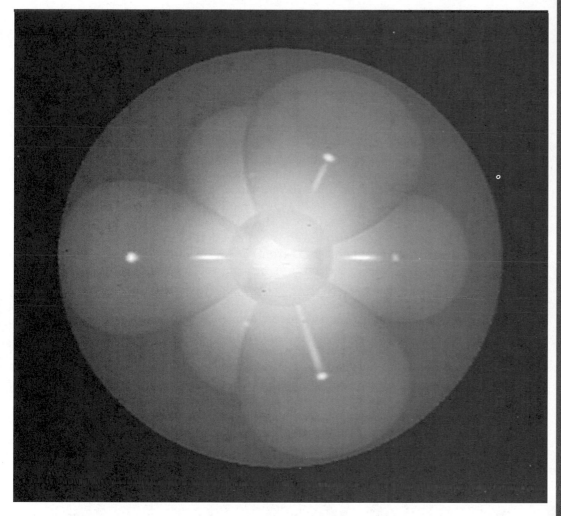

This computer-generated drawing shows the structure of an atom of neon, with the electron probability distributions surrounding the central nucleus. The inner sphere represents the 1s electrons, the outer sphere is the 2s electrons, and the lobes are the 2p electrons.

MANY- ELECTRON ATOMS

A number of difficulties arise when we attempt to construct a model that helps us to understand the structure and properties of atoms with more than one electron. For example, let's try to make an atom of helium. The nuclear charge is $Z = 2$, and we must therefore supply two electrons. As we bring in the first electron, the energy levels through which it must pass to reach the ground state are well described using our previous calculation, putting $Z = 2$. This electron eventually occupies the $1s$ state. As we bring in the second electron, it feels the *attraction* of the nucleus with $Z = 2$ and also the electrical *repulsion* of the first electron. (The first $1s$ electron, you will recall, appears as a fuzzy ball of charge that completely surrounds the nucleus; see Figure 7.10.) This second electron thus feels a net attraction toward the nucleus with a force that is somewhat weaker than we would otherwise expect for $Z = 2$; there is an *effective value* of the nuclear charge seen by the second electron, with a contribution of $+2$ from the nucleus and a contribution of perhaps -1 from the first electron, giving $Z_{eff} \cong +1$. However, as the second electron is brought in, the *first* electron feels a repulsive force from the second electron, which changes the energy levels of the *first* electron.

The problem of the mutual interactions of three or more objects is an example of what physicists call the *many-body problem*. Exact, closed-form solutions to the Schrödinger equation cannot be found for such problems. For the two-electron atom, numerical solutions can be found that give accurate results, but the solutions become more difficult to obtain as Z increases. In this chapter, we consider an approximate set of energy levels for many-electron atoms, and we try to understand some of the properties of atoms (chemical, electrical, magnetic, optical, etc.) based on those energy levels.

8.1 THE PAULI EXCLUSION PRINCIPLE

Before we consider the energy levels, let us first consider the question of how the electrons occupy the energy levels in a many-electron atom. We might expect that all Z electrons will eventually cascade down to the lowest energy level, the $1s$ state. The properties of this kind of atom should vary rather smoothly with respect to its neighbors having $Z \pm 1$ electrons. Indeed, certain of the properties of atoms, such as the energies of the emitted X rays, show this smooth variation. However, other properties do not vary in this way and thus are not consistent with this model of all electrons in the same level. For example, neon (with $Z = 10$) is an *inert gas*; it is practically unreactive and does not form chemical compounds under most conditions. Its neighbors, fluorine ($Z = 9$) and sodium ($Z = 11$), are among the most reactive of the elements and will under most conditions combine with other substances, sometimes violently. As another example, nickel ($Z = 28$) is strongly magnetic (ferromagnetic) and, for a metal, does not have a particularly large electrical conductivity. Copper ($Z = 29$) is an excellent electrical conductor but is not magnetic. Such wide variations in properties between neighboring elements suggest that it is not correct to assume that all electrons occupy the same energy level.

The rule that prevents all of the electrons in an atom from falling into the 1s level was proposed by Wolfgang Pauli in 1925, based on a study of the transitions that are present, and those that are expected but *not* present, in the emission spectra of atoms. Simply stated, the *Pauli exclusion principle* is as follows:

No two electrons in a single atom can have the same set of quantum numbers (n, l, m_l, m_s).

The Pauli principle is the most important rule governing the structure of atoms, and no study of the properties of atoms can be attempted without a thorough understanding of this principle.

Let us illustrate how the Pauli principle works in the case of helium $(Z = 2)$. The first electron in helium, in the 1s ground state, has quantum numbers $n = 1$, $l = 0$, $m_l = 0$, $m_s = +\frac{1}{2}$ or $-\frac{1}{2}$. The second electron can have the same n, l, and m_l, but it cannot have the same m_s, since then the exclusion principle would be violated. Thus if the first 1s electron has $m_s = +\frac{1}{2}$, the second 1s electron must have $m_s = -\frac{1}{2}$. Now suppose we construct an atom of lithium $(Z = 3)$. Just as with helium, the first two electrons will have quantum numbers $(n, l, m_l, m_s) = (1, 0, 0, +\frac{1}{2})$ and $(1, 0, 0, -\frac{1}{2})$. Since the third electron, according to the exclusion principle, cannot have the same set of quantum numbers as the first two, it *cannot go into the $n = 1$ level*, since there are only two different sets of quantum numbers available in the $n = 1$ level, and both of those sets have already been used. The third electron must therefore go into one of the $n = 2$ levels; experiments indicate that the next level available of the two $n = 2$ levels (2s or 2p) is the 2s level, so the third electron might have quantum numbers $(n, l, m_l, m_s) = (2, 0, 0, +\frac{1}{2})$ or $(2, 0, 0, -\frac{1}{2})$. The fourth electron, in the case of beryllium $(Z = 4)$, would have the same n, l, and m_l, but the opposite m_s value as the third electron. When we reach boron, with $Z = 5$, the fifth electron cannot go into the 2s state, since we have already assigned both possible sets of quantum numbers in that level; the fifth electron then goes into one of the 2p levels. We might therefore expect that the properties of boron, with a 2p electron, would be different from the properties of lithium or beryllium, which have only 2s electrons.

It is this process of first using up all of the possible quantum numbers for one level, and then placing electrons in the next level, which accounts for the variations in the chemical and physical properties of the elements.

Wolfgang Pauli (1900–1958, Switzerland). His exclusion principle gave the basis for understanding atomic structure, but he also contributed to the development of quantum theory, to the theory of nuclear beta decay, and to the understanding of symmetry in physical laws.

8.2 ELECTRONIC STATES IN MANY-ELECTRON ATOMS

Figure 8.1 illustrates the result of an approximate calculation of the ordering of the filling of energy levels in many-electron atoms as the atomic number

Z increases. The 1*s* level remains the lowest energy level, and the 2*s* and 2*p* levels remain fairly close in energy. The 2*s* level always lies a bit lower in energy than the 2*p* level. (The fine structure splitting is very small on the scale of this diagram.) We can understand why the 2*s* level lies lower in energy if we recall Example 7.3. An electron in the 2*s* level spends more of its time inside the Bohr radius than an electron in the "circular" 2*p* level does. The 2*p* electron sees the nuclear electric charge $+Ze$ shielded, or "screened," by the two 1*s* electrons. The 2*s* electron, however, occasionally is found closer to the nucleus than the Bohr radius, where it feels the attraction of the *entire* nuclear charge. The 2*s* electron feels, on the average, a slightly greater force of attraction to the nucleus than the 2*p* electron does; therefore the 2*s* electron is more tightly bound and lies lower in energy.

A more extreme example of the *tighter binding* of the *penetrating orbits* occurs for the *n* = 3 levels. The 3*s* electron penetrates the inner orbits (it has a large probability density at small *r*; see Figure 7.10), and the 3*p* electron penetrates almost as much. The 3*d* electron has negligible penetration of the inner orbits. As a result, the 3*s* and 3*p* levels are more tightly bound and therefore lower in energy than the 3*d* level. A similar effect occurs for the *n* = 4 levels—the tighter binding of the 4*s* and 4*p* electrons pulls their energy levels down so low that they almost coincide with the 3*d* level, as shown in Figure 8.1. The 3*d* and 4*s* levels are very close in energy—for some atoms the 3*d* level is lower and for some atoms the 4*s* is lower. This small energy difference is an important factor that contributes to the large electrical conductivity of copper, as we discuss later in this chapter.

The tighter binding of the penetrating *s* and *p* orbits also pulls the 5*s* and 5*p* levels down close to the 4*d* level, and similarly causes the 6*s* and 6*p* levels to appear at roughly the same energy as the 5*d* and 4*f* levels.

As we learned in the case of the hydrogen atom, orbits with the same value of *n* all lie at about the same average distance from the nucleus. (The electrons in the penetrating orbits spend some of their time closer to the nucleus than the nonpenetrating orbits, but also some of their time further from the nucleus; the average distance from the nucleus of the penetrating orbits is then about the same as the average distance from the nucleus of the nonpenetrating orbits. See Problem 17 in Chapter 7 for a verification of this for the hydrogen atom.) The set of orbits with a certain value of *n*, with about the same average distance from the nucleus, is known as an atomic *shell*. The atomic shells are designated by letter, as follows:

$$\begin{array}{cccccc} n & 1 & 2 & 3 & 4 & 5 \\ \text{Shell} & K & L & M & N & O \end{array}$$

The levels with a certain value of *n* and *l* (for instance, 2*s* or 3*d*) are known as *subshells*. The number of electrons that can be placed in each subshell is $2(2l + 1)$. The $(2l + 1)$ factor comes from the number of different m_l values for each *l*, since m_l can take the values 0, ±1, ±2, ±3, . . . , ±*l*. The extra factor of 2 comes from the two different m_s values; for each m_l, we can have $m_s = +\frac{1}{2}$ or $m_s = -\frac{1}{2}$. According to this scheme, the 1*s* subshell has a capacity of $2(2 \times 0 + 1) = 2$ electrons; the 3*d* subshell has a capacity of $2(2 \times 2 + 1) = 10$ electrons. (Note that this capacity doesn't depend

Penetrating orbits

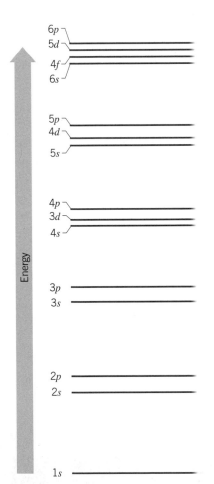

FIGURE 8.1   Atomic subshells, in order of increasing energy. The energy groupings are not to scale, but are merely representative of the relative energies of the subshells.

on n; *any d* subshell has a capacity of 10 electrons.) Table 8.1 shows the ordering and capacity of the subshells.

It is important to keep in mind exactly what is represented by Figure 8.1 and Table 8.1. They give the order of filling of the energy levels, and so they represent only the "outer" or valence electrons. For example, the first 18 electrons fill the levels up through $3p$, and the energy levels (subshells) available to the 19th electron in potassium ($Z = 19$) or calcium ($Z = 20$) are well described by Figure 8.1. However, the energy levels appropriate to the 19th electron in a heavy element such as lead ($Z = 82$) would be very different. In this case it is more correct to describe the atom in terms of shells—all of the $n = 3$ states (the M shell) are grouped together, as are all of the $n = 4$ states (the N shell), and so forth. When we discuss the *inner* structure of the atom, as in the case of X rays, the ordering of Figure 8.1 is not correct, and it is more appropriate to group the levels by shells, as we do in Section 8.5.

8.3 THE PERIODIC TABLE

Figure 8.2 shows the periodic table, which is an orderly array of the chemical elements, listed in order of increasing atomic number Z and arranged in such a way that the vertical columns, called *groups*, contain elements with rather similar physical and chemical properties. In this section we discuss the way in which the filling of electronic subshells helps us understand the arrangement of the periodic table. In the next section we examine some of the physical and chemical properties of the elements.

In attempting to understand the ordering of subshells and the periodic table, we must follow two rules for filling the electronic subshells:

1. The capacity of each subshell is $2(2l + 1)$. (This is of course just another way of stating the Pauli exclusion principle.)
2. The electrons occupy the lowest energy states available.

To indicate the electron configuration of each element, we use a notation in which the identity of the subshell and the number of electrons in it are listed. The identity of the subshell is indicated in the usual way, and the number of electrons in that subshell is indicated by a superscript. Thus hydrogen has the configuration $1s^1$, for one electron in the $1s$ shell, and helium has the configuration $1s^2$. Helium has both a filled subshell (the $1s$) and a closed major shell (the K shell) and thus is an extraordinarily stable and inert element. With lithium ($Z = 3$), we begin to fill the $2s$ subshell; lithium has the configuration $1s^2 2s^1$. With beryllium ($Z = 4$, $1s^2 2s^2$) the $2s$ subshell is full, and the next element must begin filling the $2p$ subshell (boron, $Z = 5$, $1s^2 2s^2 2p^1$). The $2p$ subshell has a capacity of six electrons, and with neon ($Z = 10$, $1s^2 2s^2 2p^6$) both the $2p$ subshell and the L shell ($n = 2$) are complete.

The next row (or *period*) begins with sodium ($Z = 11$, $1s^2 2s^2 2p^6 3s^1$), and the $3s$ and $3p$ subshells are filled in much the same way as the $2s$ and

TABLE 8.1 FILLING OF ATOMIC SUBSHELLS

n	l	Subshell	Capacity $2(2l + 1)$
1	0	$1s$	2
2	0	$2s$	2
2	1	$2p$	6
3	0	$3s$	2
3	1	$3p$	6
4	0	$4s$	2
3	2	$3d$	10
4	1	$4p$	6
5	0	$5s$	2
4	2	$4d$	10
5	1	$5p$	6
6	0	$6s$	2
4	3	$4f$	14
5	2	$5d$	10
6	1	$6p$	6
7	0	$7s$	2
5	3	$5f$	14
6	2	$6d$	10

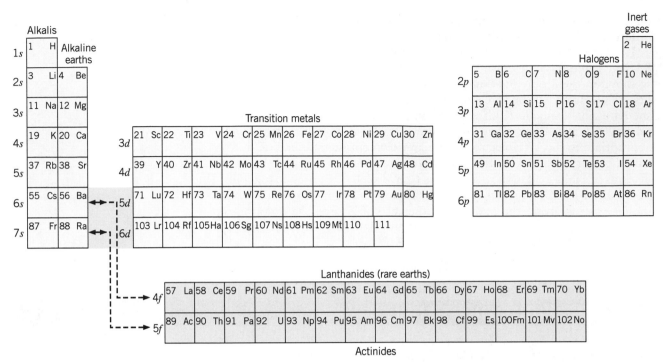

FIGURE 8.2 Periodic table of the elements.

$2p$ subshells, ending with the inert gas argon ($Z = 18$, $1s^2 2s^2 2p^6 3s^2 3p^6$). The elements of the third row (period) are chemically similar to the corresponding elements of the second row (period), and so are written directly under them. The next electron would ordinarily go into the $3d$ level. However, the highly penetrating orbit of the $4s$ electron causes the $4s$ level to appear at a slightly lower energy than the $3d$ level, so the $4s$ subshell normally fills first. The configurations of potassium ($Z = 19$) and calcium ($Z = 20$) are therefore respectively $1s^2 2s^2 2p^6 3s^2 3p^6 4s^1$ and $1s^2 2s^2 2p^6 3s^2 3p^6 4s^2$. These elements have properties similar to, and therefore appear directly under, the corresponding elements with one and two s-subshell electrons in the second and third periods.

We now begin to fill the $3d$ subshell. Since there is no $1d$ or $2d$ subshell (why not?), we would expect the first element with a d-subshell configuration to have rather different chemical properties than the elements we have placed previously; thus it should not appear in any of our previously occupied groups (columns), and so we begin a new group with scandium ($Z = 21$, $1s^2 2s^2 2p^6 3s^2 3p^6 4s^2 3d^1$). The $3d$ subshell eventually closes with zinc ($Z = 30$, $1s^2 2s^2 2p^6 3s^2 3p^6 4s^2 3d^{10}$). Along the way there are some minor variations; the most important is copper, with $Z = 29$. For this case the $3d$ level lies slightly lower than the $4s$ level, and so the $3d$ subshell fills before the $4s$, resulting in the configuration $1s^2 2s^2 2p^6 3s^2 3p^6 3d^{10} 4s^1$. As we will see in the next section, this configuration is responsible for the large electrical conductivity of copper.

In the next series of elements, the $4p$ subshell is filled, from gallium ($Z = 31$) to the inert gas krypton ($Z = 36$). When we move to the next period, we again fill the 5s subshell before the 4d subshell, and the series of 10 elements corresponding to the filling of the 4d subshell is written directly under the series that had unfilled configurations in the 3d subshell. (Silver, with $Z = 47$, corresponds exactly to copper in the fourth period, with the 4d subshell filling before the 5s.) After the completion of the 4d subshell, the 5p subshell is filled, ending with the inert gas xenon ($Z = 54$).

The next period begins with cesium and barium filling the 6s subshell, and lanthanum beginning the 5d subshell. As was the case in the previous periods, the 5d and 6s lie at almost the same energy; since the 6s is at a slightly lower energy, it fills first. However, there is yet another subshell at about the same energy as the 6s and 5d—the 4f subshell, which now begins to fill, from cerium to lutetium. This series of elements, called the lanthanides or rare earths, is usually written separately in the periodic table, since there have been no other f-subshell elements under which to write them. The 4f subshell has a capacity of 14 electrons, and so there are 14 elements in the lanthanide series. Once the 4f subshell is complete, we return to filling the 5d subshell, writing those elements in the groups under the corresponding 3d and 4d elements, and then complete the period with the filling of the 6p subshell, ending with the inert gas radon ($Z = 86$). The seventh period is filled much like the sixth, with a series known as the actinides, written under the lanthanides, corresponding to the filling of the 5f subshell.

Table 8.2 lists the electronic configurations of some of the elements.

What is most remarkable about this scheme is that the arrangement of the periodic table was known well before the introduction of atomic theory. The elements were organized into groups and periods based on their physi-

Atomic theory and the periodic table

TABLE 8.2 ELECTRONIC CONFIGURATIONS OF SOME ELEMENTS

H	$1s^1$	Y	[Kr] $5s^2 4d^1$
He	$1s^2$	Mo	[Kr] $5s^1 4d^5$
Li	$1s^2 2s^1$	Ag	[Kr] $5s^1 4d^{10}$
Be	$1s^2 2s^2$	In	[Kr] $5s^2 4d^{10} 5p^1$
B	$1s^2 2s^2 2p^1$	Xe	[Kr] $5s^2 4d^{10} 5p^6$
Ne	$1s^2 2s^2 2p^6$	Cs	[Xe] $6s^1$
Na	[Ne] $3s^1$	La	[Xe] $6s^2 5d^1$
Al	[Ne] $3s^2 3p^1$	Ce	[Xe] $6s^2 5d^1 4f^1$
Ar	[Ne] $3s^2 3p^6$	Pr	[Xe] $6s^2 4f^3$
K	[Ar] $4s^1$	Gd	[Xe] $6s^2 5d^1 4f^7$
Sc	[Ar] $4s^2 3d^1$	Dy	[Xe] $6s^2 4f^{10}$
Cr	[Ar] $4s^1 3d^5$	Yb	[Xe] $6s^2 4f^{14}$
Mn	[Ar] $4s^2 3d^5$	Lu	[Xe] $6s^2 5d^1 4f^{14}$
Cu	[Ar] $4s^1 3d^{10}$	Re	[Xe] $6s^2 5d^5 4f^{14}$
Zn	[Ar] $4s^2 3d^{10}$	Au	[Xe] $6s^1 5d^{10} 4f^{14}$
Ga	[Ar] $4s^2 3d^{10} 4p^1$	Hg	[Xe] $6s^2 5d^{10} 4f^{14}$
Kr	[Ar] $4s^2 3d^{10} 4p^6$	Tl	[Xe] $6s^2 5d^{10} 4f^{14} 6p^1$
Rb	[Kr] $5s^1$	Rn	[Xe] $6s^2 5d^{10} 4f^{14} 6p^6$

A symbol in brackets [ ] means that the atom has the configuration of the previous inert gas plus the additional electrons listed.

cal and chemical properties by Dmitri Mendeleev in 1859; understanding that organization in terms of atomic levels is a great triumph for the atomic theory. What remains is to interpret the chemical and physical properties based on the atomic levels.

8.4 PROPERTIES OF THE ELEMENTS

In this section we briefly study the way our knowledge of atomic structure helps us to understand the physical and chemical properties of the elements. Our discussion is based on the following two principles:

1. Filled subshells are normally very stable configurations. An atom with one electron beyond a closed shell will readily give up that electron to another atom to form a chemical bond. Similarly, an atom lacking one electron from a closed shell will readily accept an additional electron from another atom in forming a chemical bond.

2. Filled subshells do not normally contribute to the chemical or physical properties of an atom. Only the electrons in the unfilled subshells need be considered. (X-ray energies are an exception to this rule.)

We consider a number of different physical properties of the elements, and try to understand those properties based on atomic theory.

1. **Atomic Radii.**   We have already learned that the radius of an atom is not a precisely defined quantity, since the electron probability density determines the "size" of an atom. The radii are similarly hard to define experimentally, and in fact different kinds of experiments may give different values for the radii. One way of defining the radius is by means of the spacing between the atoms in a crystal containing that element. Figure 8.3 shows how such typical atomic radii vary with Z.

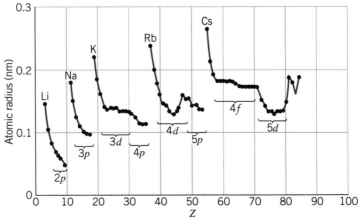

FIGURE 8.3   Atomic radii, determined from ionic crystal atomic separations. These radii are different from the mean radii of the electron cloud for free atoms.

2. **Ionization Energy**. The minimum energy needed to remove an electron from an atom is known as its *ionization energy*. For example, hydrogen has an ionization energy of 13.6 eV. Helium has an ionization energy of 24.6 eV for the first electron and 54.4 eV for the second electron. Table 8.3 gives the ionization energies of some of the elements, and Figure 8.4 shows the variation of ionization energy with atomic number Z.

3. **Electrical Resistivity**. In bulk materials, an electric current flows when a potential difference (voltage) is applied across the material. The current i and voltage V are related according to the expression $V = iR$, where R is the electrical resistance of the material. If the material is uniform with length L and cross-sectional area A, then the resistance is

$$R = \rho \frac{L}{A}$$

The *resistivity* ρ is characteristic of the kind of material and is measured in units of ohm·m. A good electrical conductor has a small resistivity ($\rho = 1.7 \times 10^{-8}$ ohm·m for copper); a poor conductor has a large resistivity ($\rho = 2 \times 10^{15}$ ohm·m for sulfur). From the atomic point of view, current depends on the movement of relatively loosely bound electrons, which can be removed from their atoms by the applied potential difference, and also on the ability of the electrons to travel from one atom to another. Thus elements with s electrons, which are the least tightly bound and which also travel farthest from the nucleus, are expected to have small resistivities.

Figure 8.5 shows the variation of electrical resistivity with atomic number.

4. **Magnetic Susceptibility**. When a material is placed in a magnetic field of intensity B, the material becomes "magnetized" and acquires a magnetization M, which for many materials is proportional to B:

$$\mu_0 M = \chi B$$

TABLE 8.3 IONIZATION ENERGIES [IN eV] OF NEUTRAL ATOMS OF SOME ELEMENTS

H	13.60	Ar	15.76
He	24.59	K	4.34
Li	5.39	Cu	7.72
Be	9.32	Kr	14.00
Ne	21.56	Rb	4.18
Na	5.14	Au	9.22

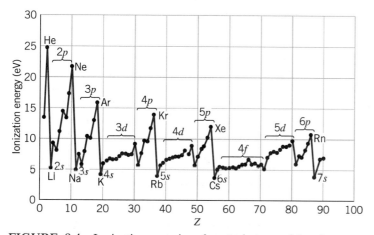

FIGURE 8.4 Ionization energies of neutral atoms of the elements.

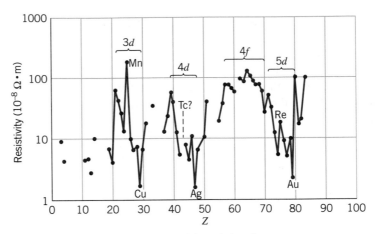

FIGURE 8.5 Electrical resistivities of the elements.

where χ is a dimensionless constant called the *magnetic susceptibility*. (Materials for which $\chi > 0$ are known as *paramagnetic*, and those for which $\chi < 0$ are called *diamagnetic*; materials that remain permanently magnetized even when B is removed are known as *ferromagnetic* and χ is undefined for such materials.)

From the atomic point of view, the magnetism of atoms depends on the **L** and **S** of the electrons in unfilled subshells, since the atomic magnetic moments $\boldsymbol{\mu}_L$ and $\boldsymbol{\mu}_S$ are proportional to **L** and **S** (recall Table 7.2). This effect is responsible for paramagnetic susceptibilities and occurs in all atoms in which **L** or **S** is nonzero. Diamagnetism is caused by the following effect: when a varying magnetic field occurs in an area bounded by an electric circuit, an *induced current* flows in the circuit; the induced current sets up a magnetic field which tends to *oppose* the changes in the applied field (Lenz's law). In the atomic physics case, the electric circuit is the circulating electron, and the induced current consists of a slight speeding up or slowing down of the electron in its orbit when a magnetic field is applied. This produces a contribution to the magnetization of the material which is opposite to the applied field **B**, and so the diamagnetic contribution to χ is negative.

Figure 8.6 shows the magnetic susceptibilities of the elements.

Just by examining Figures 8.3 to 8.6, you can see the remarkable regularities in the properties of the elements. Notice especially how similar the properties of the different sequences of elements are—for example, the electrical resistivity of the *d*-subshell elements or the magnetic susceptibility of the *p*-subshell elements. We now look at how the atomic structure is responsible for these properties.

Inert Gases

The inert gases occupy the last column of the periodic table. Since they have only filled subshells, the inert gases do not generally combine with other elements to form compounds; these elements are very reluctant to

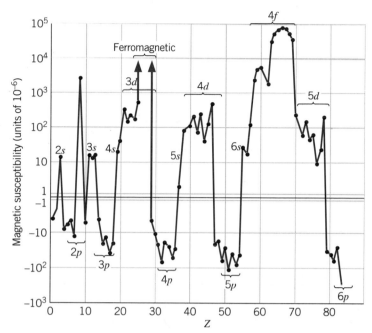

FIGURE 8.6   Magnetic susceptibilities of the elements.

give up or to accept an electron. At room temperature they are monatomic gases, and since their atoms don't like to join with each other, the boiling points are very low (typically −200°C). Their ionization energies are much larger than those of neighboring elements, because of the extra energy needed to break open a filled subshell.

p-Subshell Elements

The elements of the column (group) next to the inert gases are the halogens (F, Cl, Br, I, At). These atoms lack one electron from a closed shell and have the configuration np^5. Since a closed p subshell is a very stable configuration, these elements readily form compounds with other atoms that can provide an extra electron to close the p subshell. The halogens are therefore extremely reactive.

As we move across the series of six elements in which the p subshell is being filled, the atomic radius *decreases*. This "shrinking" occurs because the nuclear charge is increasing and pulling all of the orbits closer to the nucleus. Notice from Figure 8.3 that the halogens have the smallest radii within each p subshell series. (The ionic crystal radii of the inert gases are not known.)

As we increase the nuclear charge, the p electrons also become more tightly bound; from Figure 8.4 we see how the ionization energy increases systematically as the p subshell is filled.

From Figure 8.6 we see that each p subshell series is diamagnetic, with a characteristic negative magnetic susceptibility.

s-Subshell Elements

The elements of the first two columns (groups) are known as the alkalis (configuration ns^1) and alkaline earths (ns^2). The single s electron makes the alkalis quite reactive; the alkaline earths are similarly reactive, in spite of the filled s subshell. This occurs because the s electron wave function can extend rather far from the nucleus, where the electron is screened (by $Z - 1$ other electrons) from the nuclear charge and therefore not tightly bound. (Notice from Figure 8.3 that the ns^1 and ns^2 configurations give the largest atomic radii, and from Figure 8.4 that they have the smallest ionization energies.) For the same reasons, the ns^1 and ns^2 elements are relatively good electrical conductors. From Figure 8.6 we see that these elements are paramagnetic; for $l = 0$, there is no diamagnetic contribution to the magnetism.

Transition Metals

The three rows of elements in which the d subshell is filling (Sc to Zn, Y to Cd, Lu to Hg) are known as the *transition metals*. Many of their chemical properties are determined by the "outer" electrons—those whose wave functions extend furthest from the nucleus. For the transition metals, these are always s electrons, which have a larger mean radius than the d electrons. (Remember that the mean radius depends mostly on n; the s electrons of the transition metals have a larger n than the d electrons.) As the atomic number increases across the transition metal series, we add one d electron and one unit of nuclear charge; the net effect on the s electron is very small, since the additional d electron screens the s electron from the additional nuclear charge. The chemical properties of the transition metals are therefore very similar, as the small variation in radius and ionization energy shows.

The electrical resistivity of the transition metals shows two interesting features: a sharp rise at the center of the sequence, and a sharp drop near the end (Figure 8.5). The sharp drop near the end of the sequence indicates the small resistivity (large conductivity) of copper, silver, and gold. If we filled the d subshell in the expected sequence, copper would have the configuration $4s^2 3d^9$; however the filled d subshell is more stable than a filled s subshell, and so one of the s electrons is transferred to the d subshell, resulting in the configuration $4s^1 3d^{10}$. This relatively free, single s electron makes copper an excellent conductor. Silver ($5s^1 4d^{10}$) and gold ($6s^1 5d^{10}$) behave similarly.

At the center of the sequence of transition metals there is a sharp rise in the resistivity; apparently a half-filled shell is also a stable configuration, and so Mn ($3d^5$) and Re ($5d^5$) have larger resistivities than their neighbors. (The element Tc, with configuration $4d^5$, is radioactive and is not found in nature; its resistivity is unknown.) A similar rise in resistivity is seen at the center of the $4f$ sequence.

The transition metals have similar paramagnetic susceptibilities, due to the large orbital angular momentum of the d electrons and also to the large *number* of *d*-subshell electrons that can couple their *spin* magnetic moments. These two effects are large enough to overcome the diamagnetism of the orbital motion. It is the *d* electrons, with their large angular momentum, which are also responsible for the ferromagnetism of iron, nickel, and cobalt. As soon as the *d* subshell is filled, however, the orbital and spin magnetic moments no longer contribute to the magnetic properties (all of the m_l and m_s values, positive as well as negative, are taken); for this reason, copper and zinc are diamagnetic, not paramagnetic like their transition metal neighbors.

Lanthanides (Rare Earths)

The rare earths are rather similar to the transition metals in that an "inner" subshell (the 4*f*) is being filled after an "outer" subshell (the 6*s*). For the same reasons discussed above, the chemical properties of the rare earths should be rather similar, since they are determined mainly by the 6*s* electron; the radii and ionization energies show that this is true.

Because of the larger orbital angular momentum of *f*-subshell electrons ($l = 3$) and also because of the larger *number* of *f*-subshell electrons that can align their spin magnetic moments, the paramagnetic susceptibilities of the rare earths are even larger than those of the transition metals. Even the ferromagnetism of the rare earths is substantially stronger than that of the iron group. Generally, we think of iron as the most magnetic of the elements. If we magnetize a piece of iron, the internal magnetic field (within the piece of iron) is about 28 T. Magnetized holmium metal, a rare earth, has an internal magnetic field of 800 T, roughly 30 times that of iron! Most of the other rare earths have similar magnetic properties. (The rare earth metals do not reveal their "ferromagnetic" properties at room temperature, but must be cooled to lower temperatures. Holmium must be cooled to 20 K to reveal its magnetic properties.)

Actinides

The actinide series of elements should have chemical and physical properties similar to those of the rare earths. Unfortunately, most of the actinide elements (those beyond uranium) are radioactive and do not occur in nature. They are artificially produced elements and are available only in microscopic quantities. We are thus unable to determine many of their bulk properties.

8.5 X RAYS

X rays, as we discussed in Chapter 3, are electromagnetic radiations with wavelengths approximately from 0.01 to 10 nm (energies approximately from 100 eV to 100 keV). In Chapter 3 we discussed the *continuous* X-ray spectrum emitted by accelerated electrons. In this section we are concerned with the *discrete* X-ray line spectra emitted by atoms.

X rays are emitted in transitions between the normally filled lower energy levels of an atom. The inner electrons are so tightly bound that the energy spacing of these levels is about right for the emission of photons in the X-ray range of wavelengths. The outer electrons are relatively weakly bound, and the spacing between these outer levels is only a few electron-volts; transitions between these levels give photons of energies of a few eV, corresponding to the visible region of the spectrum. These "optical" transitions are discussed in the next section.

Since all of the inner shells of an atom are filled, X-ray transitions do not occur between these levels under normal circumstances. For example a 2p electron cannot make a transition to the 1s subshell, because all atoms beyond hydrogen have filled 1s subshells. In order to observe such a transition, we must remove an electron from the 1s subshell. This can be done by bombarding the atom with electrons (or other particles) accelerated to an energy sufficient to knock loose a 1s electron following a collision. (This requires accelerating voltages on the order of 10,000 V.)

Once we have removed an electron from the 1s subshell, an electron from a higher subshell will rapidly make a transition to fill that vacancy, emitting an X-ray photon in the process. The energy of the photon is equal to the energy difference of the initial and final atomic levels of the electron that makes the transition.

We previously defined a notation in which the $n = 1$ shell is known as the K shell. When we remove a 1s electron, we are creating a vacancy in the K shell. The X rays that are emitted in the process of filling this vacancy are known as *K-shell X rays*, or simply *K X rays*. (These X rays are emitted in transitions which come *from* the $L, M, N, \ldots$ shells, but they are known by the vacancy that they fill, not by the shell from which they originate.) The K X ray that originates with the $n = 2$ shell (L shell) is known as the K_α X ray, and the K X rays originating from higher shells are known as K_β, K_γ, and so forth. Figure 8.7 illustrates these transitions.

It is also possible that the bombarding electrons can knock loose an electron from the L shell, and electrons from higher levels will drop down to fill this vacancy. The photons emitted in these transitions are known as L X rays. The lowest-energy X ray of the L series is known as L_α, and the other L X rays are labeled in order of increasing energy as shown in Figure 8.7.

It is possible to have an L X ray emitted directly following the K_α X ray. A vacancy in the K shell can be filled by a transition from the L shell, with the emission of the K_α X ray. However, the electron that made the jump from the L shell left a vacancy there, which can be filled by an electron from a higher shell, with the accompanying emission of an L X ray.

In a similar manner, we label the other X-ray series by M, N, and so forth. Figure 8.8 shows a sample X-ray spectrum emitted by silver.

We have not considered the energy differences of the subshells within the major shells. For example, the L_α X ray could originate from one of the $n = 3$ levels (3s, 3p, 3d) and end in one of the $n = 2$ levels (2s, 2p). The energies of these different transitions will be slightly different, so that

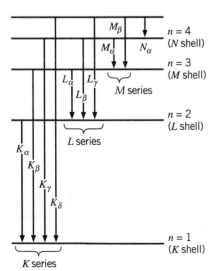

FIGURE 8.7   X-ray series.

there will be many L_α X rays, but their energy differences are very small compared, for example, with the energy difference between the L_α and L_β X rays. In fact, in many applications, we don't even see this small energy splitting.

Let us consider the K_α X ray in more detail. An electron in the L shell is screened by the two $1s$ electrons, and so it sees an effective nuclear charge of $Z_{\text{eff}} \cong Z - 2$. When one of those $1s$ electrons is removed in the creation of a K-shell vacancy, only the remaining $1s$ electron shields the L shell, and so $Z_{\text{eff}} \cong Z - 1$. (In this calculation, we neglect the small screening effect of the outer electrons; their probability densities are not zero within the L-shell orbits, but they are sufficiently small that their effect on Z_{eff} can be neglected.) The K_α X ray can thus be analyzed as a transition from the $n = 2$ level to the $n = 1$ level in a one-electron atom with $Z_{\text{eff}} = Z - 1$. Using Equations 6.36 and 6.42 for the Bohr atom, we can find the frequency of the K_α transition in an atom of atomic number Z to be

$$\nu = \frac{3cR_\infty}{4}(Z-1)^2 \tag{8.1}$$

Moseley's law

If we plot $\sqrt{\nu}$ as a function of Z, we obtain a straight line with slope $(3cR_\infty/4)^{1/2}$. Figure 8.9 is an example of such a plot. (Incidently, this result is independent of our assumption regarding the exact value of the screening correction. That is, we could have written $Z_{\text{eff}} = Z - k$, where k is some unknown number, probably close to 1. The only change in our plot would be in the intercept. We would still have a straight line with the same slope.)

This method gives us a powerful and yet simple way to determine the atomic number Z of an atom, as was first demonstrated in 1913 by the British physicist H. G. J. Moseley, who measured the K_α (and other) X-ray energies of the elements and thus determined their atomic numbers. Moseley was the first to demonstrate the type of linear relationship shown in Figure 8.9; such graphs are now known as Moseley plots. His discovery provided the first direct means of measuring the atomic numbers of the elements. Previously, the elements had been ordered in the periodic table according to increasing mass. Moseley found certain elements listed out of order, in which the element of higher Z happened to have the smaller mass

Henry G. J. Moseley (1887–1915, England). His work on X-ray spectra provided the first link between the chemical periodic table and atomic physics, but his brilliant career was cut short when he died on a World War I battlefield.

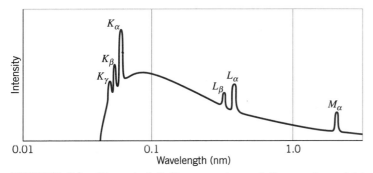

FIGURE 8.8 Characteristic X-ray spectrum of silver, such as might be produced by 30 keV electrons striking a silver target. The continuous distribution is a bremsstrahlung spectrum.

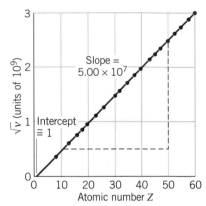

FIGURE 8.9 Square root of frequency versus atomic number for K_α X rays. The slope of the line is $5.00 \times 10^7 \, \text{s}^{-1/2}$, in good agreement with the expected value $(3cR_\infty/4)^{1/2} = 4.97 \times 10^7 \, \text{s}^{-1/2}$. The intercept is close to 1, as expected.

(for example, cobalt and nickel or iodine and tellurium). He also found gaps corresponding to yet undiscovered elements; for example, the naturally radioactive element technetium ($Z = 43$) does not exist in nature and was not known at the time of Moseley's work, but Moseley showed the existence of such a gap at $Z = 43$.

Moseley's work was of great importance in the development of atomic physics. Working in the same year as Rutherford and Bohr, Moseley not only provided a direct confirmation of the Rutherford-Bohr model, he also made a strong link between the periodic table, which was previously a rather arbitrary classification scheme of the elements, and atomic theory.

EXAMPLE 8.1

Compute the energy of the K_α X ray of sodium ($Z = 11$).

SOLUTION

The energy can be found with the help of Equation 8.1

$$\Delta E = h\nu = \frac{3hcR_\infty}{4}(Z-1)^2$$

$$= \frac{3(1240 \, \text{eV} \cdot \text{nm})(1.097 \times 10^7 \, \text{m}^{-1})}{4}(10)^2 = 1.020 \, \text{keV}$$

The measured value is 1.040 keV. The small discrepancy may be due to the screening correction in Z_{eff}, which is not exactly equal to 1.

EXAMPLE 8.2

Some measured X-ray energies in silver ($Z = 47$) are K_α: 21.990 keV and K_β: 25.145 keV. The binding energy of the K electron in silver is 25.514 keV. From these data, find: (a) the energy of the L_α X ray, and (b) the binding energy of the L electron.

SOLUTION

(a) From Figure 8.7, we see that the energies are related by:

$$\Delta E(L_\alpha) + \Delta E(K_\alpha) = \Delta E(K_\beta)$$

or

$$\Delta E(L_\alpha) = \Delta E(K_\beta) - \Delta E(K_\alpha) = 25.145 \, \text{keV} - 21.990 \, \text{keV} = 3.155 \, \text{keV}$$

(b) Again from Figure 8.7, we see that

$$\Delta E(K_\alpha) = E(L) - E(K)$$

or

$$E(L) = E(K) + \Delta E(K_\alpha) = -25.514 \, \text{keV} + 21.990 \, \text{keV} = -3.524 \, \text{keV}$$

The binding energy of the L electron is therefore 3.524 keV.

8.6 OPTICAL SPECTRA

As mentioned in the previous section, when we excite or remove one of the *outer* electrons, the resulting transitions fall in the visible range of the spectrum, and are thus known as *optical* transitions. The binding energies of the outer electrons in a typical atom are of the order of several electron-volts, and so it takes relatively little energy to remove an outer electron and produce an optical transition. In fact, it is the absorption and reemission of light by these outer electrons that are responsible for the colors of material objects (although in solids the electron energy levels are usually very different from those in isolated atoms). In contrast with X-ray spectra, which vary slowly and smoothly from one element to the next, optical spectra can show large variations between neighboring elements, especially those that correspond to filled subshells.

Beyond hydrogen, the easiest energy-level diagrams to understand are those of the alkali metals (Li, Na, K, Rb, Cs, Fr), which have a single s electron outside an inert core. Many of the excited states then correspond to the excitation of this single electron, and the resulting spectra are very similar to the spectrum of hydrogen, since the nuclear charge of $+Ze$ is shielded by the other $(Z-1)$ electrons. Figure 8.10 shows the energy levels of Li and Na along with some of the emitted transitions, which follow the same $\Delta l = \pm 1$ rule as the transitions in hydrogen.

The ground-state configuration of lithium is $1s^2 2s^1$ and the ground-state configuration of sodium is $1s^2 2s^2 2p^6 3s^1$. The excited states in both cases can be obtained by moving the outer electron to a higher state. For example, the first excited state of Li is $1s^2 2p^1$, with the $2s$ electron moving to the $2p$ level. (The energy necessary to accomplish this can be provided by various means, such as by absorption of a photon or by passage of an electric current through the material as in a gas discharge tube.) The excited electron in the $2p$ state rapidly drops back to the $2s$ state, with the emission of a photon of wavelength 670.8 nm. Since the inert core doesn't participate in this excitation or emission, we can ignore all but the outer electron in studying the levels and transitions in the alkali elements.

The ground-state configuration of helium is $1s^2$. We can produce an excited state by moving one of these electrons up to a higher level, and so some possible excited-state configurations might be $1s^1 2s^1$, $1s^1 2p^1$, $1s^1 3s^1$, and so forth. Photons are emitted when the excited electron drops back to the $1s$ level. The $\Delta l = \pm 1$ selection rule for transitions once again limits those that can occur. Figure 8.11 shows a portion of the energy level diagram for helium.

The phenomenon of *fluorescence* is responsible for the appearance of objects under the so-called "black light," which is a source of ultraviolet radiation. Photons in the ultraviolet region, invisible to the human eye, have higher energies than those in the visible region, and hence if an ultraviolet photon is absorbed by an atom, the outer electron (which is responsible for the optical transitions) can be excited to high levels. These electrons make transitions back to their ground state, accompanied by the emission of photons in the visible region. Objects seen in ultraviolet light often show colors in the blue or violet end of the spectrum that are not

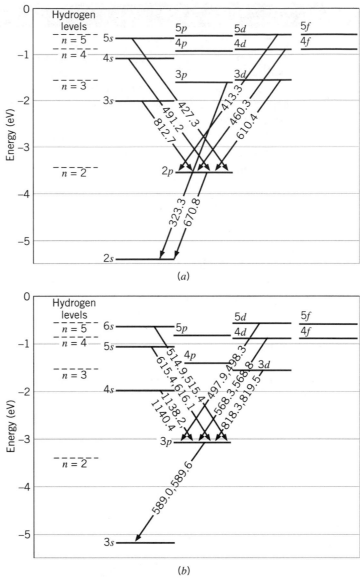

(a)

(b)

FIGURE 8.10 (a) Energy level diagram of lithium, showing some transitions (labeled with wavelength values in nanometers) in the optical region. The corresponding energies of hydrogen are included for comparison. (b) Energy level diagram of sodium. The fine-structure splitting of the 3p state makes each transition involving that state into a closely spaced doublet. (Because the fine-structure splitting increases with Z, it is negligible in lithium.)

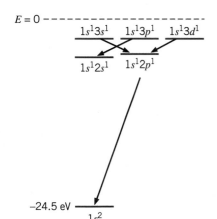

FIGURE 8.11 A small portion of the energy level diagram of helium. Note the $\Delta l = \pm 1$ transitions.

present when the objects are viewed in sunlight. We can understand this effect by considering the composition of sunlight and the optical excited states of a hypothetical atom shown in Figure 8.12. The intensity of sunlight is concentrated in the center of the visible spectrum, in the yellow region; very little intensity is present in the red or blue ends of the visible spectrum. The "yellow" photons have enough energy to excite the hypothetical atom to levels 1 and 2 shown in Figure 8.12, but not enough to reach level 3 or 4. However, the higher-energy ultraviolet photons have sufficient energy to reach the higher levels, so the light emitted by the atom has a stronger blue component when that atom is excited by ultraviolet light than when excited by sunlight. (See Color Plate 5.)

*8.7 ADDITION OF ANGULAR MOMENTA

The properties of an alkali atom such as sodium are determined primarily by the single outer electron; if that electron has quantum numbers (n, l, m_l, m_s) then the entire atom behaves as if it had those same quantum numbers. In atoms with several electrons outside of filled subshells, this is not the case. For example, the electronic configuration of carbon $(Z = 6)$ is $1s^2 2s^2 2p^2$. To find the angular momentum of carbon, we must combine the angular momenta of the two $2p$ electrons.

Suppose we have an atom with two electrons outside of filled subshells. These electrons have quantum numbers $(n_1, l_1, m_{l1}, m_{s1})$ and $(n_2, l_2, m_{l2}, m_{s2})$. The total orbital angular momentum of the atom is determined by the vector sum of the orbital angular momenta of the two electrons:

$$\mathbf{L} = \mathbf{L}_1 + \mathbf{L}_2 \tag{8.2}$$

Each vector is related to its corresponding angular momentum quantum number by

$$|\mathbf{L}| = \sqrt{L(L + 1)}\,\hbar \qquad |\mathbf{L}_1| = \sqrt{l_1(l_1 + 1)}\,\hbar \qquad |\mathbf{L}_2| = \sqrt{l_2(l_2 + 1)}\,\hbar$$

where L is the total orbital angular momentum quantum number of the atom. These vectors do not add like ordinary vectors, but have special addition rules associated with quantized angular momentum. These rules enable us to find L and its associated magnetic quantum number M_L.

1. The maximum value of the total orbital angular momentum quantum number is

$$L_{max} = l_1 + l_2$$

2. The minimum value of the total orbital angular momentum quantum number is

$$L_{min} = |l_1 - l_2|$$

3. The permitted values of L range from L_{min} to L_{max} in integer steps:

$$L = L_{min}, L_{min} + 1, L_{min} + 2, \ldots, L_{max}$$

4. The z component of the total angular momentum vector is found from the sum of the z components of the individual vectors:

$$L_z = L_{1z} + L_{2z}$$

or

$$M_L = m_{l1} + m_{l2} \tag{8.3}$$

* This is an optional section that may be skipped without loss of continuity.

Total orbital angular momentum L

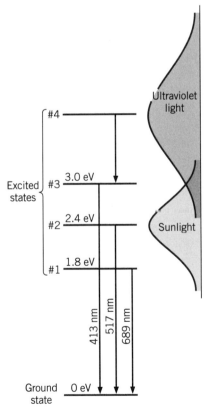

FIGURE 8.12 Excited states of a hypothetical atom. Only excited states 1 and 2 can be easily reached by exposure to sunlight; exposure to ultraviolet light populates state 4, which in turn populates state 3. Under ultraviolet light, a stronger blue or violet (413 nm) color is revealed than under sunlight.

The permitted values of M_L range from $-L$ to $+L$ in integer steps:

$$M_L = -L, -L + 1, \ldots, -1, 0, +1, \ldots, L - 1, L$$

An identical set of rules holds for coupling the spin angular momentum vectors to give the total spin angular momentum **S**. For two electrons, each of which has $s = \frac{1}{2}$, the total spin quantum number S can be 0 or 1.

All filled subshells have $L = 0$ and $S = 0$, so we don't need to consider filled subshells in analyzing the angular momentum of an atom. For this reason, filled subshells ordinarily do not contribute to the magnetic properties of atoms.

For coupling more than two electrons, couple the first two to give the maximum and minimum values of L. Then couple each allowed L to the third angular momentum to find the largest maximum and smallest minimum.

EXAMPLE 8.3

Find the total orbital and spin quantum numbers for carbon.

SOLUTION

Carbon has two $2p$ electrons outside filled subshells. Each of these electrons has $l = 1$. According to the rules for adding angular momenta, we have

$$L_{max} = 1 + 1 = 2, \qquad L_{min} = |1 - 1| = 0$$

Thus $L = 0$, 1, or 2. For the spin angular momentum, we have

$$S_{max} = \tfrac{1}{2} + \tfrac{1}{2} = 1, \qquad S_{min} = |\tfrac{1}{2} - \tfrac{1}{2}| = 0$$

and so $S = 0$ or 1. Some combinations of L and S might be forbidden by the Pauli principle. For example, to obtain $L = 2$, the two electrons must both have $m_l = +1$. The two electrons must therefore have different values of m_s, so $S = 1$ is not allowed when $L = 2$.

EXAMPLE 8.4

Find the total orbital and spin quantum numbers for nitrogen.

SOLUTION

Nitrogen has three $2p$ electrons, each with $l = 1$, outside filled subshells. If we add the first two, we get $L_{max} = 2$ and $L_{min} = 0$, as in Example 8.3, so that $L = 0$, 1, or 2. We now couple the third $l = 1$ electron to each of these values to find the largest maximum and smallest minimum, which give

$$L_{max} = 2 + 1 = 3, \qquad L_{min} = |1 - 1| = 0$$

and so $L = 0$, 1, 2, or 3. For the spin vectors, we again couple the first two to give $S_{max} = 1$ and $S_{min} = 0$. Adding the third $s = \frac{1}{2}$ electron, we have

$$S_{max} = 1 + \tfrac{1}{2} = \tfrac{3}{2}, \qquad S_{min} = |0 - \tfrac{1}{2}| = \tfrac{1}{2}$$

The resulting values of S are $\frac{1}{2}$ and $\frac{3}{2}$ (from the minimum to the maximum in integer steps). Once again, the Pauli principle may forbid certain combinations of L and S. The state with $L = 3$ cannot exist at all, because all three electrons must have $m_l = +1$, and assigning m_s quantum numbers will then result in two electrons with the same m_l and m_s, which is forbidden by the Pauli principle.

The two $2p$ electrons of carbon can combine to give $L = 0$, 1, or 2 and $S = 0$ or 1. The ground state of carbon will be identified by only one particular choice of L and S. How do we know which of these combinations will be the ground state? The rules for finding the ground state quantum numbers are known as *Hund's rules*:

1. First find the maximum value of M_S consistent with the Pauli principle. *Hund's rules* Then

$$S = M_{S,\mathrm{max}}$$

2. Next, for that M_S, find the maximum value of M_L consistent with the Pauli principle. Then

$$L = M_{L,\mathrm{max}}$$

In the case of carbon, the maximum value of M_S is $+1$, obtained when the two valence electrons both have $m_s = +\frac{1}{2}$. Thus $S = 1$. With only two electrons in the $2p$ shell, the Pauli principle places no restrictions on S; in fact, three electrons in the $2p$ shell can be assigned $m_s = \frac{1}{2}$. Our next task is to find the maximum value of M_L. The maximum value of m_l for the first electron is $+1$. The second electron cannot also have $m_l = +1$, because that would give both electrons the same set of quantum numbers, in violation of the Pauli principle. The maximum value of m_l for the second electron is 0, so $M_{L,\mathrm{max}} = +1$ and $L = 1$. The ground state of carbon is therefore characterized by $S = 1$ and $L = 1$.

EXAMPLE 8.5

Use Hund's rules to find the ground-state quantum numbers of nitrogen.

SOLUTION

The electronic configuration of nitrogen is $1s^2 2s^2 2p^3$. We begin by maximizing M_S for the three $2p$ electrons. Since three electrons in the p subshell are permitted by the Pauli principle to have $m_s = +\frac{1}{2}$, the maximum value of M_S is $\frac{3}{2}$, and therefore S is $\frac{3}{2}$. Each of the three electrons has quantum numbers $(2, 1, m_l, +\frac{1}{2})$. To maximize M_L we assign the first electron the maximum value of m_l, namely $+1$. The maximum value of m_l left for the second electron is 0, and the third electron must therefore have $m_l = -1$. The total M_L is $1 + 0 + (-1) = 0$, so $L = 0$. Thus $L = 0$, $S = \frac{3}{2}$ are the ground-state quantum numbers for nitrogen.

EXAMPLE 8.6

Find the ground-state L and S of oxygen ($Z = 8$).

SOLUTION

The electronic configuration of oxygen is $1s^2 2s^2 2p^4$. Since only three electrons in the p subshell can have $m_s = +\frac{1}{2}$, the fourth must have $m_s = -\frac{1}{2}$, so $M_{S,\max} = \frac{1}{2} + \frac{1}{2} + \frac{1}{2} + (-\frac{1}{2}) = 1$, and it follows that $S = 1$. To find L, we note that, as for nitrogen, the three electrons with $m_s = +\frac{1}{2}$ have $m_l = +1$, 0, and -1, and we maximize M_L by giving the fourth electron $m_l = +1$. Thus $M_{L,\max} = +1$, and $L = 1$.

Let us look now at the energy levels of helium. The ground-state configuration of helium is $1s^2$. Both electrons are s electrons, with $l = 0$, and so the only possible value of L is zero. Since both electrons have $m_l = 0$, the Pauli principle requires that the spin of the two electrons be opposite, so that one has $m_s = +\frac{1}{2}$ and the other has $m_s = -\frac{1}{2}$. The *only* possible total M_S is therefore zero, so the ground state of helium has $L = 0$ and $S = 0$. The first excited state has configuration $1s^1 2s^1$. Since both electrons still have $l = 0$, we must again have $L = 0$. However, the total spin S can now be 0 or 1, since the Pauli principle does not restrict m_s in this case—the two electrons already have different principal quantum numbers n, and so there is nothing to prevent them from having the same m_s. There are, therefore, *two* "first excited states" of helium, one with $L = 0$ and $S = 0$, and another with $L = 0$ and $S = 1$. (Both of these states have configuration $1s^1 2s^1$.) A state with $S = 0$ is called a *singlet* state (because there is only a single possible M_S value), and a state with $S = 1$ is called a *triplet* state (because there are three possible M_S values: $+1, 0, -1$). The classification of states into singlet and triplet is important when we consider the *selection rules* for transitions between states; these selection rules tell us which transitions are allowed (and therefore likely) to occur and which are not. The selection rules, which involve both L and S, are

L and S selection rules

$$\Delta L = 0, \pm 1 \qquad (8.4)$$

$$\Delta S = 0 \qquad (8.5)$$

(There are no selection rules for n.) Of course, the selection rule $\Delta l = \pm 1$ *for the single electron that makes the transition* still applies. For the two $1s^1 2s^1$ states of helium, the Δl rule does not permit either state to make transitions to the $1s^2$ ground state ($2s$ to $1s$ would be $\Delta l = 0$), and in addition, the ΔS rule forbids the triplet ($S = 1$) states from decaying to the $S = 0$ ground state. These transitions can thus only occur by violating these selection rules, and since that is a very unlikely event, the transitions occur with very low probability. Energy levels that have a low probability of decay must "live" for a long time before they decay; such states are known as *metastable* states.

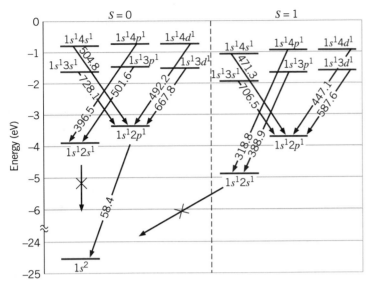

FIGURE 8.13 Energy level diagram for helium. The states are grouped into singlets ($S = 0$) and triplets ($S = 1$). Some of the transitions in the optical and ultraviolet regions are shown. Transitions marked with an X would violate the $\Delta l = \pm 1$ selection rule.

Figure 8.13 shows the energy levels and transitions in helium. The singlet and triplet levels are grouped separately, since transitions between singlet and triplet levels would violate the $\Delta S = 0$ selection rule.

Figure 8.14 shows the energy-level diagram of carbon. Notice the increasing complexity of the diagram, compared with the alkali metals and even with helium. This follows from the coupling of two electrons, both of whose l values may be different from zero. We have already discussed how the $2p^2$ configuration can give $L = 0$, 1, or 2 and $S = 0$ or 1. Only one of these ($L = 1$, $S = 1$) is the ground state of carbon; the others are excited states. More excited states can be obtained by promoting one of the $2p$ electrons to a higher level, giving configurations of $2p^1 3s^1$ ($L = 1$, $S = 0$ or 1), $2p^1 3p^1$ ($L = 0$, 1, or 2; $S = 0$ or 1), $2p^1 3d^1$ ($L = 1$, 2, or 3; $S = 0$ or 1) and so forth. Imagine the difficulty of analyzing the energy level diagram of the rare earths or actinides, which have unfilled f subshells ($l = 3$) with as many as 14 electrons!

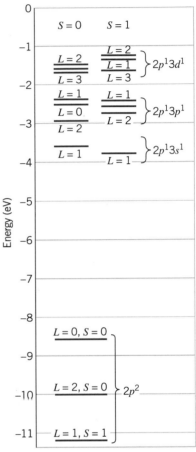

FIGURE 8.14 Energy level diagram for carbon. Each group of levels is labeled with the electron configuration. Each individual level is labeled with the total L and S.

8.8 LASERS

There are three means by which radiation can interact with the energy levels of atoms (depicted in Figure 8.15). The first two we have already discussed. In the first kind of interaction, an atom in an excited state makes a transition to a lower state, with the emission of a photon. (In all the examples we consider here, the photon energy is equal to the energy differ-

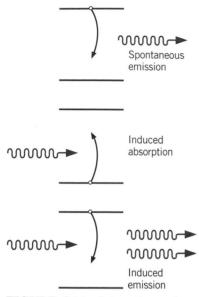

FIGURE 8.15 Interactions of radiation with atomic energy levels.

ence of the two atomic states.) This is *spontaneous emission*, which we represent as

$$atom* \longrightarrow atom + photon$$

where the asterisk indicates an excited state.

The second interaction, *induced absorption*, is responsible for absorption spectra and resonance absorption. An atom in the ground state absorbs a photon (of the proper energy) and makes a transition to an excited state. Symbolically:

$$atom + photon \longrightarrow atom*$$

The third interaction, which is responsible for the operation of the laser, is *induced (or stimulated) emission*. In this process, an atom is in an excited state. A passing photon of just the right energy (again, equal to the energy difference of the two levels) induces the atom to emit a photon and make a transition to the lower, or ground, state. (Of course, it would eventually have made that transition left on its own, but it makes it *sooner* after being prodded by the passing photon.) Symbolically,

$$atom* + photon \longrightarrow atom + 2 \ photons$$

The significant detail is that the two photons that emerge are traveling in *exactly the same direction* with *exactly the same energy*, and the associated electromagnetic waves are *perfectly in phase* (*coherent*).

Suppose we have a collection of atoms, all in the same excited state, as shown in Figure 8.16. A photon passes the first atom, causing induced emission and resulting in two photons. Each of these two photons causes an induced emission process, resulting in four photons. This process continues, doubling the number of photons at each step, until we build up an intense beam of photons, all coherent and moving in the same direction. In its simplest interpretation, this is the basis of operation of the laser. (The word *laser* is an acronym for *L*ight *A*mplification by *S*timulated *E*mission of *R*adiation.)

This simple model for a laser will not work, for several reasons. First, it is difficult to keep a collection of atoms in their excited states until

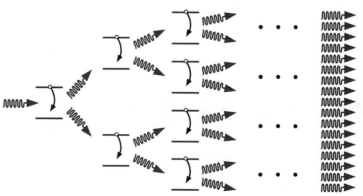

FIGURE 8.16 Buildup of intense beam in a laser. Each emitted photon interacts with an excited atom and produces two photons.

they are stimulated to emit the photon (we don't want any *spontaneous emission*). A second reason is that atoms that happen to be in their ground state undergo absorption and thus *remove* photons from the beam as it builds up.

To solve these problems, we must achieve a *population inversion*—in a collection of atoms, there must be more atoms in the upper state than in the lower state. This is called an "inversion" because under normal conditions at thermal equilibrium, the lower state always has the greater population. The "inversion" is thus an unnatural situation that must be achieved by artificial means, since it is essential for the operation of the laser.

The first laser, which was constructed by T. H. Maiman in 1960, was based on a three-level atom (Figure 8.17). The laser medium is a solid ruby rod, in which the chromium atoms are responsible for the action of the laser. The atoms, originally in the ground state, are "pumped" into the excited state by an external source of energy (a burst of light from a flash lamp that surrounds the ruby rod). The excited state decays very rapidly (by spontaneous emission) to a lower excited state, which is a metastable state—the atom remains in that level for a relatively long time, perhaps 10^{-3} s, compared with 10^{-8} s for the short-lived states. The transition from the metastable state to the ground state is the "lasing" transition, resulting from stimulated emission by a passing photon.

If the pumping action is successful, there are more atoms in the metastable state than in the ground state, and we have achieved a population inversion. However, as the lasing transition occurs, the population of the ground state is increased, thereby upsetting the population inversion. This excess of population in the ground state allows absorption of the lasing transition, thereby removing photons that might contribute to the lasing action.

The four-level laser illustrated in Figure 8.18 relieves this remaining difficulty. The ground state is pumped to an excited state that decays rapidly to the metastable state, as with the three-level laser. The lasing transition proceeds from the metastable state to yet another excited state, which in turn decays *rapidly* to the ground state. *The atom in its ground state thus cannot absorb at the energy of the lasing transition*, and we have a workable laser. Because the lower short-lived state decays rapidly, its population is always smaller than that of the metastable state, which maintains the population inversion.

The helium-neon laser is an example of a four-level laser. A mixture of helium and neon gas (about 90 percent helium) is contained in a narrow tube, as shown in Figure 8.19. An electrical current in the gas "pumps" the helium from its ground state to the excited state at an energy of about 20.6 eV. You will recall that this is a metastable state of helium—the atom remains in that state for a relatively long time because a $2s$ electron is not permitted to return to the $1s$ level by photon emission. Occasionally, an excited helium atom collides with a ground-state neon atom. When this occurs, the 20.6 eV of excitation energy may be transferred to the neon atom, since neon happens to have an excited state at 20.6 eV, and the helium atom returns to its ground state. Symbolically,

$$\text{helium*} + \text{neon} \longrightarrow \text{helium} + \text{neon*}$$

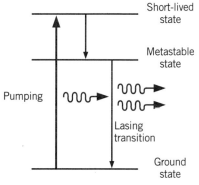

FIGURE 8.17   A three-level atom.

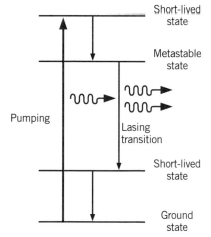

FIGURE 8.18   A four-level atom.

Helium-neon laser

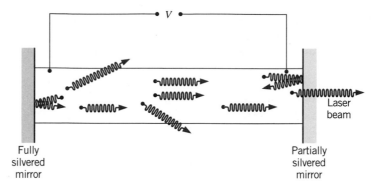

FIGURE 8.19 Schematic diagram of a He-Ne laser.

where the excited state is indicated by the asterisk. The excited state of neon corresponds to removing one electron from the filled $2p$ subshell and promoting it to the $5s$ subshell. From there it decays to the $3p$ level and eventually returns to the $2p$ ground state. Figure 8.20 illustrates this sequence of events and the level schemes. (The level shown with a dashed line, the neon $3s$ level, is not important for the basic operation of the laser, but it is necessary as an intermediate step in the return to the neon ground state, since the $\Delta l = 0$ transition $3p \rightarrow 2p$ is not allowed, but the sequence $3p \rightarrow 3s \rightarrow 2p$ is permitted.)

At any given time, there are more neon atoms in the $5s$ state than in the $3p$ state, since the good energy matchup of the $5s$ state with the helium excited state gives a much higher probability of the $5s$ state in neon being excited. This provides the population inversion that is needed for the laser.

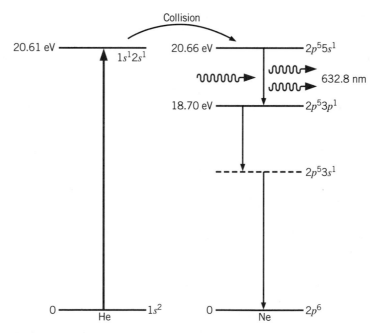

FIGURE 8.20 Sequence of transitions in a He-Ne laser.

Occasionally a neon atom in the $5s$ state emits a photon (at a wavelength of 632.8 nm) parallel to the axis of the tube. This photon causes stimulated emission by other atoms, and a beam of coherent (in-phase) radiation eventually builds up traveling along the tube axis. Mirrors are carefully aligned at the ends of the tube to help in the formation of the coherent wave, as it bounces back and forth between the two ends of the tube, causing additional stimulated emission. One of the mirrors is only partially silvered, allowing a portion of the beam to escape through one end.

The laser is not a particularly efficient device; the small helium-neon lasers you have probably seen used for laboratory or demonstration experiments have a light output of perhaps a few milliwatts; the electric power required to operate such a device may be of the order of 10 to 100 W, and thus the efficiency (power out $\div$ power in) of such a device is only about 10^{-4} to 10^{-5}. It is the *coherence* and *directionality* of the laser beam and its *energy density* that make the laser such a useful device (see Color Plate 6)—its power can be concentrated in a beam only a few millimeters in diameter and thus even a small laser can deliver 100 to 1000 W/m². Larger lasers in the megawatt (10^6 W) range are presently readily available, and research laboratories are developing 100 terawatt (10^{14} W) lasers for special applications. These powerful lasers do not operate continuously, but are instead *pulsed*, producing short (perhaps 10^{-8} s) pulses at rates of order 100 Hz. (Such a pulse is, in fact, an excellent example of a wave packet.)

SUGGESTIONS FOR FURTHER READING

Some additional basic features of the vector representation of atomic states are in:

K. W. Ford, *Classical and Modern Physics* (Lexington, Xerox College Publishing, 1974), Volume 3, Chapter 24.

Two slightly more advanced works including more detail are:

H. Semat and J. R. Albright, *Introduction to Atomic and Nuclear Physics* (New York, Holt, Rinehart and Winston, 1972).
J. C. Willmott, *Atomic Physics* (Chichester, Wiley, 1975).

A still higher level reference is:

H. G. Kuhn, *Atomic Spectra* (New York, Academic Press, 1969).

Finally, three classic works that include some introductory and advanced material covering almost all aspects of atomic structure:

A. C. Candler, *Atomic Spectra and the Vector Model* (Cambridge, Cambridge University Press, 1937).
G. Herzberg, *Atomic Spectra and Atomic Structure* (New York, Prentice-Hall, 1937).
H. E. White, *Introduction to Atomic Spectra* (New York, McGraw-Hill, 1934).

Many reference works, both popular and technical, are available about lasers. A good introductory work is:

B. A. Lengyel, *Lasers* (New York, Wiley, 1971).

Two popular articles by the 1981 Nobel prize recipient (for his work with lasers):

A. W. Schawlow, "Optical Masers," *Scientific American* **204**, 52 (June 1961).
A. W. Schawlow, "Laser Light," *Scientific American* **219**, 120 (September 1968). The entire September 1968 issue is devoted to light and includes other articles on lasers.

QUESTIONS

1. Continue Figure 8.1 upward, showing the next two major groups. What will be the atomic number of the next inert gas below Rn? What will be the structure of the eighth row (period) of the periodic table? Where do you expect the first *g* subshell to begin filling? What properties would you expect the *g*-subshell elements to have? What will be the atomic number of the second inert gas below radon?

2. Why do the 4*s* and 3*d* subshells appear so close in energy, when they belong to different principal quantum numbers *n*?

3. Would you expect element 107 to be a good conductor or a poor conductor? How about element 111? Do you expect element 112 to be paramagnetic or diamagnetic?

4. Zirconium frequently is present as an impurity in hafnium metal. Why?

5. Do you expect ytterbium (Yb) to become ferromagnetic at sufficiently low temperatures? What type of magnetic behavior would be expected at ordinary temperatures for polonium (Po)? For francium (Fr)?

6. As we move across the series of transition metal or rare earth elements, we add electrons to the *d* or *f* subshells. In chemical compounds, many of these elements show valence states of +2 or +3, which correspond to removing two *s* electrons. Explain this apparent paradox.

7. Why do the rare earth (lanthanide) elements have such similar chemical properties? What property might you use to distinguish lanthanide atoms from one another?

8. Explain why the Bohr theory gives a poor accounting of optical transitions but does well in predicting the energies of X-ray transitions.

9. What can you conclude about the electronic configuration of an atom that has both $L = 0$ and $S = 0$ in the ground state?

10. Suppose we do a Stern-Gerlach experiment using an atom that has angular momentum quantum numbers L and S in its ground state. Into how many components will the beam split? Do you expect them to be equally spaced?

11. What is the degeneracy of a state of total orbital angular momentum L that has $S = 0$? What is the degeneracy of a state of total spin angular momentum

S that has *L* = 0? What is the *total* degeneracy of a state in which both *L* and *S* are nonzero?

12. What *L* and *S* values must an atom have in order to show the *normal* Zeeman effect? Does this apply only to the ground state or to excited states also? Can an atom show the normal Zeeman effect in some transitions and the anomalous Zeeman effect in other transitions? Could the same atom even show no Zeeman effect at all in some transitions?

13. Based on the rules for coupling electron *l* and *s* values to give the total *L* and *S*, explain why filled subshells don't contribute to the magnetic properties of an atom.

14. If an atom in its ground state has *S* = 0, can you infer whether it has an even or an odd number of electrons? What if *L* = 0?

15. The *L* atomic shell actually contains three distinct levels: a 2*s* level and two 2*p* levels (a fine-structure doublet). If we look carefully at the K_α X ray under high resolution, we see two, not three, different components. Explain this discrepancy.

16. The K_α energies computed using Equation 8.1 are about 0.1 percent low for *Z* = 20, 1 percent low for *Z* = 40, and 10 percent low for *Z* = 80. Why does the simple theory fail for large *Z*? Could it be because the screening effect has not been handled correctly and that Z_{eff} is not *Z* − 1? If not, can you suggest an alternative reason?

17. The first excited state in sodium is a fine-structure doublet; the wavelengths emitted in the decay of these states are 589.59 nm and 589.00 nm, a difference of 0.59 nm. The excited 4*s* state in sodium (see Figure 8.10) decays to the 3*p* doublet with the emission of radiation at the wavelengths 1138.15 nm and 1140.38 nm, a difference of 2.23 nm. Explain how the 3*p* fine structure can give a wavelength difference of 0.59 nm in one case and 2.23 nm in the other case.

18. Suppose we had a three-level atom, like that of Figure 8.17, in which the metastable state were the higher excited state; the lasing transition would then be the upper transition. Does this atom solve the problem of absorption of the lasing transition? Would such an atom make a good laser?

19. How does a laser beam differ from a point source of light? Contrast the change in beam intensity with distance from the source for a laser and a point source.

20. Explain what is meant by a population inversion and why it is necessary for the operation of a laser.

21. How could you demonstrate that laser light is coherent? What would be the result of the same experiment using an ordinary monochromatic source? A white light source?

PROBLEMS

1. (a) List all elements with a p^3 configuration. (b) List all elements with a d^7 configuration.

2. Give the electronic configuration of (a) P; (b) V; (c) Sb; (d) Pb.

3. Use the ionization energy of sodium to find the effective nuclear charge seen by its outermost electron. Repeat the calculation for potassium. How do your results compare with the expectations based on the Bohr theory?

4. Derive Equation 8.1.

5. A certain element emits a K_α X ray of wavelength 0.1940 nm. Identify the element.

6. Compute the K_α X ray energies of calcium ($Z = 20$), zirconium ($Z = 40$), and mercury ($Z = 80$). Compare with the measured values of 3.69 keV, 15.8 keV, and 70.8 keV. (See Question 16).

7. Draw a Moseley plot, similar to Figure 8.9, for the K_β X rays using the following energies in keV:

Ne 0.858	Mn 6.51	Zr 17.7
P 2.14	Zn 9.57	Rh 22.8
Ca 4.02	Br 13.3	Sn 28.4

    Determine the slope and compare with the expected value. (Equation 8.1 applies only to K_α X rays; you will need to derive a similar equation for the K_β X rays.) Determine the z-axis intercept and give its interpretation.

8. Repeat Problem 7 for the L_α X rays (energies in keV):

Mn 0.721	Rh 2.89
Zn 1.11	Sn 3.71
Br 1.60	Cs 4.65
Zr 2.06	Nd 5.72

    Give interpretations of the slope and intercept.

9. In an X-ray tube electrons strike a target after being accelerated through a potential difference V. Estimate the minimum value of V required to observe the K_α X rays of copper.

10. Chromium has the electron configuration $4s^1 3d^5$ beyond the inert argon core. What are the ground-state L and S values?

11. Use Hund's rules to find the ground-state L and S of:
    (a) Ce, configuration $[Xe]6s^2 4f^1 5d^1$
    (b) Gd, configuration $[Xe]6s^2 4f^7 5d^1$
    (c) Pt, configuration $[Xe]6s^1 4f^{14} 5d^9$

12. Using Hund's rules, find the ground-state L and S of (a) fluorine ($Z = 9$); (b) magnesium ($Z = 12$); (c) titanium ($Z = 22$); (d) iron ($Z = 26$).

13. A certain excited state of an atom has the configuration $4d^1 5d^1$. What are the possible L and S values?

14. Assuming a magnetic moment of one Bohr magneton, find the effective magnetic field that produces the fine-structure splitting of the $3p$ state of sodium. (See Problem 17 or Figure 8.10 for the splittings.)

15. Using the wavelengths given in Figure 8.10, compute the energy difference between the $3d$ and $4d$ states in lithium; do the same for sodium. Compare those values with the corresponding $n = 4$ to $n = 3$ energy difference in hydrogen. Why is the agreement so good, considering the different values of Z?

16. (a) Using the information for lithium given in Figure 8.10, compute the energy difference of the $3p$ and $3d$ states. (b) Compute the energy of the $3s$, $4s$, and $5s$ states above the ground state. (c) The ionization energy of lithium in its ground state is 5.39 eV. What is the ionization energy of the $2p$ state? Of the $3s$ state?

17. Consider the following transitions from different fine-structure doublets in sodium (all wavelengths in nm):

$$3p\text{–}3s: 588.995, 589.592$$
$$4p\text{–}3s: 330.303, 330.241$$
$$5p\text{–}3s: 285.307, 285.286$$
$$6p\text{–}3s: 268.047, 268.038$$

Compute the splitting, in eV, of these four p states. Determine the dependence of this splitting on the quantum number n, and compare with the simple theory of Section 7.8.

18. Using the wavelengths given in Figure 8.13, compute the energy difference between the $1s^14p^1$ and $1s^13p^1$ singlet $(S = 0)$ states in helium. Compare this energy difference with the value expected using the Bohr model, assuming that the p electron is screened by the s electron. Repeat the calculation for the $3d$ and $4d$ triplet $(S = 1)$ states.

19. (a) List the six possible sets of quantum numbers (n, l, m_l, m_s) of a $2p$ electron. (b) Suppose we have an atom such as carbon, which has two $2p$ electrons. Ignoring the Pauli principle, how many different possible combinations of quantum numbers of the two electrons are there? (c) How many of the possible combinations of part (b) are eliminated by applying the Pauli principle? (d) Suppose carbon is in an excited state with configuration $2p^13p^1$. Does the Pauli principle restrict the choice of quantum numbers for the electrons? How many different sets of quantum numbers are possible for the two electrons?

20. Use the degeneracies of the states with all possible total L and S to find how many different levels the $2p^13p^1$ excited state of carbon includes. (See Figure 8.14.) Compare this result with the result of counting the individual m_l and m_s values from Problem 19(d). (See also Question 11.)

21. (a) What is the longest wavelength of the absorption spectrum of lithium? (b) What is the longest wavelength of the absorption spectrum of helium? In what region of the spectrum does this occur? (c) What are the *shortest* wavelengths in the absorption spectra of helium and lithium? In what region of the electromagnetic spectrum are these?

22. A small helium-neon laser produces a light beam with an average power of 3.5 mW and a diameter of 2.4 mm. (a) How many photons per second are emitted by the laser? (b) What is the amplitude of the electric field of the light wave? Compare this result with the electric field at a distance of 1 m from an incandescent light bulb that emits 100 W of visible light.

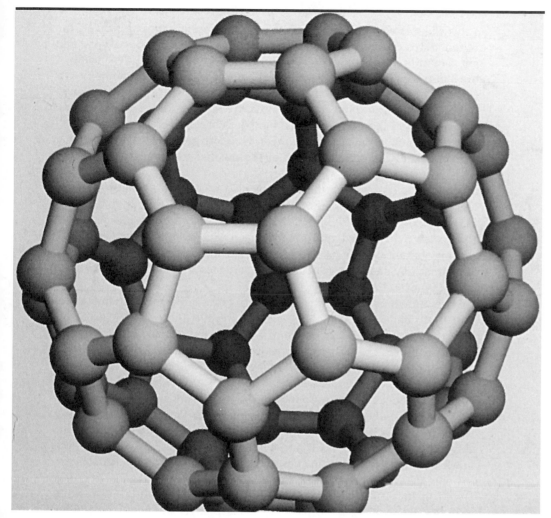

Molecules range from the simple with only two atoms to very complex organic molecules such as DNA. The photo shows a computer model of the molecule C_{60}, known as a "buckminsterfullerene" or "buckyball."

MOLECULAR STRUCTURE

In this chapter we consider the combination of atoms into molecules, the excited states of molecules, and the ways that molecules can absorb and emit radiation. From a variety of experiments we learn that the spacing of atoms in molecules is of the order of 0.1 nm, and that the binding energy of an atom in a molecule is of the order of electron-volts. This spacing and binding energy are about what we expect for electronic orbits, and we therefore expect that the forces that bind molecules together originate with the electrons. The negatively charged electrons provide the binding that overcomes the Coulomb repulsion of the positively charged nuclei of the atoms in the molecule.

However, we recognize that there is also a Coulomb repulsion of the *electrons*, and it is not apparent why stable molecules form at all. The key to this problem is the existence of the spatial probability densities of atomic orbits, such as we calculated for hydrogen and illustrated in Chapter 7. These probability densities are frequently not spherically symmetric, and very often may show overwhelming preferences for one spatial direction over another.

When atoms are brought together to form molecules, atomic states change into molecular states. These states are filled in order of increasing energy by the valence electrons contributed by the atoms of the molecule. Just as in the case of atoms, *it is the probability densities of the occupied molecular states that determine the nature of molecular bonds and the structure and properties of molecules.*

Just as we began to study atomic physics by looking at the simplest atom, we begin our study of molecular physics with the simplest molecule, H_2^+, the singly ionized hydrogen molecule. We next turn to other simple molecules, such as H_2 and NaCl, and finally we look at how our previous knowledge of atomic wave functions can help us to understand the molecular states that form the basis of organic chemistry.

We conclude this chapter with a discussion of the rotational and vibrational excited states of molecules. The electromagnetic radiations emitted in transitions between the states give a distinctive signature of the molecule and its structure. *Molecular spectroscopy*, the study of these radiations, finds application in such diverse areas as identification of atmospheric pollutants and the search for life in outer space.

9.1 THE HYDROGEN MOLECULE ION

What complicates the molecular structure is also what complicates atomic structure—there are too many electrons present for us to be able to write down and solve the dynamical equations that govern the structure of the atom or molecule. We therefore use the same tactic to study molecular structure that we used for atomic structure: we begin with a molecule that has only one electron. Such a molecule is H_2^+, *the hydrogen molecule ion*, which results when we remove an electron from a molecule of ordinary hydrogen, H_2.

Before we turn to the wave mechanical properties of H_2^+, let us try to guess what it is that holds this simple molecule together. We first realize that it is *not* correct to think of H_2^+ as an atom of hydrogen (proton plus electron) joined to a second proton. Since the atom of hydrogen in such a combination is electrically neutral, there is no electrostatic Coulomb force to hold the two pieces together. In this kind of molecule, at least, it is apparently not correct to identify the electron as belonging exclusively to one or the other of the components. *The electron must somehow be shared between the two parts.* The electron must spend a significant part of its time in the region between the two protons. In the language of quantum mechanics, the electron's probability density must have a large value in that region.

As we learned in Chapter 7, an electron in the ground state of hydrogen has an energy of -13.6 eV, a wave function $\psi = (\pi a_0^3)^{-1/2} e^{-r/a_0}$, where a_0 is the Bohr radius, and a probability density proportional to ψ^2. Figure 9.1 shows the electron wave functions for two protons separated by a large distance. As we bring the two protons closer together, the wave functions begin to overlap, and we must combine them according to the rules of quantum mechanics—first add the wave functions, then square the result to find the combined probability density. (Note that this gives a very different result from first squaring, then adding.)

We can combine these two wave functions in two different ways, depending on whether they have the same signs or opposite signs. The absolute sign of a wave function is arbitrary. When we calculate the normalization constant of a wave function, we actually compute its square. We could choose either the positive or the negative root; for convenience we usually choose the positive one. When we calculate probability densities ψ^2 for a single wave function, the choice of sign becomes irrelevant. However, when we combine different wave functions, their *relative* signs determine whether the two functions add or subtract, which can result in very different probability densities.

Consider the two different combinations of wave functions shown in Figure 9.2. In one case, the two wave functions have the same sign, and in the other case they have opposite signs. This difference in sign has a substantial effect on the probability distributions, which are shown in Figure 9.3. The probability density corresponding to $\psi_1 + \psi_2$ (Figure 9.3a) has relatively large values in the region between the two protons. This suggests a concentration of negative charge between the protons, which can supply enough Coulomb attraction to form a stable molecule. The probability density corresponding to $\psi_1 - \psi_2$ (Figure 9.3b), however, has a vanishing probability density midway between the protons and a small density of negative charge

FIGURE 9.1   The electron wave functions for two hydrogen atoms separated by a large distance.

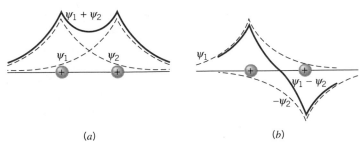

FIGURE 9.2   The overlap of two hydrogenic wave functions. In (*a*), the two wave functions have the same sign, while in (*b*) they have opposite signs.

in this region. There is not enough negative charge to overcome the Coulomb repulsion of the protons, and as a result this combination of wave functions does not lead to the formation of a stable molecule.

Let us consider the energy in the hydrogen molecule ion as a function of the separation distance R between the two protons. The Coulomb potential energy of the protons is $U_p = e^2/4\pi\varepsilon_0 R$; this function is plotted in Figure 9.4. To find the electron energy as a function of R, let us first consider the case when the two protons are very far apart. In this case the electron is in the ground state orbit about one of the protons, for which $E = -13.6$ eV. As we bring the protons together, the electron becomes more tightly bound (since it is attracted by both protons) and its energy becomes more negative. As $R \to 0$, the system approaches a single atom with $Z = 2$. For the wave function $\psi_1 + \psi_2$ (Figure 9.2a), the combined wave function has a maximum at $R = 0$ and resembles the ground-state wave function of an atom with $Z = 2$. Recalling the result from Chapter 6 for the electron energy in a hydrogen-like atom,

$$E_n = (-13.6 \text{ eV})\frac{Z^2}{n^2}$$

where n is the principal quantum number, we find the energy of a ground-state electron for $Z = 2$ to be -54.4 eV. The energy corresponding to the sum of the two wave functions, which we label E_+, therefore has the value -13.6 eV at large R and approaches the value -54.4 eV at small R. The result of an exact calculation of E_+ is shown in Figure 9.4.

For the combination corresponding to the difference between the two wave functions, the energy is once again -13.6 eV for large R. As $R \to 0$, the combined wave function approaches 0 (Figure 9.2b). The lowest energy

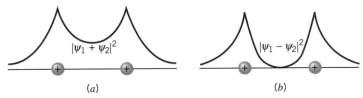

FIGURE 9.3   The probability densities corresponding to the two wave functions of Figure 9.2.

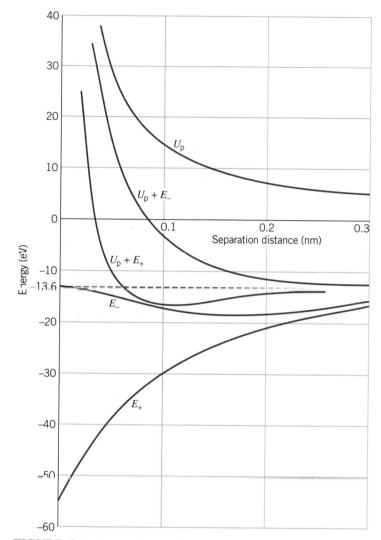

FIGURE 9.4 Dependence of energy on separation distance for H_2^+.

level with a wave function that vanishes at $R = 0$ is the $2p$ state, for which the energy in a $Z = 2$ hydrogen-like atom is -13.6 eV. The energy E_- corresponding to the wave function $\psi_1 - \psi_2$ therefore has the value -13.6 eV for both large R and small R. Its exact form is shown in Figure 9.4.

The total energy of the hydrogen molecule ion is the sum of the proton energy U_p and the electron energy E_+ or E_-. These two sums are also plotted in Figure 9.4. You can see that the combination $U_p + E_-$ has no minimum and therefore no stable bound state. The wave function $\psi_1 - \psi_2$ does not lead to a stable configuration for the hydrogen molecule ion, just as we originally suspected.

The sum $U_p + E_+$ gives a minimum at an equilibrium separation $R_{eq} = 0.106$ nm and an energy of -16.3 eV. The binding energy B of H_2^+ is the energy necessary to take apart the ion into H and H^+ and corresponds to the depth of the potential valley of $U_p + E_+$ in Figure 9.4:

Binding energy of H_2^+ $$B = E(H + H^+) - E(H_2^+) = -13.6 \text{ eV} - (-16.3 \text{ eV}) = 2.7 \text{ eV}$$

Note that we have defined molecular binding energy as the energy difference between the separate components (H and H^+) and the combined system (H_2^+).

It is interesting to note that the stability is achieved at $R_{eq} \cong 2a_0$. You will recall from Chapter 7 that the radial probability density for the 1s state of hydrogen has its maximum value at $r = a_0$. Thus the stable configuration of the H_2^+ ion is such that the maximum in the radial probability density for a single H atom would fall exactly in the middle of the molecule! This is once again consistent with our expectations for the structure of H_2^+—the electron must spend most of its time between the two protons.

In summary, from our study of this simple molecule we have learned that an important feature of molecular bonding concerns the *sharing* of a single electron by two atoms of the molecule. This sharing is responsible for the stability of the molecule. With this in mind we can now add a second electron and consider the H_2 molecule.

EXAMPLE 9.1

Suppose a spherically symmetric negative charge of diameter R_{eq} were centered between two protons at the H_2^+ equilibrium configuration, so that the two protons just touch the surface of the sphere. What charge on the sphere is required to give the system a binding energy of 2.7 eV?

SOLUTION

Let the negative charge on the sphere be some fraction δ of the charge of the electron. Then the potential energy between the sphere and either of the protons is

$$U_e = -\frac{1}{4\pi\varepsilon_0}\frac{\delta e^2}{R_{eq}/2}$$

The total energy of the system includes the repulsion of the protons and the attraction of the negative sphere for each of the protons:

$$E = U_p + 2U_e = \frac{1}{4\pi\varepsilon_0}\frac{e^2}{R_{eq}} - 2\frac{1}{4\pi\varepsilon_0}\frac{\delta e^2}{R_{eq}/2}$$

$$= \frac{1}{4\pi\varepsilon_0}\frac{e^2}{R_{eq}}(1 - 4\delta)$$

Inserting the value of $R_{eq} = 0.106$ nm and $E = -B = -2.7$ eV, we can solve to find

$$\delta = 0.30$$

This quantity of charge is roughly consistent with the fraction of $|\psi|^2$ that appears between the two protons in Figure 9.3a.

9.2 THE H₂ MOLECULE AND THE COVALENT BOND

Suppose we have two hydrogen atoms separated by a very large distance. Associated with each atom there is a $1s$ electronic state, at an energy of -13.6 eV, since the atoms are so far apart that there is no interaction between the electrons. As we bring the atoms closer together to form a H₂ molecule, the electron wave functions begin to overlap, so that the electrons are "shared" between the two atoms. As we have seen in the previous section, this can occur in such a way that the two electron wave functions *add* in the region between the two protons, giving a stable molecule, or *subtract*, leading to no stable molecule. The separate, individual electronic states of the atoms now become *molecular* states. Notice that, as shown in Figure 9.5, the number of states does not change as the separation R is reduced. When the atoms are separated by a large distance, there are two states, each at -13.6 eV, so the total energy at $R = \infty$ is -27.2 eV. When the separation is reduced, there are still two states, but now at different energies. One state corresponds to the sum of the two wave functions and leads to a stable H₂ molecule; the other state corresponds to the difference of the two wave functions and does not give a stable molecule. The molecular state that leads to a stable molecule is known as a *bonding* state, and the one that does not lead to a stable molecule is an *antibonding* state.

Bonding and antibonding states

As we found previously for H₂⁺, in order to form a molecule, the electron probability distribution must be large in the region between the two protons. In the case of H₂, this is true for both electrons, and it is certainly our expectation, based on the Pauli principle, that for the two electrons *both* to occupy the *molecular* state leading to the large probability in the central region, their spins must be oppositely directed; that is, one must have spin "up" and the other spin "down."

The sharing of electrons in a molecule such as H₂ is the origin of the *covalent* bond; this type of bonding occurs most frequently in molecules containing two identical atoms, and it is therefore also called *homopolar* or *homonuclear* bonding.

Covalent bond

Let us summarize the essential features of covalent bonding:

1. As two atoms are brought together, the electrons interact and the separate atomic states and energy levels are transformed into molecular states.

2. In one of the molecular states, the electron wave functions overlap in such a way as to give a *lower* energy than the separated atoms had; this is the bonding state that leads to the formation of stable molecules.

3. The other molecular state (the antibonding state) has an increased energy relative to the separated atoms and does not lead to the formation of stable molecules.

4. The restrictions of the Pauli principle apply to molecular states just as they do to atomic states; each molecular state has a maximum occupancy of two electrons, corresponding to the two different orientations of electron spin.

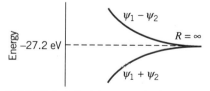

FIGURE 9.5 Energy of different combinations of wave functions in H₂.

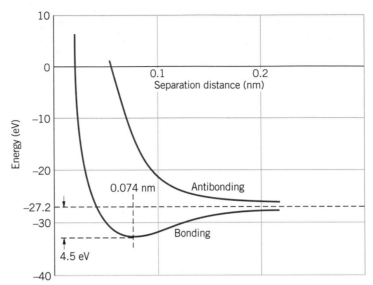

FIGURE 9.6   Bonding and antibonding in H_2.

The energy of the bonding state for H_2 is shown in Figure 9.6; as you can see, there is a minimum with $E = -31.7$ eV at $R = 0.074$ nm. When we separate the H_2 molecule into two neutral H atoms, the total energy is 2×-13.6 eV $= -27.2$ eV, and so the *molecular binding energy* of H_2 is -27.2 eV $- (-31.7$ eV$) = 4.5$ eV.

Comparing Figures 9.4 and 9.6 you can immediately see the effect of adding an additional electron to H_2^+: the binding energy is greater (the molecule is more tightly bound), and the nuclei are drawn closer together. Both of these effects are due to the presence of the increased electron density in the region between the two protons.

We can also understand why He does not form the molecule He_2—as two He atoms are brought together, the bonding and antibonding states are formed in much the same way as with H_2. The He_2 molecule would have four electrons, two in the bonding state and two in the antibonding state, and the net effect is that no stable molecule forms. (However, He_2^+ is stable, with two bonding and only *one* antibonding electrons. The binding energy of He_2^+ is 3.1 eV and the separation is 0.108 nm, remarkably close to the corresponding values of H_2^+.)

9.3 OTHER COVALENT BONDING MOLECULES

Other hydrogenlike systems with a single *s* electron also form stable molecules through covalent bonding; for example, see the values for Li_2 and Na_2 given in Table 9.1. Atoms with valence electrons in *p* states can also

TABLE 9.1 PROPERTIES OF *s*-BONDED MOLECULES

Molecule	Dissociation Energy (eV)*	Equilibrium Separation (nm)
H_2	4.52	0.074
Li_2	1.10	0.267
Na_2	0.80	0.308
K_2	0.59	0.392
LiNa	0.91	0.281
KNa	0.66	0.347
LiH	2.43	0.160
Rb_2	0.47	0.422
NaRb	0.61	0.359
Cs_2	0.43	0.450
NaH	2.09	0.189

* "Dissociation energy" is equivalent to "binding energy" as a measure of the strength of a chemical bond. Either term indicates the energy needed to break the bond.

form diatomic molecules through covalent bonds—oxygen and nitrogen, for example. Since there are three atomic *p* states, there will be six molecular states, and the classification of levels can become quite tedious, but we can understand the structure of molecules composed of atoms with *p* electrons based on the geometry of atomic *p* states. We discuss three applications of this type of covalent bonding: *pp* homopolar molecules, *sp* directed bonds, and *sp* hybrid states. First we review some of the properties of atomic *p* states.

In Chapter 7 we solved the Schrödinger equation for the H atom and showed the spatial probability distributions for the various possible electronic wave functions. Of course, these solutions for hydrogen will *not* be correct for other atoms, but the essential features of the geometry of the atomic states remains correct. Let us concentrate our attention on the *p* states, of which there are three, corresponding to $m_l = -1, 0,$ and $+1$. The probability distributions corresponding to these m_l values were shown in Figure 7.9. We can imagine these distributions to have a sort of "figure eight" shape with two distinct lobes of large probability. In the $m_l = 0$ case, the "figure eight" has its long axis along the *z* axis, and the two lobes of maximum probability occur in the $+$ and $-$ *z* directions. In the $m_l = \pm 1$ cases, the "figure eight" lies in the *xy* plane, and we can think of it as rotating rapidly about the *z* axis, counterclockwise for $m_l = +1$ and clockwise for $m_l = -1$, giving again two lobes of large probability, which now fall along a line that has any orientation in the *xy* plane. For our purposes here, it is not as convenient to use the m_l notation as it is to use a different notation in which we assign each of the three possible *p* states a label that gives the direction in space corresponding to the lobes of maximum probability. Thus p_z is the state with regions of large electron probability along the *z* axis, and similarly for p_x and p_y. Figure 9.7 shows a *schematic* representation of these probability distributions. (The p_z state corresponds exactly to $m_l = 0$; p_x and p_y correspond to *mixtures* of $m_l = +1$ and $m_l =$

p_x, p_y, p_z States

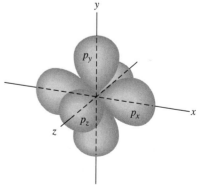

FIGURE 9.7 Probability distributions of three different *p* electrons.

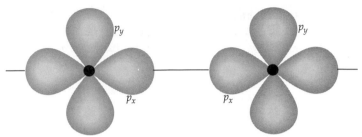

FIGURE 9.8   Two atoms with p electrons. The p_z probability distribution, which extends perpendicular to the page, is not shown.

−1.) We consider the structure of molecules containing p electrons based on this simple model of the three mutually perpendicular p states p_x, p_y, and p_z.

pp Covalent Bonds

Consider what happens when we bring together two p-shell atoms, whose probability distributions are each similar to Figure 9.7. We assume that the atoms approach along the x axis, as in Figure 9.8. As the atoms are brought together, the p_x states overlap (Figure 9.9a), giving (if the two wave functions add) an increased negative charge density between the two nuclei and contributing to the bonding of the atoms in the molecule. There is a much weaker overlap between the p_y states (Figure 9.9b) and also between the p_z states (which are not shown in the figure). Because the overlap of the p_y states is not along the line connecting the nuclei, there are components to the binding force that oppose one another, and only a much smaller resultant force acts along the line connecting the nuclei (Figure 9.9b). In addition, there is less overlap of the p_y states. The net result is that the p_y states (and also the p_z states) are less effective in binding the molecule than the p_x states.

This somewhat oversimplified model suggests that the p_x state should have a much greater bonding effect (and also a greater antibonding effect) than the p_y and p_z states. It also suggests that the bonding and antibonding effects of p_y states should be the same as those of p_z states.

Let us now consider the energies of the molecular states as a function of the nuclear separation distance R. We assume that we are dealing with two atoms, having valence electrons in the $2p$ shell. Each atom has a $1s$ *atomic* state which is filled by two electrons. When these states overlap, the result is $1s$ bonding and antibonding *molecular* states just as in the case of H_2. There is a total of four $1s$ electrons in the molecule, and with two in each state the $1s$ bonding and antibonding molecular states are filled to capacity. The same is true of the $2s$ states. The *atomic* $2s$ levels form bonding and antibonding *molecular* states, and since each atom has a filled $2s$ shell, the four $2s$ electrons fill both the bonding and antibonding molecular states.

Since the atoms have partially filled $2p$ shells, the final molecular bonding depends critically on the molecular $2p$ states. For each *atomic* p state (p_x, p_y, p_z) there are corresponding bonding and antibonding *molecular* states.

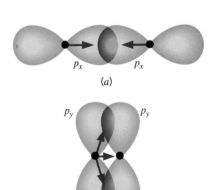

FIGURE 9.9   (a) Overlap of p_x probability distributions. The vectors indicate the force on the nuclei due to overlap. (b) Overlap of p_y probability distributions. The off-axis forces give a smaller resultant force along the axis.

However, the bonding and antibonding effects of these states are not equivalent, as Figure 9.9 illustrates. The p state that happens to lie along the line of approach (p_x) has a significantly greater effect that those that lie off the line of approach (p_y, p_z). The p_x bonding state must therefore lie lower in energy than the p_y and p_z bonding states, and the p_x antibonding state must lie higher in energy than the p_y and p_z antibonding states. Figure 9.10 illustrates the energies of the molecular states.

Let us consider three diatomic molecules composed of atoms with $2p$ valence electrons: N_2, O_2, and F_2. Nitrogen has seven electrons; the $1s$ and $2s$ shells are filled, and three electrons occupy the $2p$ shell. In the N_2 molecule, there are therefore six $2p$ electrons to fill the $2p$ molecular states, each of which can hold two electrons. The three lowest $2p$ molecular levels are filled. Since all three of these are bonding states, N_2 forms a very stable diatomic molecule (binding energy = 9.8 eV, separation = 0.11 nm). Because of this stability, nitrogen in the form of N_2 is rather unreactive under most circumstances.

Oxygen has four $2p$ electrons, and so the O_2 molecule must have eight electrons in the $2p$ molecular levels. These not only fill up the three bonding levels, as with N_2, but also the lowest ($2p_y$ or $2p_z$) *antibonding* level. We therefore expect O_2 to be *less* stable than N_2, since it has three bonding and one antibonding molecular states, and experience is consistent with this expectation. The binding energy of O_2 (5.1 eV) is less than that of N_2, and the nuclear separation is greater (0.12 nm). The O_2 molecule is less stable than the N_2 molecule, and the O_2 molecular bonds can be broken by relatively modest chemical reactions, as, for example, the oxidation of metals exposed to air.

Fluorine has five $2p$ electrons, so the F_2 molecule has ten electrons in the $2p$ states, filling three bonding and now *two* antibonding levels. This makes F_2 even less stable than O_2; the energy necessary to dissociate F_2 into two F atoms is only 1.6 eV, and fluorine gas reacts quite violently with many substances. In fact, since the photons of visible light have energies in excess of 1.6 eV (typically 2 to 4 eV), F_2 is normally unstable and is broken down into F atoms by exposure to light. This process is called *photodissociation*.

Although their molecular level diagrams may be more complicated, we expect all molecules based on p-shell atoms similar to N_2, O_2, and F_2 to behave in similar ways. Unfortunately, the other np^3 atoms (P, As, Sb, Bi) and np^4 atoms (S, Se, Te, Po) are solids at ordinary temperatures, and their molecular properties can normally not be observed, since the interaction of the atoms in a solid is usually more important than molecular bonding effects (as we will see in Chapter 11). However, the other halogens (Cl, Br, I, At) have properties similar to F_2—relatively weakly bound, diatomic molecules, which are quite reactive.

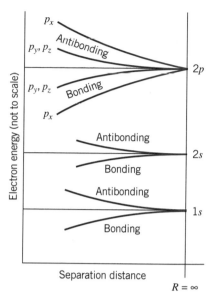

FIGURE 9.10 Bonding and antibonding $2p$ states.

sp Molecular Bonds

It is often the case that a stable molecule is formed from two atoms, one with an *s*-state valence electron and the other with one or more *p*-state

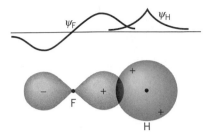

FIGURE 9.11   Overlap of *s* and *p* wave functions.

valence electrons. Consider, for example, the HF molecule. The electron wave function of H we have described previously, and the F atom has five electrons in the *p* shell, so of the three 2*p* atomic states, two will each have their capacity of two electrons, and the third will have a single electron. We ignore the four paired electrons, which do not significantly affect the molecular bonding, and concentrate instead on the single unpaired *p* electron. The two-lobed probability distribution corresponds to a two-lobed *p*-state wave function, in which the signs of ψ are opposite for the two lobes. The 1*s* wave function of H has only one sign (Figure 9.11). If, as the H and F atoms approach from a large distance, the H wave function happens to have the same sign as the nearer lobe of the F wave function, an increased electron probability results, and hence a *bonding sp* state is formed. It is also possible to have *antibonding sp* states, which result from the H and F wave functions having opposite signs.

Table 9.2 gives dissociation energies and nuclear separation distances for some *sp*-bonded diatomic molecules.

Let us now consider the water molecule, H_2O. Oxygen has eight electrons, four of which occupy the 2*p* shell. When we place these electrons in the *2p atomic* states, we begin with one electron each in the p_x, p_y, and p_z states, and then the fourth 2*p* electron must pair with one of the first three. An oxygen atom therefore has two unpaired 2*p* electrons, each of which can form a bond with the 1*s* electron of H to form a molecule of H_2O. Figure 9.12 shows a representation of the electron probability distributions we might expect for an oxygen atom and for a molecule of H_2O. Such a molecule has *directed* bonds, which have a fixed, measurable relative direction in space. The expected angle between the two bonds is 90°; this angle can be measured experimentally by, for example, measuring the electric dipole moment of the atom, and the result, 104.5°, is somewhat larger than we expect. This discrepancy can be interpreted as arising from the Coulomb repulsion of the two H atoms, which tends to spread the bond angle somewhat.

Directed bonds

As another example, consider the NH_3 (ammonia) molecule. With $Z = 7$, the nitrogen atom has three unpaired *p* electrons, one each in the p_x, p_y, and p_z atomic states. Each of these can form a bond with a H atom

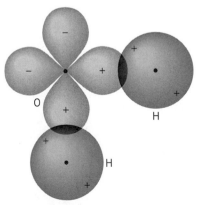

FIGURE 9.12 Overlap of electronic wave functions in H_2O.

TABLE 9.2   PROPERTIES OF *sp*-BONDED MOLECULES

Molecule	Dissociation Energy (eV)	Equilibrium Separation (nm)
HF	5.90	0.092
HCl	4.48	0.128
HBr	3.79	0.141
HI	3.10	0.160
LiF	5.98	0.156
LiCl	4.86	0.202
NaF	4.99	0.193
NaCl	4.26	0.236
KF	5.15	0.217
KCl	4.43	0.267

to form the NH_3 molecule, and we expect to find three mutually perpendicular *sp* bonds (Figure 9.13). The measured bond angle is 107.3°, again indicating some repulsion between the H atoms.

Table 9.3 lists some bond angles measured for other molecules that have *sp* directed bonds. As you can see, the bond angle does indeed approach 90° in many cases. Based on the discussion given above, you should be able to explain why this happens as the Z of the central atom increases.

sp Hybrid States

One example of a 2*p* atom we have so far not considered is carbon, and for a special reason: carbon forms a great variety of different types of molecular bonds, with a resulting diversity in the type and complexity of molecules containing carbon. It is this diversity that is the basis for the many kinds of *organic molecules* that can form, based on various kinds of carbon molecular bonds, and so an understanding of the physics of carbon molecular bonds is essential to the understanding of many fundamental questions of structure and processes in molecular biology.

Carbon, with six electrons, has the configuration $1s^2 2s^2 2p^2$, so we expect carbon under ordinary circumstances to show a valence of 2, with the two 2*p* electrons contributing to the structure, and we might therefore expect to form stable molecules such as CH_2, with directed *sp* bonding (similar to H_2O) and a bond angle of roughly 90°. Instead, what forms is CH_4 (methane) in a tetrahedral structure (Figure 9.14), with four equivalent bonds. For another example, the elements of the third column of the periodic table (boron, aluminum, gallium . . .) have the outer configuration $ns^2 np$ ($n = 2$ for boron, $n = 3$ for aluminum, etc.), and we expect these elements to form compounds as if they had a single valence electron. We therefore expect halides such as BCl or GaF, oxides such as B_2O or Al_2O, nitrides such as B_3N or Al_3N, hydrides such as BH or GaH, and so forth. Instead we find that boron, aluminum, and gallium generally behave as if they had three valence electrons, and form compounds such as BCl_3, Al_2O_3, AlN, and B_2H_6. Furthermore, the three valence electrons seem to be equivalent; there seems to be no way, for example, to associate two of the valence electrons with *s* states and one with a *p* state. The bonds formed by the three electrons make angles of 120° with one another.

It is the effect of *sp hybridization* that is responsible for the valence of three (rather than one) in boron and four (rather than two) in carbon. The four bonds in CH_4 are *equivalent* and *identical*, which would *not* be expected if we had two *ss* bonds and two *sp* bonds; similarly in BF_3 or BCl_3, the three bonds are identical and are clearly *not* identified with two *sp* bonds and one *pp* bond.

The normal meaning of *hybrid* is an offspring resulting from the union of parents of different types, in which the offspring is not exactly like either parent, but retains some of the attributes of each. In the case of molecules, hybridization refers to a process by which the states can no longer be identified as either *s* or *p* states, but rather are mixtures of *s* and *p* states. The formation of *sp* hybrids is normally as follows:

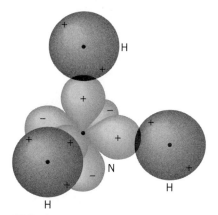

FIGURE 9.13 Overlap of electronic wave functions in NH_3.

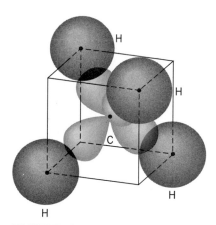

FIGURE 9.14 Tetrahedral arrangement of molecular bonds in CH_4.

sp Hybridization

TABLE 9.3 BOND ANGLES OF *sp* DIRECTED BONDS

Molecule	Bond Angle
H_2O	104.5°
H_2S	93.3°
H_2Se	91.0°
H_2Te	89.5°
NH_3	107.3°
PH_3	93.3°
AsH_3	91.8°
SbH_3	91.3°

1. In an atom with the configuration $2s^2 2p^n$, one of the $2s$ electrons is excited to the $2p$ shell, giving a configuration $2s2p^{n+1}$.
2. The hybrid states are formed by taking equal mixtures of the wave functions representing the 2s state and each of the 2p states. For example, in the case of boron, the configuration of $2s^2 2p$ is converted to $2s2p^2$. Assuming the $2p$ states to be $2p_x$ and $2p_y$, the resultant *hybrid* wave functions can be represented as different combinations of ψ_{2s}, ψ_{2p_x}, and ψ_{2p_y}, such as

$$\psi = \psi_{2s} + \psi_{2p_x} + \psi_{2p_y}$$

Other combinations can be formed by subtracting, instead of adding, the individual wave functions.

Illustrations of the probability distributions expected for sp, sp^2, and sp^3 hybrids are shown in Figure 9.15. Keep in mind that these do not yet represent *molecular* states—they are merely the contribution of one of the atoms to the bonding electronic distributions of the molecule.

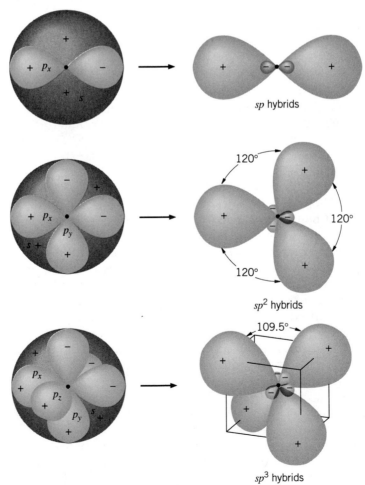

sp hybrids

sp^2 hybrids

sp^3 hybrids

FIGURE 9.15 Probability distributions in sp, sp^2, and sp^3 hybrids.

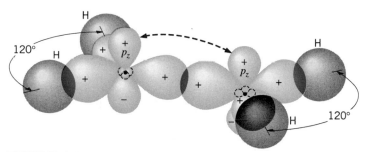

FIGURE 9.16 Molecular bonding in C_2H_4. For clarity the p_z overlap is not shown; the dashed line indicates the bond formed by the unhybridized p_z states.

The tetrahedral structure of CH_4 is therefore merely the result of the symmetrical spatial arrangement of the four sp^3 hybrid states of C, with each hybrid state bonded to one H. The bond angle of such a symmetrical tetrahedron is 109.5°, in good agreement with the measured bond angles of CH_4 and other sp^3 hybrids, shown in Table 9.4.

It is also possible for only two $2p$ electrons in the $2s2p^3$ configuration of carbon to become involved in hybrid states; the third $2p$ electron then is available, for example, to form ordinary pp molecular states. The ethylene molecule C_2H_4 is an example of such a structure. The three sp^2 hybrids form bond angles of 120°; each carbon atom has two of its three sp^2 hybrid states joined to a H atom, and the third is joined to the other carbon. The unhybridized p electron also forms a bond between the two carbons, in a manner similar to that shown in Figure 9.9b for the "off-axis" pp molecular states. Figure 9.16 shows a representation of the bonding in C_2H_4. There are two bonds between the pair of carbons, one from the sp^2 hybrid and the other from the unhybridized p_z state. Another example of sp^2 hybrids in carbon is benzene (C_6H_6) in which each carbon is joined to one H and two other carbons by the sp^2 hybrids, with again one unhybridized p orbital available to bond the carbons. The basic structure of benzene is a ring of carbon atoms, as shown in Figure 9.17, with the expected angle of 120°

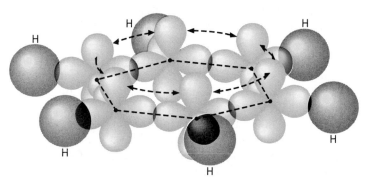

FIGURE 9.17 Molecular bonding in benzene. As in Figure 9.16, the overlap of the unhybridized p_x states is not shown, but the dashed arrows indicate the p_z bonds.

TABLE 9.4 BOND ANGLES OF sp^3 HYBRIDS

Molecule	Bond Angle
CCl_4	109.5°
C_2H_6	109.3°
C_2Cl_6	109.3°
$CClF_3$	108.6°
CH_3Cl	110.5°
$SiHF_3$	108.2°
SiH_3Cl	110.2°
$GeHCl_3$	108.3°
GeH_3Cl	110.9°

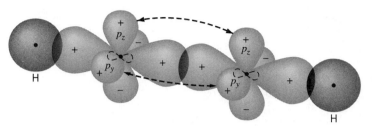

FIGURE 9.18   Molecular bonding in C_2H_2. The carbons are joined by three bonds, one from the p_x hybrid and two from the unhybridized p_y and p_z states.

between the hybrid states, which gives the molecule its characteristic hexagonal shape (see Color Plate 7 and Figure 4.14).

Finally, it is also possible for carbon to form sp hybrids, leaving two unhybridized p states. Acetylene (C_2H_2) is an example of such a molecule, with the two carbons now joined by three bonds, one from the sp hybrid and two from the unhybridized p states. Figure 9.18 shows the bonding arrangement in C_2H_2.

This variety of bonds entered into by carbon is the basis for the varied properties of organic molecules, from the simple ones we have studied here, to the complex ones that form the basis of living things. However, it is not only carbon that shows sp hybridization, but other atoms as well. (Indeed, the failure of NH_3 to conform to the expected 90° bond angle could be blamed on sp hybridization rather than on the repulsion of the H atoms.) It is also possible to have $3s$-$3p$ hybrids (silicon) and $4s$-$4p$ hybrids (germanium). It is this hybridization that gives these materials, like carbon, a valence of 4 and a symmetrical bonding arrangement, which are partly responsible for the usefulness of Si and Ge as semiconducting materials, as we will learn in Chapter 11. It is also interesting to speculate on the possibility of a new type of organic chemistry, including new life forms, based on Si or Ge, rather than C (see Color Plate 8).

9.4   IONIC BONDING

In covalent bonding, as we have seen, the bonding electrons do not belong to any particular atom in the molecule, but rather are shared among the atoms. It is also possible to form a molecule that results from the extreme opposite case in which valence electrons are not shared but instead spend all of their time in the neighborhood of only one of the atoms of the molecule.

Consider the ionic molecule NaCl. Suppose we have a neutral sodium atom ($1s^2 2s^2 2p^6 3s^1$), with one $3s$ electron outside of a closed shell, and a neutral chlorine atom ($1s^2 2s^2 2p^6 3s^2 3p^5$), which lacks one electron from a filled $3p$ shell. To remove the outer electron from Na requires 5.14 eV, the *ionization energy* of Na, and we are left with a positively charged Na^+ ion. If we then attach that electron to the Cl atom, creating a negatively charged

Cl⁻ ion, the energy *released* is 3.61 eV, the *electron affinity* of Cl. The energy is released because the filled $3p$ shell is an especially stable configuration, which is energetically very favorable. Thus, if we borrow 5.14 eV to ionize the Na, we get back immediately 3.61 eV by attaching the electron to the Cl. We can get back the remaining 1.53 eV (= 5.14 eV − 3.61 eV) by moving the Na⁺ and Cl⁻ close enough together that their Coulomb potential energy is 1.53 eV. The separation distance corresponding to this potential energy is found from the potential energy equation, $U = e^2/4\pi\varepsilon_0 R$:

Electron affinity

$$R = \frac{e^2}{4\pi\varepsilon_0}\frac{1}{U} = \frac{1.44 \text{ eV·nm}}{1.53 \text{ eV}} = 0.941 \text{ nm}$$

That is, as long as the Cl⁻ and Na⁺ are closer together than 0.941 nm, the Coulomb attraction will supply enough energy to overcome the difference between the ionization energy of Na and the electron affinity of Cl. Put another way, Na⁺ and Cl⁻ *ions* separated by less than 0.941 nm have a more stable configuration than neutral Na and Cl *atoms.*

It would therefore seem that the closer together we push the Na⁺ and Cl⁻ ions, the more tightly bound they become. However, if they are pushed too close together, the filled $2p$ shell of Na⁺ and the filled $3p$ shell of Cl⁻ will begin to overlap. Because electrons are identical particles, when two electron wave functions overlap it is no longer possible to identify which electron came from Na and which from Cl. There is no "Na electron" or "Cl electron"—all electrons are alike. Thus if a $2p$ electron from Na and a $3p$ electron from Cl begin to overlap, they will occasionally *both* try to behave as $2p$ electrons in Na and also occasionally as $3p$ electrons in Cl. But this is not possible—Na⁺ already has a full $2p$ electron shell, and Cl⁻ has a full $3p$ shell; because of the Pauli principle, no additional electrons are permitted in either shell. Forcing the two ions closer together therefore requires that electrons be pushed from the $2p$ or $3p$ shells into a higher shell, in order to "make room" for the overlapping electrons. Since this takes energy, we must therefore add energy to the Na⁺ + Cl⁻ system in order to reduce the separation of the ions. We can imagine this energy as a sort of "potential energy" of repulsion, which increases rapidly as we try to force the ions close together.

In summary, when the ions are far apart, they attract one another and when they are too close together, they repel one another; in between there must be an equilibrium position where the attractive and repulsive forces are balanced. It is this equilibrium position that determines the size of an ionic molecule.

Figure 9.19 shows the energy of the NaCl molecule as a function of the separation distance between the nuclei. Taking the energy zero point to refer to the neutral atoms, we can separate the molecular energy into three terms: the constant $\Delta E = 1.53$ eV, which represents the difference in energy between the ions and the neutral atoms; the Coulomb attraction U_C between the ions; and the Pauli "repulsion" U_R, which is represented in Figure 9.19 as a potential that rises rapidly for small R and falls to 0 for values of R that exceed the sum of the ionic radii, at which point the electron shells no longer overlap and the Pauli repulsion does not occur. The sum of these three terms gives the molecular energy of NaCl, which shows a minimum

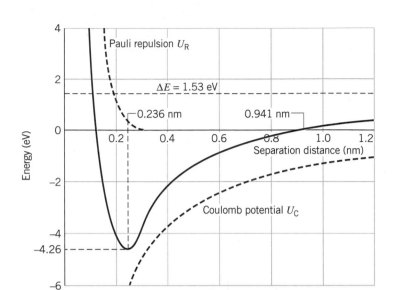

FIGURE 9.19   Molecular energy in NaCl. The "zero" of the energy scale represents neutral Na and Cl atoms. The solid curve is the sum of the three contributions to the molecular energy.

at an equilibrium separation of 0.236 nm. At that distance the energy is −4.26 eV, so the binding energy or dissociation energy (the energy needed to split the molecule into neutral atoms) is 4.26 eV.

Table 9.5 shows the equilibrium separation and dissociation energy of several ionic molecules.

Remember that we are concerned here with isolated molecules and not with collections of molecules in solids. When we speak of NaCl in this section, we mean not the solid salt but rather a gas of NaCl molecules. The spacing between atoms in a solid can be very different from the spacing in a molecule.

TABLE 9.5   PROPERTIES OF SOME IONIC DIATOMIC MOLECULES

Molecule	Dissociation Energy (eV)	Equilibrium Separation (nm)
NaCl	4.26	0.236
NaF	4.99	0.193
NaH	2.08	0.189
LiCl	4.86	0.202
LiH	2.47	0.239
KCl	4.43	0.267
KBr	3.97	0.282
RbF	5.12	0.227
RbCl	4.64	0.279

EXAMPLE 9.2

(a) What is the value of the Pauli repulsion energy of NaCl at the equilibrium separation? (b) Estimate the value of the Pauli repulsion energy at a separation of 0.1 nm.

SOLUTION

(a) At the equilibrium separation distance, the Coulomb energy is

$$U_C = -\frac{1}{4\pi\varepsilon_0}\frac{e^2}{R_{eq}} = -\frac{1.44\ \text{eV}\cdot\text{nm}}{0.236\ \text{nm}} = -6.10\ \text{eV}$$

and the Pauli repulsion energy can be found from the molecular energy E:

$$U_R = E - U_C - \Delta E = -4.26\ \text{eV} - (-6.10\ \text{eV}) - 1.53\ \text{eV} = 0.31\ \text{eV}$$

(b) From Figure 9.19 we estimate $E = +4.0$ eV at $R = 0.1$ nm. The Coulomb energy at $R = 0.1$ nm is -14.4 eV, and the repulsion energy is

$$U_R = E - U_C - \Delta E = +4.0\ \text{eV} - (-14.4\ \text{eV}) - 1.53\ \text{eV} = 16.9\ \text{eV}$$

Note how rapidly the repulsion energy increases at small R.

Covalent and ionic bonding represent two extreme cases, one in which electrons are shared between two atoms and the other in which the electrons always are associated with one of the atoms. How can we decide whether two atoms will join by covalent or ionic bonding? The answer depends on the willingness of the atoms to share their electrons, or equivalently the degree to which one atom dominates the other in its desire to have the valence electrons all to itself.

For homonuclear diatomic molecules, we expect a purely covalent bond; since the two atoms are exactly alike, neither can dominate and the electron is completely shared between them. For heteronuclear molecules, however, the situation is quite different. There can be no purely ionic or purely covalent bond. Since the two atoms have different atomic numbers and different electronic configurations, the wave function of a valence or bonding electron, even if shared, will not be exactly the same near one atom as it is near the other. The electron must therefore spend more time near one atom than near the other; this means that one atom may have a slight excess of negative charge and the bond, even if we think of it as being covalent, will also have a small ionic character. Conversely, a "pure" ionic bond has a small covalent character; we learned in Chapter 7 that electron wave functions do not suddenly drop off to zero amplitude, but instead fall exponentially to zero. Therefore, even in an ionic molecule like NaCl, the wave function of the electron that was transferred to the Cl⁻ ion is not zero at the location of the Na⁺ ion; the wave function may indeed have a very small amplitude at the Na⁺ ion, but it is not zero. The electron thus spends some of its time, even if very little, being shared between the atoms, and the bonding, while mostly ionic, has a small covalent character as well.

Linus Pauling (1901–1994, United States). Although he is generally considered to be a chemist, his work on molecular bonds transcends the traditional chemistry-physics boundary. He was awarded two Nobel prizes, the chemistry prize in 1954 for his work on bonding and the peace prize in 1963 for his efforts toward a ban on the testing of nuclear weapons.

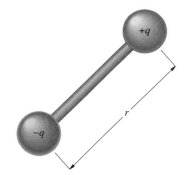

FIGURE 9.20 An electric dipole.

An empirical way to characterize the relative ionic character of a molecule is in terms of the *electric dipole moment* of the molecule. An electric dipole, as illustrated in Figure 9.20, consists of two charges $+q$ and $-q$ separated by a distance r. The electric dipole moment of this arrangement is defined as

$$p = qr \qquad (9.1)$$

In a purely covalent molecule, there is no excess charge on either atom, so in effect $q = 0$ and the dipole moment is expected to be zero. In an ionic molecule, on the other hand, a net positive charge resides on one atom and a net negative charge on the other, so the dipole moment is not zero.

If NaCl were a purely ionic molecule, we would expect a dipole moment of

$$p = eR_{eq} = (1.60 \times 10^{-19} \text{ C})(0.236 \times 10^{-9} \text{ m}) = 3.78 \times 10^{-29} \text{ C·m}$$

Fractional ionic character

The measured electric dipole moment of NaCl is 3.00×10^{-29} C·m, which is 0.79 or 79% of the maximum value corresponding to a purely ionic bond. Because NaCl is only partially ionic, the measured dipole moment is smaller than the full ionic value. By taking the ratio between the measured electric dipole moment and the maximum value calculated from Equation 9.1, we can determine a measure of the fractional ionic character of the molecular bond.

Electronegativity

Is there a property of atoms that allows us to predict whether they will join with one another in ionic or covalent bonds? One property that has some predictive value is the *electronegativity*, which can be roughly defined as the capability of an atom to attract electrons when it forms chemical bonds. The electronegativity of an atom can be computed from the sum of the energy cost of removing an electron (the ionization energy) and the energy gain in adding an electron (the electron affinity). Table 9.6 shows the electronegativity values for some elements.

If the two atoms in a molecule have equal electronegativities, they have equal tendencies to attract electrons and should therefore form covalent bonds. If the two electronegativities are very different, one atom will have a greater tendency to attract electrons; as a result, there will be a net negative charge on one atom and a net positive charge of equal magnitude on the other. The molecular bonding will then have at least a partial ionic character.

TABLE 9.6 ELECTRONEGATIVITIES OF SOME ELEMENTS

H	2.1		
Li	1.0	F	4.0
Na	0.9	Cl	3.0
K	0.8	Br	2.8
Rb	0.8	I	2.5
Cs	0.7		

Source: L. Pauling, *The Chemical Bond* (Ithaca, Cornell University Press, 1967).

We therefore expect a rough relationship between the fractional ionic character (determined by comparing the measured electric dipole moment with its maximum expected value) and the magnitude of the *difference* between the electronegativity values. Figure 9.21 shows the fractional ionic characters plotted against the electronegativity differences. Although the points are scattered, you can see that there is indeed a direct relationship between the quantities: molecules with small electronegativity differences have only a small ionic character and therefore form covalent bonds, while molecules with large electronegativity differences form mostly ionic bonds.

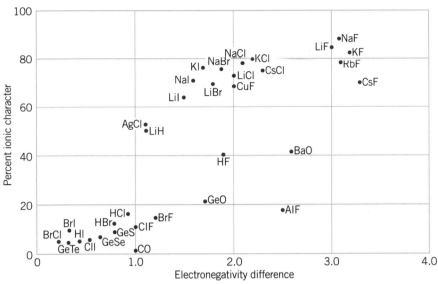

FIGURE 9.21   The fractional ionic character of bonds in diatomic molecules.

9.5   MOLECULAR VIBRATIONS

A molecule can absorb and emit electromagnetic energy just as an atom does; for example, the molecule H_2 can absorb a photon of the right energy to excite one of the $1s$ electrons to an excited $2p$ state, and it will then return to the ground state by emitting a photon. The energies of such photons are typically of the same order as those from electronic transitions in atoms, 1 to 10 eV. We have discussed such optical transitions (in the visible region of the electromagnetic spectrum) for atomic systems in Chapter 8, and since the optical transitions in molecules are very similar, we will not discuss them any further.

The absorption or emission of an optical photon changes the electronic state of motion in a molecule. There are, however, other ways for molecules, but not atoms, to absorb and emit electromagnetic radiation. It is possible to change the state of motion of the individual atoms themselves, by making them vibrate relative to one another or rotate about the center of mass of the atom. Just like the electronic motion, the vibrational and rotational motions are *quantized*—vibrational energy and rotational energy can be emitted or absorbed only in discrete bundles of a certain size. In the remainder of this chapter, we study the alternative ways a molecule can absorb or emit electromagnetic energy by changing its rotational or vibrational state of motion. In these discussions, the electrons are completely ignored—both vibration and rotation depend on the mass of the vibrating or rotating object, and the electrons have too little mass to be significant.

We begin with the vibrational motion of a molecule. The two atoms in the molecule oscillate so that the center of mass remains fixed (Figure 9.22). (We consider only diatomic molecules, in order to simplify the discus-

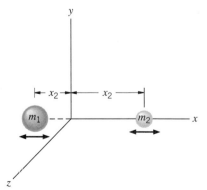

FIGURE 9.22   A vibrating diatomic molecule.

sions, but the same general principles hold for molecules with more than two atoms.) In Chapter 5 we studied the wave mechanics of the simple harmonic oscillator, and we found that, when the potential energy is given by $U = \frac{1}{2}kx^2$, the energy levels are

Energies of vibration

$$E = h\nu(N + \tfrac{1}{2}) \qquad N = 0, 1, 2, \ldots \qquad (9.2)$$

where ν is the same as the classical oscillation frequency:

$$\nu = \frac{1}{2\pi} \sqrt{k/m} \qquad (9.3)$$

The vibrational energy levels are illustrated in Figure 9.23.

A molecule can absorb or emit energy by changing its state of vibration. When a photon of energy $h\nu$ is absorbed, the vibrational quantum number N increases by one unit. An excited molecule can lose energy by emitting a photon of energy $h\nu$, in which case N decreases by one unit. In either absorption or emission, there is a *selection rule* that indicates which transitions are permitted to occur:

Vibrational selection rule

$$|\Delta N| = 1 \qquad (9.4)$$

The only transitions that can occur are those that change N by one unit. These are indicated in Figure 9.23. Notice that all of the transitions have the same energy $h\nu$.

In order to calculate the frequencies of the emitted transitions using Equation 9.3, we must know the mass m and the vibrational force constant k. Since both atoms in the molecule are participating in the vibration, it is not clear which mass to use. Let the masses of the two atoms be m_1 and m_2. As they both pass through their equilibrium positions, the total energy of the molecule is

$$E_T = \tfrac{1}{2}m_1 v_1^2 + \tfrac{1}{2}m_2 v_2^2$$

$$= \frac{p_1^2}{2m_1} + \frac{p_2^2}{2m_2} \qquad (9.5)$$

In a frame of reference in which the center of mass of the molecule is fixed, $p_1 = p_2$ and the energy can be written (with $p = p_1 = p_2$) as

$$E_T = \frac{1}{2}p^2 \left(\frac{1}{m_1} + \frac{1}{m_2}\right) = \frac{1}{2}p^2 \left(\frac{m_1 + m_2}{m_1 m_2}\right) = \frac{p^2}{2m} \qquad (9.6)$$

where

$$m = \frac{m_1 m_2}{m_1 + m_2} \qquad (9.7)$$

That is, the energy of the system is the same as if it were a single mass m, moving with momentum p. This m is a sort of effective mass of the whole molecule and is known as the *reduced mass*. It is the mass we should use in calculating the vibrational frequency.

Notice that $m = \frac{1}{2}m_1$ when $m_1 = m_2$, as in a homonuclear molecule—the effective mass is half the mass of an individual atom. Whenever one mass

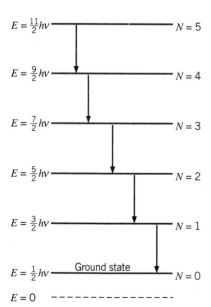

FIGURE 9.23   Energy levels and transitions in a vibrating molecule.

is much greater than the other, the reduced mass has a value nearly equal to that of the lighter mass. This is consistent with our expectation, since the inertia of the heavier mass reduces its tendency to move, and most of the vibrational motion is done by the lighter mass.

The calculation of the vibrational force constant is illustrated in the following example.

EXAMPLE 9.3

Find the vibrational frequency and photon energy for H_2.

SOLUTION

To find the frequency from Equation 9.3, we must know the force constant k. To estimate k, we consider the molecule to behave like a simple harmonic oscillator in the vicinity of its equilibrium separation R_{eq}, and we fit a parabolic oscillator potential energy $U = \frac{1}{2}kx^2$ to the molecular energy in that region. Figure 9.24 shows the region of the energy minimum and a parabola that approximates the curve near the minimum. The equation of the parabola is

$$E - E_{min} = \frac{1}{2}k(R - R_{eq})^2 \qquad (9.8)$$

The constant k can be estimated from the graph by finding the value of $R - R_{eq}$ that is necessary for a certain value of $E - E_{min}$. As shown in the figure, when $E - E_{min} = 0.50$ eV, the value of $R - R_{eq}$ is $\frac{1}{2}(0.034$ nm$) = 0.017$ nm. Solving for k, we obtain

$$k = \frac{2(E - E_{min})}{(R - R_{eq})^2} = \frac{2(0.50\,\text{eV})}{(0.017\,\text{nm})^2} = 36 \times 10^{20}\,\text{eV/m}^2$$

The reduced mass of molecular hydrogen is half the mass of a hydrogen atom. We can now calculate the vibrational frequency:

$$\nu = \frac{1}{2\pi}\sqrt{\frac{k}{m}} = \frac{1}{2\pi}\sqrt{\frac{kc^2}{mc^2}}$$

$$= \frac{1}{2\pi}\sqrt{\frac{(36 \times 10^{20}\,\text{eV/m}^2)(9.0 \times 10^{16}\,\text{m}^2/\text{s}^2)}{(0.5)(1.008\,\text{u})(931.5\,\text{MeV/u})}}$$

$$= 1.3 \times 10^{14}\,\text{Hz}$$

The corresponding photon energy is

$$E = h\nu = (4.14 \times 10^{-15}\,\text{eV·s})(1.3 \times 10^{14}\,\text{Hz}) = 0.54\,\text{eV}$$

The wavelength of this radiation is 2.3 μm, which is in the infrared region of the spectrum. Molecular vibrations typically give photons in the infrared.

Note that the parabola gives a reasonable approximation to the actual energy curve only for energies up to about 1 eV above the minimum, which corresponds to the excited state with $N = 2$. If we were to excite the H_2 molecule to energies greater than 1 eV above its ground state, we would expect to see deviations from the behavior predicted by the simple harmonic

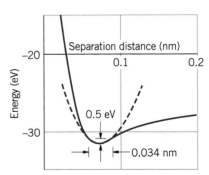

FIGURE 9.24 Fitting a parabola (dashed line) to the energy minimum of H_2.

oscillator. In particular, all the transitions would no longer have the same energy, and changes other than $\Delta N = \pm 1$ may occur. In other molecules, the harmonic oscillator approximation may remain valid for larger vibrational quantum numbers, but eventually it will fail in all cases at sufficiently large N.

Summarizing our conclusions from this section for the vibrational motion of molecules, we expect the simple harmonic oscillator to give a reasonable description of the motion near the energy minimum, with a sequence of emission or absorption photons all of energy $h\nu$ corresponding to changes of one unit in the vibrational quantum number. The emitted or absorbed radiations are generally in the infrared region of the spectrum.

9.6 MOLECULAR ROTATIONS

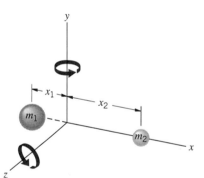

FIGURE 9.25 A rotating diatomic molecule.

A second way that a molecule can change its state of motion when it absorbs or emits radiation is by rotating about the center of mass. Consider the diatomic molecule shown in Figure 9.25. The coordinate system has as its origin the center of mass of the molecule, so that $x_1 m_1 = x_2 m_2$. The rotational kinetic energy is

$$K = \tfrac{1}{2} I \omega^2 \tag{9.9}$$

where I is the rotational inertia of the molecule and ω is its angular velocity. We can also write the kinetic energy in terms of the angular momentum $|\mathbf{L}| = I\omega$

$$K = \frac{|\mathbf{L}|^2}{2I} \tag{9.10}$$

The rotational inertia of the molecule is

$$I = m_1 x_1^2 + m_2 x_2^2 \tag{9.11}$$

We can write this in terms of the reduced mass $m = m_1 m_2/(m_1 + m_2)$ and the equilibrium separation $R_{eq} = x_1 + x_2$ as

$$I = m R_{eq}^2 \tag{9.12}$$

The kinetic energy is the total energy of the system which must be quantized, since there is no potential energy. Without going through the solution to the Schrödinger equation for this kind of rotational motion, we can guess at the value of the quantized energies. For the hydrogen atom, the magnitude of the angular momentum $\mathbf{L}$ is given in Equation 7.11 by $|\mathbf{L}| = \sqrt{l(l + 1)}\hbar$ where $l = 0, 1, 2, 3, \ldots$, and we expect therefore that the energies of the rotating diatomic molecule are

Energies of rotation

$$E = \frac{|\mathbf{L}|^2}{2I} = \frac{L(L + 1)\hbar^2}{2m R_{eq}^2} \tag{9.13}$$

where L is a quantum number that takes the values 0, 1, 2, . . . , just like the orbital angular momentum of the H atom.

As we excite the molecule to a high rotational state, it drops back toward the ground state by emitting photons corresponding to rotational transitions. The selection rule for these photons is

$$|\Delta L| = 1 \tag{9.14}$$

Rotational selection rule

The energy of a photon is then the energy difference between two adjacent levels:

$$\Delta E = E_{L+1} - E_L$$

$$= \frac{(L+1)(L+2)\hbar^2}{2mR_{eq}^2} - \frac{L(L+1)\hbar^2}{2mR_{eq}^2}$$

$$= (L+1)\frac{\hbar^2}{mR_{eq}^2} \tag{9.15}$$

You will recall that the transitions among the *vibrational* excitations all had the same energy; however, the energies of the *rotational* transitions depend on L. Figure 9.26 shows a typical sequence of rotational levels and the transitions between them.

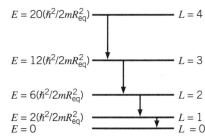

$E = 20(\hbar^2/2mR_{eq}^2)$ — $L = 4$

$E = 12(\hbar^2/2mR_{eq}^2)$ — $L = 3$

$E = 6(\hbar^2/2mR_{eq}^2)$ — $L = 2$

$E = 2(\hbar^2/2mR_{eq}^2)$ — $L = 1$
$E = 0$ — $L = 0$

FIGURE 9.26 Energy levels and transitions in a rotating diatomic molecule.

EXAMPLE 9.4

Calculate the energies and wavelengths of the three lowest radiations emitted by molecular H_2.

SOLUTION

The photon energy equals the energy difference between the levels, which is given by Equation 9.15. To evaluate this energy, we must first determine the rotational quantity $\hbar^2/mR_{eq}^2$. For H_2, the reduced mass m is half the mass of a hydrogen atom. Then

$$\frac{\hbar^2}{mR_{eq}^2} = \frac{\hbar^2 c^2}{mc^2 R_{eq}^2} = \frac{(hc)^2}{4\pi^2 mc^2 R_{eq}^2}$$

$$= \frac{(1240\,\text{eV·nm})^2}{4\pi^2(0.5 \times 1.008\,\text{u} \times 931.5\,\text{MeV/u})(0.074\,\text{nm})^2}$$

$$= 0.0152\,\text{eV}$$

Equation 9.15 now gives the energies directly, and the corresponding wavelengths are found from $\lambda = hc/\Delta E$.

$L = 1$ to $L = 0$	$\Delta E = 0.0152\,\text{eV}$	$\lambda = 81.6\,\mu\text{m}$
$L = 2$ to $L = 1$	$\Delta E = 0.0304\,\text{eV}$	$\lambda = 40.8\,\mu\text{m}$
$L = 3$ to $L = 2$	$\Delta E = 0.0456\,\text{eV}$	$\lambda = 27.2\,\mu\text{m}$

The emitted energies form a sequence $\Delta E, 2\,\Delta E, 3\,\Delta E, \ldots$, and the emitted wavelengths are $\lambda, \lambda/2, \lambda/3, \ldots$

The emitted rotational transitions are (like the vibrational transitions) in the infrared region, but the wavelengths are 1–2 orders of magnitude larger than those of the vibrational transitions. This region of the spectrum corresponds to the far infrared and microwave radiations.

EXAMPLE 9.5

Figure 9.27 shows a portion of the rotational absorption spectrum of a molecule. Determine the rotational inertia of the molecule.

SOLUTION

Each peak in the absorption spectrum corresponds to the molecule moving from one level of Figure 9.26 to the next higher level after absorbing a photon of the proper energy or frequency. The frequencies can be found from Equation 9.15:

$$\nu = \frac{\Delta E}{h} = (L+1)\frac{\hbar}{2\pi I}$$

where Equation 9.12 has been used to replace mR_{eq}^2 with the rotational inertia I. This gives the frequency of each peak in Figure 9.27, but doesn't allow us to find I since we don't know the L value for the peaks. We can avoid this problem by calculating the difference in frequency $\Delta \nu$ between adjacent peaks L and $L+1$:

$$\Delta \nu = (L+2)\frac{\hbar}{2\pi I} - (L+1)\frac{\hbar}{2\pi I} = \frac{\hbar}{2\pi I}$$

or, estimating the spacing $\Delta \nu$ between the peaks of Figure 9.27 as 6.2×10^{11} Hz,

$$I = \frac{\hbar}{2\pi \, \Delta \nu} = \frac{1.05 \times 10^{-34}\,\text{J} \cdot \text{s}}{2\pi(6.2 \times 10^{11}\,\text{Hz})}$$

$$= 2.7 \times 10^{-47}\,\text{kg} \cdot \text{m}^2$$

$$= 0.016\,\text{u} \cdot \text{nm}^2$$

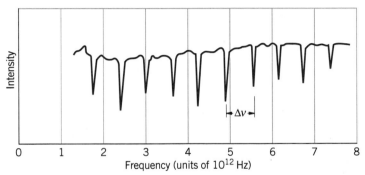

FIGURE 9.27   A molecular absorption spectrum.

This value would correspond to, for example, a molecule with a reduced mass of 1 u (such as one consisting of hydrogen combined with a much heavier atom) and an equilibrium separation of 0.13 nm.

Summarizing this section, we have seen that the rotational motion of a molecule results in a sequence of energy levels that are not equally spaced. The rotational energy levels are spaced 1–2 orders of magnitude closer than the vibrational levels. The emitted radiations form an ascending series of energies (or a descending series of wavelengths) in the far infrared or microwave region of the spectrum. The radiations are restricted by the selection rule that permits the rotational quantum number L to change by only one unit.

9.7 MOLECULAR SPECTRA

A complex molecule can absorb or emit energy in a variety of ways, as indicated schematically by the energy-level diagram of Figure 9.28. The emitted or absorbed energy can change the electronic state (in analogy with the change of energy levels of electrons in atoms). The energy necessary to make this change is of the order of eV, corresponding to photons in the visible range of the spectrum. Within the energy minimum of any electronic state, there are vibrational and rotational states. The vibrational states are evenly spaced and have energy spacings of typically 0.1–1 eV. The rotational states are not evenly spaced, and they have a smaller energy spacing of typically 0.01–0.1 eV. Because the rotational spacing is much smaller than the vibrational spacing, it is convenient to consider each vibrational state as providing the basis on which a sequence of rotational states is built. We discuss here only the vibrational and rotational structure, not the electronic excited states.

Figure 9.29 shows a detail of the rotational and vibrational structure. The states are labelled with the vibrational quantum number N and the rotational quantum number L. Transitions between the levels must satisfy both the rotational and vibrational selection rules

$$|\Delta N| = 1 \quad and \quad |\Delta L| = 1 \quad (9.16)$$

Consider the state with quantum numbers $N = 1$ and $L = 4$, as shown in Figure 9.29. The molecule cannot make a transition to the next lowest rotational state ($N = 1$, $L = 3$), because it would violate the vibrational selection rule $|\Delta N| = 1$. All transitions must simultaneously satisfy *both* selection rules. The molecule can make a transition to the state with $N = 0$ and $L = 3$ or to the state with $N = 0$ and $L = 5$. Absorption transitions must also satisfy these selection rules.

Combined vibrational and rotational selection rule

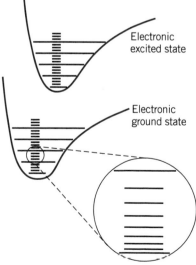

Electronic excited state

Electronic ground state

FIGURE 9.28 Molecular electronic, vibrational, and rotational energy levels.

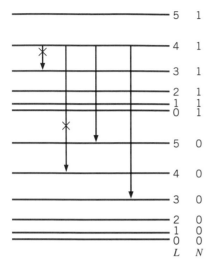

	5	1
	4	1
	3	1
	2	1
	1	1
	0	1
	5	0
	4	0
	3	0
	2	0
	1	0
	0	0
	L	*N*

FIGURE 9.29 Combined rotational and vibrational energy levels. Transitions marked with an X violate the selection rules and are not allowed. Note that the rotational spacing is much smaller than the vibrational spacing, so a sequence of rotational states is built on each vibrational state.

Energies of absorbed photons

The energy of the state with quantum numbers N and L can be written as the sum of the vibrational and rotational terms:

$$E_{NL} = \left(N + \frac{1}{2}\right)h\nu + \frac{L(L+1)\hbar^2}{2mR_{eq}^2} \tag{9.17}$$

where

$$N = 0, 1, 2, \ldots \quad \text{and} \quad L = 0, 1, 2, \ldots$$

Since the vibrational term is so much larger than the rotational term, the *emission* wavelengths in the spectrum will usually correspond to $N \to N - 1$, while $L \to L \pm 1$; the *absorption* wavelengths will be those for transitions in which N increases by one unit.

For absorption from initial state N, L to a final state $N + 1, L \pm 1$ the possible photon energies are

$$\Delta E = E_{N+1, L\pm1} - E_{NL}$$

$$= \left[\left(N + \frac{3}{2}\right)h\nu + \frac{(L \pm 1)(L \pm 1 + 1)\hbar^2}{2mR_{eq}^2}\right]$$

$$- \left[\left(N + \frac{1}{2}\right)h\nu + \frac{L(L+1)\hbar^2}{2mR_{eq}^2}\right]$$

$$\Delta E = h\nu + \frac{\hbar^2}{mR_{eq}^2}(L+1) \quad \text{for } L \to L + 1 \tag{9.18}$$

$$\Delta E = h\nu - \frac{\hbar^2}{mR_{eq}^2}(L) \quad \text{for } L \to L - 1 \tag{9.19}$$

Figure 9.30 shows the expected spectrum of absorption photons. Starting from the center and increasing to the right is a series of photons whose energies are given by Equation 9.18:

$$h\nu + \hbar^2/mR_{eq}^2, h\nu + 2\hbar^2/mR_{eq}^2, h\nu + 3\hbar^2/mR_{eq}^2, \ldots$$

Also starting from the center but decreasing to the left is a series of photons whose energies are given by Equation 9.19:

$$h\nu - \hbar^2/mR_{eq}^2, h\nu - 2\hbar^2/mR_{eq}^2, h\nu - 3\hbar^2/mR_{eq}^2, \ldots$$

Note that the photon of energy $h\nu$ is missing at the center of the spectrum; such a photon would correspond to a "pure" vibrational transition which would violate the rotational selection rule.

We can compare this ideal spectrum with the real spectrum of Figure 9.31, which shows the absorption transitions of the molecule HCl. Although the basic structure of Figure 9.30 is present, there are a number of differences, which we can explain based on the structure of the molecule.

1. *The transitions are not equally spaced.* We expect all of the transitions of the spectrum to be separated by a constant energy $\hbar^2/mR_{eq}^2$ but, as you can see, this is not the case. The explanation for this effect lies with our assumption that the moment of inertia of the molecule is constant. As the molecule rotates, there is an apparent "centrifugal force," which tends

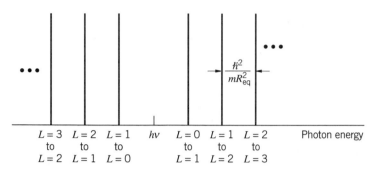

FIGURE 9.30 Expected sequences of absorption transitions between combined rotational and vibrational states.

to increase the separation of the atoms, increasing the rotational inertia and decreasing the rotational energy. As Figure 9.31 shows, this effect becomes more severe as L increases.

2. *The heights of the peaks are quite different.* The heights of the peaks give the intensity of the transitions, and the intensity of any transition is proportional to the population of the particular level from which that transition originates. The populations of the levels decrease as the energy increases, according to the Maxwell-Boltzmann distribution factor $e^{-E/kT}$. The populations also increase with increasing L according to the factor $2L + 1$, which gives the angular momentum degeneracy of each level; in effect, the more substates there are in each level, the greater its population. We can therefore write the population of a level of energy E_{NL} as

$$p(E_{NL}) = (2L + 1)e^{-E_{NL}/kT} = (2L + 1)e^{-[(N+1/2)h\nu + L(L+1)\hbar^2/2mR_{eq}^2]/kT} \quad (9.20)$$

where we have used E_{NL} from Equation 9.17.

For the first few levels, the populations increase as L increases, because the exponential factor does not differ much from 1. However, as the energy continues to increase, the rapidly decreasing exponential factor begins to dominate and the populations decrease accordingly, becoming negligible for $L > 10$.

The level with maximum population, which corresponds to the most intense peak in the spectrum, can be found from Equation 9.20 by locating

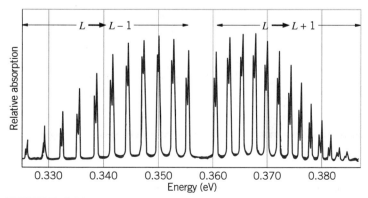

FIGURE 9.31 The molecular absorption spectrum of HCl.

its maximum, where $dp/dL = 0$. The result is

$$2L + 1 = \sqrt{\frac{4kT}{\hbar^2/mR_{eq}^2}} \tag{9.21}$$

For a measurement at room temperature ($kT \cong 0.025$ eV) and for $\hbar^2/mR_{eq}^2 \cong 0.0026$ eV (estimated from the peak spacing in Figure 9.31), we obtain $L \cong 3$ for the peak of maximum intensity, in agreement with Figure 9.31.

3. *Each peak appears to be two very closely spaced peaks.*   Chlorine consists of two types of atoms (isotopes) whose masses are roughly 35 u and 37 u. The differing masses result in slightly different vibrational and rotational energies for the two isotopes, with the heavier atom having the smaller energies. Thus the peaks for the Cl atoms with mass 37 u appear at slightly lower energies than the peaks for the Cl atoms with mass 35 u. The lighter atom occurs with about three times the abundance of the heavier atom, so the peaks for the lighter atom have the greater intensity.

Molecular spectroscopy   Just as atomic spectroscopy gives us a way to identify atoms from their characteristic emission or absorption spectrum, *molecular spectroscopy* enables us to identify molecules by the radiation they absorb or emit. Each molecule has its own characteristic "fingerprint" that can be easily recognized. It is important that this technique tells us exactly the composition of the molecule—the number of atoms of each type, the isotopic ratios, even the state of ionization of the molecule. Thus we could easily distinguish CO from CO_2, $H^{35}Cl$ from $H^{37}Cl$, H_2^+ from H_2.

As you might imagine, such a precise identification technique has many applications to areas where it is necessary to identify precisely trace amounts of molecules. Two applications in particular are of interest. The absorption spectra of our atmosphere can be used to identify trace amounts of various pollutants, and thus molecular spectroscopy helps to measure the purity of our air. Similarly, the absorption spectra of interstellar dust are used to identify the molecules which are present in interstellar space. This is the only technique we have (so far) for learning about the formation of complex molecules in our galaxy, because stars are too hot to permit molecules to exist.

Unfortunately, our atmosphere absorbs much of the infrared and microwave radiation that characterizes the spectra of these molecules, but spectrometers carried on satellites beyond the atmosphere are permitting the observation of those radiations and the identification of many varieties of molecules, including some relatively complex organic molecules. As an added benefit, when those spectrometers are aimed toward the Earth, they can measure how the infrared radiation emitted by the Earth is absorbed in its atmosphere, and thus detect the presence of various atmospheric pollutants.

EXAMPLE 9.6

(*a*) From the absorption spectrum of HCl (Figure 9.31), determine the vibrational force constant. (*b*) Determine the rotational energy spacing of the peaks and compare with the expected value for HCl.

SOLUTION

(*a*) The vibrational frequency can be found from the energy of the "missing" central transition in Figure 9.31. This energy can be read from the figure to be about 0.358 eV, and so

$$\nu = \frac{\Delta E}{h} = \frac{0.358 \text{ eV}}{4.14 \times 10^{-15} \text{ eV} \cdot \text{s}} = 8.65 \times 10^{13} \text{ Hz}$$

The reduced mass of HCl can be found from Equation 9.7 to be $m = 0.98$ u. We can now find the force constant by solving Equation 9.3 for k:

$$k = 4\pi^2 m \nu^2 = 4\pi^2 (0.98 \text{ u})(8.65 \times 10^{13} \text{ Hz})^2$$

$$= 2.89 \times 10^{29} \text{ u} \cdot \text{Hz}^2 = 2.99 \times 10^{21} \text{ eV/m}^2$$

(*b*) From the figure, the energy spacing of the peaks is estimated to be 0.0026 eV. The expected spacing (see Figure 9.30) is

$$\frac{\hbar^2}{mR_{eq}^2} = \frac{(hc)^2}{4\pi^2 (mc^2) R_{eq}^2} = \frac{(1240 \text{ eV} \cdot \text{nm})^2}{4\pi^2 (0.98 \text{ u} \times 931.5 \text{ MeV/u})(0.127 \text{ nm})^2}$$

$$= 0.00265 \text{ eV}$$

This calculation gives excellent agreement with the value estimated from the spectrum.

SUGGESTIONS FOR FURTHER READING

Some comprehensive books on molecular spectroscopy that include introductory as well as advanced material:

G. M. Barrow, *Introduction to Molecular Spectroscopy* (New York, McGraw-Hill, 1962).
M. Karplus and R. N. Porter, *Atoms and Molecules* (Menlo Park, W. A. Benjamin, 1970).
L. Pauling, *The Chemical Bond* (Ithaca, Cornell University Press, 1967).

An advanced but very detailed work:

G. Herzberg, *Molecular Spectra and Molecular Structure* (New York, Van Nostrand, 1950). Volume I covers diatomic molecules.

Two popular-level articles on the search for molecules in space:

B. E. Turner, "Interstellar Molecules," *Scientific American* **228**, 50 (March 1973).
B. Zuckerman, "Interstellar Molecules," *Nature* **268**, 491 (August 1977).

Information on the properties of molecules is tabulated in many locations; see, for example, the *Handbook of Chemistry and Physics* (Chemical Rubber Publishing Co.) or the *Journal of Physical and Chemical Reference Data.*

QUESTIONS

1. Why does H_2 have a smaller radius and a greater binding energy than H_2^+?

2. The molecule LiH has a simple electronic structure. The H atom would like to gain an electron to fill its $1s$ subshell, but the Li atom would similarly like to fill its $2s$ subshell. Based on the atomic structure of H and Li, which would you expect to dominate in the desire for an additional electron? Are the electronegativity values of Table 9.6 consistent with this?

3. In general, would you expect ss bonds or pp bonds to be stronger? Why?

4. Explain why the bond angles of the sp directed bonds (Table 9.3) approach $90°$ as the atomic number of the central atom increases.

5. How do the molecular force constants k compare with those of ordinary springs? What do you conclude from this comparison?

6. Why is it unnecessary to consider rotations of the diatomic molecule of Figure 9.25 about the x axis? Estimate the rotational inertia for rotations about the x axis; compare the typical rotational energies with corresponding values for rotations about the y or z axis.

7. Explain how the equilibrium separation in a molecule can be determined by measuring the absorption or emission lines for rotational states.

8. How would the rotational energy spacing of D_2 (a form of "heavy hydrogen"; the mass of D is twice the mass of H) compare with the rotational spacing of H_2? How would their vibrational energy spacings compare? Their equilibrium separation distances?

9. For a molecule like HCl, estimate the number of rotational levels between the first two vibrational levels.

10. Why does an atom generally absorb radiation only from the ground state, while a molecule can absorb from many excited rotational or vibrational states?

11. If a collection of molecules were all in the $N = 0$, $L = 0$ ground state, how many lines would there be in the absorption spectrum?

PROBLEMS

1. Calculate the ionization energy of H_2.

2. Calculate the Coulomb energy of KBr at the equilibrium separation distance.

3. The ionization energy of potassium is 4.34 eV; the electron affinity of iodine is 3.06 eV. At what separation distance will the KI molecule gain enough Coulomb energy to overcome the energy needed to form the K^+ and I^- ions?

4. Bond strengths are frequently given in units of kilojoules per mole. Find the molecular dissociation energy (in eV) from the following bond strengths: (a) NaCl, 410 kJ/mole; (b) Li_2, 106 kJ/mole; (c) N_2, 945 kJ/mole.

5. (a) Assuming a separation of 0.193 nm, compute the expected electric dipole moment of NaF. (b) The measured dipole moment is 27.2×10^{-30} C·m. What is the fractional ionic character of NaF?

6. The equilibrium separation of HI is 0.160 nm and its measured electric dipole moment is 1.47×10^{-30} C·m. Find the fractional ionic character of HI.

7. The equilibrium separation of BaO is 0.194 nm and the measured electric dipole moment is 26.5×10^{-30} C·m. Calculate the fractional ionic character, assuming two valence electrons.

8. An empirical function that might be used to represent the molecular energy is

$$E = \frac{A}{R^9} - \frac{B}{R}$$

where A and B are positive constants. Find A and B in terms of the equilibrium separation R_{eq} and the dissociation energy E_0. Sketch the resulting function.

9. Show that the angle between bonds in the tetrahedral carbon structure (Figure 9.14) is 109.5°.

10. Derive Equation 9.12 from Equation 9.11.

11. Find the difference in the rotational energy $\hbar^2/mR^2$ of HCl for the two different masses of Cl (35 u and 37 u).

12. Figure 9.32 shows the energy minimum of molecular NaCl, through which a parabola has been drawn. Following the methods of Example 9.3, find the effective vibrational force constant, the vibrational frequency and wavelength, and the vibrational photon energy for NaCl. In what range of the electromagnetic spectrum would such radiations be found? What is the maximum vibrational quantum number for which the parabolic approximation remains valid for NaCl?

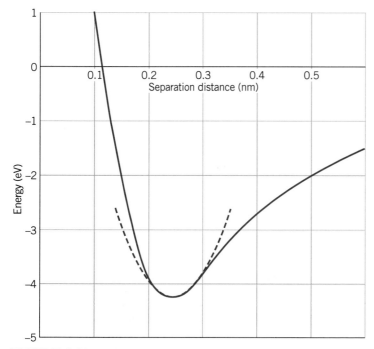

FIGURE 9.32 Problem 12.

13. Complete the following table, which compares properties of H_2 molecules when one or both of the H is replaced with an atom of deuterium (D), which is a "heavy hydrogen" atom with twice the mass of ordinary hydrogen.

Molecule	Vibrational frequency	R_{eq}	$\hbar^2/mR_{eq}^2$
H_2	1.32×10^{12} Hz	0.074 nm	0.0152 eV
HD			
D_2			

14. The actual dissociation energy of a molecule depends not only on the depth of the energy minimum but on the "zero-point" energy of the vibrational motion (see Figures 9.23 and 9.28). Given that the dissociation energy of H_2 is 4.52 eV and that the vibrational energy is 0.54 eV, find the dissociation energies of HD and D_2. (See Problem 13.)

15. Following the method of Example 9.4, calculate the energies and wavelengths of the three lowest rotational transitions emitted by molecular NaCl. In what region of the electromagnetic spectrum would these radiations be found?

16. Estimate the number of rotational states between the vibrational states for (a) H_2 (see Example 9.3); (b) HCl (see Example 9.6); (c) NaCl (see Problems 12 and 15).

17. Figure 9.29 illustrates the combined rotational-vibrational structure when the vibrational energy is much larger than the rotational energy. Make a sketch showing the reverse situation, in which the rotational energy is much larger than the vibrational energy. Use a scale in which $\hbar^2/mR_{eq}^2 = 20$ units and $hv = 2$ units; show rotational levels up to $L = 3$ and vibrational levels to $N = 3$.

18. (a) Sketch a diagram, similar to Figure 9.29, showing all possible absorption transitions from the $N = 0$ to the $N = 1$ states. Include rotational states up to $L = 5$. (b) Use the values $hv = 10$ units and $\hbar^2/2mR_{eq}^2 = \frac{1}{4}$ unit and show the energy spectrum of the absorption, including all transitions from part (a). Label each transition with the initial and final quantum numbers.

19. Based on the energy levels given by Equation 9.17, calculate the energies of the *emitted* photons.

20. (a) What is the reduced mass of the KCl molecule? (b) With an equilibrium separation of 0.267 nm, find the spacing of the transitions in the combined rotational-vibrational spectrum.

21. Figure 9.33 shows the absorption spectrum of the molecule HBr. Following the basic procedures of Section 9.7, find: (a) the energy of the "missing" transition; (b) the effective force constant k; (c) the rotational spacing $\hbar^2/mR_{eq}^2$. Estimate the value of the rotational spacing expected for HBr and compare with the value deduced from the spectrum. Why are there only single lines and not double lines as in the case of HCl?

22. Derive Equation 9.21 from Equation 9.20.

23. (a) In a collection of H_2 molecules at room temperature, what is the ratio of the number of molecules in the $N = 1$ vibrational state to the number in the $N = 0$ vibrational state? (Ignore the rotational structure for this problem.) (b) What is the ratio of the number in the $N = 2$ state to the number in the ground state?

24. Repeat Problem 23 for NaCl molecules.

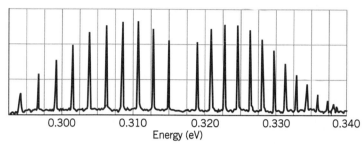

0.300   0.310   0.320   0.330   0.340
Energy (eV)

FIGURE 9.33   Problem 21.

25. (a) In a collection of H_2 molecules at room temperature, find the relative numbers of molecules in the first four rotational states of the $N = 0$ vibrational state. (b) Repeat part (a) if the temperature is 30 K. (*Hint*: Don't forget the degeneracy of the levels.)

26. Repeat Problem 25 for NaCl.

27. At what temperature would 25 percent of a collection of HCl molecules be in the first excited vibrational state? (Ignore the rotational structure.)

28. The most intense absorption line in the rotational-vibrational spectrum of CO at room temperature occurs for $L = 7$. Justify this value with a calculation. (The equilibrium separation of CO is 0.113 nm.)

29. CUPS Exercise 7.13.

30. CUPS Exercise 7.14.

31. CUPS Exercise 8.15.

32. CUPS Exercise 8.18.

Some processes in nature are determined by the statistical distribution of random events. When the number of particles in the system is large, the distribution of the possible outcomes of a measurement becomes unobservably narrow, and we can regard such processes as leading to definite measurable results. Other processes are governed by strictly deterministic laws but often seem chaotic in their outcomes. Such processes are represented by apparently irregular geometric patterns called fractals.

STATISTICAL
PHYSICS

Many physics experiments are analyzed as if the interactions take place in single, isolated events. For example, in the emission of light from the atoms of a gas at low density, the transitions of an electron in any atom are unlikely to be affected by the presence of other atoms, and so we can treat the light from a collection of many atoms as we would treat light from a single atom. Rutherford scattering and Compton scattering are similar examples of experiments that can be analyzed in the same way.

On the other hand, consider the addition of energy to a gas in a container by raising its temperature. If we add a total energy E to a gas of N atoms, we cannot predict with certainty how much energy any particular atom will acquire. On the average, each atom's energy will increase by E/N, but some atoms might acquire no additional energy at all, while others might absorb $10\,E/N$ or even $100\,E/N$.

This sharing of energy among the many parts of a system cannot be simply analyzed in terms of single isolated events. The analysis of such *cooperative* phenomena requires the techniques of *statistical physics*, in which we are concerned not with calculating the *exact* outcome of single, isolated events, but with predicting the *average* outcome of many cooperative events, based on the *statistical distribution* of the possible outcomes.

In this chapter we discuss the laws of statistical physics, and we illustrate some systems that are governed by *classical* statistics and some others that require *quantum* statistics. These statistical concepts are necessary to understand the bulk properties of matter, which are discussed briefly in this chapter and more extensively in Chapter 11.

10.1 STATISTICAL ANALYSIS

When we pass an electric current through a tube containing a gas of very low density, such as mercury vapor, light is emitted by the atoms of the gas. Electrons in individual atoms are pushed from the ground state to excited states, and they then return to the ground state with the emission of one or more photons. In the case of mercury vapor, we see individual photons corresponding to green light, blue light, orange light, and so forth. Each photon has a definite wavelength and corresponds to a transition between two levels of definite energies. Aside from the effect of the uncertainty principle, the wavelengths are "sharp." If we analyze the light with a high-resolution device such as a diffraction grating, the resulting spectrum (Figure 10.1) shows the sharpness of the spectral lines. We can understand this spectrum based on our knowledge of the excited states of a single mercury atom; as long as the density of the gas is low, the number of atoms in the tube does not affect the observed spectrum. We treat the light emitted by this collection of atoms as if the individual emissions occur singly and in isolation.

Consider now the contrasting case of the tungsten filament of an ordinary incandescent light bulb. Figure 10.2 shows the spectrum in this case, which has the continuous distribution of wavelengths that we call "white" light. All wavelengths are present, not just a finite number. Isolated tungsten atoms, like those of mercury, emit light at a finite number of discrete, well-

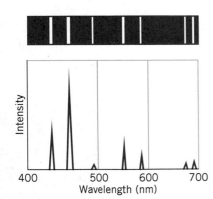

FIGURE 10.1 The line spectrum of mercury in the visible region.

COLOR PLATE 1

Electron microscope image of bacteria on the head of a pin. (Source: Dr. Tony Brain and David Parker, Science Photo Library/Photo Researchers.)

COLOR PLATE 2

A ring of 48 iron atoms forms a "corral" in which the probability density due to the wave function of the trapped electron is clearly visible. This image was obtained with a scanning tunneling microscope. (Courtesy of International Business Machines Corporation.)

COLOR PLATE 3

An electron microscope was used to move individual xenon atoms to spell out a familiar name. (Courtesy of International Business Machines Corporation.)

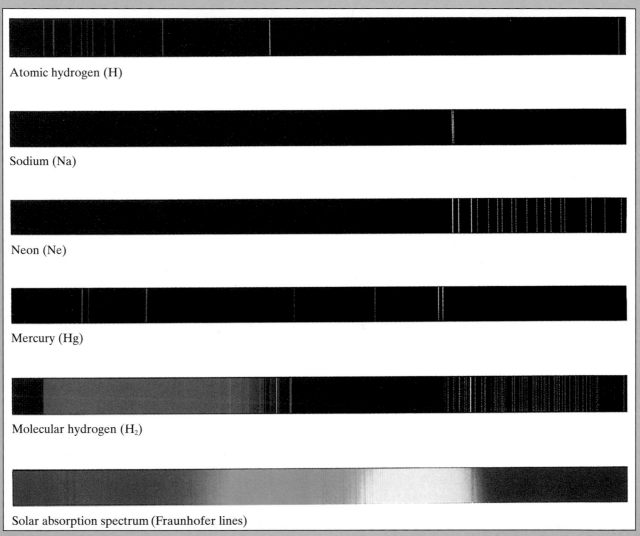

Atomic hydrogen (H)

Sodium (Na)

Neon (Ne)

Mercury (Hg)

Molecular hydrogen (H$_2$)

Solar absorption spectrum (Fraunhofer lines)

COLOR PLATE 4

Some emission and absorption line spectra. (Courtesy of Bausch and Lomb.)

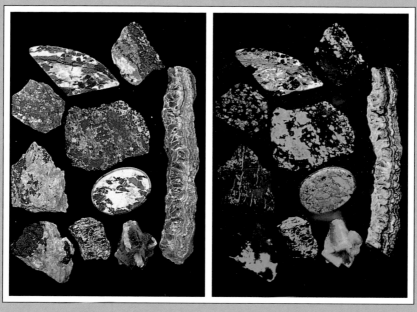

COLOR PLATE 5

Some minerals viewed in natural (left) and ultraviolet (right) light. (Courtesy Dr. Julius Weber)

COLOR PLATE 6

The muliple color laser beams shown here were generated simultaneously by 500-W dye lasers developed for use in separating isotopes of uranium. (Courtesy Textron Defense Systems, Everett, Massachusetts.)

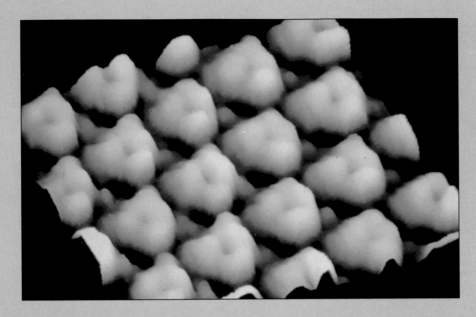

This scanning tunneling microscope image of benzene molecules shows their ringlike structure. (Courtesy of International Business Machines Corporation.)

COLOR PLATE 8

Captain Kirk and Mr. Spock encounter a horta, a rock-eating creature whose biochemistry is based on silicon rather than carbon. (Courtesy Paramount Pictures.)

A diode laser, compared in size with a grain of salt. (Courtesy AT&T Bell Laboratories.)

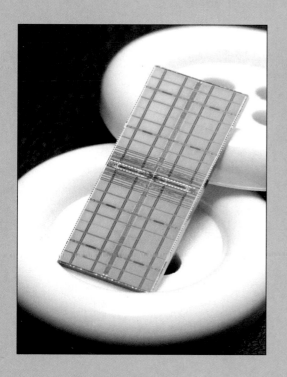

An integrated circuit, in which millions of individual circuit elements are produced on a single "chip" of silicon. This chip serves as a 16,000,000-bit random access memory component in computers. The diameter of the buttons is about 2 cm. (Courtesy of International Business Machines Corporation.)

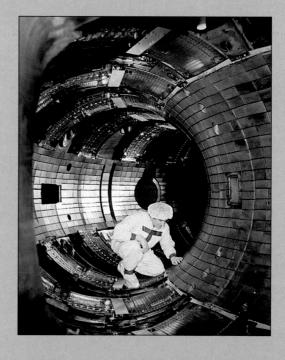

A technician inside the toroidal magnetic chamber of the Tokamak Fusion Test Reactor. (Courtesy Princeton Plasma Physics Laboratory.)

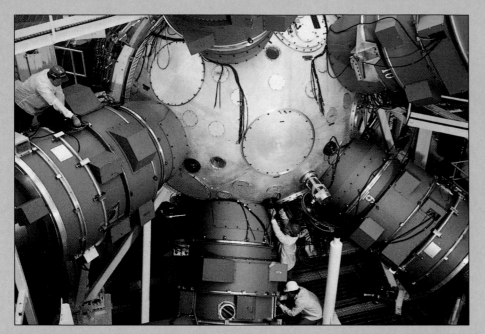

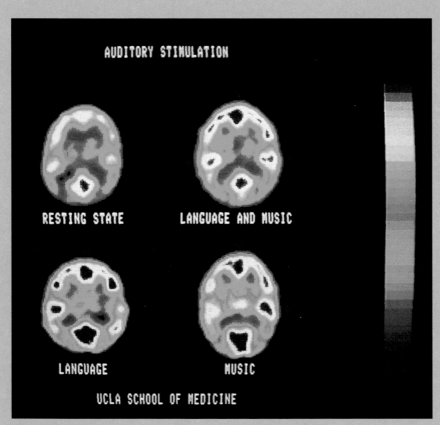

COLOR PLATE 13

Brain activity in a patient after the introduction of glucose labeled with ^{18}F. Active areas of the brain that metabolize glucose more rapidly have greater concentrations of ^{18}F and are shown in lighter colors. This image shows different areas of the brain that are active for language and music. The bright area at the bottom of each image is the visual cortex, which is also active because the patient's eyes were open. (Source: Dr. John Maziotta et al./Photo Researchers.)

COLOR PLATE 14

An aerial view of the Fermi National Accelerator Laboratory near Chicago. Beams of protons and antiprotons circulate in opposite directions around the 1-km diameter ring and collide near the upper left section of the ring. To the left of the ring can be seen the two smaller injector rings, the 16-story office building, and three experimental lines. (Courtesy Fermi National Accelerator Laboratory.)

COLOR PLATE 15

A large detector at the Fermilab collider. The black arches that have been removed to the sides are calorimeter detectors that contain photomultipliers and electronics used to measure the energies of particles that are created in collisions inside the detector. The calorimeters are pushed together so that they surround the region where the collisions of protons and antiprotons take place. (Courtesy Fermi National Accelerator Laboratory.)

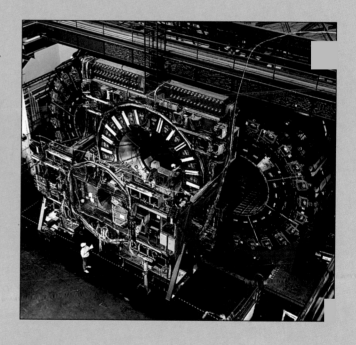

COLOR PLATE 16

Particle tracks observed in a large calorimeter detector such as that of Color Plate 15. A collision in the center of the detector creates many particles, whose trajectories can be seen here. A magnetic field causes the trajectories of charged particles to curve; the radius of curvature determines the momentum of the particle. (Courtesy Fermi National Accelerator Laboratory.)

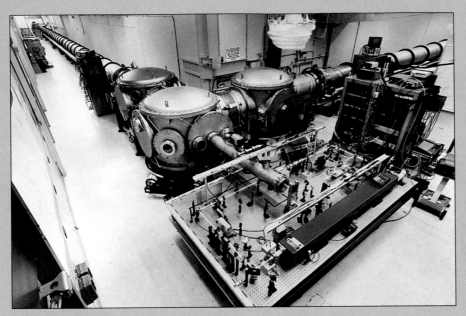

COLOR PLATE 17

The Laser Interferometer Gravity Wave Observatory. The interferometer consists of two arms, in analogy to the Michelson interferometer (Figure 2.6). Each arm contains two suspended masses. A passing gravitational wave changes the distance between the masses, which can be detected through the interference pattern of the laser light that travels along the two arms. The photograph shows a prototype with arms 40 meters long. The LIGO facility will eventually be built with arms 4 kilometers long. (Courtesy California Institute of Technology.)

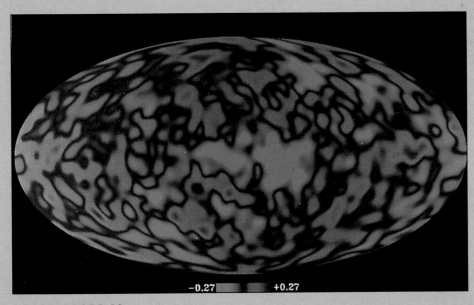

−0.27   +0.27

COLOR PLATE 18

A map of the microwave background of the sky made by the COBE satellite. The shadings represent temperature fluctuations of ± 0.00003 K. It is believed that dark matter collects in the cooler areas and then attracts hydrogen gas, resulting in the formation of galaxies. (Courtesy NASA.)

defined wavelengths, but in a solid tungsten filament the cooperative effect of other nearby atoms changes the energy levels and makes the spectrum continuous. Even though the mercury vapor and the tungsten filament may contain roughly the same number of atoms, in one case we can ignore the presence of the other atoms, while in the other case we must consider the mutual influence of many or all of the atoms in the sample.

There are two ways to approach the analysis of a complex system. The first is to specify a set of *microscopic* properties, such as the position and velocity of each atom. For even a small system containing perhaps 10^{15} atoms, this is obviously a hopeless task. The second way is to recognize that such a description is not only impossibly complex, it is also unnecessary because it provides far more detail than is useful. We can understand and predict the behavior of systems containing many particles in terms of a few *macroscopic* properties, such as the temperature or the pressure of a gas. The development of relationships between microscopic and macroscopic properties was one of the great triumps of nineteenth-century physics. In the case of a gas confined to a container, for example, kinetic theory gives the relationship between the microscopic motion of the molecules and the macroscopic temperature and pressure.

More generally, we can make a *statistical* analysis by counting the number of different arrangements of the microscopic properties of a system. For example, consider the distribution of 2 units of energy to a "gas" of four identical but distinguishable particles. Each particle can acquire energy only in integral units. How can these four particles share the 2 units of energy? One way is for one particle to have the entire 2 units. There are four different ways to accomplish this distribution, corresponding to choosing each of the four particles to take the 2 units of energy. Another way to distribute the energy is to give two different particles 1 unit each. There are six ways to accomplish this distribution (Table 10.1). Each possible energy distribution is called a *macrostate*—a state of the system that can be observed through the measurement of a macroscopic property such as the temperature. In our simple system, there are two macrostates, one in which one particle has 2 units of energy and another in which two particles each have 1 unit of energy. The different arrangements of microscopic variables corresponding to a single macrostate are called *microstates*. In our system, there are four microstates corresponding to macrostate A and six microstates corresponding to macrostate B. The number of microstates corresponding to a given macrostate is called the *multiplicity W*. For our system, $W_A = 4$ and $W_B = 6$.

One application of these statistical principles is to determine the direction of the natural evolution of a system. The *second law of thermodynamics* says, in effect, that isolated systems evolve in a direction such that the multiplicity increases.* That is, if we started our system in macrostate A and allowed the four particles to interact with one another, later we might

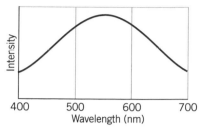

FIGURE 10.2 The continuous spectrum of an incandescent source in the visible region.

TABLE 10.1 MACROSTATES OF SIMPLE SYSTEM

| | Microstate | | | |
| | Energy of Particle | | | |
Macrostate	1	2	3	4
A	2	0	0	0
	0	2	0	0
	0	0	2	0
	0	0	0	2
B	1	1	0	0
	1	0	1	0
	1	0	0	1
	0	1	1	0
	0	1	0	1
	0	0	1	1

---

* The second law is often expressed in terms of the *entropy S* by saying that an isolated system must evolve such that $\Delta S \geq 0$. The Austrian physicist Ludwig Boltzmann (1844–1906) developed the relationship between the entropy of a system and the multiplicity of its macrostate: $S = k \ln W$, where k is the Boltzmann constant. The increase in the entropy of a system as it changes from one macrostate to another is thus equivalent to an increase in the corresponding multiplicities.

find it in macrostate B; however, if the system began in macrostate B, we would be less likely to find it later in macrostate A, because a change from B to A involves a less probable decrease in multiplicity. As the number of particles in the system increases, differences in multiplicity become greater, and changes involving decreases in multiplicity become so improbable as to be unobservable.

Implicit in this statistical analysis is the following postulate:

All microstates are equally probable.

Our system can be found with equal probability in any of the 10 microstates listed in Table 10.1. It is this postulate that allows us to assign a greater statistical weight to macrostate B; since the system can be found with equal probability in any of the six microstates of B or the four microstates of A, the relative weight of B is $6/4 = 3/2$, so in a large number of identical systems we expect to find 40 percent in macrostate A and 60 percent in macrostate B.

For another example, in the card game of poker a royal flush consisting of the A♥, K♥, Q♥, J♥, 10♥ is just as probable as the *particular* worthless hand 10♠, 8♣, 5♦, 4♥, 2♠. What makes a royal flush so special and rare is that there are only four possible royal flushes among the 2,598,960 possible poker hands, most of which are worthless. Even though a royal flush is just as likely as a *particular* worthless hand, it is much less likely than *any* worthless hand. In the language of statistical physics, the macrostate "royal flush" has fewer microstates (and is therefore less probable) than the macrostate "worthless hand" or the macrostate "one pair."

The assumption of equal probabilities of the microstates allows us to do calculations on the system. For example, suppose we have a large number of identical systems like that of Table 10.1 whose microstates are distributed randomly. Let us reach into each system and measure the energy of a particle. What is the probability that the particle will have 2 units of energy? Among the 10 *equally probable* microstates consisting of a total of 40 particles, there are four particles with 2 units of energy. Thus $p(2) = 4/40 = 0.10$, and we expect that 10 percent of our measurements would find $E = 2$. Similarly, $p(1) = 0.30$ and $p(0) = 0.60$.

The statistical analysis of a complex system gives us a way of describing the state of the system, its average properties, and its evolution in time. Since physics is often concerned with exactly these details, the statistical analysis is very useful indeed. It can be applied to systems with a small number of particles, such as a nucleus with the order of 10^2 particles, or a very large number, such as a star like the Sun with 10^{57} particles. Our next task is to determine whether there are differences between the statistical behaviors of classical and quantum particles.

10.2 CLASSICAL VERSUS QUANTUM STATISTICS

To illustrate the differences between classical and quantum statistics, let us first consider another example similar to the one of the previous section:

the distribution of a total of 6 energy units to a collection of five identical but distinguishable particles. Instead of a tabulation similar to Table 10.1, the energy distribution is illustrated in Figure 10.3. There are ten macrostates, labelled A through J. Each dot indicates a particle with a certain energy; for instance, in macrostate B there are three particles with energy $E = 0$, one with $E = 1$, and one with $E = 5$. The multiplicity of each macrostate (the number of microstates) can be worked out in tabular form, or it can be calculated directly using standard methods from permutation theory:

$$W = \frac{N!}{N_0!N_1!N_2!N_3!N_4!N_5!N_6!} \tag{10.1}$$

where N is the total number of particles and N_E is the number with energy E. As in the previous example, note that the multiplicity is greatest when the energy is shared by a larger number of particles, such as macrostate E, and smallest when the energy is concentrated in a smaller number of particles, such as macrostates A or J.

As we did previously, let us calculate the probability to measure any particular value of the energy of a particle. As before, this can be done by considering a collection of all 210 possible microstates and counting the number of times each value of the energy appears. Symbolically,

$$p(E) = \frac{\sum_{i=A}^{J} N_E W_i}{N \sum_{i=A}^{J} W_i} \tag{10.2}$$

where N_E represents the number of particles with energy E in each particular macrostate A through J. The sums are carried out over the ten macrostates. For example, a particle with 2 units of energy appears only in macrostates C, E, G, I, and J, and so the probability to find a particle with 2 units of energy is

$$p(2) = \frac{1 \times 20 + 1 \times 60 + 3 \times 10 + 2 \times 30 + 1 \times 5}{5 \times 210} = 0.167$$

TABLE 10.2 ENERGY PROBABILITIES FOR SYSTEM OF FIGURE 10.3

Energy	Probability
0	0.400
1	0.267
2	0.167
3	0.095
4	0.048
5	0.019
6	0.005

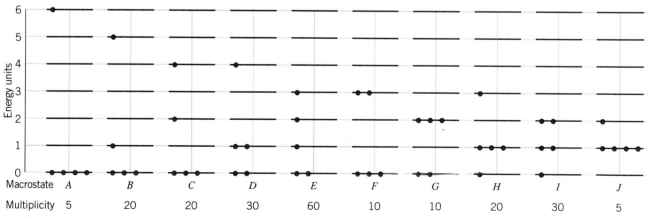

FIGURE 10.3 The macrostates of a system in which five identical particles share 6 units of energy.

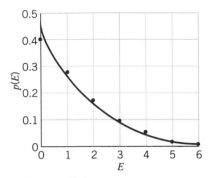

FIGURE 10.4   The probability to find a particular value of the energy in a collection of classical particles. The curve is an exponential function that closely fits the points.

TABLE 10.3   ENERGY PROBABILITIES FOR QUANTUM PARTICLES

	Probability	
Energy	Integral Spin	Spin 1/2
0	0.420	0.333
1	0.260	0.333
2	0.160	0.200
3	0.080	0.133
4	0.040	0.000
5	0.020	0.000
6	0.020	0.000

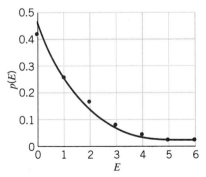

FIGURE 10.5   The probability distribution for quantum particles with integral spin. The curve is approximately exponential but rises a but more steeply at low energy than the curve of Figure 10.4.

Table 10.2 gives the probabilities associated with measuring each of the possible energies. Note that the probability decreases with increasing energy. The probabilities are plotted in Figure 10.4. The smooth curve is an exponential of the form $p \propto e^{-\beta E}$, where β is a constant chosen to fit the data. You can see that the decrease of p with increasing E is approximately exponential.

This example represents the application of classical statistics. Although real systems composed of many particles have too many macrostates and microstates to tabulate, we can analyze their properties by determining the distribution function $p(E)$ that describes how the energy is shared among the parts of a system. Later in this chapter we discuss the application of classical statistics to the distribution of energies of the molecules of a gas. Our experience with this simple example serves as a useful guide: The true classical distribution function turns out to be an exponential of the form $e^{-\beta E}$. However, there are other phenomena, such as the electrical conductivity and specific heat of metals, the behavior of liquid helium, and thermal radiation, which cannot be successfully analyzed using classical statistics. For these phenomena, we must use the methods of *quantum statistics*.

Why should classical statistics differ from quantum statistics? Two of the assumptions made in the example of this section are inconsistent with basic principles of quantum physics:

1. *In quantum physics, identical particles must be treated as indistinguishable.*   In calculating the multiplicity of the macrostates, we assumed the classical particles to be identical but distinguishable. That is, the particles are numbered 1 through 5 (or carry other distinguishing marks). In macrostate *A*, for instance, it is possible to distinguish the microstate in which particle 1 has $E = 6$ from the microstate in which particle 2 has $E = 6$, so we count each of these as separate microstates in determining the multiplicity of 5. If we regard the particles as indistinguishable quantum particles (such as electrons or photons), we can't tell the difference between these microstates. Quantum mechanics makes the observation process a part of the analysis of a system; the uncertainty principle, for example, imposes limits not on the behavior of a system but on our knowledge of its behavior. If we can't *observe* the five microstates of *A* as separate arrangements, we can't *count* them as separate arrangements. For identical quantum particles, the multiplicity of each macrostate becomes exactly 1.

2. *Quantum mechanics can impose limits on the maximum number of particles that can be assigned to any particular state.*   For example, suppose the particles in our example were electrons. The Pauli principle forbids two electrons in a system from being in the same state of motion (or having the same set of quantum numbers). Since electrons can have spin up or spin down, there can be no more than two electrons with a given energy value. For electrons, macrostates *A*, *B*, *C*, *G*, *H*, and *J* are forbidden, because in each one there are more than 2 particles with the same value of the energy.

We can repeat the calculation of the probability $p(E)$ for two cases of quantum particles: photons or alpha particles, which have integral spins and are not restricted by the Pauli principle, and electrons or protons, which have a spin of $\frac{1}{2}$ and *are* restricted by the Pauli principle to no more than 2 per energy value. In the first case, we have ten macrostates with equal multiplicities of 1, and in the second case we have three macrostates (D, E, I) again with equal multiplicities of 1. Table 10.3 gives the resulting values of $p(E)$. The probabilities for the particles with integral spins are plotted in Figure 10.5; the curve is approximately exponential but rises a bit more steeply at low energies. The values for the particles with $s = \frac{1}{2}$ are plotted in Figure 10.6; the behavior is no longer even approximately exponential; instead, it is rather flat near $E = 0$ and then drops rapidly to 0 for the higher energies.

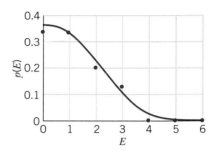

FIGURE 10.6 The probability distribution for quantum particles with spin $\frac{1}{2}$. The curve, which is drawn to characterize the points, becomes approximately flat at low energies.

The example we have analyzed in this section should be regarded only as indicating some differences between classical and quantum statistics. The true energy distribution functions $p(E)$ for many-particle systems look different from the ones obtained for a simple five-particle system.

Particles with integral spin are described by a distribution function that is approximately exponential but that rises more sharply at low energies; this difference is responsible for a number of interesting cooperative effects. This distribution function is called the *Bose-Einstein* distribution, and particles governed by this distribution are called *bosons*. Particles that obey the Pauli principle are described by a very different distribution which becomes flat at low energies, like Figure 10.6; this distribution function is called the *Fermi-Dirac* distribution, and particles with half-integral spin are known as *fermions*. Later in this chapter we develop and apply the Fermi-Dirac and Bose-Einstein distribution functions to a variety of physical situations.

Bosons and fermions

10.3 THE DISTRIBUTION OF MOLECULAR SPEEDS

Let us imagine a large container filled with a certain type of gas at a reasonably low pressure. As we have discussed in the previous section, the available energy, which we assume to be only in the form of translational kinetic energy, is distributed over the N molecules in such a way that there are many more molecules with small kinetic energies than with large kinetic energies. We can investigate this assertion in detail by measuring the velocity distribution of the molecules. As we have argued frequently in this text, it makes no sense to ask "How many molecules (per unit volume) have velocity components v_x, v_y, and v_z?", because we are not sure what it means for a molecule to have *exactly* those velocity components. (In other words, how exact is exact?) It makes more sense, and in fact is a better description of what we actually measure, to ask, "How many molecules (per unit volume) have velocities between (v_x, v_y, v_z) and $(v_x + dv_x, v_y + dv_y, v_z + dv_z)$?"

We expect that, if the box is at rest, the distribution of velocities is symmetric around zero (just as many molecules should be moving to the right as to the left) and we also expect the distribution to be such that finding a molecule with a large velocity is unlikely. (Remember the example

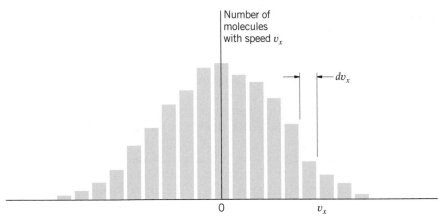

FIGURE 10.7   A possible result of measuring the distribution of one component of the velocity of molecules of a gas.

of the previous section—it was very unlikely to find one particle with all the energy.) If we were to perform an experiment in which we measured the component v_x in the interval dv_x, we might obtain a result of the form shown in Figure 10.7. We may assume that this is a representation of a smooth curve, and a familiar curve that has the proper form (and that is also generally associated with random statistical processes) is the *normal* or *Gaussian* distribution e^{-x^2}. We therefore assume the distribution to be represented by

$$f(v_x) \, dv_x = \frac{1}{A_x} e^{-bv_x^2} \, dv_x \qquad (10.3)$$

Maxwell velocity distribution

where $f(v_x) \, dv_x$ gives the number of molecules per unit volume with velocity components between v_x and $v_x + dv_x$. Equation 10.3 is the *Maxwell velocity distribution*. The constants A_x and b will be determined later.

More often we are interested not in the *velocity* distribution but rather in the *speed* distribution. In order to obtain the speed distribution we must first generalize Equation 10.3 to three dimensions:

$$f(v_x, v_y, v_z) \, dv_x \, dv_y \, dv_z = \frac{1}{A_x A_y A_z} e^{-bv_x^2} e^{-bv_y^2} e^{-bv_z^2} \, dv_x dv_y dv_z$$

$$= \frac{1}{A} e^{-b(v_x^2 + v_y^2 + v_z^2)} \, dv_x \, dv_y \, dv_z \qquad (10.4)$$

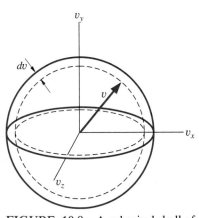

FIGURE 10.8   A spherical shell of radius v and thickness dv in the $v_x v_y v_z$ coordinate system.

A is a new constant to be determined. The speed v is $(v_x^2 + v_y^2 + v_z^2)^{1/2}$, and we wish to know the distribution of speeds $f(v) \, dv$. In order to find this distribution, we must know how dv is related to dv_x, dv_y, and dv_z, or equivalently how many different combinations of v_x, v_y, and v_z give the same value of v. One way to see how this is done is to imagine a coordinate system in which the axes are labeled v_x, v_y, v_z; the points of constant v then form the surface of a sphere of radius v, and allowing for a range dv gives a spherical shell of thickness dv, as shown in Figure 10.8. The volume of the spherical shell is $4\pi v^2 \, dv$, and replacing the "volume" element

$dv_x \, dv_y \, dv_z$ with $4\pi v^2 \, dv$, we obtain the *Maxwell speed distribution*:

$$f(v) \, dv = \frac{4\pi}{A} e^{-bv^2} v^2 \, dv \qquad (10.5)$$

Maxwell speed distribution

The function $f(v)$ is plotted in Figure 10.9.

An example of an experiment to measure the distribution of molecular speeds is shown in Figure 10.10. A small hole in the side of an "oven" allows a stream of molecules to escape; we assume the hole to be small enough so that the distribution of speeds inside the oven is not changed. The beam of molecules is made to pass through a slot in a disc attached to an axle rotating at angular velocity ω. At the other end of the axle is a second slotted disc, but the slot is displaced from the first by an angle θ. In order for a molecule to pass through both slots and strike the detector, it must travel the length L of the axle in the same time that it takes the axle to rotate by the angle θ, and thus $L/v = \theta/\omega$. Keeping L and θ fixed, we can vary ω; measuring the number of molecules striking the detector for each different value of ω enables us to measure the Maxwell speed distribution. A set of such experimental results is shown in Figure 10.11, and the remarkable agreement between the measured speed distribution and that predicted according to Equation 10.5 justifies the assumptions we have made in its derivation.

(From this example you can also see the importance of the interval dv. What we measure is always the product $f(v) \, dv$, and in this case the range of velocities is determined primarily by the width of the slots in the discs. To make dv very small, and thereby measure v "exactly," we would need to make the slots very small, so that very few molecules could get through. For the "perfect" experiment, we make $dv \to 0$ by making the slot width equal to 0, and no molecules get through the apparatus!)

The final step in the derivation of the Maxwell distribution is the evaluation of the constants A and b in Equation 10.5. The constant A is a *normalization* constant, similar to the normalization constant of quantum physics. It is chosen so that the integral over all speeds gives the total number of molecules (per unit volume) in the container:

$$\int_0^\infty f(v) \, dv = n \qquad (10.6)$$

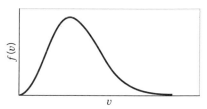

FIGURE 10.9 The Maxwell speed distribution $f(v)$.

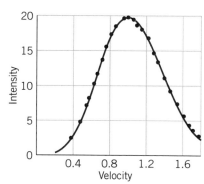

FIGURE 10.11 Result of a measurement of the distribution of atomic speeds of thallium vapor. The solid line is the Maxwell speed distribution corresponding to the oven temperature of 870 K.

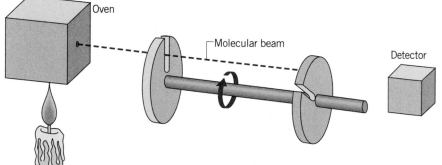

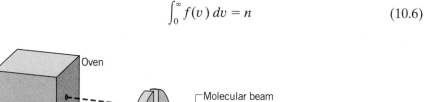

FIGURE 10.10 Apparatus to measure the distribution of molecular speeds.

where n is the number of molecules per unit volume. Since we have two constants to determine, we need a second equation, and here we turn to a result obtained from kinetic theory: the average kinetic energy per molecule of a gas in thermal equilibrium at absolute temperature T is $\frac{3}{2}kT$, where k is the Boltzmann constant. Computing the average kinetic energy per molecule is equivalent to finding the average value of $\frac{1}{2}mv^2$ over the distribution of speeds:

$$K_{av} = \frac{1}{n}\int_0^\infty (\tfrac{1}{2}mv^2) f(v)\, dv = \tfrac{3}{2}kT \tag{10.7}$$

[This expression for the average value is of the same form as Equation 5.11 for the quantum-mechanical average value; here $f(v)$ plays the role of the probability density $P(x)$.] Definite integrals of the form $\int_0^\infty x^{2j}e^{-ax^2}\, dx$ for $j = 1, 2, 3, \ldots$ can be found in many handbooks:

$$\int_0^\infty x^{2j}e^{-ax^2}\, dx = \frac{1\cdot 3\cdot 5\cdots(2j-1)}{2^{j+1}a^j}\sqrt{\frac{\pi}{a}} \tag{10.8}$$

It is left as an exercise to show that Equations 10.6, 10.7, and 10.8 give the values of the constants:

$$b = \frac{m}{2kT} \tag{10.9}$$

$$\frac{1}{A} = n\left(\frac{m}{2\pi kT}\right)^{3/2} \tag{10.10}$$

so that

$$f(v) = 4\pi n\left(\frac{m}{2\pi kT}\right)^{3/2} v^2 e^{-mv^2/2kT} \tag{10.11}$$

EXAMPLE 10.1

Find the values of the mean (average) speed, the root-mean-square speed, and the most probable speed from the Maxwell speed distribution.

SOLUTION

Since the speeds are distributed according to $f(v)$, the mean speed is

$$v_m = \frac{1}{n}\int_0^\infty v\, f(v)\, dv = \frac{4\pi}{nA}\int_0^\infty v^3 e^{-bv^2}\, dv$$

The integral can be shown to have the value $1/2b^2$, and so

$$v_m = \frac{4\pi}{2b^2 nA} = 2\pi\left(\frac{m}{2\pi kT}\right)^{3/2}\left(\frac{2kT}{m}\right)^2$$

$$= \left(\frac{8kT}{\pi m}\right)^{1/2} \tag{10.12}$$

Since one of the conditions used to evaluate A and b was that the mean kinetic energy per molecule was $\frac{3}{2}kT$, it follows that the mean value of v^2 is

$$(v^2)_{\mathrm{m}} = \frac{3kT}{m}$$

(Note that this is *not* equal to v_{m}^2). The *root-mean-square* speed v_{rms} is

$$v_{\mathrm{rms}} = \sqrt{(v^2)_{\mathrm{m}}}$$

$$= \left(\frac{3kT}{m}\right)^{1/2} \tag{10.13}$$

The *most probable* speed v_{p} can be found from $f(v)$ using standard calculus techniques to find maxima. Its value is

$$v_{\mathrm{p}} = \left(\frac{2kT}{m}\right)^{1/2} \tag{10.14}$$

The relationship of v_{p}, v_{m}, and v_{rms} is shown in Figure 10.12.

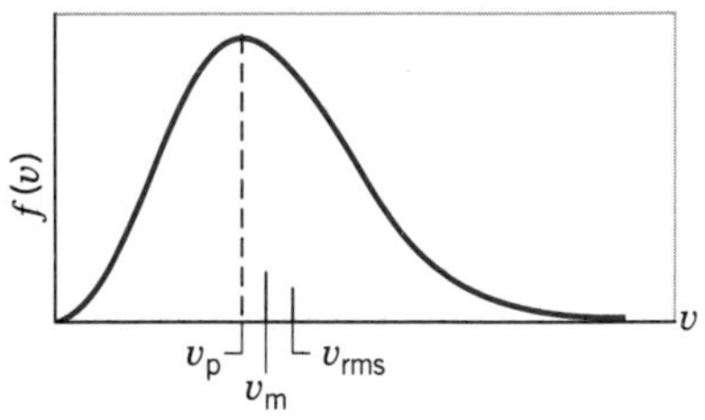

FIGURE 10.12 The relationship of the most probable speed v_{p}, the mean speed v_{m}, and the root-mean-square speed v_{rms}.

EXAMPLE 10.2

The light that reaches us from distant stars is Doppler shifted not only because the stars are moving relative to us, but also because the atoms of the stars are in rapid thermal motion. Consider a star that happens to be at rest relative to the Earth, and calculate the frequency distribution of the emitted light, assuming that only a single frequency ν_0 is emitted in the rest frame of each atom. (Assume also that the speed of thermal motion is small compared with c.)

SOLUTION

The formula for the relativistic Doppler shift is (from Chapter 2):

$$\nu = \nu_0 \sqrt{\frac{1 - u/c}{1 + u/c}}$$

Let us rewrite this as

$$\nu = \frac{\nu_0(1 - u/c)}{\sqrt{1 - u^2/c^2}}$$

and assuming $u \ll c$, the factor in the denominator does not differ appreciably from unity. The velocity u, upon which the Doppler shift depends, is the *component* v_x of the velocity along the line joining the Earth with the star, and so we must use the Maxwell *velocity* distribution, rather than the *speed* distribution

$$f(v_x)\, dv_x = \frac{1}{A_x} e^{-bv_x^2}\, dv_x$$

With

$$\nu = \nu_0 \left(1 - \frac{v_x}{c} \right)$$

we have

$$v_x = c \left(1 - \frac{\nu}{\nu_0} \right)$$

and

$$|dv_x| = \frac{c}{\nu_0} \, d\nu$$

so the distribution of observed frequencies is

$$f(\nu) \, d\nu = \frac{1}{A_x} \left(\frac{c}{\nu_0} \right) e^{-bc^2(1-\nu/\nu_0)^2} \, d\nu$$

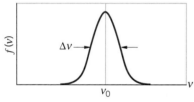

FIGURE 10.13 The distribution of frequencies in a Doppler-broadened spectral line.

This distribution is shown in Figure 10.13. Notice that, since the average value of v_x is zero (recall Figure 10.7) the average value of ν must be ν_0. The effect of the thermal motion of the atoms of the star is not to shift the spectral line, but rather to *broaden* it, and the broadening gives a direct measurement of the temperature. Let us define the broadening to be the range of frequencies $\Delta\nu$ that corresponds to the intensity of the spectral line having half its maximum value; this is called the "full width at half maximum," or FWHM. The only feature of $f(\nu)$ that depends on ν is the exponential, which is unity at $\nu = \nu_0$; $f(\nu)$ thus drops to half of its maximum value when the exponential factor has the value $\frac{1}{2}$:

$$e^{-bc^2(1-\nu/\nu_0)^2} = \frac{1}{2}$$

$$bc^2 \left(1 - \frac{\nu}{\nu_0} \right)^2 = \ln 2$$

$$1 - \frac{\nu}{\nu_0} = \pm \sqrt{\frac{\ln 2}{bc^2}}$$

Since ν represents the point at which $f(\nu)$ drops by half, the range $\Delta\nu$ is $\nu_+ - \nu_-$, where ν_+ and ν_- represent the solutions to the above expression for the positive and negative roots

$$\nu_\pm = \nu_0 \left(1 \pm \sqrt{\frac{\ln 2}{bc^2}} \right)$$

$$\Delta\nu = 2\nu_0 \sqrt{\frac{\ln 2}{bc^2}}$$

$$= (2\sqrt{\ln 2}) \, \nu_0 \sqrt{\frac{2kT}{mc^2}}$$

For a star composed of hydrogen atoms, $mc^2 \cong 938$ MeV and we can find

$$\Delta\nu = 7.14 \times 10^{-7}\nu_0\ \sqrt{T}$$

For a star like the sun, with $T \cong 6000$ K, $\Delta\nu \simeq 5.5 \times 10^{-5}\nu_0$. This small width is nevertheless relatively easy to measure with modern spectroscopic equipment, and so the *Doppler broadening* gives us a way to measure the surface temperature of stars.

Study the above example to see how we accomplished a *change of variable* to go from the velocity distribution to a frequency distribution. Notice that there were three steps involved:

1. Find an expression that relates the new variable to the velocity.
2. Differentiate the expression to find the relationship between the differentials of the two variables.
3. Substitute for the velocity and its differential in the expression for the speed or velocity distribution.

10.4 THE MAXWELL-BOLTZMANN DISTRIBUTION

If we add energy to a system of particles such as an ideal gas in which translational kinetic energy is the only form of energy, we can use Equation 10.11 and the change-of-variable techniques described at the end of the previous section to convert the speed distribution into an energy distribution. We begin with the expression relating the new variable, the energy E, to the speed:

$$E = \tfrac{1}{2}mv^2$$

$$dE = mv\ dv$$

$$dv = \frac{dE}{mv} = \frac{dE}{\sqrt{2mE}} \tag{10.15}$$

$$f(E)\ dE = 4\pi n \left(\frac{m}{2\pi kT}\right)^{3/2} \frac{2E}{m}\ e^{-E/kT} \frac{dE}{\sqrt{2mE}}$$

$$= \frac{2n}{\pi^{1/2}(kT)^{3/2}}\ E^{1/2}e^{-E/kT}\ dE \tag{10.16}$$

It is now reasonable to ask how this result, which gives the energy distribution when the energy takes the form of translational kinetic energy, can be carried over to other cases in which the energy may take different forms. The normalization factor, for example, was derived from a special assumption regarding the relationship in a gas between temperature and mean molecular kinetic energy—surely this will not be correct for other types of systems. Similarly, the factor $E^{1/2}$ in Equation 10.16 came from the calculation of the relationship between dE and the "volume" element $dv_x\, dv_y\, dv_z$ of the original velocity distribution; this will likewise not be correct in other situations. The remaining exponential factor $e^{-E/kT}$ turns out to be the only characteristic of the velocity and speed distributions that is true in general, no matter what form the energy takes. The *Maxwell-Boltzmann distribution function* is this exponential factor with its normalization constant:

Maxwell-Boltzmann distribution

$$f_{\mathrm{MB}}(E) = A^{-1}e^{-E/kT} \qquad (10.17)$$

The normalization constant A in general can depend on many factors, such as on density and temperature in the case of the molecular gas described by Equation 10.10. According to the Maxwell-Boltzmann distribution, the relative number of particles having a particular value of the energy E is:

$$p(E) = g(E)f_{\mathrm{MB}}(E) \qquad (10.18)$$

The function $g(E)$, known as the *density of states* factor, describes the number of possible ways the system can have a certain value of E. In the case of the molecular gas, the density of states factor comes from the "volume" element $dv_x\, dv_y\, dv_z$, which we modified to the "volume" of the spherical shell $4\pi v^2\, dv$, which is proportional to $E^{1/2}\, dE$, giving $g(E) \propto E^{1/2}$ for the case of the molecular gas. When we discuss discrete energy levels, such as those of an atomic system, $g(E)$ will be proportional to the *degeneracy* of the atomic levels; recall that the degeneracy gives the number of different ways an atomic system can have the same value of E, which is exactly how we define $g(E)$.

Note that the Maxwell-Boltzmann factor $f_{\mathrm{MB}}(E)$ has exactly the exponential form $e^{-\beta E}$ we expected from our numerical example of Section 10.2.

EXAMPLE 10.3

(a) In a gas of atomic hydrogen at room temperature, what is the relative population of the first excited state at $E = 10.2$ eV? How much hydrogen would be required in order to have a reasonable probability of finding one atom in the first excited state? (b) At what temperature would we expect to find $\frac{1}{10}$ of the atoms in the first excited state?

SOLUTION

(a) We recall from our study of atomic hydrogen that the degeneracy of the atomic levels is $2n^2$. Thus $g = 2$ for the ground state ($n = 1$) and $g = 8$ for the first excited state ($n = 2$). The constant A is the same for the two states, and so

$$\frac{p(E_2)}{p(E_1)} = \frac{g(E_2)}{g(E_1)} e^{-(E_2 - E_1)/kT}$$

At room temperature ($T = 293$ K), $kT = 0.0252$ eV; thus

$$\frac{p(E_2)}{p(E_1)} = \frac{8}{2} e^{-10.2\,\text{eV}/0.0252\,\text{eV}}$$

$$= 4e^{-405}$$

$$= 0.6 \times 10^{-175}$$

In order to have one atom in the excited state, we therefore require about 1.7×10^{175} atoms of hydrogen, about 3×10^{148} kg, a quantity greater than the mass of the universe!

(*b*) We now require $p(E_2)/p(E_1) = 0.100/0.900$ and solve for T:

$$0.111 = 4e^{-10.2\,\text{eV}/kT}$$

$$kT = 2.85\,\text{eV}$$

$$T = 3.30 \times 10^4\,\text{K}$$

EXAMPLE 10.4

A certain atom with total atomic spin $\frac{1}{2}$ has a magnetic moment μ. A collection of such atoms is placed in a magnetic field of strength B. (*a*) What is the ratio, at temperature T, of the number of atoms with their spins aligned along the field to those with their spins aligned opposite to the field? (*b*) By making reasonable estimates for μ and B, estimate the temperature at which this ratio becomes 1.1, that is, at which the difference between the numbers parallel and antiparallel to the field becomes 10 percent.

SOLUTION

(*a*) The energy of interaction with the magnetic field is $F = -\boldsymbol{\mu} \cdot \mathbf{B}$, and the degeneracies of the states with $m_s = +\frac{1}{2}$ and $m_s = -\frac{1}{2}$ are identical, so $g(E_1) = g(E_2)$. Therefore we have:

$$\frac{p(E_2)}{p(E_1)} = e^{-(E_2 - E_1)/kT}$$

The energy of those aligned parallel to the field is $E_1 = -\mu B$, while those aligned antiparallel have energy $E_2 = +\mu B$, and thus

$$\frac{p(E_2)}{p(E_1)} = e^{-2\mu B/kT}$$

(*b*) A typical atom has $\mu \cong \mu_B$ (one Bohr magneton) and the largest magnetic fields that can be produced in the laboratory are of order 10 T, and so

$$\mu B \cong \mu_B B = 9.27 \times 10^{-24}\,\text{J/T} \times 10\,\text{T}$$

$$= 9.27 \times 10^{-23}\,\text{J}$$

$$= 5.79 \times 10^{-4}\,\text{eV}$$

FIGURE 10.14 Example 10.4. The shaded bars on the energy levels indicate the relative populations at $T = 141$ K.

When $p(E_2)/p(E_1) = 1/1.1$ (as in Figure 10.14),

$$\frac{2\mu B}{kT} = \ln 1.1 = 9.53 \times 10^{-2}$$

$$kT = \frac{2(5.79 \times 10^{-4}\,\text{eV})}{9.53 \times 10^{-2}} = 1.22 \times 10^{-2}\,\text{eV}$$

$$T = 141\,\text{K}$$

This temperature is only a factor of 2 below room temperature, which suggests that atomic magnetic effects become important at temperatures of the order of room temperature and that atomic magnetism may be the source of ordinary magnetic effects. Since nuclear magnetic moments are about 1000 times smaller than atomic magnetic moments, they should become important at temperatures 1000 times smaller.

10.5 QUANTUM STATISTICS

As we discussed in Section 10.2, the distribution functions for the indistinguishable particles of quantum physics are different from those of classical physics. Because of the unusual behavior of quantum systems, we must have separate distribution functions for those particles, like the electron, that obey the Pauli exclusion principle, and those particles that do not. We will not derive these distribution functions, but merely state them and discuss some of their properties.

Particles that do not obey the Pauli principle are those with integral spins (0, 1, 2, . . . , in units of $\hbar$) and are known collectively as *bosons*. The

Bose-Einstein distribution

distribution function for bosons is known as the *Bose-Einstein distribution* and has the form

$$f_{\text{BE}}(E) = \frac{1}{Ae^{E/kT} - 1} \tag{10.19}$$

Note the similarity to the Maxwell-Boltzmann distribution function. In particular, for energies not too different from kT, when A is a large number, the Bose-Einstein distribution reduces directly to the Maxwell-Boltzmann distribution. Since A depends *inversely* on the density (see Equation 10.10), the Maxwell-Boltzmann distribution is a good approximation at low densities, such as for gases, while for liquids and solids of higher densities we expect the quantum statistical distributions to be necessary.

Particles of half-integral spin ($\frac{1}{2}$, $\frac{3}{2}$, . . .) which obey the Pauli principle, such as electrons or nucleons, are known as *fermions*, and their distribution

Fermi-Dirac distribution

function is the *Fermi-Dirac distribution*:

$$f_{\text{FD}}(E) = \frac{1}{Ae^{E/kT} + 1} \tag{10.20}$$

How the minor change in sign in the denominator between f_{BE} and f_{FD} gives such a radical change in the form of the distribution function is not immediately obvious, and to show the differences we need to know more about the normalization coefficient A, which is not a constant but depends on T. For the Bose-Einstein distribution, in most cases of practical interest A is either independent of T or depends so weakly on T that the exponential term $e^{E/kT}$ dominates. However, for the Fermi-Dirac distribution, A is strongly dependent on T, and the dependence is usually approximately exponential, so A is written as

$$A = e^{-E_F/kT} \qquad (10.21)$$

and the Fermi-Dirac distribution becomes

$$f_{FD}(E) = \frac{1}{e^{(E-E_F)/kT} + 1} \qquad (10.22)$$

where E_F is called the *Fermi energy*.

Let us look qualitatively at the differences between f_{BE} and f_{FD} at low temperatures. For the Bose-Einstein distribution, assuming for the moment $A = 1$, in the limit of small T the exponential factor becomes large for large E, and so $f_{BE} \to 0$ for states with large energies. The only energy levels that have any real chance of being populated are those with $E \cong 0$, for which the exponential factor approaches 1, the denominator becomes very small, and $f_{BE} \to \infty$. Thus when T is small, all of the particles in the system try to occupy the lowest energy state. This effect is known as "Bose condensation" and we will see how it has some rather unexpected consequences.

Such a situation is not possible for fermions, such as electrons. We know that the electrons in an atom, for example, do not all occupy the lowest energy state, no matter what the temperature. Let us see how the Fermi-Dirac distribution function prevents this. The exponential factor in the denominator of f_{FD} is $e^{(E-E_F)/kT}$. For values of $E > E_F$, when T is small the exponential factor becomes large and f_{FD} goes to zero, just like f_{BE}. When $E < E_F$, however, the story is very different, for then $E - E_F$ is negative, and $e^{(E-E_F)/kT}$ goes to zero for small T, so $f_{FD} \cong 1$. *The occupation probability is therefore only one per quantum state,* just as required by the Pauli principle. Even at very low temperatures, fermions do not "condense" into the lowest energy level.

In Figures 10.15 to 10.17, the three distributions f_{MB}, f_{FD}, and f_{BE} are plotted as functions of the energy E. (Note the qualitative similarity with Figures 10.4 to 10.6.) You can see, by comparing these figures, that all of the distribution functions fall to zero at large values of E; when $E \gg kT$, the occupation probability is very small, as we calculated for f_{MB} in the case of the first excited state of the hydrogen atom in Example 10.3. Notice also that, even though f_{MB} becomes large for small E, it remains finite. The Bose-Einstein distribution, f_{BE}, on the other hand, becomes infinite as $E \to 0$; this is the "condensation" effect referred to earlier, in which all of the particles try to occupy the lowest quantum state. You can see that f_{FD} never becomes larger than 1.0, just as we expect for particles that obey the

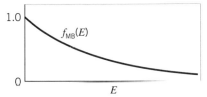

FIGURE 10.15 The Maxwell-Boltzmann distribution function.

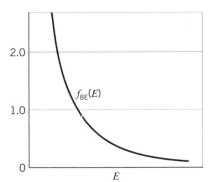

FIGURE 10.16 The Bose-Einstein distribution function.

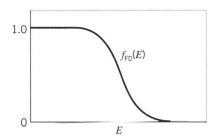

FIGURE 10.17 The Fermi-Dirac distribution function.

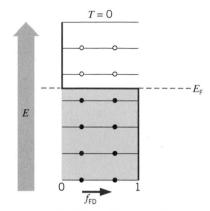

FIGURE 10.18 Filling of electronic levels according to the Fermi-Dirac distribution at $T = 0$. Open circles represent unoccupied states; filled circles represent occupied states.

Pauli principle. At $T = 0$, all energy levels up to E_F are occupied ($f_{FD} = 1.0$) and all energy levels above E_F are empty ($f_{FD} = 0$). Figure 10.18 shows a hypothetical set of energy levels and how they would be populated at $T = 0$ by a particle such as an electron, for which $g(E) = 2$. As T increases, some levels above E_F are partially occupied ($f_{FD} > 0$), while some levels below E_F are partially empty ($f_{FD} < 1$). The higher the temperature, the more "spread out" f_{FD} becomes. Figure 10.19 shows how the energy levels of a system might be populated at two different temperatures, T_1 and T_2.

From Equation 10.22 you can see that, when $E = E_F$, f_{FD} is $\frac{1}{2}$, independent of T. Thus an alternative definition of E_F is that point at which the occupation probability is exactly 0.5. The Fermi energy varies only slightly with temperature for most materials, and we can usually regard it as constant for many applications. As we will see in Section 10.7, for electrons in a metal E_F depends on the electron density of the material, which doesn't change much with temperature. For some materials, notably semiconductors, the density of *conduction* electrons can change significantly with temperature, and thus E_F in these materials is temperature dependent.

10.6 APPLICATIONS OF BOSE-EINSTEIN STATISTICS

Blackbody Radiation

As we did in our discussion of cavity radiation in Chapter 3, we treat the cavity as a box filled with electromagnetic radiation, but now we consider the radiation to be in the form of photons. For this calculation, we assume the box to be filled with a "gas" of photons. Photons have spin 1, so they are bosons and obey Bose-Einstein statistics.

The parameter A of the Bose-Einstein distribution, Equation 10.19, depends on the total number of particles described by the distribution; the analogous parameter for the Maxwell-Boltzmann distribution is given by Equation 10.10. Because photons are continuously created and destroyed as radiation is emitted or absorbed by the walls of the cavity, the total number of particles is not constant, and the parameter A loses its significance. We eliminate this factor from the Bose-Einstein distribution by setting $A = 1$ in Equation 10.19.

An important part of this calculation is the density of states function $g(E)$. In the case of the ordinary molecular gas, you will recall we determined the density of states in Equation 10.5 by considering the three-dimensional coordinate system v_x, v_y, v_z, and the "volume" of the spherical shell for this case was $4\pi v^2 \, dv$. Since all photons move with the speed of light, this process makes no sense for photons.

Instead, we recall a similar problem we discussed in Chapter 5, where we considered the allowed energies of a particle confined to a region of space in one, two, or three dimensions. In that discussion we learned that

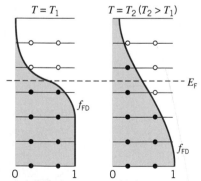

FIGURE 10.19 Filling of electronic levels according to the Fermi-Dirac distribution at $T > 0$.

requiring $\psi = 0$ at the boundaries resulted in the wave numbers k_x, k_y, and k_z taking a certain set of discrete values $\pi n_x/L$, $\pi n_y/L$, and $\pi n_z/L$, respectively. The case for electromagnetic waves in a box is similar; if the walls are made of a conducting medium, then the electric field must vanish at the boundary, and the same set of discrete wave numbers results for the electromagnetic waves. The quantization of wave number is equivalent to the quantization of momentum components, since $p_x = \hbar k_x$, $p_y = \hbar k_y$, and $p_z = \hbar k_z$. Thus

$$
\begin{aligned}
p &= \sqrt{p_x^2 + p_y^2 + p_z^2} \\
&= \hbar \sqrt{k_x^2 + k_y^2 + k_z^2} \\
&= \frac{\hbar \pi}{L} \sqrt{n_x^2 + n_y^2 + n_z^2}
\end{aligned}
\tag{10.23}
$$

For photons, $E = pc$, so

$$
E = \frac{\hbar c \pi}{L} \sqrt{n_x^2 + n_y^2 + n_z^2}
\tag{10.24}
$$

This expression gives the discrete values of E that are permitted; in order to find the density of states $g(E)$ we need to know how many of these discrete values occur between E and $E + dE$. As we did in the case of velocities in the Maxwell distribution, this calculation is easiest if we imagine a coordinate system similar to Figure 10.8, in which the axes are labelled n_x, n_y, n_z. Of course, these are not really continuous varibles, since they may take only integer values; furthermore, only positive integers are allowed, since n_x, n_y, and n_z count the nodes in the electric field. A spherical shell in this coordinate system has a "radius" $n = \sqrt{n_x^2 + n_y^2 + n_z^2}$ and a "volume" $4\pi n^2\, dn$. The number of allowed energy values is only one-eighth of this, because the positive values of n_x, n_y, and n_z occupy only one-eighth of the coordinate space. Finally, for each value of n, there are two distinct waves, corresponding to the two possible polarizations of the wave. The allowed number of states is therefore

$$
g(n)\, dn = 2 \times \tfrac{1}{8} \times 4\pi n^2\, dn
\tag{10.25}
$$

and since

$$
E = \frac{\hbar c \pi}{L} n
\tag{10.26}
$$

we find

$$
g(E)\, dE = \pi \left(\frac{L}{\hbar c \pi} \right)^3 E^2\, dE
\tag{10.27}
$$

The number of photons having energy in the range E to $E + dE$ is, according to the Bose-Einstein distribution,

$$
p(E)\, dE = g(E) f_{\mathrm{BE}}\, dE
$$

$$
p(E)\, dE = \pi \left(\frac{L}{\hbar c \pi} \right)^3 E^2 \frac{1}{e^{E/kT} - 1}\, dE
\tag{10.28}
$$

The total radiant energy carried by photons of energy E is $E\,p(E)$, and the energy density (energy per unit volume) of photons in the range E to $E + dE$ is

$$u(E)\,dE = \frac{E\,p(E)\,dE}{L^3} \tag{10.29}$$

Substituting, and converting from photon energy E to wavelength, we find

Planck blackbody formula

$$u(\lambda)\,d\lambda = \frac{8\pi hc}{\lambda^5}\frac{1}{e^{hc/\lambda kT} - 1}\,d\lambda \tag{10.30}$$

Multiplying by $c/4$, as is necessary when we go from the energy density of the radiation to the radiancy, as in Equation 3.31, we find the result that was given in Equation 3.36.

Thus the Planck theory of blackbody radiation, which was so successful in accounting for experimental results, follows directly from the Bose-Einstein distribution for photons.

Einstein Theory of Heat Capacity

In an ordinary solid, most of the physical properties originate either with the valence electrons or with the latticework of atoms. Electrical conductivity, for example, originates with the valence electrons, while the propagation of mechanical waves is due to the lattice of atoms. The heat capacities of solids have contributions from both lattice and conduction electrons; at all but the lowest temperatures, the lattice contribution is dominant and the electronic contribution can therefore be neglected.

When we heat a solid, we increase the thermal motion of the atoms; the difference between a solid and a gas is that in the solid the atoms are fixed to equilibrium positions, and their motion consists of oscillations about the equilibrium position. For a one-dimensional oscillator, we count two degrees of freedom (one for its potential energy and one for kinetic energy); according to the equipartition theorem there is an energy of $\frac{1}{2}kT$ per degree of freedom, and thus the total energy of the *three-dimensional* solid is $3kT$ ($\frac{1}{2}kT$ per degree of freedom $\times$ 2 degrees of freedom per dimension $\times$ 3 dimensions) for each atom. Since a mole has Avogadro's number of atoms, the total energy per mole is

$$E = 3N_A kT$$

$$= 3RT \tag{10.31}$$

where $N_A k = R$, the universal gas constant. The molar heat capacity C is the energy change (heat energy) per unit change in temperature $\Delta E/\Delta T$, and thus we expect

$$C = 3R \tag{10.32}$$

This classical result is known as the *law of Dulong and Petit* and predicts that the molar heat capacity should take the constant value $3R$ (24.9 J/mole·K), independent of the type of material or the temperature.

The values in Table 10.4 show that this simple result gives quite a remarkable agreement with the experimental values at ordinary temperatures. However, at low temperatures, the heat capacity begins to deviate considerably from the classical value and tends to zero as T goes to zero.

The explanation for this failure of classical physics was first given by Einstein, who assumed that the *oscillations* (not the atoms) of the solid obeyed Bose-Einstein statistics. Just as electromagnetic waves are analyzed as "particles" (quanta of electromagnetic energy, or photons) that obey Bose-Einstein statistics, so are mechanical or acoustic waves analyzed as "particles" (quanta of vibrational energy, called *phonons*) that also obey Bose-Einstein statistics. Einstein made the simplifying assumption that all of the phonons (oscillations) have the same frequency, and so there is no density of states factor to worry about. We have seen in Chapter 5 that a quantized oscillator has an energy of $\hbar\omega(n + \frac{1}{2})$. Each additional value of n represents an additional phonon; to go from a vibrational energy of $\frac{5}{2}\hbar\omega$ to $\frac{7}{2}\hbar\omega$ we must "create" a phonon of energy $\hbar\omega$. One mole of the solid contains N_A atoms with $3N_A$ phonons of oscillation, each with energy $\hbar\omega$; the probability distribution for the phonons is f_{BE}, and so the energy per mole is

$$E = 3N_A\hbar\omega \frac{1}{e^{\hbar\omega/kT} - 1} \qquad (10.33)$$

The heat capacity can be found from dE/dT:

$$C = \frac{dE}{dT} = 3N_A\hbar\omega \frac{(e^{\hbar\omega/kT})(\hbar\omega/kT^2)}{(e^{\hbar\omega/kT} - 1)^2}$$

$$C = 3R \left(\frac{\hbar\omega}{kT}\right)^2 \frac{e^{\hbar\omega/kT}}{(e^{\hbar\omega/kT} - 1)^2} \qquad (10.34)$$

Einstein formula for heat capacity

Where T is small, the exponential term dominates, and $C \propto e^{-\hbar\omega/kT}$, so indeed C approaches 0 for small T, in agreement with experiment. Figure 10.20 compares the experimental values of C with the behavior predicted by Equation 10.34. As you can see, the agreement is reasonably good. (The vibrational energy $\hbar\omega$ is an adjustable parameter of the theory and takes different values for different materials.)

TABLE 10.4   HEAT CAPACITIES OF COMMON METALS

Metal	Room Temp (300 K)		Low Temp (25 K)	
	$J/g \cdot K$	$J/mole \cdot K$	$J/g \cdot K$	$J/mole \cdot K$
Al	0.904	24.4	0.0175	0.473
Ag	0.235	25.3	0.0287	3.10
Au	0.129	25.4	0.0263	5.18
Cu	0.387	24.6	0.016	1.0
Fe	0.450	25.1	0.0075	0.42
Pb	0.128	26.5	0.0681	14.1
Sn	0.222	26.3	0.058	6.9

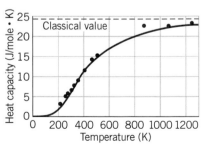

FIGURE 10.20 The Einstein theory of the heat capacity of diamond. The dots are the data points and the solid curve is Equation 10.34 with $\hbar\omega = 0.114$ eV.

In this calculation we have oversimplified by assuming all of the phonons to have the same frequency. The proper calculation, which was first done by Debye, assumes a distribution of phonon frequencies with a density of states given by an expression of the same form as that for the "photon gas" of blackbody radiation; the predicted low temperature behavior is then $C \propto T^3$, in much better agreement with experiment than the result of the Einstein calculation. Nevertheless, even our simple calculation shows that the behavior of C can be found by applying quantum statistics to the vibrations of a solid.

Liquid Helium

One of the most remarkable substances we can study in the laboratory is liquid helium. Here are some of its properties:

1. Helium gas is the most inert of the inert gases. Under normal conditions, it forms no compounds, and it has the lowest boiling point, 4.18 K, of any material.

2. Below 4.18 K, helium behaves much like an ordinary liquid. As the helium boils, the escaping gas forms bubbles, like a boiling pot of water. As the liquid is cooled further, a sudden transition occurs at a temperature of 2.18 K, the violent boiling stops, and the liquid becomes absolutely still. (Evaporation continues, but only from the surface.)

3. As the liquid is cooled below 2.18 K, the specific heat and the thermal conductivity both increase suddenly and discontinuously. Figure 10.21 shows the specific heat as a function of temperature. The form of the figure looks rather like the Greek letter λ, and so the transition point at 2.18 K has become known as the *lambda point*. The thermal conductivity rises at the λ point by a factor of perhaps 10^6.

4. When a liquid flows through a narrow capillary tube, its *viscosity* causes some resistance to the flow. When liquid helium flows through a capillary tube, its viscosity drops at the lambda point by a factor of perhaps 10^6. Below the lambda point helium flows easily through narrow capillary tubes (or other narrow constrictions), through which it would not flow above the lambda point.

5. Below the lambda point, liquid helium has the power to seemingly defy gravity, flowing up and over the walls of its container. The helium forms a thin film which lines the walls of the container, the remaining liquid is then drawn up by the film like a siphon, and the helium can be seen dripping from the bottom of the container, as in the photograph of Figure 10.22.

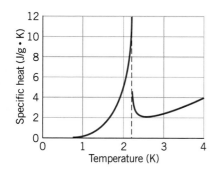

FIGURE 10.21 The specific heat of liquid helium. The discontinuity at 2.18 K is the lambda point.

All of these strange properties occur because liquid helium obeys Bose-Einstein statistics. Ordinary helium has two electrons filling the $1s$ shell, so the total angular momentum of the electrons is zero. It happens also that the helium nucleus (alpha particle) also has a spin of zero. Therefore the total spin of the atom (electron spin + nuclear spin) is zero, and a

(*a*) (*b*) (*c*)

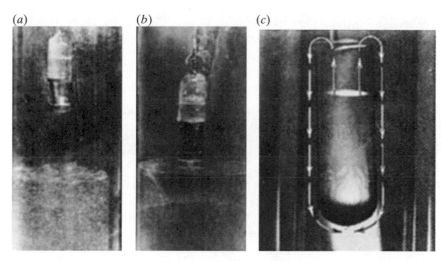

FIGURE 10.22 Some properties of liquid helium. In (*a*), the liquid is shown boiling violently at a temperature above 2.18 K. Liquid helium is also present in the small container, held in by a porous plug at the bottom. In (*b*), the helium has been cooled below 2.18 K. The boiling stops, and the viscosity essentially vanishes, allowing the helium to leak out of the small container through the very fine pores of the plug. In (*c*) the liquid helium escapes from the container by flowing up and over the walls and can be seen dripping from the bottom.

helium atom behaves like a boson. At 2.18 K, a *change of phase* occurs in the helium liquid. Above the lambda point, helium behaves like an ordinary liquid; this phase is called He I. The phase that occurs below the lambda point is called He II, and its special properties are due to its *superfluid* component. As the temperature is decreased from the lambda point toward absolute zero, the relative concentration of the normal fluid decreases and that of the superfluid increases. The unusual properties of liquid helium are all caused by the superfluid component, which is also known as a *quantum liquid.* Because the helium atoms obey Bose-Einstein statistics, the Pauli principle does not prevent all of the atoms from being in the same quantum state. This begins to happen at the lambda point, and is an example of the effect referred to as Bose condensation. We can think of the superfluid as being a single quantum state made up of a very large number of atoms; the atoms behave in a cooperative way, giving the superfluid its unusual properties.

By way of comparison, if we try the same kinds of experiments with the rarer isotope of helium, ^{3}He, the behavior is very different. Although ^{3}He has zero atomic spin, just like ^{4}He, it has only three particles rather than four in its nucleus, and its *nuclear* spin is $\frac{1}{2}$. The total atomic (electronic + nuclear) spin is therefore $\frac{1}{2}$, and ^{3}He behaves like a fermion and obeys the Fermi-Dirac distribution. Since the Pauli principle prevents more than one fermion from occupying any quantum state, no Bose condensation is expected for ^{3}He, and indeed none is observed until ^{3}He is cooled to about

0.002 K. At this point the weak coupling of two ^{3}He to form a boson occurs, and the ^{3}He pairs can experience Bose condensation. (A similar effect is responsible for superconductivity; see Section 11.7.)

10.7 APPLICATION OF FERMI-DIRAC STATISTICS

In a metal, the valence electrons are not very strongly bound to individual atoms, and consequently travel rather freely throughout the volume of the metal. We can treat these electrons as a "gas" that obeys the Fermi-Dirac distribution. The density of states function can be derived in a way similar to that for the photon gas. (Even the factor of 2, to account for the two different spin states, is the same.) The only difference is that, rather than using $E = pc$, we use the appropriate relationship for nonrelativistic electrons, $E = p^2/2m$. After some manipulations, we find

$$g(E)\, dE = \frac{8\sqrt{2}\pi L^3 m^{3/2}}{h^3} E^{1/2}\, dE \tag{10.35}$$

and therefore the number of electrons with energy between E and $E + dE$ is

$$p(E)\, dE = \frac{8\sqrt{2}\pi L^3 m^{3/2}}{h^3} \frac{E^{1/2}\, dE}{e^{(E-E_F)/kT} + 1} \tag{10.36}$$

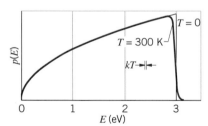

FIGURE 10.23 The number of occupied electronic levels at $T = 0$ and $T = 300$ K, according to the Fermi-Dirac distribution. The Fermi energy E_F is chosen to be 3.0 eV.

This function is plotted in Figure 10.23. You can see immediately why we neglected the influence of the electrons when we computed the specific heat of a metal and considered only the lattice contribution. When we add heat to a metal, only a very few of the electrons near E_F absorb the energy and change their state of motion; the vast majority of the electrons are not affected, and thus the influence of the electrons on the specific heat is indeed negligible.

We can find a numerical value for E_F by assuming that the sample contains a total number N of these free electrons:

$$N = \int_0^\infty p(E)\, dE \tag{10.37}$$

At $T = 0$, the Fermi-Dirac distribution function has the value 1 for $E < E_F$ and 0 for $E > E_F$, so the integral reduces to

$$N = \frac{8\sqrt{2}\pi L^3 m^{3/2}}{h^3} \int_0^{E_F} E^{1/2}\, dE \tag{10.38}$$

Evaluating the integral and solving the resulting expression for E_F, we obtain

Fermi energy at T = 0

$$E_F = \frac{h^2}{2m}\left(\frac{3N}{8\pi V}\right)^{2/3} \tag{10.39}$$

where $V = L^3$, the volume of the metal.

We can also find the mean energy of the electrons

$$E_m = \frac{1}{N} \int_0^\infty E p(E) \, dE \tag{10.40}$$

and it is left as an exercise to show that

$$E_m = \tfrac{3}{5} E_F \tag{10.41}$$

EXAMPLE 10.5

Compute the Fermi energy E_F for sodium.

SOLUTION

Each sodium atom contributes one valence electron to the metal, and so the number of electrons per unit volume, N/V, is equal to the number of sodium atoms per unit volume. This can in turn be found from the density of sodium and the mass of a sodium atom:

$$\frac{N}{V} = \text{atoms per unit volume} = \frac{\rho N_A}{M}$$

$$= \frac{(0.971 \text{ g/cm}^3)(6.02 \times 10^{23} \text{ atoms/mole})}{23.0 \text{ g/mole}}$$

$$= 2.54 \times 10^{22} \text{ cm}^{-3}$$

$$= 2.54 \times 10^{28} \text{ m}^{-3}$$

$$E_F = \frac{h^2}{2m} \left(\frac{3}{8\pi} \frac{N}{V} \right)^{2/3}$$

$$- \frac{(hc)^2}{2mc^2} \left(\frac{3}{8\pi} \, 2.54 \times 10^{28} \text{ m}^{-3} \right)^{2/3}$$

$$= \frac{(1240 \text{ eV·nm})^2}{2(0.511 \times 10^6 \text{ eV})} (2.09 \times 10^{18} \text{ m}^{-2}) \left(\frac{10^{-9} \text{ m}}{\text{nm}} \right)^2$$

$$= 3.15 \text{ eV}$$

The average energy of the valence electrons is $\tfrac{3}{5} E_F$ or 1.89 eV. *Even at the absolute zero of temperature*, the electrons still have quite a large average energy.

From Figure 10.23 we see that the change in $p(E)$ between $T = 0$ and $T = 300$ K (room temperature) is relatively small, and so these values for E_F and E_m are approximately correct at room temperature.

The meaning of these numbers is as follows. Instead of isolated atoms with individual energy levels, we consider the metal to be a single system with a very large number of energy levels (at least as far as the valence electrons are concerned). Electrons fill these energy levels, in accordance

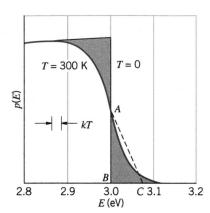

FIGURE 10.24   Detail of Figure 10.23 near E_F.

with the Pauli principle, beginning at $E = 0$. By the time we add 2.54×10^{22} valence electrons to 1 cm³ of sodium, we have filled energy levels up to $E_F = 3.15$ eV; all levels below E_F are filled and all levels above E_F are empty. Electrons have an almost continuous energy distribution (the levels are discrete, but they are very close together) from $E = 0$ to $E = E_F$, with an average energy of 1.89 eV. At $T = 300$ K, a relatively small number of electrons is excited from below E_F to above E_F; the range over which electrons are excited is of order $kT \cong 0.025$ eV, so that only electrons within 0.025 eV of E_F are affected by the change from $T = 0$ to $T = 300$ K.

The change in the number of occupied electron states from $T = 0$ to $T = 300$ K is represented in greater detail by the shaded area of Figure 10.24. If we approximate that area by triangle ABC, the number of electrons n_{ex} excited from below E_F to above E_F is approximately $\frac{1}{2}(AB)(BC)$. The height AB is $p(E_F)$; since $f_{ED}(E_F) = \frac{1}{2}$, it follows that $AB = p(E_F) = \frac{1}{2}g(E_F)$. On the scale of Figure 10.24, the base BC of the triangle is about $3kT$ so

$$n_{ex} \cong \frac{1}{2} \times \frac{g(E_F)}{2} \times 3kT \qquad (10.42)$$

Combining Equations 10.35 and 10.39, we find

$$g(E_F) = \tfrac{3}{2}N\frac{1}{E_F} \qquad (10.43)$$

and therefore the fraction of electrons that changes state when the metal is heated from $T = 0$ to a temperature T is about

$$\frac{n_{ex}}{N} \cong \frac{9}{8}\frac{kT}{E_F} \qquad (10.44)$$

Electronic heat capacity

At room temperature $kT \cong 0.025$ eV, and since E_F has values of several eV, the fraction of electrons that are excited is less than 1 percent. Assuming those electrons to contribute to the heat capacity like an ideal gas ($C_V = \frac{3}{2}R$), the net electronic contribution to the heat capacity is of order $0.01R$, and is therefore negligible at ordinary temperatures compared with the lattice contribution of $3R$.

At what temperatures would we expect the electronic contribution to the heat capacity to become important? Since the fraction of electrons that contribute to C depends on T, by raising T we should be able to increase this fraction. If we increase the temperature until $kT \cong E_F$, we should notice the electrons contributing to the heat capacity, but this requires temperatures of the order of 10^4 K, well beyond the boiling point of metals! If we go down to very low temperatures, we find that the electronic contribution to the heat capacity is still proportional to T, but the lattice contribution, as discussed in the previous section, is proportional to T^3. At sufficiently low temperatures, the lattice contribution can become smaller than the electronic contribution. At very low temperatures, we do in fact observe $C \propto T$, rather than $C \propto T^3$, thus verifying the contribution of the electrons to the heat capacity.

Although the Fermi-Dirac distribution is an accurate and realistic description of the behavior of fermions, we have oversimplified by our use of the free-electron theory. Electrons are not really free, even in conductors. Even though the valence electrons move relatively freely, they see a periodic potential caused by the atoms of the lattice, and this makes significant changes in the density of states function $g(E)$. These changes, and their effect on the behavior of metals, insulators, and semiconductors, are discussed in Chapter 11.

Similar calculations using the Fermi-Dirac distribution determine the radius of white dwarf and neutron stars (see Problems 23 and 24).

SUGGESTIONS FOR FURTHER READING

An excellent introduction to statistical physics, including examples of arranging and sorting problems similar to the one in Section 10.2:

D. McLachlan Jr., *Statistical Mechanical Analogies* (Englewood Cliffs, Prentice-Hall, 1968).

More extensive treatments of statistical physics, at about the same mathematical level as this chapter:

N. Ashby and S. C. Miller, *Principles of Modern Physics* (San Francisco, Holden-Day, 1970). Chapter 2 discusses probability, and Chapters 10 and 11 cover classical and quantum statistics.

F. Reif, *Statistical Physics*, Volume 5 of the Berkeley Physics Series (New York, McGraw-Hill, 1967). Covers only classical statistics.

Advanced texts, with many applications of classical and quantum statistics:

C. Kittel, *Thermal Physics* (New York, Wiley, 1969).

P. M. Morse, *Thermal Physics* (New York, Benjamin, 1965).

R. D. Reed and R. R. Roy, *Statistical Physics for Students of Science and Engineering* (Scranton, Intext, 1971).

The properties of liquid helium are discussed in a very general and highly readable book:

K. Mendelssohn, *The Quest for Absolute Zero* (New York, McGraw-Hill, 1966).

For additional work on the properties of ^3He, see:

N. D. Mermin and D. M. Lee, "Superfluid Helium 3," *Scientific American* **235**, 56 (December 1976).

QUESTIONS

1. Suppose a container filled with a gas moves with constant velocity v. How is the Maxwell velocity distribution different for such a gas, compared with the same container of gas at rest?

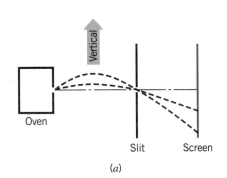

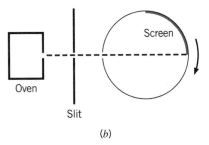

FIGURE 10.25   Question 3.

2. The population inversion necessary for the operation of a laser is sometimes called a "negative temperature." What is the meaning of a negative temperature? Does it have a physical interpretation?

3. Figure 10.25 shows two different experimental arrangements used to measure the distribution of molecular speeds. Based on the figures, explain how each apparatus might operate, and try to guess at how the observed distribution of molecules might appear. Where do the fastest molecules land? The slowest?

4. How would Figure 10.7 change if the temperature of the gas were increased?

5. How is the speed distribution of a gas at temperature T different from that at temperature $2T$? The energy distribution? Sketch the speed and energy distributions at the two temperatures.

6. Consider a mixture of two gases of molecular masses m_1 and $m_2 = 2m_1$ in thermal equilibrium at temperature T. How do their speed distributions differ? How do their energy distributions differ?

7. It is generally more convenient, wherever possible, to use Maxwell-Boltzmann statistics rather than quantum statistics. Under what circumstances can a quantum system be described by Maxwell-Boltzmann statistics?

8. Suppose we had a gas of hydrogen *atoms* at relatively high density. Do the atoms behave as fermions or as bosons? Would a gas of deuterium (heavy hydrogen) atoms behave any differently? (*Hint*: The nuclear spin is $\frac{1}{2}$ for hydrogen and 1 for deuterium.)

9. The early universe contained a large density of neutrinos (massless spin-$\frac{1}{2}$ particles that travel at the speed of light). Which statistical distribution would be needed to describe the properties of the neutrinos?

10. Would you expect the photoelectric effect to depend on the temperature of the surface of the metal? Explain.

11. Estimate the mean kinetic energy of the "free" electrons in a metal if they obeyed Maxwell-Boltzmann statistics. How does this compare with the result of applying Fermi-Dirac statistics? Why is there such a difference?

PROBLEMS

1. Calculate the probabilities listed in Table 10.2.

2. In a collection of 32 molecules, the speeds are distributed as follows:

Number of molecules	Range of speed (m/s)	Number of molecules	Range of speed (m/s)
1	0.0–0.1	4	0.5–0.6
3	0.1–0.2	3	0.6–0.7
5	0.2–0.3	2	0.7–0.8
6	0.3–0.4	2	0.8–0.9
5	0.4–0.5	1	0.9–1.0

Sketch this distribution and find the most probable speed, the mean speed, and the rms speed.

3. Carry through the calculation of the constants b and A for the Maxwell speed distribution, using Equations 10.6, 10.7, and 10.8.

4. Show that the most probable speed v_p of the Maxwell speed distribution is $(2kT/m)^{1/2}$.

5. For a gas that obeys the Maxwell distribution, find $\Delta v = \sqrt{(v^2)_m - (v_m)^2}$. What is the meaning of this quantity?

6. Equation 10.16 gives the distribution of kinetic energies in a gas. Find:

   (a) The most probable kinetic energy K_p.

   (b) The mean kinetic energy, $K_m = n^{-1}\int_0^\infty Kf(K)dK$.

   (c) The rms kinetic energy, $K_{rms} = [n^{-1}\int_0^\infty K^2 f(K)dK]^{1/2}$.

   (d) The variation in the kinetic energy, $\Delta K = [(K^2)_m - (K_m)^2]^{1/2}$.

   Interpret each of these quantities.

7. Consider a collection of N noninteracting atoms with a single excited state at energy E. Assume the atoms obey Maxwell-Boltzmann statistics, and take both the ground state and the excited state to be nondegenerate. (a) At temperature T, what is the ratio of the number of atoms in the excited state to the number in the ground state? (b) What is the average energy of an atom in this system? (c) What is the total energy of the system? (d) What is the heat capacity of this system?

8. Suppose we have a gas in thermal equilibrium at temperature T. Each molecule of the gas has mass m. (a) What is the ratio of the number of molecules at the Earth's surface to the number at height h (with potential energy mgh)? (b) What is the ratio of the density of the gas at height h to the density at the surface? (c) Would you expect this simple model to give an adequate description of the Earth's atmosphere?

9. A collection of noninteracting hydrogen atoms is maintained in the $2p$ state in a magnetic field of strength 5.0 T. (a) At room temperature, find the fraction of the atoms in the $m_l = +1$, 0, and -1 states. (b) If the $2p$ state made a transition to the $1s$ state, what would be the relative intensites of the three normal Zeeman components? Ignore any effects of electron spin.

10. The following method is used to measure the molecular weight of very heavy molecules. A liquid containing the molecules is spun rapidly in a centrifuge, which establishes a variation in the density of the liquid. The density is measured, such as by absorption of light, to determine the molecular weight. Assign a fictitious "centrifugal" force to act on the molecules and show that the density varies as $\rho = \rho_0 e^{m\omega^2 x^2/2kT}$ where ω is the angular velocity of the centrifuge and x measures the distance along the centrifuge tube.

11. One mole of helium gas at a temperature of 293 K is confined to a cubical container whose sides are 12.0 cm long. (a) Find the mean kinetic energy K_m. (*Hint:* Use v_{rms} from Example 10.1.) (b) How many molecules have energies in an interval of width $0.01 K_m$ centered on K_m?

12. Find the blackbody energy spectrum $u(E)$, and convert to the wavelength spectrum $u(\lambda)$, Equation 10.30.

13. (a) From the blackbody energy density $u(E)$, the energy per unit volume per unit energy interval at the energy E, find the number of photons of energy E

per unit volume per unit energy interval. (b) Show that the total number of photons per unit volume at temperature T is $N = 8\pi(kT/hc)^3 \int_0^\infty x^2 dx/(e^x - 1)$. (c) The value of the integral is about 2.404. How many photons per cubic centimeter are there in a cavity filled with radiation at $T = 300$ K? At $T = 3$ K?

14. From the blackbody energy spectrum $u(E)$, find the *energy* at which the maximum radiation is emitted. Compare this result with Wien's displacement law (see Chapter 3) and account for any differences.

15. A blackbody is radiating at a temperature of 2.50×10^3 K. (a) What is the total energy density of the radiation? (b) What fraction of the energy is emitted in the interval between 1.00 and 1.05 eV? (c) What fraction is emitted between 10.00 and 10.05 eV?

16. (a) Derive Equation 10.39 from Equation 10.38. (b) Derive Equation 10.41 from Equation 10.40.

17. Compute the Fermi energy E_F and the average electron energy for copper.

18. Calculate the Fermi energy for magnesium, assuming two free electrons per atom.

19. In sodium metal at room temperature, compute the energy difference between the points at which the Fermi-Dirac occupation probability has the values 0.1 and 0.9. What do you conclude about the "sharpness" of the distribution?

20. In sodium metal (see Example 10.5), calculate the number of electrons per unit volume at room temperature in an interval of width $0.01 E_F$ centered on the mean energy E_m.

21. A certain metal has a Fermi energy of 3.00 eV. Find the probability to find an electron with energy between 5.00 eV and 5.10 eV for (a) $T = 295$ K; (b) $T = 2500$ K.

22. Assume that the nucleons in a nucleus obey Fermi-Dirac statistics. Find the Fermi energy and the average energy of the protons or neutrons in the nucleus of a calcium atom, which contains 20 protons and 20 neutrons and has a radius of 4.1×10^{-15} m. Are these values reasonable?

23. After a star like our Sun stops producing energy by hydrogen fusion, it collapses to a white dwarf star, with a radius about the same as the Earth's. Take the mass of the Sun to be 2.00×10^{30} kg and its radius after collapse to be 6.40×10^3 km, and calculate the Fermi energy of the electrons in such a white dwarf. Assume one electron for every two protons or neutrons.

24. When a star is somewhat more massive than the Sun, it will collapse to a neutron star, which can be regarded as a giant atomic nucleus, composed of neutrons. For a star about twice the Sun's mass, the radius of the neutron star will be about 10 km. Find the Fermi energy of the neutrons in such a neutron star.

This scanning electron microscope photograph shows salt crystals, which are built up from a lattice of sodium and chloride ions. In the absence of impurities, an exact cubic crystal is formed.

Imperfections in the lattice can produce distortions of the shape, but the basic cubic structure remains.

SOLID-STATE PHYSICS

In this chapter we study the way atoms or molecules combine to make solids. In particular, we discuss how the principles of quantum mechanics are essential in understanding the properties of solids.

At first thought, there seem to be so many different solids that to classify them and form some general rules for their properties would appear to be a hopeless task. The book you are reading is made of paper and cloth, held together by a glue made from resins, once liquid and now solid. Your desk might be made of wood, metal, or plastic; your chair might be made of similar materials, and might perhaps be covered with cloth, leather, or plastic fabric and contain fiber or synthetic foam padding. Around you, there might be many books and papers, pencils made of wood and metal and graphite, rubber erasers, metal and plastic pens, and a calculator with a plastic body containing semiconductors and a liquid crystal display. Looking through the glass window, you see trees, bricks, concrete, metal, selected by humans or by nature for strength, utility, or attractiveness. Each of these solids has a characteristic color, texture, strength, hardness, or ductility; it has a certain measurable electrical conductivity, thermal conductivity, magnetic susceptibility, and melting point; it has certain characteristic emission or absorption spectra in the visible, infrared, ultraviolet, or other regions of the electromagnetic spectrum.

It is a fair generalization to say that all of these properties depend on two features of the structure of the material: the type of atoms or molecules of which the substance is made, and the way those atoms or molecules are joined or stacked together to make the solid. It is the formidable task of the *solid-state* (or *condensed-matter*) *physicist* or *physical chemist* to try to relate the structure of materials to their observed physical or chemical properties.

In this chapter we concentrate our discussions on materials in which the atoms or molecules occupy regular or periodic sites; this structure is called a *lattice*, and materials with this structure are called *crystals*. Crystalline materials include many metals, chemical salts, and semiconductors. One property that distinguishes crystals is their *long-range order*—once we begin constructing the lattice in one location, we determine the placement of atoms that are quite far away. In this respect, the crystal is like a brick wall, in which the bricks are stacked in a periodic array and the placement of a brick is predetermined by the original arrangement of bricks far away compared with the size of a brick. (By contrast, *amorphous* materials such as glass or paper have no long-range order, and their structure is more similar to a pile of bricks than a brick wall.)

Quantum mechanics plays a fundamental role in determining all properties of the solid: mechanical, electrical, thermal, magnetic, optical, and so forth. To illustrate the application of quantum mechanics to the study of solids, in this chapter we consider primarily the electrical properties of solids, especially their electrical conduction.

11.1 IONIC SOLIDS

In our study of ionic bonding in molecules, we learned that the cohesive forces in ionic molecules originate from the electrostatic attraction between a closed-shell ion, such as Na^+, and another closed-shell ion, such as Cl^-. Such materials are expected to form solids readily, because a Na^+ ion can simultaneously attract many Cl^- ions to itself, thereby building up a solid structure. Ionic bonding is a particularly strong type of bonding, and since the forces are electrostatic in origin, we might suppose that the more negative ions we surround a positive ion with, the more stable and strong will be the solid. (Covalent bonds, on the other hand, involve specific electron wave functions and so are limited in the number of near neighbors that can participate in the bonding.)

Ionic solids are crystalline, rather than amorphous, because we can pack ions together more efficiently in a regular array than in a random arrangement. (We can also stack bricks more efficiently in a regular array than in a random pile. In the regular array, there are more bricks per unit volume, and their average separation is smaller.)

The simplest type of crystal lattice is the *cubic* lattice, in which we imagine the atoms to be placed at the corners of a succession of cubes which cover the volume of the crystal. Figure 11.1 shows how this would look if we imagine the atoms to stack like hard spheres. This type of stacking is not the most efficient, because there are large gaps at the center of each face of the cube, and also in the middle of the cube itself. We get a better stacking arrangement, which has more atoms per unit volume, if we place another sphere in the gap at the center of each face of the cube, or else at the center of the body of the cube (even though in both cases the spheres at the corners are no longer in direct contact with one another). These two lattices are known as *face-centered cubic* (fcc) and *body-centered cubic* (bcc) and are illustrated in Figures 11.2 and 11.3.

Cubic lattice

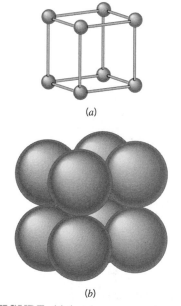

(a)

(b)

FIGURE 11.1 Arrangement of atoms in a simple cubic crystal. Part (a) shows the relationship of the bonds and part (b) shows the packing of atoms.

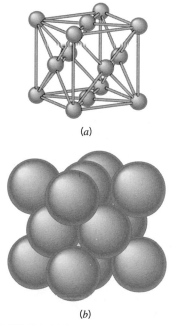

(a)

(b)

FIGURE 11.2 Arrangement of atoms in a face-centered cubic crystal.

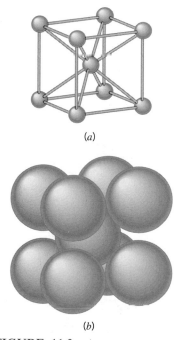

(a)

(b)

FIGURE 11.3 Arrangement of atoms in a body-centered cubic crystal.

The fcc lattice gives a slightly more efficient packing (more atoms per unit volume) and so it is usually the most stable structure. However, atoms do not stack like hard spheres, and often the bcc structure is preferred. These two crystal types, fcc and bcc, also occur for materials other than ionic solids, and in fact some materials change their structure from fcc to bcc as the temperature is changed.

A common material that has the fcc lattice structure is NaCl, and for that reason the fcc lattice is often called *NaCl structure*. In order to have the atoms attract one another, we must alternate Na^+ and Cl^- ions, as is shown in Figure 11.4. Notice that a given Na^+ ion is attracted by 6 close Cl^- neighbors, and does not "belong to" any single Cl^- ion. *It is therefore wrong to consider ionic solids as being composed of molecules.*

A typical bcc structure is CsCl, as shown in Figure 11.5, and so the bcc lattice is often known as the *CsCl structure*. In this case each ion is surrounded by 8 neighbors of the opposite charge.

The forces that hold ionic solids together are, like the forces that hold ionic molecules together, of electrostatic origin. Each Na^+ ion in the NaCl structure is surrounded by 6 Cl^- ions at a distance R. At the slightly larger distance of $R\sqrt{2}$, each Na^+ ion sees 12 Na^+ ions, and at the still greater distance of $R\sqrt{3}$ there are 8 Cl^- ions. We can continue in this way to add the alternating attractive and repulsive contributions to find the total Coulomb potential energy:

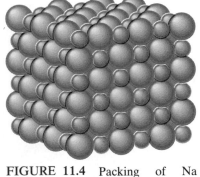

FIGURE 11.4 Packing of Na (small spheres) and Cl (large spheres) in a crystal of NaCl.

$$U_C = -6\frac{e^2}{4\pi\varepsilon_0}\frac{1}{R} + 12\frac{e^2}{4\pi\varepsilon_0}\frac{1}{R\sqrt{2}} - 8\frac{e^2}{4\pi\varepsilon_0}\frac{1}{R\sqrt{3}} + \cdots$$

$$= -\frac{e^2}{4\pi\varepsilon_0}\frac{1}{R}\left(6 - \frac{12}{\sqrt{2}} + \frac{8}{\sqrt{3}} - \cdots\right) \qquad (11.1)$$

or

$$U_C = -\alpha\frac{e^2}{4\pi\varepsilon_0}\frac{1}{R} \qquad (11.2)$$

where α, called the *Madelung constant*, is the factor in parentheses in Equation 11.1:

$$\alpha = 6 - \frac{12}{\sqrt{2}} + \frac{8}{\sqrt{3}} - \cdots \qquad (11.3)$$

This quantity depends only on the geometry of the lattice and is evaluated by summing the slowly converging series of alternating positive and negative terms. The result is

$$\alpha = 1.7476 \qquad \text{(fcc or NaCl lattice)}$$

For the bcc lattice, a similar calculation gives

$$\alpha = 1.7627 \qquad \text{(bcc or CsCl lattice)}$$

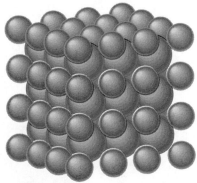

FIGURE 11.5 Packing of Cs (large spheres) and Cl (small spheres) in a crystal of CsCl.

As in the case of ionic molecules, the electrostatic force is opposed by a repulsive force due to the Pauli principle which keeps the filled subshells from overlapping and thus keeps the atoms from getting too close together.

The repulsive potential energy can be approximated as

$$U_R = AR^{-n} \qquad (11.4)$$

where A gives the strength of the potential energy and n determines how rapidly it increases at small R. For most ionic crystals, n is in the 8–10 range. The total energy of an ion in the lattice is the sum of the Coulomb and repulsive potential energies:

$$E = -\alpha \frac{e^2}{4\pi\varepsilon_0}\frac{1}{R} + \frac{A}{R^n} \qquad (11.5)$$

The energies are illustrated in Figure 11.6. Note that, as in the case of ionic molecules, there is a stable minimum in the energy which determines both the equilibrium separation R_0 and the binding energy. To find this minimum, we set dE/dR to zero, which gives

$$A = \frac{\alpha e^2 R_0^{n-1}}{4\pi\varepsilon_0 n} \qquad (11.6)$$

The binding energy B of an ion in the crystal is the depth of the energy well at $R = R_0$, the equilibrium separation between nearest-neighbor ions. Substituting Equation 11.6 into Equation 11.5 and evaluating the resulting equation at $R = R_0$, we obtain

$$B = -E(R_0) = \frac{\alpha e^2}{4\pi\varepsilon_0 R_0}\left(1 - \frac{1}{n}\right) \qquad (11.7)$$

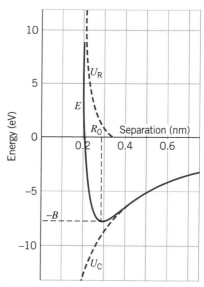

FIGURE 11.6 Contributions to the energy of an ionic crystal. Values are shown for NaCl.

Ionic binding energy

From thermodynamic measurements, it is possible to determine the bulk *cohesive energy* of a solid. In effect, the cohesive energy of an ionic solid is defined as the energy necessary to dismantle the solid into individual ions. Some measured values of the cohesive energies and nearest-neighbor spacings are given in Table 11.1. The value of the exponent n is determined from compressibility data.

The cohesive energy of a bulk sample can be calculated by multiplying the binding energy for a single ion, determined from Equation 11.7, by the number of ions in the sample, *except* that such a calculation would count each ion twice.* In one mole of an ionic solid, there are Avogadro's number N_A of positive ions and also N_A negative ions, for a total of $2N_A$ ions per mole. The relationship between the molar cohesive energy C and the ionic binding energy B is then

$$C = \frac{1}{2}(B)(2N_A) = BN_A \qquad (11.8)$$

where the factor of $\frac{1}{2}$ corrects for the problem of double counting of the ions.

The following examples illustrate the relationship between cohesive energy of the bulk solid and the binding energy per ion pair.

---

* Consider ions A and B. If we use Equation 11.7 to compute the binding energy of ion A, the result includes the interaction of ion A with *all* the ions of the solid, including ion B. Similarly, the binding energy of ion B calculated from Equation 11.7 includes the interaction of B with A. If we calculated the total binding energy of the solid by adding together the binding energies of ions A and B, we would be including the interaction between A and B twice.

TABLE 11.1   PROPERTIES OF IONIC CRYSTALS

Crystal	Nearest-Neighbor Separation (nm)	Cohesive Energy (kJ/mol)	n	Structure
LiF	0.201	1030	6	fcc
LiCl	0.257	834	7	fcc
NaCl	0.281	769	9	fcc
NaI	0.324	682	9.5	fcc
KCl	0.315	701	9	fcc
KBr	0.330	671	9.5	fcc
RbF	0.282	774	8.5	fcc
RbCl	0.329	680	9.5	fcc
CsCl	0.356	657	10.5	bcc
CsI	0.395	600	12	bcc
MgO	0.210	3795	7	fcc
BaO	0.275	3029	9.5	fcc

EXAMPLE 11.1

(a) Determine the experimental value of the binding energy of an ion pair in the NaCl lattice. (b) Compare the experimental value of the ionic binding energy with the calculated value.

SOLUTION

(a) From Equation 11.8, we have

$$B = \frac{C}{N_A} = \frac{769 \times 10^3 \text{ J/mol}}{(6.02 \times 10^{23} \text{ ions/mol})\,(1.60 \times 10^{-19} \text{ J/eV})} = 7.98 \text{ eV}$$

(b) The calculated value of the ionic binding energy is obtained from Equation 11.7:

$$B = \frac{\alpha e^2}{4\pi\varepsilon_0 R_0}\left(1 - \frac{1}{n}\right) = \frac{(1.7476)\,(1.44 \text{ eV} \cdot \text{nm})}{0.281 \text{ nm}}(0.889) = 7.96 \text{ eV}$$

The agreement between the experimental and calculated values is very good.

EXAMPLE 11.2

How much energy *per neutral atom* would be needed to take apart a crystal of NaCl?

SOLUTION

If we supply an energy of C to a mole of NaCl, we obtain N_A Na$^+$ ions and N_A Cl$^-$ ions. To convert these to neutral atoms, we must remove an electron from each Cl^-, which costs us the electron affinity of Cl (3.61 eV), and then we must attach that electron to the Na^+, which returns the ionization energy of Na (5.14 eV). The net cost per pair of Na and Cl atoms is

$$7.98\,eV + 3.61\,eV - 5.14\,eV = 6.45\,eV$$

Since expending this much energy gives two neutral atoms (Na and Cl), the net cost per atom is half that amount, or 3.23 eV.

Based on the high cohesive energies of ionic solids given in Table 11.1, we can explain the following properties of the ionic solids:

1. They form relatively stable and hard cubic crystals.
2. They are poor electrical conductors, since there are no free electrons available.
3. They have high vaporization temperatures. Since the cohesive energies are of the order of several electron-volts, in order to vaporize the solid we must heat it to a temperature high enough so that $kT \sim 1$ eV, or $T \sim 12{,}000$ K. This is quite a high estimate; actual values are more like 1000 to 2000 K, still a relatively high temperature.
4. They are transparent to visible radiation. A photon of visible light will interact with a material only if it can excite an electron from its ground state to an excited state. Since the ionic solids all have filled shells, it is necessary to excite an electron from one shell to the next. The energy required for this is typically much larger than the energies of photons of visible light, and so the photons pass right through ionic solids.
5. Ionic solids absorb strongly in the infrared. We can show this by considering the *force* on an ion in a solid. We can estimate this force from the derivative of the total potential energy, which is given by Equation 11.5:

Properties of ionic solids

$$F = -\frac{dU}{dR} = -\frac{\alpha e^2}{4\pi\varepsilon_0 R^2}\left(1 - \frac{R_0^{n-1}}{R^{n-1}}\right) \qquad (11.9)$$

Suppose we displace the atom by a small distance x from R_0:

$$F = -\frac{\alpha e^2}{4\pi\varepsilon_0 (R_0 + x)^2}\left(1 - \frac{R_0^{n-1}}{(R_0 + x)^{n-1}}\right)$$

$$= -\frac{\alpha e^2}{4\pi\varepsilon_0 R_0^2}\frac{1}{(1 + x/R_0)^2}\left(1 - \frac{1}{(1 + x/R_0)^{n-1}}\right)$$

$$\cong -\frac{\alpha e^2}{4\pi\varepsilon_0 R_0^2}(n-1)\frac{x}{R_0} \qquad (11.10)$$

keeping only terms to first order in the small quantity x/R_0. The ion thus experiences a restoring force $F = -kx$ with:

$$k \cong \frac{\alpha e^2}{4\pi\varepsilon_0 R_0^3}(n-1) \qquad (11.11)$$

and oscillates with a frequency $\nu = (1/2\pi)\sqrt{k/m}$. For NaCl, we estimate the force constant to be:

$$k = \frac{8(1.75)\,(1.44\,\text{eV}\cdot\text{nm})}{(0.281\,\text{nm})^3} = 909\,\text{eV/nm}^2$$

and the corresponding frequency is:

$$\nu = \frac{1}{2\pi}\sqrt{\frac{(909\,\text{eV/nm}^2)\,(9\times10^{16}\,\text{m}^2/\text{s}^2)\,(10^{18}\,\text{nm}^2/\text{m}^2)}{23\,\text{u}\times931.5\,\text{MeV/u}\times10^6\,\text{eV/MeV}}}$$

$$= 9.83\times10^{12}\,\text{Hz}$$

(We have used the mass of a Na atom for this estimate.) Photons of this frequency have a wavelength of 30.5 μm, in the infrared region. A beam of infrared radiation of wavelength 30.5 μm will set the Na ions oscillating, and in the process the radiation is absorbed.

6. Ionic solids are usually soluble in polar liquids such as water. The water molecule has an electric dipole moment, which can exert an attractive force on the charged ions, breaking the ionic bonds and dissolving the ionic solid.

11.2 COVALENT SOLIDS

As we discussed in Chapter 9, carbon forms molecules by covalent bonding of its four outer electrons in sp^3 hybrid orbits. Such bonds are highly *directional*, and we have seen how it is possible to calculate the angle between the bonds based on the symmetry of the bonding configuration. Solid carbon, in the form of diamond, is an example of a solid in which the interatomic forces are also of a covalent nature. As in a molecule, the four equivalent sp^3 hybrid states participate in covalent bonds, and since they are equivalent they must make equal angles with one another. The manner in which this is done is shown in Figure 11.7. A central carbon atom is covalently bound to four other carbons that occupy four corners of a cube as shown. The angle between the bonds is 109.5°, as it was in the covalently bonded molecules.

Figure 11.8 illustrates how the solid structure characteristic of diamond is constructed of such bonds. Each carbon has four close neighbors with which it shares electrons in covalent bonds. The basic structure is known as *tetrahedral*, and many compounds have a similar structure as a result of covalent bonding. Table 11.2 shows some of these compounds. The cohesive energy is the energy required to dismantle the solid into individual atoms. The structure is also known as the *zinc sulfide* or *zinc blende* structure.

Some of the covalent solids listed in Table 11.2 have bond energies larger than those of ionic solids. Substances such as diamond and silicon carbide are particularly hard. Other covalent solids with structures similar to carbon are silicon and germanium; the structure of these solids is responsible for their behavior as semiconductors.

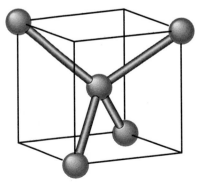

FIGURE 11.7 The tetrahedral structure of carbon.

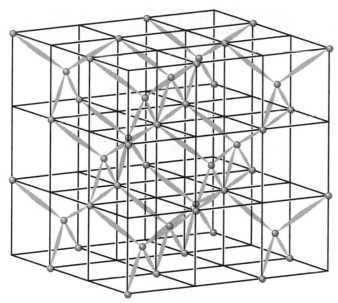

FIGURE 11.8   The lattice structure of diamond.

The covalent solids do not have the same similarity of characteristics that ionic solids do, and so we cannot make the same generalizations. Carbon, in the diamond structure, has a large bond energy and is therefore very hard and transparent to visible light; germanium and tin have similar structures, but are metallic in appearance and highly reflective. Carbon (as diamond) has an extremely high melting point (4000 K); germanium and tin melt at much lower temperatures more characteristic of ordinary metals. Of course, these differences depend on the actual bond energy in the solid, which in turn depends on the type of atoms of which the solid is made. Those solids with large bond energies are hard, have high melting points, are poor electrical and thermal conductors, and are transparent to visible light. Those solids with small bond energies may have very different properties.

TABLE 11.2   SOME COVALENT SOLIDS

Crystal	Nearest-Neighbor Distance (nm)	Cohesive Energy (kJ/mol)
ZnS	0.235	609
C (diamond)	0.154	710
Si	0.234	447
Ge	0.244	372
Sn	0.280	303
CuCl	0.236	921
GaSb	0.265	580
InAs	0.262	549
SiC	0.189	1185

11.3   OTHER SOLID BONDS

Metallic Bonds

The valence electrons in a metal are usually rather loosely bound, and frequently the electronic shells are only partially filled, so that metals tend not to form covalent bonds. The basic structure of metals is a "sea" or "gas" of approximately free electrons surrounding a lattice of positive ions. The metal is held together by the attractive force between each individual metal ion and the electron gas.

The most common crystal structures of metallic solids are fcc, bcc, or a third type known as *hexagonal close-packed* (*hcp*). The hcp structure is shown in Figure 11.9; like the fcc structure, it is a particularly efficient way of packing atoms together. Some metals and their characteristics are shown in Table 11.3. The cohesive energy of metal bonds tends to fall in the range 100–400 kJ/mol (1–4 eV/atom), making the metals less strongly bound than ionic or covalent solids. As a result, many metals have relatively low melting points (some below a few hundred °C). The relatively free electrons in the metal interact readily with photons of visible light, so metals are not transparent. The free electrons are responsible for the high electrical and thermal conductivity of metals. Since metallic bonds don't depend on any particular sharing or exchange of electrons between specific atoms, the exact nature of the atoms of the metal is not as important as it is in the case of ionic or covalent solids; as a result we can make many kinds of metallic alloys by mixing together different metals in varying proportions.

Molecular Solids

None of the solids we have discussed so far can be considered as composed of individual molecules. It is, however, possible for molecules to exert forces on one another and to bind together in solids. Since molecules have

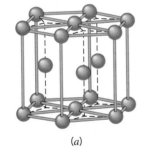

(a)

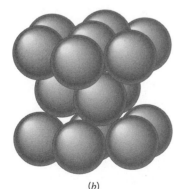

(b)

FIGURE 11.9 Arrangement of atoms in a hexagonal close-packed crystal.

TABLE 11.3   STRUCTURE OF METALLIC CRYSTALS

Metal	Crystal Type	Nearest-Neighbor Distance (nm)	Cohesive Energy (kJ/mol)
Fe	bcc	0.248	418
Li	bcc	0.304	158
Na	bcc	0.372	107
Cu	fcc	0.256	337
Ag	fcc	0.289	285
Pb	fcc	0.350	196
Co	hcp	0.251	424
Zn	hcp	0.266	130
Cd	hcp	0.298	112

electrons that are already shared in *molecular* bonds, there are no available electrons to participate in ionic, covalent, or metallic bonds with other molecules. Molecular solids are held together by much weaker forces, which generally depend on the *electric dipole moments* of the molecules. These forces between one molecule and another are much weaker than the internal forces that hold a molecule together; thus a molecule *can* retain its identity in a molecular solid.

The ionic, covalent, and metallic solids are all held together by ordinary $1/R^2$ Coulomb forces; in ionic solids, the force is between positive and negative ions, while in covalent and metallic solids, it is between the positive ions and the shared electrons. In a molecular solid, the molecules are electrically neutral, and so no Coulomb forces are possible. It is, however, possible for the dipole moment of one molecule to exert an attractive force on the dipole moment of another. The dipole cohesive force, which is proportional to $1/R^3$, is in general weaker than the Coulomb force. Molecular solids are therefore more weakly bound and have lower melting points than ionic, covalent, or metallic solids, because it takes less thermal energy to break the bonds of a molecular solid.

Electric dipole forces are important in molecular bonding for all molecules that have permanent electric dipole moments (*polar molecules*). We can imagine such forces as occurring when the positive end of one dipole exerts an attractive force on the negative end of the next, as shown in Figure 11.10.

The water molecule is an example of a case in which such forces occur. The oxygen atom in water tends to attract all of the electrons of the molecule and so looks like the negative end of the dipole; the two "bare" protons are the positive ends of the dipole, and each can attract the negative oxygen of adjacent water molecules (Figure 11.11). Although it is not obvious from this oversimplified explanation, it is this sort of bonding that causes the characteristic hexagonal crystal structure of ice, and is therefore responsible for the beautiful hexagonal patterns of snowflakes. When bonding of this sort involves hydrogen atoms, as it does in water, it is known as *hydrogen bonding*.

It is also possible to have dipole forces exerted between atoms or molecules that have no permanent dipole moments. For example, consider an atom of an inert gas such as neon. The atom has filled electron shells, so it is spherically symmetric and has no permanent electric dipole moment. However, quantum mechanical fluctuations* can produce an instantaneous electric dipole moment, which induces a dipole moment by polarizing a neighboring atom and produces an attractive force of the kind shown in Figure 11.10. This attractive force, which is known as the *van der Waals force*, is responsible for the bonding in certain molecular solids (as well as for such physical effects as surface tension and friction). Examples of solids that are bound by the van der Waals force include those composed of the inert gases (Ne, Ar, Kr, and Xe), symmetric molecules such as CH_4 and $GeCl_4$, halogens, and other gases such as H_2, N_2, and O_2.

---

* These fluctuations are too rapid to be observed in the laboratory. Measurements give only the *average* value of this fluctuating dipole moment, which is zero.

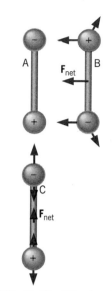

FIGURE 11.10 Dipole A exerts attractive forces on dipoles B and C. Each charge is both attracted and repelled, and the net effect is attraction.

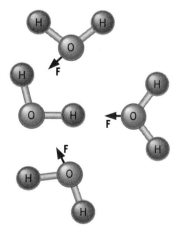

FIGURE 11.11 Dipole forces between water molecules.

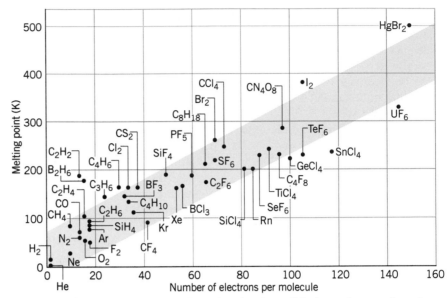

FIGURE 11.12 The melting points of molecular solids depend approximately on the number of electrons per molecule.

The van der Waals force is extremely weak; it falls off with separation distance like R^{-7}. In inert gas crystals, the nearest neighbor distance is 0.3–0.4 nm, but the cohesive energies are typically only 10 kJ/mol or 0.1 eV/atom. Solids bound by these weak forces have low melting points, because little thermal energy is required to break the bonds. In fact, since the induced dipole moment of an atom should be approximately proportional to the *total number* of electrons in the atom, we might expect that the melting points of nonpolar molecular solids should be roughly proportional to the number of electrons in each molecule. Figure 11.12 shows this relationship; although the properties of the individual solids cause considerable scatter of the points, there is rough relationship of the sort expected.

11.4 BAND THEORY OF SOLIDS

When two identical atoms, such as sodium, are very far apart, the electronic levels in one are not affected by the presence of the other, and we may think of the atoms as being isolated. The 3s electron of each atom has a single energy with respect to its nucleus. As we bring the atoms closer together, the electron wave functions start to overlap, and the interaction between the atoms causes two different 3s levels to form, depending on whether the two wave functions add or subtract. This is the effect that is responsible for molecular binding, which we discussed in Section 9.2. A representation of the energy levels is shown in Figure 11.13.

As we bring together a large number of atoms to form a solid, the same sort of effect occurs. When the sodium atoms are far apart, all 3s electrons have the same energy, and as we begin to move them together, the energy levels begin to "split." The situation for five atoms is shown in Figure 11.14. There are now five energy levels that result from the five overlapping electron wave functions. As the number of atoms is increased to the very large numbers that characterize an ordinary piece of metal (perhaps 10^{22} atoms), the levels become so numerous and so close together that we can no longer distinguish the individual levels, as shown in Figure 11.15. We can regard the N atoms as forming an almost continuous *band* of energy levels. Since those levels were identified with the 3s atomic levels of sodium, we refer to the 3s *band*.

Each energy band has a total of N individual levels. Each level can hold $2 \times (2l + 1)$ electrons (corresponding to the two different orientations of the electron spin and the $2l + 1$ orientations of the electron orbital angular momentum) so that the capacity of each band is $2(2l + 1)N$ electrons.

Figure 11.16 shows a more complete representation of the energy bands in sodium metal. The 1s, 2s, and 2p bands are each full; the 1s and 2s bands each contain $2N$ electrons and the 2p band contains $6N$ electrons. The 3s band *could* accommodate $2N$ electrons as well; however, each of the N atoms contributes only one 3s electron to the solid, and so there is a total of only N 3s electrons available. The 3s band is therefore half full. Above the 3s band is a 3p band, which could hold $6N$ electrons, but which is completely empty.

The situation we have described is the ground state of sodium metal. When we add energy to the system (thermal or electrical energy, for example), the electrons can move from the filled states to any of the empty states. In this case, electrons from the partially full 3s band can absorb a small amount of energy and move to empty 3s states within the 3s band, or they can absorb a larger amount of energy and move to the 3p band.

We can describe this situation in a more correct way if we recall our discussions of quantum statistics from Chapter 10. Electrons are described by the Fermi-Dirac distribution. At a temperature of $T = 0$ K, all electron

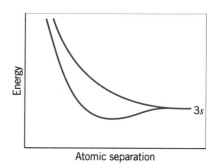

FIGURE 11.13 Splitting of 3s levels when two atoms are brought together.

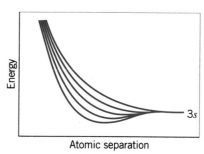

FIGURE 11.14 Splitting of 3s levels when five atoms are brought together.

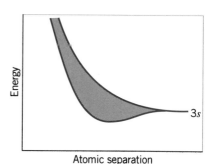

FIGURE 11.15 Formation of 3s band by a large number of atoms.

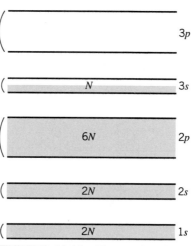

FIGURE 11.16 Energy bands in sodium metal.

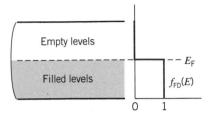

FIGURE 11.17   Detail of a half-filled band, at $T = 0$, showing the Fermi-Dirac distribution on the right.

Energy gap, valence and conduction bands

levels below the Fermi energy E_F are filled and all levels above the Fermi energy are empty, as shown in Figure 11.17. In the case of sodium, the Fermi energy would be in the middle of the $3s$ band, since all electron levels below that energy are occupied. At higher temperatures the Fermi energy gives the level at which the occupation probability is 0.5; the Fermi energy does not change significantly as we increase the temperature, but the occupation probability of the levels above E_F is no longer zero. Figure 11.18 shows a situation in which the thermal excitation of electrons leads to a small population of the $3p$ band.

Sodium is an example of a substance that is a good electrical conductor. When we apply a very modest potential difference, of the order of 1 V, electrons can easily absorb energy because there are N unoccupied states within the $3s$ band, all within an energy of 1 eV. Electrons absorb energy as they are accelerated by the applied voltage, and they are therefore free to move as long as there are many unoccupied states within the accessible energy range. In sodium there are N relatively free electrons that can easily move to N unoccupied energy states, and sodium is therefore a good conductor.

A material in which one band is completely full and the next higher completely empty, on the other hand, is a poor conductor. We refer in this case to the *gap* between the energy bands, and the Fermi energy lies somewhere in the gap as shown in Figure 11.19. Once again, at $T = 0$, all states below E_F are filled. As the temperature is raised, the shape of the Fermi-Dirac distribution function changes; if the energy gap E_g between the bands is large compared with kT, there will be few electrons thermally excited from the lower band, called the *valence band*, to the upper band, called the *conduction band*. This situation is shown in Figure 11.20. There are many electrons in the valence band available for electrical conduction, but there are very few empty states for them to move through, so they do not contribute to the electrical conductivity. There are many empty states in the conduction band, but at ordinary temperatures there are so few electrons in that band that their contribution to the electrical conductivity is also very small. These substances are classified as *insulators* and in general they have two properties: a large energy gap (a few electron-volts) between the valence and conduction bands, and a Fermi level that is in the gap

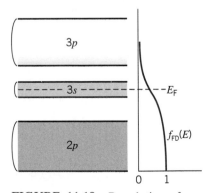

FIGURE 11.18   Population of energy bands in sodium at $T > 0$. The $2p$ band is no longer completely full and the $3p$ band is no longer completely empty.

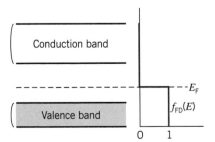

FIGURE 11.19   Band structure in which E_F lies in the gap between bands.

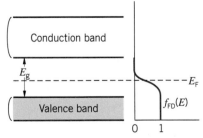

FIGURE 11.20   When $E_g \gg kT$, the conduction band is still unpopulated. This situation is characteristic of an insulator.

between the bands (i.e., a filled valence band and an empty conduction band).

When a material has the basic structure of an insulator, but with a much smaller energy gap (1 eV or less) the behavior becomes quite different, and these materials are known as *semiconductors*. Figure 11.21 shows a representation of such a substance at ordinary temperatures. There are now many electrons in the conduction band, and of course many empty states accessible to them, so that they can conduct relatively easily. There are also many empty states in the valence band, so that some of the electrons in the valence band can also contribute to the electrical conductivity by moving about through those states. We consider these two mechanisms of electrical conduction in detail in Section 11.8. For now we note two characteristic properties of semiconductors that relate directly to the band structure as shown in Figure 11.21. (1) Because thermal excitation across the gap is relatively probable, the electrical conductivity of semiconductors depends more strongly on temperature than the electrical conductivity of insulators or conductors. (2) It is possible to alter the structure of these materials, by adding impurities in very low concentration, in such a way that the Fermi energy changes and may move up toward the conduction band or down toward the valence band. This process, known as *doping*, can have a great effect on the conductivity of the semiconductor.

In the examples we have discussed so far, it is not apparent why the band theory is so useful in understanding the properties of a solid. Sodium, for example, is expected to be a good conductor based on its atomic properties alone (a relatively loosely bound $3s$ electron); on the other hand, solid xenon has only filled atomic shells and should be a poor conductor. These conclusions follow either from simple atomic theory or from band theory. However, there are many cases in which atomic theory leads to wrong predictions while band theory gives correct results. We consider two examples. (1) Magnesium has a filled $3s$ shell, and on the basis of atomic theory alone we expect it to be a poor electrical conductor. It is, however, a very good electrical conductor. (2) The $2p$ shell of carbon has only two electrons of the maximum number of six. Carbon should therefore be a relatively good conductor; instead it is an extremely poor conductor. We can understand both of these materials based on the unusual way the bands of these solids behave when the atoms are close enough so that the band gap disappears and the bands overlap. In magnesium (Figure 11.22), for example, the (filled) $3s$ and (empty) $3p$ bands overlap, and the result is a single band with a capacity of $2N + 6N = 8N$ levels. Only $2N$ of those are filled, and so magnesium behaves like a material with a single band filled only to one-fourth its capacity. Magnesium is therefore a very good conductor. In carbon, the extreme overlap of the electronic wave functions at close range first causes mixing of the $2s$ and $2p$ bands, in a way similar to magnesium; a single band is created with a capacity of $8N$ electrons (Figure 11.23). As the atoms approach still closer, the band divides into two separate bands, each with a capacity of $4N$ electrons. Since carbon has four valence electrons (two $2s$ and two $2p$), the lower $4N$ states are completely filled and the upper $4N$ states of the conduction band are completely empty. Carbon is therefore an insulator. Germanium and silicon have the same

Semiconductors

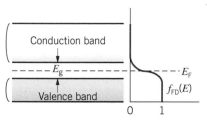

FIGURE 11.21 Band structure of a semiconductor. The gap is much smaller than in an insulator, so there is now a small population of the conduction band.

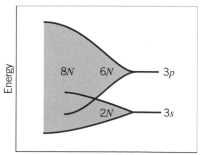

FIGURE 11.22 Band structure in magnesium. The $3s$ and $3p$ bands overlap, forming a single band.

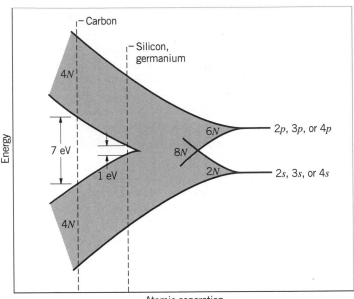

FIGURE 11.23 Band structure of carbon, silicon, and germanium. The combined *ns* + *np* band splits into two bands. Carbon is an insulator because the gap is large; Ge and Si are semiconductors because their energy gaps are smaller.

type of structure as carbon, but their equilibrium separation is greater, and the gap between the valence and conduction bands is smaller, about 1 eV; it is this feature that causes Ge and Si to be semiconductors.

Summary of band theory   In this section we have discussed a conceptual approach to the band theory of solids. Let us summarize the main features of this theory:

1. When many atoms are brought together to form a solid, the interactions between the atoms cause the discrete energy levels of the isolated atoms to "smear out" into bands.

2. The properties of the bands are determined by the atomic properties of the isolated atoms and by the equilibrium separations of the atoms in the solid.

3. The properties of the solid are determined by the occupation of the bands, by the spacing between the bands, and by the relative location of the Fermi energy.

*11.5 JUSTIFICATION OF BAND THEORY

The band theory of solids has had great success in accounting for the properties of metals, insulators, and semiconductors. In the last section we

---

* This is an optional section that may be skipped without loss of continuity.

suggested that energy bands form when atoms are brought close together so that their electron wave functions overlap. In this section we consider a different approach to band theory that is based on the quantum mechanics of an electron moving through a lattice of ions. In analogy with solutions to the Schrödinger equation discussed in Chapter 5, in which an electron in a potential energy well shows discrete energy levels, we will see that an electron in a periodic potential energy provided by a lattice of ions can show energy bands.

To simplify the problem we consider only a one-dimensional lattice of ions (Figure 11.24). The electron is represented by a de Broglie wave traveling through the lattice. The interaction between the electron and the lattice can be represented as a scattering problem, similar to Bragg scattering (Section 3.1). The Bragg condition for scattering is

$$2d \sin \theta = n\lambda \qquad (n = 1, 2, 3, \cdots) \tag{11.12}$$

where d is the atomic spacing and θ is the angle of incidence measured from the plane of atoms (*not* from the normal). In a two-dimensional lattice, the incident wave can be scattered in many different directions, depending on the plane where we imagine the reflection to occur (recall Figure 3.6); in one dimension, however, only one possible reflection can occur—the incident wave can be reflected back in the opposite direction. We can use the Bragg condition for this case, with $d = a$ (the spacing between the ions or atoms of the lattice) and $\theta = 90°$ (the angle between the "reflecting plane" and the incident wave), and so

$$2a \sin 90° = n\lambda \tag{11.13}$$

and using $k = 2\pi/\lambda$, we find

$$k = n\frac{\pi}{a} \tag{11.14}$$

For wave numbers that do not satisfy this condition, the electron propagates freely through the lattice and behaves like a free particle whose energy is only kinetic:

$$E = \frac{p^2}{2m} = \frac{\hbar^2 k^2}{2m} \tag{11.15}$$

Since there are no restrictions on k, all values of E are allowed. The relationship between E and k is a parabola, as shown in Figure 11.25.

For wave numbers that satisfy the Bragg condition, the reflected and incident waves add to produce standing waves, which always result when we superpose two waves of equal wavelengths traveling in opposite directions. Depending on the phase difference between the waves, their amplitudes can add or subtract, so two different possible standing waves can result. Their probability densities are shown in Figure 11.26. For one of the waves (ψ_1), the electrons are more likely to be found close to the positive ions; these electrons are more tightly bound to the lattice—the energy of the electron is a bit lower than that of the free electron, due to the negative potential energy between the electron and the ions. Electrons represented by the other wave (ψ_2) are most likely to be found in the region between

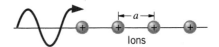

FIGURE 11.24   One-dimensional Bragg scattering. The only possible scattering is a reflection back along the original direction.

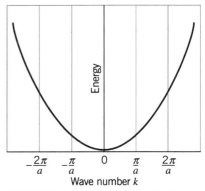

FIGURE 11.25   The parabolic relationship between energy and wave number for a free particle.

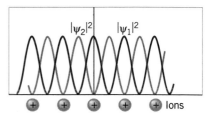

FIGURE 11.26   Probability densities of two different standing waves.

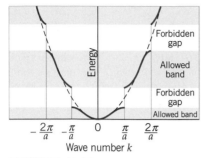

FIGURE 11.27 The relationship between energy and wave number for a one-dimensional lattice. The dashed curve is the free-particle parabola. The solid curves represent waves scattered by the lattice.

the ions; they are less tightly bound, so their energies are a bit above those of the unscattered electrons (for which the probability density is flat, so they are equally likely to be found at any location).

The resulting dependence of the energy of the electrons on the wave number k is illustrated by the S-shaped curve segments in Figure 11.27. For wave numbers that are far from satisfying the Bragg condition (that is, values of k that are not close to $n\pi/a$), the curve segments overlap the dashed parabola representing the free particle. Close to the wave numbers that satisfy the Bragg condition, however, the energy deviates from that of the free particle, a bit below the parabola for the more tightly bound electrons that spend more time near the ions and a bit above the parabola for the less tightly bound electrons that are more likely found between the ions.

Notice from Figure 11.27 that, even though all values of k are permitted, there are certain allowed bands of energy values separated by forbidden gaps. An electron traveling in this lattice is permitted to have energies only in the regions corresponding to the allowed bands. This indicates how a periodic array of atoms results in energy bands.

Kronig-Penney Model

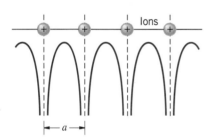

FIGURE 11.28 Potential energy due to ions in a one-dimensional lattice.

For an even more rigorous approach, we can solve the Schrödinger equation in one dimension for a periodic lattice. Figure 11.28 shows a representation of the potential energy for an electron in a one-dimensional periodic lattice of positive ions. The problem can be greatly simplified if we replace the Coulomb potential energy with a square well, as shown in Figure 11.29. Each well has a width b and a depth U_0. By taking the limit as $b \to 0$ and $U_0 \to -\infty$ such that the product bU_0 remains constant, we get a fair approximation to the Coulomb potential. This representation of the lattice is called the *Kronig-Penney model*.

The solution to the Schrödinger equation for this problem, which we shall not discuss in detail, can be represented as the product of two functions: a free-particle wave function such as $\cos kx$ and a periodic function $u(x)$ that has the same period as the lattice:

$$\psi(x) \sim u(x) \cos kx \qquad (11.16)$$

where $u(x) = u(x + a)$. Halfway between the ions, $u(x) \approx 1$ and the electron behaves like a free particle. It is only in the region near the ions where $u(x)$ differs significantly from 1.

From this point the solution proceeds in analogy with the method for solving the Schrödinger equation discussed in Chapter 5. We substitute the wave function of Equation 11.16 into the Schrödinger equation, and when we apply the periodic boundary condition we obtain an expression that determines the allowed values of the energy of the electron. In this case the resulting equation is

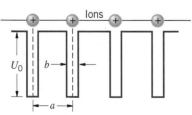

FIGURE 11.29 A simplified version of the potential energy of Figure 11.28.

$$\cos ka = \frac{maU_0 b}{\hbar^2} \frac{\sin \alpha a}{\alpha a} + \cos \alpha a \qquad (11.17)$$

where

$$\alpha = \sqrt{\frac{2mE}{\hbar^2}} \qquad (11.18)$$

Let us see how this expression gives energy bands rather than discrete energy values.

Let the right side of Equation 11.17 be represented as a function $f(\alpha a)$. The solid curve of Figure 11.30 shows $f(\alpha a)$ as a function of αa. The graph is drawn with f on the horizontal axis and αa on the vertical axis. Note from Figure 11.30 that for certain values of αa, $f(\alpha a)$ becomes greater than $+1$ or less than -1. However, the left side of Equation 11.17 is a cosine function and must *always* be between -1 and $+1$. Any value of αa that allows the right side of Equation 11.17 to fall outside the range from -1 to $+1$ must therefore be forbidden. Because α depends on E, these forbidden values of α result in the forbidden energy gaps, as shown in Figure 11.30.

We have discussed two different one-dimensional calculations that lead to allowed energy bands separated by forbidden gaps: one based on the scattering of electron waves in a periodic lattice and another based on solving the Schrödinger equation for a periodic potential energy that approximates the Coulomb potential. While neither of these simple calculations gives an accurate representation of the behavior of electrons moving in a real three-dimensional solid, they do serve to give us an indication of the origin of allowed bands and forbidden gaps, and they remind us once again that the wave behavior of electrons has important observable consequences, in this case the electrical conduction in solids.

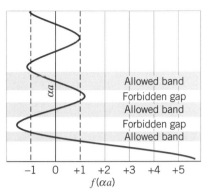

FIGURE 11.30 Allowed energy bands from one-dimensional model of lattice. The solid curve represents the right side of Equation 11.17. The allowed bands correspond to those regions where the curve lies between $+1$ and -1.

11.6 ELECTRONS IN METALS

Each atom of the element copper has a loosely bound valence electron. When copper atoms are brought together to form solid copper, these electrons become part of a "sea" or "gas" of nearly free electrons that can easily move throughout the metal. Other metals behave similarly. We can therefore, in many instances, regard certain properties of a metal as determined by the properties of the electron gas. In the ideal gas model of an ordinary gas, molecules move freely and experience forces only when they scatter from other molecules. In the electron gas model, the electrons move freely and experience forces only when they scatter from the ion cores. (This is the basis of the Kronig-Penney model discussed in the previous section.) The motion of the electrons is constrained by two factors: The distribution of energies is determined by the Fermi-Dirac function (in contrast to an ordinary gas, which obeys Maxwell-Boltzmann statistics), and some of the energy values are forbidden by the band theory. Within the allowed bands, the motion of the electrons is restricted only by the Fermi-Dirac distribution. With these assumptions, we can study many of the properties of metals, such as electrical conduction, heat capacity, and heat conduction.

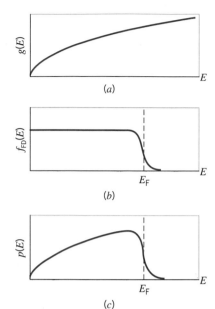

FIGURE 11.31 (*a*) The density of states factor, Equation 10.35. (*b*) The Fermi-Dirac distribution function, Equation 10.22. (*c*) The electron energy distribution, determined from the product of (*a*) and (*b*).

Figure 11.31 summarizes the main details of the Fermi-Dirac energy distribution as it might be applied to electrons in solids. The distribution of electrons $p(E)$ is determined by the product of the density of states factor, Equation 10.35, and the Fermi-Dirac factor, Equation 10.22. At $T = 0$, all states above the Fermi energy E_F are empty and all states below E_F are occupied. For temperatures greater than 0, E_F identifies the point at which the Fermi-Dirac factor has the value $\frac{1}{2}$. The difference between $p(E)$ at $T = 0$ and at room temperature was illustrated in Figure 10.23; only a small number of electrons near E_F are affected by the temperature increase.

We calculated the Fermi energy at $T = 0$ by using Equation 10.38, in which the integral of $p(E)$ over all energies gives the total number of electrons N. The same procedure can be used to find E_F at any temperature:

$$N = \int_0^\infty p(E)\, dE = \frac{8\sqrt{2}\,\pi V m^{3/2}}{h^3} \int_0^\infty \frac{E^{1/2}\, dE}{e^{(E-E_F)/kT} + 1} \qquad (11.19)$$

In principle, we can evaluate the integral and solve for E_F, as we did to obtain Equation 10.39. However, the integral cannot be evaluated in closed form. The solution can be approximated as

$$E_F(T) \approx E_F(0)\left[1 - \frac{\pi^2}{12}\left(\frac{kT}{E_F(0)}\right)^2 \right] \qquad (11.20)$$

Here $E_F(T)$ represents the Fermi energy at temperature T and $E_F(0)$ represents the Fermi energy at $T = 0$. At room temperature, $kT = 0.025$ eV, and for most metals the Fermi energy is a few eV, so the change in the Fermi energy between 0 K and room temperature is only about 1 part in 10^4. We can therefore regard the Fermi energy as a constant for our applications, and we will represent it simply as E_F. Table 11.4 shows the Fermi energies of some metals.

Electrical Conduction

When an electric field **E** is applied to a metal, a current flows in the direction of the field. The flow of charges is described in terms of a *current density* **j**, the current per unit cross-sectional area. In an ordinary metal, the current density is proportional to the applied electric field:

$$\mathbf{j} = \sigma \mathbf{E} \qquad (11.21)$$

where the proportionality constant σ is the *electrical conductivity* of the material. We would like to understand the conductivity in terms of the properties of the metal.

The free electrons in our electron gas experience a force $\mathbf{F} = -e\mathbf{E}$ and a corresponding acceleration $-e\mathbf{E}/m$. We observe that in a conductor the current is constant in time, so the increase in velocity from the electric field must be opposed, in this case by collisions with the lattice. This model of conduction in metals views the electrons as accelerated by the field only for short intervals, following which they are slowed by collisions. The net result is that the electrons acquire on the average a steady *drift velocity*

TABLE 11.4 FERMI ENERGIES OF SOME METALS

Metal	E_F (eV)
Ag	5.50
Au	5.53
Ca	4.72
Cs	1.52
Cu	7.03
Li	4.70
Mg	7.11
Na	3.15

v_d, given by the acceleration times the average time τ between collisions:

$$\mathbf{v}_d = \frac{-e\mathbf{E}}{m}\,\tau \qquad (11.22)$$

The magnitude of the current density is determined by the number of charge carriers and their average speed:

$$\mathbf{j} = -ne\mathbf{v}_d \qquad (11.23)$$

where n is the density of electrons available for conduction. Substituting for the drift velocity, we obtain

$$\mathbf{j} = \frac{ne^2\tau}{m}\,\mathbf{E} \qquad (11.24)$$

and the conductivity is therefore

$$\sigma = \frac{ne^2\tau}{m} \qquad (11.25)$$

If we knew the average time τ between collisions, we could calculate the conductivity of the metal. Figure 11.32a represents the Fermi distribution of electron velocities in the metal. Like the Fermi energy distribution, it is flat at $v = 0$ and it falls to zero near the Fermi velocity v_F, but (unlike the energy distribution) it has positive and negative branches since the electrons can in general move in either direction. When an electric field is applied, all electrons acquire on the average an additional velocity component equal to the drift velocity v_d, which shifts the entire velocity distribution to the left (opposite the field direction), as shown in Figure 11.32b. Even though the entire distribution shifts, the net effect of applying the electric field is centered on a small number of electrons in the vicinity of v_F. The electric field causes some electron states near v_F in the direction of $\mathbf{E}$ to become unoccupied, while an equal number of states near v_F in a direction opposite to $\mathbf{E}$ become occupied. To understand the conduction process, it is sufficient to focus our attention on the electrons that are moving with speeds close to v_F.

The unknown factor in Equation 11.25 is the time between collisions, which we can express as

$$\tau = \frac{l}{v_d} \qquad (11.26)$$

where l is the *mean free path* of the electrons, the average distance the electrons travel between collisions.

EXAMPLE 11.3

The measured conductivity of copper at room temperature is 5.88×10^7 $\Omega^{-1}\text{m}^{-1}$, and the Fermi energy of copper is 7.03 eV. (a) Find the average time between collisions of the conduction electrons. (b) Find the mean free path of electrons in copper.

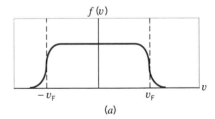

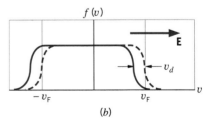

FIGURE 11.32 (a) The Fermi-Dirac velocity distribution function. (b) When an electric field is applied, the distribution shifts as the electrons are accelerated in a direction opposite the field.

SOLUTION

(a) The density of charge carriers in copper can be determined from its density ρ and its molar mass M:

$$n = \frac{\rho N_A}{M} = \frac{(8.96 \text{ g/cm}^3)(6.02 \times 10^{23} \text{ atoms/mole})}{63.5 \text{ g/mole}}$$

$$= 8.49 \times 10^{28} \text{ atoms/m}^3$$

Solving Equation 11.25 for τ, we obtain

$$\tau = \frac{\sigma m}{ne^2} = \frac{(5.88 \times 10^7 \, \Omega^{-1}\text{m}^{-1})(9.11 \times 10^{-31} \text{ kg})}{(8.49 \times 10^{28} \text{ atoms/m}^3)(1.60 \times 10^{-19} \text{ C})^2} = 2.46 \times 10^{-14} \text{ s}$$

(b) The Fermi velocity is

$$v_F = \sqrt{\frac{2E_F}{m}} = \sqrt{\frac{2(7.03 \text{ eV})(1.60 \times 10^{-19} \text{ J/eV})}{9.11 \times 10^{-31} \text{ kg}}} = 1.57 \times 10^6 \text{ m/s}$$

and so the mean free path is

$$l = v_F \tau = (1.57 \times 10^6 \text{ m/s})(2.46 \times 10^{-14} \text{ s}) = 3.87 \times 10^{-8} \text{ m} = 38.7 \text{ nm}$$

It may seem somewhat surprising that the mean free path is so large, compared with the spacing between copper atoms in the lattice (0.26 nm). On the average, an electron passes about 150 copper atoms before making a collision. However, recall that the Fermi energy in a conductor lies near the middle of the conduction band (see Figure 11.17), and that electrons near the middle of an allowed band are not scattered at all by the lattice (Figure 11.27). In a perfectly periodic lattice, the conductivity would be infinite—the electrons would not scatter at all!

In a real metallic lattice, there are two contributions to the scattering of electrons, and therefore to the conductivity: the atoms are in thermal motion and therefore do not occupy exactly the positions of a perfectly arranged lattice, and lattice imperfections and impurities cause deviations from the ideal lattice. The first effect is temperature dependent and dominates at high temperatures; the second effect is independent of temperature and dominates at low temperatures. Figure 11.33 represents the resistivity (the inverse of conductivity) of sodium metal as a function of the temperature.

Heat Capacity

When a small quantity of heat dQ is transferred to a gas, its internal energy is raised by an equivalent amount dE (assuming no external work is done), and the temperature correspondingly increases by dT. The heat capacity is then $C = dE/dT$. To determine the heat capacity for the electron gas, we must know how the energy of the electrons varies with temperature. The mean or average electron energy was found at $T = 0$ using Equation 10.40. A similar calculation can be used to find the temperature dependence

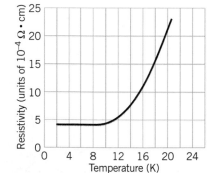

FIGURE 11.33 Dependence of electrical resistivity of sodium metal on temperature.

of E_m:

$$E_m = \frac{1}{N} \int_0^\infty E \, p(E) \, dE \tag{11.27}$$

Following the same procedure as Equation 11.19, we can also find an approximate result for the temperature dependence of E_m:

$$E_m(T) = E_m(0) \left[1 + \frac{5\pi^2}{12} \left(\frac{kT}{E_F} \right)^2 \right] \tag{11.28}$$

Here $E_m(0) = \frac{3}{5} E_F$, as in Equation 10.41. Once again, the ratio kT/E_F is about 10^{-2}, so the change in the electron energy with increasing temperature is small. The electronic contribution to the heat capacity then follows directly by differentiating Equation 11.28:

$$C = \frac{dE}{dT} = N_A \frac{dE_m}{dT} = \frac{\pi^2}{2} R \frac{kT}{E_F} \tag{11.29}$$

Here we have used the total energy E of a mole of electrons ($E = N_A E_m$) to find the molar heat capacity. This calculation agrees with the crude estimate of Section 10.7: The electronic contribution to the molar heat capacity is of the order of $0.01R$, which is negligible compared with the room-temperature lattice contribution of $3R$.

Thermal Conductivity

The expression for the thermal conductivity of a gas, obtained from classical kinetic theory, is

$$K = \frac{1}{3} C v l \tag{11.30}$$

where C is the heat capacity per unit volume, v is the molecular speed, and l is the mean free path. Converting Equation 11.29 from molar heat capacity to heat capacity per unit volume (by substituting nk for $R = N_A k$, where n is the density of electrons), we can write the thermal conductivity for the electron gas as

$$K = \frac{1}{3} \frac{\pi^2}{2} nk \frac{kT}{E_F} v l \tag{11.31}$$

We expect that, in analogy with the electrical conductivity, only the electrons near the Fermi energy contribute to the thermal conductivity, so we can take $v = v_F$. Writing $E_F = \frac{1}{2} m v_F^2$ and using Equation 11.26 for the time τ between collisions, we obtain

$$K = \frac{\pi^2}{3} \frac{nk^2 \tau T}{m} \tag{11.32}$$

Note that the quantity $n\tau/m$ appears in both this expression for the thermal conductivity and Equation 11.25 for the electrical conductivity. This should not be surprising, because both the transport of heat and the transport of

TABLE 11.5 LORENZ NUMBER OF METALS [UNITS OF 10^{-8} W·Ω/K²]

Metal	0°C	100°C
Ag	2.31	2.37
Au	2.35	2.40
Cd	2.42	2.43
Cu	2.23	2.33
Pb	2.47	2.56
Sn	2.52	2.49
Zn	2.31	2.33

electric charge through the metal depend on the actions of the electron gas. If we take the ratio of these two quantities, all details of the properties of the material cancel:

$$\frac{K}{\sigma} = \frac{\pi^2 k^2}{3e^2} T \tag{11.33}$$

This remarkable result says that the ratio of the thermal and electrical conductivities depends only on the temperature, and should be the same for all metals at a given temperature, since all of the properties relating to a specific metal (l and v) cancel from the ratio. This result is known as the *Wiedemann-Franz* law; the constant $\pi^2 k^2/3e^2$ is known as the Lorenz number and has the theoretical value 2.44×10^{-8} W·Ω/K². Table 11.5 gives some experimental values of the Lorenz number, determined from the ratio K/σ, for various metals at different temperatures, and it is indeed quite nearly constant, both for different metals and for different temperatures.

This excellent agreement once again lends confidence to the use of the electron gas model coupled with the Fermi-Dirac distribution to explain properties of metals.

11.7 SUPERCONDUCTIVITY

At low temperatures, the resistivity of a metal (the inverse of its conductivity) is nearly constant. As the temperature of a material is lowered, the lattice contribution to the resistivity decreases while the impurity contribution remains approximately constant, and as we approach $T = 0$ K the resistivity should approach a constant value. Many metals, known as *normal* metals, behave in this way, as illustrated in Figure 11.33.

The behavior of another class of metals is quite different. These metals behave normally as the temperature is decreased, but at some critical temperature T_c (which depends on the properties of the metal), the resistivity drops suddenly to zero, as shown in Figure 11.34. These materials are known as *superconductors*. The resistivity of a superconductor is not merely very small at temperatures below T_c; it vanishes! Such materials can conduct electric currents even in the absence of an applied voltage, and the conduction occurs with no i^2R (joule heating) losses.

Superconductivity has been observed in 28 elements at ordinary pressures, in several additional elements at high pressure, and in hundreds of compounds and alloys. Since the original discovery of superconductivity in Hg in 1911, the focus of research has been to search for materials with the highest possible critical temperature, because many possible large-scale applications of superconductivity are presently impractical owing to the high cost of keeping materials below their critical temperatures. Table 11.6 summarizes some superconducting materials and their critical temperatures. You can see that before 1986, progress in raising the critical temperature was very slow, but since 1986 dramatic and rapid increases in T_c have been achieved.

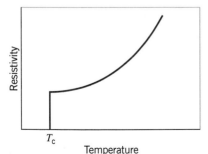

FIGURE 11.34 Resistivity of a superconductor.

TABLE 11.6 SOME SUPERCONDUCTING MATERIALS

Material	T_c (K)	Gap (meV)	Year
Zn	0.85	0.24	1933
Al	1.18	0.34	1937
Sn	3.72	1.15	1913
Hg	4.15	1.65	1911
Pb	7.19	2.73	1913
Nb	9.25	3.05	1930
Nb_3Sn	18.1		1954
Nb_3Ge	23.2		1973
$La_xBa_{2-x}CuO_4$	30		1986
$La_xSr_{2-x}CuO_5$	40		1986
$YBa_2Cu_3O_7$	95		1987
$Tl_2Ba_2Ca_2Cu_3O_{10}$	125		1988

Conspicuously absent from the list of superconductors are the best metallic conductors (Cu, Ag, Au), which suggests that superconductivity is *not* caused by a good conductor getting better but instead must involve some fundamental change in the material. In fact, superconductivity results from a kind of paradox: ordinary materials can be good conductors if the electrons have a relatively weak interaction with the lattice, but superconductivity results from a *strong* interaction between the electrons and the lattice.

Consider an electron moving through the lattice. As it moves, it attracts the positive ions and disturbs the lattice, much as a boat moving through water creates a wake. These disturbances propagate as lattice vibrations, which can then interact with another electron. In effect, two electrons interact with one another through the intermediary of the lattice; the electrons move in correlated pairs that do not lose energy by interacting with the lattice. (These pairs are not necessarily traveling together through the lattice; they may be separated by a large distance.) In the absence of a net current, the members of a pair have opposite momenta; when a net current is established, both members of the pair acquire a slight increase in momentum in the same direction, and this motion is responsible for the current.

According to the successful BCS theory of superconductivity,* below the critical temperature there is a small energy gap E_g in the occupation probability of electrons in a superconductor (Figure 11.35). Below the gap, the electrons form pairs, which are known as *Cooper pairs*. Once a single Cooper pair forms, it is energetically favorable for other pairs to form, so the change from the normal state above T_c to the superconducting state below T_c is quite sudden. (The paired electrons can be regarded as zero-spin bosons, which can "condense" into a common ground state.)

When a superconductor is cooled below T_c, the gap opens and Cooper pairs begin to form. As the material is cooled further, the gap widens.

---

* The theory of superconductivity was developed in 1957 by John Bardeen, Leon N. Cooper, and J. Robert Schrieffer, who were awarded the 1972 Nobel prize in physics for their work. Bardeen also shared the 1956 Nobel prize for his research on semiconductors and his development of the transistor.

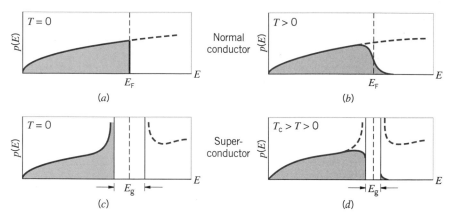

FIGURE 11.35   The occupation probability for electrons in (*a*) a normal conductor at $T = 0$; (*b*) a normal conductor at $T > 0$; (*c*) a superconductor at $T = 0$; and (*d*) a superconductor at $0 < T < T_c$. In (*c*), there is an energy gap of width E_g, and states displaced from within the gap pile up on either side of the gap. At higher temperatures, as in (*d*), the gap is narrower, and there are empty states below the gap and filled states above the gap. As the temperature is raised to T_c, the gap width becomes 0 and the occupation probability of the superconductor in (*d*) approaches that of the normal conductor in (*b*). The gap width is exaggerated in this figure; generally, $E_g \sim 10^{-3}$ eV.

Values of the energy gap listed in Table 11.5 correspond to the limiting case as $T \to 0$. The energy gaps, which can be regarded as representing the binding energy of a Cooper pair, are very small, of the order of 10^{-3} eV. At $T = 0$, all states below the gap are occupied, and all electrons are paired. When $0 < T < T_c$, there are some unoccupied states below the gap, and some states above the gap are occupied by normal (unpaired) electrons.

Beginning in 1986, a new class of superconductors was discovered with unusually high values of T_c. In the 75 years from the discovery of superconductivity in 1911 until 1986, the highest T_c had gone from 4 K to about 23 K. In 1986, several materials were discovered with T_c in the range $30 - 40$ K. By 1987 it was up to 93 K, and it rose to 125 K in 1988. Crossing the boundary at 77 K is important, because it means that cooling can be accomplished with liquid nitrogen instead of liquid helium, which costs nearly an order of magnitude more than liquid nitrogen. The rapid increase in T_c has led to the hope that it might be possible to develop materials that are superconductors at room temperature. Such materials could enable the transmission of electric power over long distances without resistive losses.

The high-T_c superconductors are oxides of copper in combination with other elements. They are ceramics, which means that they are rather brittle and not easily formed into wires to carry current. The crystal structure is characterized by planes of copper and oxygen between planes of the other elements. It seems likely that the superconductivity occurs in the copper oxide planes, but it is not yet clear that the complete explanation for these new superconductors is given by the BCS theory.

Superconducting materials have many applications that take advantage of their abilities to carry electrical currents without resistive losses. Electromagnets can be constructed that carry large currents and therefore produce large magnetic fields (of order 5 to 10 T). If we use superconducting wires, currents as large as 100 A can be carried by very fine wires, of order 0.1 mm diameter, and thus such magnets can be constructed in a smaller space, using less material, than would be possible with ordinary conductors. Figure 11.36 illustrates such a piece of apparatus. Starting from zero, the current is increased until the magnetic field **B** reaches some desired value. At that point the switch S is closed and the return current, following the path of least resistance, flows through the switch rather than through the power supply. The power supply can then be turned off, and in principle the current will continue to circulate through the coil and switch forever; in practice, observations over the course of years have shown no decrease of the current.

11.8 INTRINSIC AND IMPURITY SEMICONDUCTORS

A semiconductor is a material with an energy gap E_g of order 1 eV between the valence band and the conduction band. At $T = 0$, all states in the valence band are full and all states in the conduction band are empty; recall that the Fermi-Dirac distribution is a step function at $T = 0$ and gives an occupation probability of exactly 1 for all states below E_F and exactly 0 for all states above E_F. As the temperature is raised, however, some states above E_F are occupied and some states below E_F are empty. At room temperature, the relationship between the Fermi energy, the valence and conduction bands, and the electron energy distribution might be as we pictured in Figure 11.21.

Although the value of the room-temperature Fermi-Dirac distribution function is nearly zero in the conduction band (in fact, it is so close to zero that it cannot be seen on the scale of Figure 11.21) it is not *exactly* zero; Figure 11.37 shows a greatly magnified view of $f_{FD}(E)$ near the bottom of the conduction band. The value of $E - E_F$ is about 0.5 eV if E_F lies near the middle of the 1 eV energy gap, and therefore $E - E_F \gg kT$, since at room temperature $kT \sim 0.025$ eV. The "1" in the denominator of the Fermi-Dirac distribution is therefore negligible, and $f_{FD}(E)$ is approximately exponential, as shown in Figure 11.37. Assuming the Fermi energy to lie near the middle of the gap, the occupation probability near the bottom of the conduction band is of order $e^{-E_g/2kT} \cong 10^{-9}$. Thus one atom in 10^9 contributes an electron to the electrical conductivity; compare this with a metal in which essentially *every* atom contributes an electron to the conductivity. (On the other hand, consider an insulator, which has a band structure very similar to that of a semiconductor, except the energy gap is perhaps 5 eV instead of 1 eV. This small difference in the size of the energy gap has an enormous effect on the occupation probability of the conduction band at room temperature: $e^{-E_g/2kT} \cong 10^{-44}$. Thus in an ordinary sample

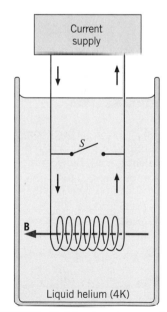

FIGURE 11.36 A superconducting coil immersed in liquid helium can carry a current indefinitely without resistive losses.

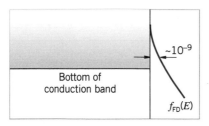

FIGURE 11.37 The tail of the Fermi-Dirac distribution function near the bottom of the conduction band. On the scale of this diagram, the "1" of $f_{FD}(E)$ is 1000 km off to the right and E_F is about 1 m below the bottom of the page.

Holes

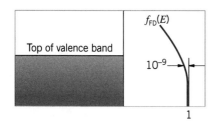

FIGURE 11.38 The Fermi-Dirac distribution function near the top of the valence band, showing the small fraction of empty states.

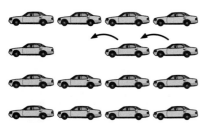

FIGURE 11.39 The motion of cars to the left, each filling the vacant space, is equivalent to the motion of the vacant space to the right.

containing of order 10^{20} atoms, there may be 10^{11} conduction electrons in a semiconductor, 10^{20} in a conductor, and none in an insulator.)

Figure 11.38 shows the corresponding region near the top of the valence band. If there are a few filled states in the conduction band, there must be a few *empty* states in the valence band, and the Fermi-Dirac distribution is just a tiny bit smaller than 1; in fact it is approximately $1 - e^{(E-E_F)/kT}$. This number is about $1 - 10^{-9}$, based on our discussion for the electrons in the conduction band. (Since all the electrons in the conduction band came originally from the valence band, the number of electrons in the conduction band is *exactly* equal to the number of vacancies in the valence band. Thus the value of $(E - E_F)$ for the valence band must be exactly equal to the value of $-(E - E_F)$ for the conduction band, and so the Fermi energy must lie exactly at the center of the gap.)

When we apply an electric field to a semiconductor, the electrons in the conduction band try to follow the field and therefore give an electric current. The electrons in the valence band also try to follow the field, but in order to do so each electron must move from a filled state to one of the very few vacancies. If we follow the motion of the electrons in the valence band as they hop from one vacancy to the next, we see an apparent motion of the vacancies in the opposite direction. (Imagine a large parking lot, with cars parked bumper to bumper, and with a single empty spot as in Figure 11.39. As we watch first one car moving to the empty spot, then the next car filling the spot vacated by the first, and so on, there is an apparent movement of the vacancy opposite to the motion of the cars.) These vacancies are known as *"holes"* and it is convenient to represent the motion of electrons in the valence band as the motion of holes in the opposite direction. The current in a semiconductor therefore consists of two parts: the negatively charged electrons in the conduction band and the positively charged holes in the valence band. Although the number of electrons in the conduction band is equal to the number of holes in the valence band, the two contributions to the current are in general not equal, since the electrons in the conduction band move more easily than the electrons in the valence band which produce the motion of the holes. Typically, the contribution of the electrons to the current at room temperature is about two to four times the contribution of the holes.

The material we have been describing thus far is an *intrinsic* semiconductor and is characterized by several features: (1) the number of electrons in the conduction band is equal to the number of holes in the valence band; (2) the Fermi energy lies at the middle of the gap; (3) the electrons contribute most to the current, but the holes are important also; (4) about 1 electron in 10^9 contributes to the conduction.

Because only 1 electron in 10^9 contributes to the conductivity of an intrinsic semiconductor, the presence of impurities which cannot be removed from the material at the level of just 1 part in 10^9 can significantly alter the conductivity of the semiconductor in a way that is not easily controllable. However, if impurities with known properties are *deliberately* introduced into the semiconductor in carefully controlled amounts, their contribution to the conductivity can be precisely determined. At impurity levels of only 1 part in 10^6 or 10^7, the impurity contribution to the conductivity dominates the intrinsic contribution.

Such materials are known as *impurity* semiconductors, and the process of introducing the impurity is known as *doping*. Impurity semiconductors can be of two varieties: those in which the impurity contributes additional electrons to the conduction band and those in which the impurity contributes additional holes to the valence band.

Let us consider a material such as silicon or germanium, in which there are four valence electrons in hybrid orbitals. In the band theory view, these fill the $4N$ states of the valence band; in the atomic view the lattice is constructed so that each Ge or Si atom has four neighbors with which it shares an electron, and so all electrons participate in covalent bonding (Figure 11.40). Now suppose we replace one of the Si or Ge atoms with an atom that has five valence electrons, such as phosphorous, arsenic, or antimony. Four of the five electrons form covalent bonds with the neighboring Si or Ge atoms, but the fifth electron is relatively weakly bound to the impurity atom and can be easily detached to contribute to the conductivity (Figure 11.41).

On an energy level diagram, the energies of these loosely bound electrons appear as discrete levels in the energy gap just below the conduction band, as in Figure 11.42. The energy needed for such electrons to enter the conduction band is relatively small, about 0.01 eV in Ge and 0.05 eV in Si, and so excitations can occur easily at room temperature ($kT \sim 0.025$ eV). These energy levels are known as *donor states* and the impurity is known as a *donor*, because electrons are "donated" to the conduction band. A semiconductor that has been doped with donor impurities is known as *n-type* semiconductor, because the conductivity is due mostly to the negative electrons.

A valence-3 atom, such as boron, aluminum, gallium or indium, can also be used as an impurity. When such an atom replaces a Si or Ge atom in the lattice, all of its three electrons form covalent bonds with neighboring Si or Ge atoms. But since the impurity atom is surrounded by four atoms in the lattice, one of the surrounding Si or Ge atoms has an unbound electron (Figure 11.43). Since the completing of the four pairs of covalent

Impurity semiconductors

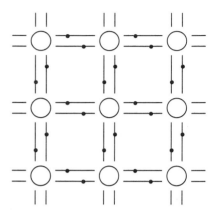

FIGURE 11.40 Covalent bonding in Ge or Si. Each atom shares four electrons in covalent bonds.

Donor states

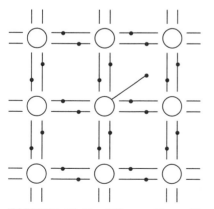

FIGURE 11.41 When a Ge or Si atom is replaced with a valence-5 atom, there is an extra electron that does not participate in covalent bonds.

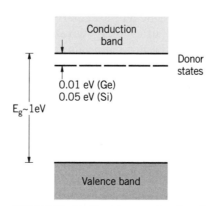

FIGURE 11.42 Energy levels of donor atoms.

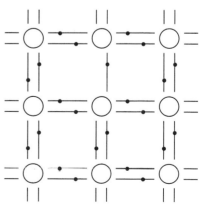

FIGURE 11.43 If a Ge or Si atom is replaced by a valence-3 atom, one covalent bond is lacking.

bonds is energetically very favorable, an electron is easily captured to complete the symmetry of the lattice. This creates a hole in the valence band and therefore contributes to the conductivity. The energy level diagram is shown in Figure 11.44. These impurity atoms form discrete levels, just above the valence band, known as *acceptor states*. At room temperature, electrons are readily excited from the valence band to the acceptor states.

Acceptor states

A material that has been doped with acceptor impurities is known as *p-type* semiconductor, because the conductivity is due mostly to the positively charged holes. (Remember that *n*-type and *p*-type materials are both *electrically neutral* since they are made from neutral atoms. The designations *n* and *p* refer only to the charge carriers, not to the material itself. If, for example, we remove some electrons from *n*-type material, it becomes positively charged.)

At $T = 0$ the Fermi level in *n*-type semiconductors lies between the donor states and the conduction band (remember, all states below E_F are full and all above E_F are empty; at $T = 0$ the donor states are all occupied). In *p*-type semiconductors, the Fermi level at $T = 0$ lies between the valence band and the acceptor states. As the temperature is raised, the thermal excitation of electrons from the valence band to the conduction band (as in an intrinsic semiconductor) causes the Fermi level to move toward the center of the energy gap, as shown in Figure 11.45. For low doping levels and at a high enough temperature, the material may behave like an intrinsic semiconductor.

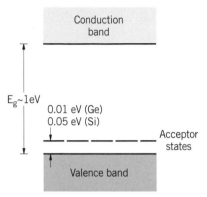

FIGURE 11.44 Energy levels of acceptor atoms.

11.9 SEMICONDUCTOR DEVICES

The *p-n* Junction

When a *p*-type semiconductor is placed in contact with an *n*-type semiconductor (Figure 11.46) electrons flow from the *n*-type material into the *p*-type material, until equilibrium is established. This equilibrium occurs when the Fermi levels in the two substances become identical. The electrons cannot travel very far from the junction region, because the semiconductor does not conduct particularly well.

The resulting energy level diagram is shown in Figure 11.47. The region between the two materials is known as the *depletion* region, because it has been somewhat depleted of charge carriers. Electrons from the donor states of the *n*-type material fill up the holes of the acceptor states of the *p*-type material. In this region the donor states *do not* provide electrons for the conduction band and the acceptor states *do not* provide holes in the valence band.

Actually, these devices are not made by bringing two different materials into contact, but rather by doping one side of a material so that it becomes *n* type and the other side so that it becomes *p* type. The doping is carefully controlled, and typically depletion layers have a thickness of the order of 1 μm.

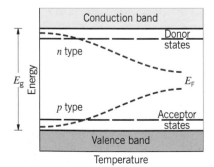

FIGURE 11.45 In a semiconductor the Fermi level changes with temperature.

The excess electrons that have entered the *p*-type material give that side of the depletion region a negative charge, which tends to repel additional electrons from the *n* region. There is a corresponding positive charge in the *n* region. These charges are associated with the fixed ions in those regions; the acceptor atoms in the *p* region acquire an electron and become fixed sites of negative charge, while the donor atoms in the *n* region lose an electron and become fixed sites of positive charge.

In equilibrium, enough negative charge builds up to stop the flow of electrons completely, and there is a net electric field $\mathbf{E}_0$ in the depletion region that results in a force (in the opposite direction) on the electrons that prevents any further flow of charge. Equivalently, there is a potential difference V_0 between the *n*-type and *p*-type regions; for electrons to flow from the *n* region to the *p* region, they must climb the energy barrier of height eV_0.

In the tail of the Fermi distribution of electrons in the conduction band of the *n* region, there will be a small number of electrons with enough energy to climb the energy barrier and enter the *p* region, where they recombine with holes (that is, they "fall" from the conduction band of the *p* region into the valence band). This gives the *diffusion* or *recombination* contribution to the current, which is directed from the *p* region to the *n* region (the current direction always being opposite to the direction of electron flow).

Even though holes provide the dominant contribution to the conduction in the *p* region, there are also electrons which provide a smaller contribution to the current. Electrons are thermally excited from the valence band in the *p* region to the conduction band, where they are accelerated by the electric field into the *n* region. This gives the *drift* or *thermal* contribution to the current, which is opposite to the diffusion current. At equilibrium, the two contributions to the current cancel one another, so that the net current is zero.

Figure 11.47 represents the energy levels and currents associated only with the electrons. There are also hole contributions to the diffusion and drift currents, which also cancel one another.

Let us now apply an external voltage V_{ext} across the junction so that the *p*-type material is made more positive than the *n* material; that is, we connect the + terminal of a battery to the *p* side of the junction and the − terminal of the battery to the *n* side (Figure 11.48). The effect of the battery is to *lower* the energy hill by an amount eV_{ext}. (The vertical axis shows electron energy, and a potential difference of V_{ext} gives an electron energy of $-eV_{ext}$.) This situation is called a *forward voltage* or *forward biasing*. The forward bias causes the depletion region to become narrower, because the battery pulls electrons out of the *p* region and injects them back into the *n* region. Because the energy hill is lower, more electrons can diffuse from the *n* region into the *p* region, so the diffusion current is considerably increased. (That is, there are more electrons in the *n* region in the tail of the Fermi distribution with energies above the bottom of the conduction band of the *p* region.) The drift current, however, is unaffected by the presence of the battery or the height of the hill. There is now a net current through the junction in the forward direction.

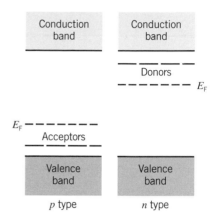

FIGURE 11.46 *n*-type and *p*-type semiconductors before being placed in contact.

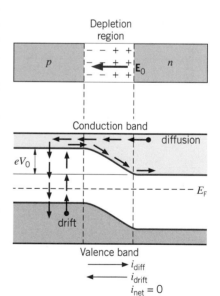

FIGURE 11.47 A *p-n* junction. The electric field in the depletion region inhibits additional flow of electrons. Energetic electrons in the tail of the Fermi distribution contribute to the diffusion current, while thermal excitation gives an equal and opposite drift current.

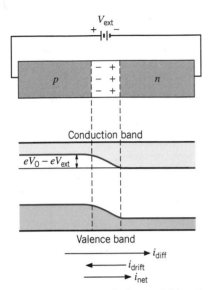

FIGURE 11.48 A forward-biased *p-n* junction. The potential hill is smaller, the diffusion current is larger, and there is a net forward current.

Now we reverse the battery connections (Figure 11.49), a situation known as *reverse voltage* or *reverse biasing*. This *raises* the hill by the amount eV_{ext}, *widens* the depletion region (because the battery pulls more electrons from the *n* region and injects them into the *p* region), and *decreases* the diffusion current. The drift current is again unchanged, so now there is a relatively small net current in the reverse direction.

Figure 11.50 shows the upper tail of the Fermi-Dirac distribution of the electrons extending into the conduction band of the *n*-type region. Only those electrons in the portion of the tail above the energy E_0 of the bottom of the conduction band of the *p*-type region can flow back across into the *p*-type region and it is these electrons that produce the recombination current. The number of electrons in that tail above the energy E_0 is approximately

$$N_1 = ne^{-(E_0 - E_F)/kT} \tag{11.33}$$

where n is some proportionality factor, and where we have approximated the Fermi-Dirac function as an exponential by neglecting the 1 in the denominator. (Since $E_0 - E_F \gtrsim 1$ eV and $kT \cong 0.025$ eV, this is an excellent approximation.) The recombination current is proportional to N_1, and since the thermal and recombination currents are equal, the thermal current is also proportional to N_1. Applying V_{ext} changes the level E_0 to $E_0 - eV_{ext}$, and the number of electrons in the tail above $E_0 - eV_{ext}$ is

$$N_2 = ne^{-(E_0 - eV_{ext} - E_F)/kT} \tag{11.34}$$

The recombination current is now proportional to N_2; applying the bias did not change the thermal current, so it is still proportional to N_1. The net current is given by the difference:

$$i \propto N_2 - N_1 = ne^{-(E_0 - E_F)/kT}(e^{eV_{ext}/kT} - 1) \tag{11.35}$$

We can rewrite this expression as

$$i = i_0(e^{eV_{ext}/kT} - 1) \tag{11.36}$$

This function is plotted in Figure 11.51, and it is immediately obvious why such *p-n* junctions, also known as *diodes*, have the property of *rectifying* varying currents. When the applied voltage is such that the junction is

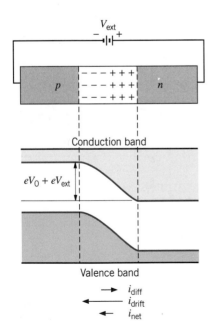

FIGURE 11.49 A reverse-biased *p-n* junction. The potential hill is larger, the diffusion current is smaller, and there is a small net reverse current.

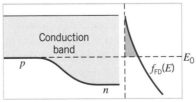

FIGURE 11.50 The recombination current depends on the density of electrons in the tail of $f_{FD}(E)$ above the energy E_0 of the bottom of the conduction band in the *p*-type material.

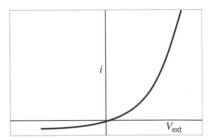

FIGURE 11.51 Current-voltage characteristics of an ideal *p-n* junction.

forward biased, a large forward current can flow. (When $V_{ext} = 1$ V, $i \cong 2 \times 10^{17} i_0$.) When the applied voltage is such that the junction is reverse biased, only a very small current can flow. (When $V_{ext} = -1$ V, $i \cong -i_0$.) Even very small forward voltages can produce large forward currents; even very large reverse voltages can produce only small reverse currents. This ability to rectify a varying voltage is illustrated in Figure 11.52.

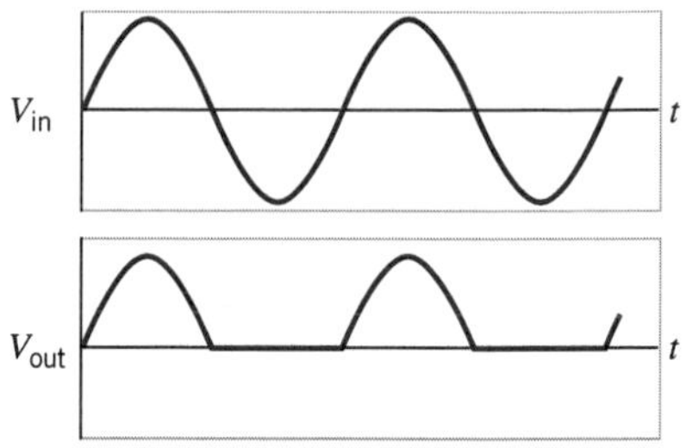

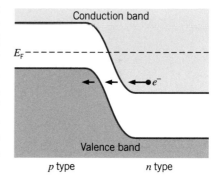

FIGURE 11.52 A *p-n* junction can rectify an ac voltage.

The Tunnel Diode

When the *p* and *n* regions are very heavily doped, the depletion layer becomes much narrower, perhaps 10 nm, and the energy diagram might look like Figure 11.53. When a small forward bias is applied, there is now a third contribution to the current—an electron from the conduction band of the *n* region can "tunnel" through the forbidden region directly into the valence band of the *p* region. This process of course depends on the wave nature of the electron and is an example of the type of barrier penetration we have discussed previously, in Section 5.7. The narrow depletion layer makes the process possible. The wavelength of an electron near the Fermi surface is about 1 nm, and if the thickness of the depletion layer were many orders of magnitude larger than this, tunneling would be unlikely to occur.

As the forward voltage is increased, the potential hill is lowered, and soon it no longer becomes possible for an electron to "tunnel" directly through the forbidden region. For a voltage of a few tenths of a volt, the "tunneling" current becomes zero. At this point the tunnel diode behaves like an ordinary diode. Figure 11.54 illustrates the characteristic current-voltage relationship for a tunnel diode.

Tunnel diodes are useful in electric circuits as high-speed elements, because the characteristics of the device can change as rapidly as the bias voltage can be changed. They can also be used as switches. If we were to pass current through the tunnel diode so that we were on the peak of the characteristic curve, a small increase in the current would cause the voltage to jump suddenly to a much larger value.

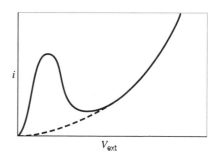

FIGURE 11.53 A *p-n* junction under heavy doping. Electrons can tunnel across the narrow gap.

Photodiodes

A photodiode is a *p-n* junction whose operation involves the emission or absorption of light. These devices operate on principles similar to ordinary atoms. An electron in the valence band may absorb a photon and make a transition to the conduction band. Since photons of visible light have energies of order 2 to 3 eV, a semiconductor with its gap of order 1 eV is just right for such a transition. Conversely, an excited electron from the conduction band can drop back down to the valence band, emitting a photon in the process.

A common device that emits visible light is the LED, or light-emitting diode. An external current supplies the energy necessary to excite electrons to the conduction band, and when the electrons fall back down to recombine with holes, a photon is emitted. The energy is of course equal to the

FIGURE 11.54 Current-voltage characteristics of a tunnel diode. The dashed curve shows the characteristics of an ordinary *p-n* junction diode.

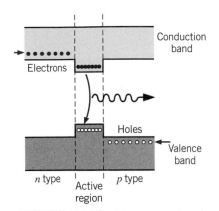

FIGURE 11.55 The energy bands in a diode laser. The active region has a smaller gap than the *n*-type and *p*-type regions on either side.

difference in energy of the electronic states. A common material for such devices, which are used for indicator lights, calculator displays, and so forth, is gallium arsenide phosphide, GaAsP, which emits photons in the red region of the spectrum.

Figure 11.55 shows another application of the emission of light by a semiconductor, in this case a *diode laser* or *semiconductor laser*. A thin layer of semiconducting material is sandwiched between *n*-type and *p*-type regions having a slightly larger energy gap. Electrons are injected from an external circuit into the *n*-type material, from which they diffuse into the middle layer. The electrons are prevented from diffusing into the *p*-type layer by a potential barrier, so they tend to concentrate in the middle layer. In a similar fashion, holes are injected into the *p*-type layer and again concentrate in the middle layer. This creates a population inversion similar to that described in our discussion of the laser in Section 8.8. An electron drops into the valence band accompanied by the emission of a photon; that photon then induces other transitions leading to the avalanche of photons that gives the lasing action.

The physical construction of a typical diode laser is illustrated in Figure 11.56. The lasing material is a narrow (0.2 μm) layer of GaAs, and the *p*-type and *n*-type layers are GaAlAs a few μm in thickness. The ends of the material are cleaved to create mirror-like surfaces that reflect a portion of the light wave, enhancing stimulated emission in the active region. This device emits at a wavelength of 840 nm, in the near infrared region. Diode lasers at this wavelength are commonly used in communication to send signals along optical fibers. By varying the materials of the laser, it is possible to obtain visible radiation.

Diode lasers are of small size (Color Plate 9), and they consume very little power (typically 10 mW, compared with the standard HeNe laser that may consume several watts). As a result, diode lasers can be powered by batteries. Efficiencies of the order of 20% are possible (that is, 20% of the electrical power supplied to the device appears as radiation in the beam), compared with 0.1% in the HeNe laser. The light signal can be turned on or off in switching times that are characteristic of semiconductors (< 100 ps), and thus we have a device that can rapidly modulate the beam.

The reverse case occurs when a beam of photons of sufficient energy is incident on the depletion region of a *p-n* junction. When this occurs, the absorption of the photon gives the electron enough energy to move to the conduction band, and in the process a hole is created in the valence band. The large electric field in the depletion region (such devices are usually operated under reverse bias) immediately forces the hole toward the *p* side and the electron toward the *n* side. A current is generated when the electron travels through an external circuit to recombine with the hole in the *p*-type material. Devices such as this have two common applications: In a silicon solar cell, the purpose is to create electric current from sunlight, so the current in the external circuit is used to supply power to an electric device. In various kinds of photon detecting devices, the purpose is to detect and measure the intensity of a beam of photons incident on the material. Such devices are used as light meters for cameras, as well as for detectors of high-energy photons such as X rays and gamma rays.

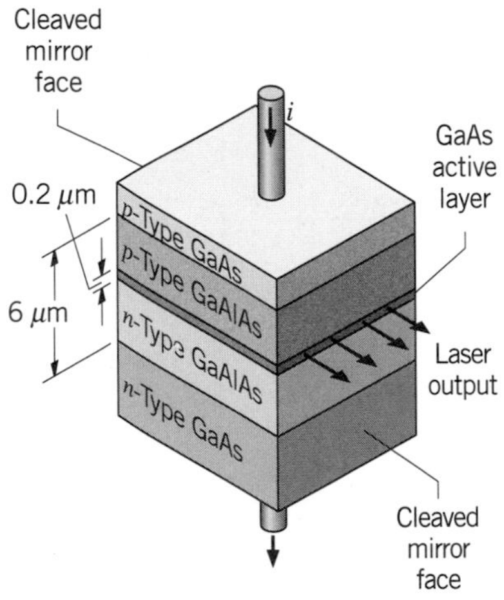

FIGURE 11.56 A diode laser. The lasing action occurs in the thin GaAs layer.

Junction Transistors

A transistor consists of a thin region of one type of semiconductor sandwiched between two thicker regions of the other type. According to the type of semiconductor used, such transistors are known as *npn* or *pnp* transistors. There is in principle no difference between the two types, other than the reversing of all bias voltages.

An unbiased *npn* transistor is illustrated in Figure 11.57. As with the *p-n* junction, electrons and holes travel across the two regions of contact until equilibrium is established. Electrical contact is made with each of the three regions. The central region is known as the base (B), and the other regions are known as the emitter (E) and collector (C). It is sometimes helpful to think of a transistor as two *p-n* junctions back to back, although the narrowness of the base has important consequences. In typical operation, the EB junction is forward biased and the BC junction is reverse biased, as shown in Figure 11.58. As with the *p-n* junction, the flow of electrons from E to B is enhanced by the forward biasing. Once into the base, electrons diffuse through its narrow region and are accelerated to the collector by the voltage across the BC junction. Not all of the electrons from the emitter leave the collector; a small fraction may leave through the base into the external circuit.

Transistors are found in so many different applications that to list them would fill many volumes. We mention briefly two applications that can be understood with reference to Figure 11.58. The continuous flow of current from C to E can be interrupted if the EB junction is not forward biased. The transistor can thus serve as a switch, in which the application of a dc voltage level to the base can turn the current on or off. This on-off switching has found application in logic circuits and especially in digital computers. A second application takes advantage of the small current leaving the base. Suppose the transistor is operating in such a way that a constant 5 percent of the current leaves the base while the other 95 percent goes through the emitter. If we are pulling 10 mA of current through C, then 0.5 mA emerges from B while 9.5 mA goes to E. Let us now attach to the base a device that draws an additional 0.1 mA through the base; that is, the base current rises to 0.6 mA. To keep the base current at a constant 5 percent of the collector current, the collector current must rise to 12 mA, and so 11.4 mA must emerge from the emitter. Thus a 0.1-mA variation in the current at the base caused a 1.9-mA variation at the emitter, and the transistor acts like an amplifier with a current gain of 19.

Almost every aspect of daily life is affected in some way by the transistor, by high-speed digital computers made possible by the transistor, and by the integrated circuit, a collection of transistors reduced to such a small size that thousands of transistors fit into the space formerly occupied by one (Color Plate 10). From pocket calculators to microwave ovens to automobiles to aircraft, the integrated circuit has become an essential component of our society. This is only one example of the unexpected directions in which basic research may lead—all of these practical applications followed from experiments studying the properties of solids.

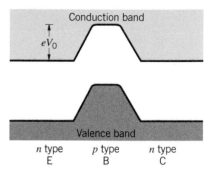

FIGURE 11.57 A *npn* junction transistor.

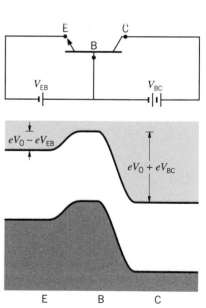

FIGURE 11.58 Typical biasing arrangement of *npn* transistor.

SUGGESTIONS FOR FURTHER READING

More detailed and comprehensive books on solid-state physics:

R. H. Bube, *Electrons in Solids: An Introductory Survey,* 3rd ed. (San Diego, Academic Press, 1992).

A. J. Dekker, *Solid State Physics* (Englewood Cliffs, Prentice-Hall, 1957).

A. Hart-Davis, *Solids: An Introduction* (London, McGraw-Hill, 1975).

C. Kittel, *Introduction to Solid State Physics,* 6th ed. (New York, Wiley, 1986).

T. L. Martin and W. F. Leonard, *Electrons and Crystals* (Belmont CA, Wadsworth, 1970).

M. N. Rudden and J. Wilson, *Elements of Solid State Physics,* 2nd ed. (Chichester, Wiley, 1993).

R. T. Sanderson, *Chemical Bonds and Bond Energy* (New York, Academic Press, 1976).

High-temperature superconductors are discussed in R. J. Cava, "Superconductors beyond 1-2-3," *Scientific American* **263**, 42 (August 1990).

The entire September 1977 issue of *Scientific American* is devoted to integrated circuits and includes articles on their manufacture and applications.

Bulk properties of solids are tabulated in many references, including the *Handbook of Chemistry and Physics* (Chemical Rubber Publishing Co.) and the *American Institute of Physics Handbook* (New York, McGraw-Hill, 1963).

QUESTIONS

1. Compare the equilibrium separations and binding energies of ionic *solids* (Table 11.1) with those of the corresponding ionic *molecules* (Table 9.5). Account for any systematic differences.

2. How should Equation 11.7 be modified to be valid for MgO and BaO?

3. Assuming that its other properties don't also change with temperature, at what temperature would you expect carbon to begin to behave like a semiconductor?

4. Would you expect the Wiedemann-Franz law to apply to semiconductors? To insulators?

5. (a) Why does the electrical conductivity of a metal decrease as the temperature is increased? (b) How would you expect the conductivity of a semiconductor to change with temperature?

6. Why is it that only the electrons near E_F contribute to the electrical conductivity?

7. Would you expect silicon to behave like an insulator at a low enough temperature? Would it behave like a conductor at a high enough temperature?

8. What determines the drift speed of an electron in a metal? What determines the Fermi speed?

9. Do the superconducting elements have any particular electronic structure or configuration in common?

10. Three different materials have filled valence bands and empty conduction bands, and the Fermi energy lies in the middle of the gap. The gap energies are 10 eV, 1 eV, and 0.01 eV. Classify the electrical properties of these materials at room temperature and at 3 K.

11. In what way does a *p-n* junction behave as a capacitor?

12. Semiconductors are sometimes called "nonohmic" materials. Why?

13. If a semiconductor is doped at a level of one impurity atom per 10^9 host atoms, what is the average spacing between the impurity atoms?

14. Explain the processes that contribute to the current in a forward-biased *p-n* junction. Do the same for a reverse-biased junction.

15. What limits the response time of a *p-n* junction when the external voltage is varied? Why does a tunnel diode not have the same limits?

16. The energy gap E_g is 0.72 eV for Ge and 1.10 eV for Si. At what wavelengths will Ge and Si be transparent to radiation? At what wavelengths will they begin to absorb significantly?

17. Why is a semiconductor better than a conductor for applications as a solar cell or photon detector? Would an insulator be even better?

PROBLEMS

1. Consider the packing of hard spheres, as represented in Figure 11.1. The corners of the basic cube are the centers of the eight spheres. (a) What fraction of the volume of each sphere is inside the volume of the basic cube? (b) Let r be the radius of each sphere and let a be the length of a side of the cube. Express a in terms of r. (c) What fraction of the volume of the cube is taken up by the portions of the spheres? This fraction is called the *packing fraction*.

2. Compute the packing fractions (see Problem 1) of the fcc and bcc structures (Figures 11.2 and 11.3). Which structure fills the space most efficiently?

3. Derive Equations 11.6 and 11.7.

4. By summing the contributions for the attractive and repulsive Coulomb potential energies, show that the Madelung constant has the value 2 ln 2 for a one-dimensional "lattice" of alternating positive and negative ions.

5. Calculate the first 3 contributions to the electrostatic potential energy of an ion in the CsCl lattice.

6. Use the values of the cohesive energies of the ionic solids from Table 11.1, along with the other values from the listed references, and see how the boiling points of the ionic solids vary with the cohesive energy. Boiling points may be found in the *Handbook of Chemistry and Physics*.

7. (a) Find the binding energy per ion pair in CsCl from the cohesive energy. (b) Find the binding energy per ion pair in CsCl from Equation 11.7. (c) Find the binding energy per atom for CsCl. The ionization energy of Cs is 3.89 eV.

8. Repeat Problem 7 for LiF. The ionization energy of Li is 5.39 eV, and the electron affinity of F is 3.45 eV.

9. Calculate the Coulomb energy and the repulsion energy for NaCl at its equilibrium separation.

10. Calculate the wavelength at which CsI will absorb energy by means of the vibrational motion of the ions.

11. The electric field of a dipole is proportional to $1/r^3$. Assuming that the induced dipole moment of molecule B is proportional to the electric dipole field of molecule A, show that the van der Waals force is proportional to r^{-7}. (*Hint*: Calculate the potential energy of dipole B.)

12. The density of sodium is 0.971 g/cm^3 and its molar mass is 23.0 g. In the bcc structure, what is the distance between sodium atoms?

13. Copper has a density of 8.96 g/cm^3 and molar mass of 63.5 g. Calculate the center-to-center distance between copper atoms in the fcc structure.

14. Calculate the binding energy per atom for metallic Na and Cu.

15. (a) In copper at room temperature, what is the electron energy at which the Fermi-Dirac distribution has the value 0.1? (b) How does $E - E_F$ compare with kT?

16. Calculate the de Broglie wavelength of an electron with energy E_F in copper, and compare the value with the atomic separation in copper.

17. What is the probability per unit volume in sodium that states between 0.10 eV and 0.11 eV above the Fermi energy are occupied at room temperature (293 K)?

18. From the Fermi energy for Mg, find the number of free electrons per atom. The molar mass of Mg is 24.3 g, and its density is 1.74 g/cm^3.

19. At what temperature would the Fermi energy of Au be reduced by 1%? Compare this temperature with the melting point of Au (1337 K).

20. Compute the Fermi energy for zinc, which has two valence electrons per atom. The molar mass of zinc is 91.22 g, and its density is 6.51 g/cm^3.

21. Calculate the expected value of the Lorenz number.

22. Use the Wiedemann-Franz ratio to calculate the thermal conductivity of copper at room temperature. The electrical conductivity is 5.88×10^7 Ω^{-1} m^{-1}.

23. We can understand the behavior of donor impurities with the following rough calculation. When we replace an atom of silicon with an atom of phosphorous, the outer electron of phosphorous is screened and sees $Z_{\text{eff}} \cong 1$. (a) Compute the binding energy of the electron, assuming that silicon has a dielectric constant of 12 that effectively reduces the electric field experienced by the electron. Compare this value with the value shown in Figure 11.42. (b) Compute the radius of the electron's orbit. How many atoms does it encounter in a sphere of that radius? The lattice spacing of Si is 0.234 nm.

24. When a material such as germanium is used as a photon detector, an incoming photon makes many interactions and excites many electrons across the gap between the valence and the conduction band. (a) ^{137}Cs emits a 662-keV gamma ray. How many electrons are excited across the 0.72-eV gap of germanium by the absorption of this gamma ray? (b) The number N calculated in part (a) is subject to statistical fluctuations of $\sqrt{N}$. Compute the variation in N and

the fractional variation in N. (c) What is the corresponding variation in the measured energy of the gamma ray? This result is the experimental resolution of the detector.

25. Assuming the energy gap in intrinsic silicon is 1.1 eV and that the Fermi energy lies at the middle of the gap, calculate the occupation probability at 293 K of (a) a state at the top of the valence band and (b) a state at the bottom of the conduction band. (c) How do the occupation probabilities change if the temperature is increased by 100 K?

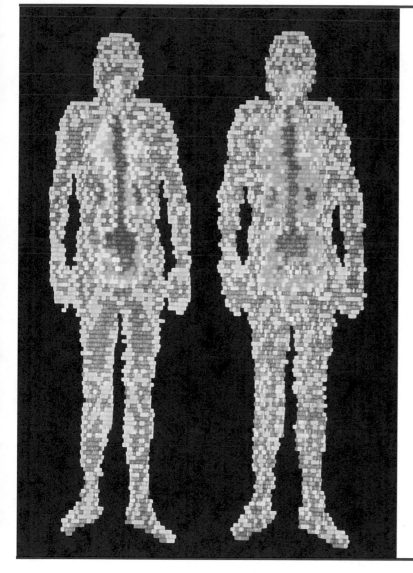

Radioactive isotopes have proven to be valuable tools for medical diagnosis. The photo shows full-body front (right) and back gamma-ray scans of a man injected with a radioactive isotope. The isotope tends to concentrate in the bones, so the images reveal a higher intensity of radiation from the spine, ribs, and pelvis.

NUCLEAR STRUCTURE AND RADIOACTIVITY

When we probe deep inside the atom, we find at its center the nucleus, occupying only 10^{-15} of the volume of the atom, but providing the electrical force that holds the atom together. Were it not for the Coulomb attraction provided by the nucleus, the mutual repulsion of the electrons would cause the atom to fly apart. What keeps the nucleus itself from flying apart, under the repulsive force of the positive charges? A positive charge of magnitude e at the surface of the nucleus experiences an electrostatic repulsive force that gives a potential energy of roughly 20 MeV. In order to keep that positive charge inside the nucleus, the nuclear force must provide a binding energy in excess of 20 MeV—thousands of times larger than typical atomic binding energies!

There are many similarities between atomic structure and nuclear structure, which will make our study of the properties of the nucleus somewhat easier. Nuclei are subject to the laws of quantum physics. They have ground and excited states and emit photons (known as *gamma rays*) in transitions between the excited states. Just like atomic states, nuclear states can be labeled by their angular momentum.

There are, however, two major differences between the study of atomic and nuclear properties. In atomic physics, the electrons experience the force provided by an external agent, the nucleus; in nuclear physics, there is no such external agent. The constituents of the nucleus move about under the influence of a force that *they themselves* provide. The mutual interactions of the electrons on one another have an influence on the atomic level scheme, but it is a relatively small influence, and we saw how we could understand a great deal of atomic structure based primarily on the study of the interaction between one electron and the nucleus; we treat the effect of the other electrons as a minor perturbation. In nuclear physics, the mutual interaction of the nuclear constituents is just what provides the nuclear force, so we cannot treat this many-body problem as a perturbation. We therefore cannot avoid the mathematical difficulties in the nuclear case, as we did in the atomic case.

The second problem associated with nuclear physics is that we cannot write the nuclear force in a simple form like the Coulomb force or the gravitational force. There is no closed-form analytical expression that can be written to describe the mutual forces of the nuclear constituents.

In spite of these difficulties, we can learn a great deal about the properties of the nucleus by studying the interactions between different nuclei, the radioactive decay of nuclei, and the properties of some nuclear constituents. In this chapter and the next we describe these studies and how we learn about the nucleus from them.

12.1 NUCLEAR CONSTITUENTS

The work of Rutherford, Bohr, and their contemporaries showed that the positive charge of the atom is confined in a very small nuclear region at the center of the atom, that the nucleus has a charge of $+Ze,$ and that the nucleus provides most (99.9 percent) of the atomic mass. It was also known that the masses of the atoms were very nearly integer multiples of the mass of hydrogen, the lightest atom; a glance at Appendix B confirms this observation. We call this integer A, the *mass number*. It is therefore reasonable to suppose that the hydrogen nucleus is composed of a fundamental unit of positive charge (a correct assumption), and that all heavier nuclei contain an integral number A of these positive units (an incorrect assumption, as we shall see). This fundamental unit is the *proton*, with an electric charge of $+e$ and a mass equal to the atomic mass of hydrogen, less the electronic mass and binding energy. If a nucleus of mass number A contained A protons, it would have a nuclear charge of Ae rather than Ze; since $A > Z$ for all atoms heavier than hydrogen, this model gives too much positive charge to the nucleus. This difficulty was removed by the *proton-electron model*, in which it was postulated (again incorrectly) that the nucleus also contained $(A - Z)$ electrons. Under these assumptions, the nuclear mass would be about A times the mass of the proton (since the mass of the electrons is negligible) and the nuclear electric charge would be $A(+e) + (A - Z)(-e) = Ze$, in agreement with experiment. However, this model leads to several difficulties. First, as we discovered in Chapter 4 (see Example 4.7), the presence of electrons in the nucleus is not consistent with the uncertainty principle, which would require those electrons to have unreasonably large ($\sim$19 MeV) kinetic energies. A more serious problem concerns the total *intrinsic spin* of the nucleus.

From measurements of the *very* small effect of the nuclear magnetic moment on the atomic transitions (called the *hyperfine splitting*), we know that the proton has an intrinsic spin of $\frac{1}{2}$, just like the electron. Consider an atom of deuterium, sometimes known as "heavy hydrogen." It has a nuclear charge of $+e$, just like ordinary hydrogen, but a mass of two units, twice that of ordinary hydrogen. The proton-electron nuclear model would then require that the deuterium *nucleus* contain two protons and one electron, giving a net mass of two units and a net charge of one. Each of these three particles has a spin of $\frac{1}{2}$, and the rules for adding angular momenta in quantum mechanics would lead to a spin of deuterium of either $\frac{1}{2}$ or $\frac{3}{2}$. However, the measured total spin of deuterium is 1. For these and other reasons, the hypothesis that electrons are a nuclear constituent must be discarded.

The resolution of this dilemma came in 1932 with the discovery of the *neutron*, a particle of roughly the same mass as the proton (actually about 0.1 percent more massive) but having no electric charge. According to the *proton-neutron* model, a nucleus consists of Z protons and $(A - Z)$ neutrons, giving a total charge of Ze and a total mass of roughly A times the mass of the proton, since the proton and neutron masses are roughly the same.

The proton and neutron are, except for their electric charges, very similar to one another, and so they are classified together as *nucleons*. Some properties of the two nucleons are listed in Table 12.1.

The chemical properties of any element depend on its atomic number Z, but not on its mass number A. It is possible to have two different nuclei, with the same Z but with different A (that is, with different numbers of neutrons). Atoms of these nuclei are identical in all their chemical properties, differing only in mass and in those properties that depend on mass.

TABLE 12.1   PROPERTIES OF THE NUCLEONS

Name	Symbol	Charge	Mass	Rest Energy	Spin
Proton	p	$+e$	1.007276 u	938.28 MeV	$\frac{1}{2}$
Neutron	n	0	1.008665 u	939.57 MeV	$\frac{1}{2}$

Isotopes   Nuclei with the same Z but different A are called *isotopes*. Hydrogen, for example, has three isotopes: ordinary hydrogen ($Z = 1$, $A = 1$), deuterium ($Z = 1$, $A = 2$), and tritium ($Z = 1$, $A = 3$). All of these are indicated by the chemical symbol H. When we discuss nuclear properties it is important to distinguish among the different isotopes. We do this by indicating, along with the chemical symbol, the atomic number Z, the mass number A, and the *neutron number $N = A - Z$* in the following format:

$$_{Z}^{A}X_{N}$$

where X is any chemical symbol. Since the chemical symbol and the atomic number Z give the same information, it is not necessary to include both of them in the isotope label. Also, if we specify Z then we don't need to specify *both* N and A. It is sufficient to give only the chemical symbol and A. The three isotopes of hydrogen would be indicated as $_{1}^{1}H_{0}$, $_{1}^{2}H_{1}$, and $_{1}^{3}H_{2}$, or more compactly as ^{1}H, ^{2}H, and ^{3}H.

In Appendix B you will find a list of isotopes and some of their properties.

EXAMPLE 12.1

Give the symbol for the following: (*a*) The isotope of helium with mass number 4. (*b*) The isotope of tin with 66 neutrons. (*c*) An isotope with mass number 235 that contains 143 neutrons.

SOLUTION

(*a*) From the periodic table, we find that helium has $Z = 2$. Since $A = 4$, $N = A - Z = 2$. Thus the symbol would be $_{2}^{4}He_{2}$ or ^{4}He.

(*b*) Again from the periodic table, we know that for tin (Sn), $Z = 50$. Since we are given $N = 66$, then $A = Z + N = 116$. The symbol is $_{50}^{116}Sn_{66}$ or ^{116}Sn.

(*c*) Given that $A = 235$ and $N = 143$, we know that $Z = A - N = 92$. From the periodic table, we find that this element is uranium, and so the proper symbol for this isotope is $_{92}^{235}U_{143}$ or ^{235}U.

12.2 NUCLEAR SIZES AND SHAPES

It is as difficult to define precisely the radius of a nucleus as it is the radius of an atom. The probability distribution for atomic electrons makes the atom appear like a "fuzzy ball" of charge with no sharp boundary; the charge distribution does not stop at any clearly defined point. We can, however, take the average or most probable orbital radius of the outermost electron to represent the radius of an atom.

The nucleus must be treated in much the same way, although there are not neutron or proton orbits that we can use for this purpose. Most nuclei are rather spherical (although some are slightly stretched or flattened) and the radial variation of the nuclear density is approximately represented in Figure 12.1. From a variety of experiments, we know many remarkable features of the nuclear density. We have already discussed how strong the nuclear force must be to keep the Coulomb repulsion from blowing the nucleus to pieces. We might expect that this force would cause the protons and neutrons to congregate at the center of the nucleus, giving an increasing density in the central region. However, Figure 12.1 shows that this is not the case—the density remains quite constant. Some mechanism keeps the nucleus from collapsing toward the center. This fact gives one important clue about a property of the nuclear force, as we discuss in Section 12.4. Another interesting feature of Figure 12.1 is that the density of a nucleus seems not to depend on the mass number A—very light nuclei, such as ^{12}C, have roughly the same central density as very heavy nuclei, such as ^{209}Bi. Stated another way, the number of protons and neutrons per unit volume is approximately constant over the entire range of nuclei:

$$\frac{\text{number of neutrons and protons}}{\text{volume of nucleus}} = \frac{A}{\frac{4}{3}\pi R^3} \cong \text{constant}$$

Thus

$$A \propto R^3$$

This suggests a proportionality between the nuclear radius R and the cube root of the mass number:

$$R \propto A^{1/3}$$

or, defining a constant of proportionality R_0,

$$R = R_0 A^{1/3} \tag{12.1}$$

The constant R_0 must be determined by experiment, and a typical experiment might be to scatter charged particles (alpha particles or electrons, for example) from the nucleus and to infer the radius of the nucleus from the distribution of scattered particles. From such experiments, we know the value of R_0 is approximately 1.2×10^{-15} m. (The exact value depends, as in the case of atomic physics, on exactly how we define the radius, and values of R_0 usually range from 1.0×10^{-15} m to 1.5×10^{-15} m.) The length 10^{-15} m is 1 femtometer (fm), but physicists often refer to this length as one fermi, in honor of the Italian-American physicist, Enrico Fermi.

Nuclear radii

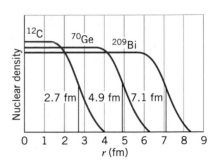

FIGURE 12.1 The radial dependence of the nuclear charge density.

EXAMPLE 12.2

Compute the approximate nuclear radii of carbon ($A = 12$), germanium ($A = 70$), and bismuth ($A = 209$).

Enrico Fermi (1901–1954, Italy–United States). There is hardly a field of modern physics to which he did not make contributions in theory or experiment; perhaps Newton was the only other scientist with such skills in both theoretical and experimental work. He developed the statistical laws for spin-$\frac{1}{2}$ particles and in the 1930s he proposed a theory of beta decay that is still used today. He was the first to demonstrate the transmutation of elements by neutron bombardment (for which he received the 1938 Nobel prize), and he directed the construction of the first nuclear reactor.

SOLUTION

Using Equation 12.1, we obtain:

Carbon: $R = (1.2 \text{ fm})A^{1/3} = (1.2 \text{ fm})(12)^{1/3} = 2.7 \text{ fm}$
Germanium: $R = (1.2 \text{ fm})A^{1/3} = (1.2 \text{ fm})(70)^{1/3} = 4.9 \text{ fm}$
Bismuth: $R = (1.2 \text{ fm})A^{1/3} = (1.2 \text{ fm})(209)^{1/3} = 7.1 \text{ fm}$

As you can see from Figure 12.1, these values define the mean radius, the point at which the density falls to half the central value.

EXAMPLE 12.3

Compute the density of a typical nucleus, and find the resultant mass if we could manufacture a nucleus with a radius of 1 cm.

SOLUTION

$$\rho = \frac{m}{V} = \frac{Am_\mathrm{p}}{\frac{4}{3}\pi R^3} = \frac{Am_\mathrm{p}}{\frac{4}{3}\pi R_0^3 A} = \frac{1.67 \times 10^{-27} \text{ kg}}{\frac{4}{3}\pi(1.2 \times 10^{-15} \text{ m})^3} = 2 \times 10^{17} \text{ kg/m}^3$$

The mass of our hypothetical nucleus would be

$$m = \rho V = \left(2 \times 10^{17} \frac{\text{kg}}{\text{m}^3}\right)\left(\frac{4}{3}\pi\right)(0.01 \text{ m})^3 = 8 \times 10^{11} \text{ kg}$$

about the mass of a 1-km sphere of ordinary matter!

The result of Example 12.3 shows the great density of what physicists call *nuclear matter*. Although examples of such nuclear matter in bulk are not found on Earth (a sample of nuclear matter the size of a large building would have a mass as great as that of the entire Earth), they are found in certain extremely hot stars, in which the high temperature "boils off" the electrons from atoms, leaving only the bare nuclei, which then form an aggregate nucleus out of the entire stellar material.

One way of measuring the size of a nucleus is to scatter charged particles, such as alpha particles, as in Rutherford scattering experiments. As long as the alpha particle is outside the nucleus, the Rutherford scattering formula holds, but when the distance of closest approach is less than the nuclear radius, deviations from the Rutherford formula occur. Figure 12.2 shows the results of a Rutherford scattering experiment in which such deviations are observed. (Problem 5 suggests how a value for the nuclear radius can be inferred from these data.)

Other scattering experiments can also be used to measure the nuclear radius. Figure 12.3 shows a sort of "diffraction pattern" that results from the scattering of energetic electrons by a nucleus. In each case the first diffraction minimum is clearly visible. (The intensity at the minimum doesn't fall to zero because the nuclear density doesn't have a sharp edge, as illustrated in Figure 12.1.) For scattering of radiation of wavelength λ by a circular disc of diameter D, the first diffraction minimum should appear at an angle of $\theta = \sin^{-1}(1.22\ \lambda/D)$. This is analogous to diffraction by a

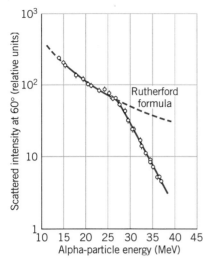

FIGURE 12.2 Deviations from Rutherford formula in scattering from ^{208}Pb are observed for alpha-particle energies above 27 MeV.

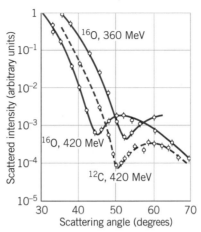

FIGURE 12.3 Diffraction of 360-MeV and 420-MeV electrons by ^{12}C and ^{16}O nuclei.

slit of width a, for which the first diffraction minimum appears at the angle $\theta = \sin^{-1}(\lambda/a)$. At an electron energy of 420 MeV, the observed minima for ^{16}O and ^{12}C give a radius of 2.6 fm for ^{16}O and 2.3 fm for ^{12}C (see Problem 6), in good agreement with the values 3.0 fm and 2.7 fm computed from Equation 12.1.

12.3 NUCLEAR MASSES AND BINDING ENERGIES

Suppose we have a proton and an electron at rest separated by a large distance. The total energy of this system is the total rest energy of the two particles, $m_pc^2 + m_ec^2$. Now we let the two particles come together to form a hydrogen atom in its ground state. In the process, several photons are emitted, the *total* energy of which is 13.6 eV. The total energy of this system is the rest energy of the hydrogen atom, m_Hc^2, plus the total photon energy, 13.6 eV. Conservation of energy demands that the total energy of the system of isolated particles must equal the total energy of atom plus photons:

$$m_ec^2 + m_pc^2 = m_Hc^2 + 13.6 \text{ eV}$$

or

$$m_ec^2 + m_pc^2 - m_Hc^2 = 13.6 \text{ eV}$$

That is, the rest energy of the combined system (the hydrogen atom) is less than the rest energy of its constituents by 13.6 eV. This energy difference is the *binding energy* of the atom. We can regard the binding energy as either the "extra" energy we obtain when we assemble an atom from its components or else the energy we must supply to disassemble the atom into its components.

Nuclear binding energies are calculated in a similar way. Consider, for example, the nucleus of deuterium, $^2_1H_1$, which is composed of one proton and one neutron. The nuclear binding energy of deuterium is the difference between the total rest energy of the constituents and the rest energy of their combination:

$$B = m_n c^2 + m_p c^2 - m_D c^2 \tag{12.2}$$

where m_D is the mass of the deuterium nucleus. In using mass tables to do these calculations, it is important to remember that the tabulated masses are *atomic masses*, not nuclear masses. To calculate nuclear binding energies, however, we must use nuclear masses. The relationship between atomic and nuclear masses is:

$$m_{atom}c^2 = m_{nucleus}c^2 + Zm_{electron}c^2 + total\ electron\ binding\ energy \tag{12.3}$$

Nuclear rest energies are of the order of 10^9 to 10^{11} eV, total electron rest energies are of the order of 10^6 to 10^8 eV, and electron binding energies are of the order of 1 to 10^5 eV. Thus the last term of Equation 12.3 is very small compared with the other two terms, and we can safely neglect it to the accuracy we need for these calculations.

The *nuclear* mass of hydrogen (the proton mass) is the *atomic* mass of hydrogen (1.007825 u) less the mass of one electron. The *nuclear* mass of deuterium is the *atomic* mass of deuterium (2.014102 u) less the mass of one electron. Substituting for the *nuclear* masses that appear in Equation 12.2, we can find the binding energy in terms of the *atomic* masses:

$$B = m_n c^2 + [m(^1H) - m_e]c^2 - [m(^2H) - m_e]c^2$$

$$= [m_n + m(^1H) - m(^2H)]c^2$$

Notice that the electron mass cancels in this calculation, as it would in any such calculation, since the constituents include Z hydrogen atoms (with Z electrons) and the atom of atomic number Z also includes Z electrons. We can therefore generalize this equation to give the total binding energy of any nucleus $^A_Z X_N$:

Nuclear binding energy

$$B = [Nm_n + Zm(^1_1H_0) - m(^A_Z X_N)]c^2 \tag{12.4}$$

The masses that appear in Equation 12.4 are *atomic* masses. For deuterium, we then have

$$B = (1.008665\ u + 1.007825\ u - 2.014102\ u)(931.5\ MeV/u)$$

$$= 2.224\ MeV$$

Here we use $c^2 = 931.5$ MeV/u to convert mass units to energy units.

EXAMPLE 12.4

Find the total binding energy B and also the binding energy per nucleon B/A for $^{56}_{26}Fe_{30}$ and $^{238}_{92}U_{146}$.

SOLUTION

From Equation 12.4, for $^{56}_{26}$Fe$_{30}$ with $N = 30$ and $Z = 26$,

$$B = (30 \times 1.008665\ \text{u} + 26 \times 1.007825\ \text{u} - 55.934939\ \text{u})(931.5\ \text{MeV/u})$$

$$= 492.3\ \text{MeV}$$

$$\frac{B}{A} = (492.3\ \text{MeV})/56 = 8.791\ \text{MeV per nucleon}$$

For $^{238}_{92}$U$_{146}$,

$$B = (146 \times 1.008665\ \text{u} + 92 \times 1.007825\ \text{u} - 238.050785\ \text{u})(931.5\ \text{MeV/u})$$

$$= 1802\ \text{MeV}$$

$$\frac{B}{A} = (1802\ \text{MeV})/238 = 7.571\ \text{MeV per nucleon}$$

Example 12.4 gives us insight into an important aspect of nuclear structure. The values of B/A we calculated show us that the nucleus ^{56}Fe is *relatively* more tightly bound than is the nucleus ^{238}U—the binding energy *per nucleon* is greater for ^{56}Fe than for ^{238}U. Alternatively, this calculation shows that, given a large supply of protons and neutrons, we would release more energy by assembling those nucleons into nuclei of ^{56}Fe than we would by assembling them into nuclei of ^{238}U.

Repeating this calculation for the entire range of nuclei, we obtain the results shown in Figure 12.4. The binding energy per nucleon starts at small

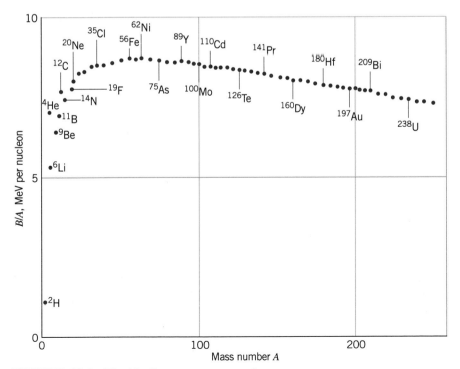

FIGURE 12.4 The binding energy per nucleon.

values (0 for the proton and neutron, 1.11 MeV for deuterium), rises to a maximum of 8.795 MeV for ^{62}Ni, and then falls to values of 7.5 MeV for the heavy nuclei.

The shape of Figure 12.4 is determined primarily by three factors: (1) a constant term, which originates because nucleons interact only with their nearest neighbors (as we discuss in the next section), (2) a sharp decrease for light nuclei (which have relatively more surface nucleons and therefore fewer near neighbors than heavy nuclei), and (3) a gradual decrease for heavy nuclei due to the Coulomb repulsion of the nuclear protons.

Figure 12.4 suggests that we can liberate energy from the nucleus in two different ways. If we split a heavy nucleus into two lighter nuclei, energy is released, because the binding energy per nucleon is greater for the two lighter fragments than it is for the original nucleus. This process is known as *nuclear fission*. Alternatively, we could combine two light nuclei into a heavier nucleus; again, energy is released when the binding energy per nucleon is greater in the final nucleus than it is in the two original nuclei. This process is known as *nuclear fusion*. We consider fission and fusion in greater detail in Chapter 13.

12.4   THE NUCLEAR FORCE

Our clues to the structure of the hydrogen atom came from studies of transitions between its excited states. The Balmer series and the Ritz combination principle provided information that helped us to understand the forces that hold the atom together. Moreover, understanding atomic structure was helped in hydrogen by the simplicity of its structure—a single electron revolving about the nucleus, with no other electrons to complicate the structure.

This experience suggests that we should begin our study of the nuclear force by looking at the simplest system in which that force operates—the deuterium nucleus, which consists of one proton and one neutron. We hope to learn something about the nuclear force from the photons emitted in transitions between the excited states of this nucleus. Unfortunately, our hope is not fulfilled—deuterium has *no excited states*. When we bring a proton and an electron together to form a hydrogen atom, a whole spectrum of photons is emitted as the electron drops into its ground state; from this spectrum we learn the energies of the excited states. When we bring a proton and a neutron together to form a deuterium nucleus, only one photon (of energy 2.224 MeV) is emitted as the system drops directly into its ground state.

Even though we can't use the excited states of deuterium, we can learn about the nuclear force in the proton-neutron system by scattering neutrons from protons as well as by doing a variety of different experiments with heavier nuclei. From these experiments we have learned the following characteristics of the nuclear force:

1. It is a very different kind of force from electromagnetism, gravitation, or the other kinds of forces we commonly encounter. It is also the strongest of the known forces, and so it is sometimes known as the *strong* force.

2. The strong nuclear force has a very short range—the region in which the force acts is limited to nuclear dimensions (10^{-15} m or so). There are two major pieces of evidence for this short range. The first comes from our study of the density of nuclear matter. As we add nucleons to the nucleus, the central density remains roughly constant (Figure 12.1). This suggests that each added nucleon feels a force only from its nearest neighbors, and *not* from all the other nucleons in the nucleus. In this respect, a nucleus behaves somewhat like a crystal, in which each atom interacts primarily with its nearest neighbors, and additional atoms make the crystal larger but don't change its density.

   The second bit of evidence for the short range comes from Figure 12.4. Since the binding energy per nucleon is roughly constant, total nuclear binding energies are roughly proportional to A. For a force with long range (such as the gravitational and electrostatic forces, which have infinite range) the binding energy would be roughly proportional to A^2. (For example, because each of the Z protons in a nucleus feels the repulsion of the other $Z - 1$ protons, the total electrostatic energy of the nucleus is proportional to $Z(Z - 1)$, which is roughly Z^2 for large Z.)

   Figure 12.5 illustrates the dependence of the nuclear binding energy on the separation distance between the nucleons. The binding energy is relatively constant for separation distances less than about 1 fm, and it is zero for separation distances much greater than 1 fm.

3. The nuclear force between any two nucleons does not depend on whether the nucleons are protons or neutrons—the n-p nuclear force is the same as the n-n nuclear force, which is in turn the same as the p-p nuclear force.

A successful model for the origin of this short-range force is the *exchange force*. Suppose we have a neutron and a proton in the nucleus. The neutron emits a particle, and it also exerts a strong attractive force on that particle. The nearby proton, if the emitted particle happens to be close to it, can also exert a strong force on the particle, perhaps strong enough to absorb the particle. The proton can then emit a particle that can be absorbed by the neutron. The proton and neutron each exert a strong force on the exchanged particle, and thus they appear to exert a strong force on each other. The situation is similar to that shown in Figure 12.6, in which two people play catch with a ball to which each is attached by a spring. Each player exerts a force on the ball, and the effect is as if each exerted a force on the other.

How can a neutron of rest energy $m_n c^2$ emit a particle of rest energy mc^2 and still remain a neutron, without violating conservation of energy? The answer to this question can be found from the uncertainty principle, $\Delta E \, \Delta t \sim \hbar$. We don't know that energy has been conserved unless we measure it, and we can't measure it more accurately than the uncertainty ΔE in a time interval Δt. We can therefore "violate" energy conservation

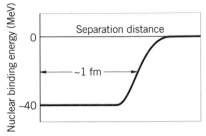

FIGURE 12.5 Dependence of nuclear binding energy on the separation distance of nucleons.

Exchange force

FIGURE 12.6 An attractive exchange force.

by an amount ΔE for a time interval of at most $\Delta t = \hbar/\Delta E$. The amount by which energy conservation is violated in our exchange force model is mc^2, the rest energy of the exchanged particle. This particle can thus exist only for a time interval (in the laboratory frame) of at most

$$\Delta t = \frac{\hbar}{mc^2} \qquad (12.5)$$

The longest distance this particle can possibly travel in the time Δt is $x = c\,\Delta t$, since it can't move faster than the speed of light. We can therefore obtain a relationship between the range of the exchange force and the rest energy of the exchanged particle:

$$x = c\,\Delta t = c\left(\frac{\hbar}{mc^2}\right)$$

or

$$mc^2 = \frac{\hbar c}{x} \qquad (12.6)$$

Inserting into this expression an estimate for the range of the nuclear force of 10^{-15} m or 1 fm, we can estimate the rest energy of the exchanged particle (using the convenient value $\hbar c \cong 200$ MeV·fm):

$$mc^2 \cong 200 \text{ MeV}$$

The exchanged particle cannot be observed in the laboratory during the exchange, for to do so would violate energy conservation. However, if we provide energy to the nucleons from an external source (for example, by causing a nucleus to absorb a photon), the "borrowed" energy can be repaid and the particle can be observed. When we carry out this experiment, the nucleus is found to emit pi mesons (pions), which have a rest energy of 140 MeV, remarkably close to our estimate of 200 MeV. Many observable properties of the nuclear force have been successfully explained by a model based on the exchange of pions. We discuss the properties of pions in Chapter 14.

12.5 RADIOACTIVE DECAY

One of the remarkable properties of certain nuclei is their ability to transform themselves spontaneously from one value of Z and N to another. Other nuclei are stable and do not decay into different nuclei. Usually, for each A value there are one or two stable nuclei. All other nuclei with that A value are unstable and undergo some sort of decay process, until stability is reached.

Figure 12.7 shows a plot of all the known nuclei, with stable nuclei indicated by dark shading. For light stable nuclei, the neutron and proton numbers are roughly equal. However, for the heavy stable nuclei, the factor $Z(Z-1)$ in the Coulomb repulsion energy grows rapidly, so extra neutrons

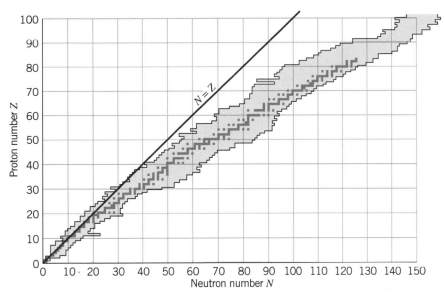

FIGURE 12.7 Stable nuclei are shown in dark; known radioactive nuclei are in light shading.

are required to supply the additional binding energy needed for stability. For this reason, all heavy stable nuclei have $N > Z$.

There are no stable nuclei with $A = 5$ or 8. The alpha particle ${}_{2}^{4}He_{2}$ is a particularly stable nucleus ($B/A = 7.07$ MeV); a nucleus with $A = 5$, such as ${}_{2}^{5}He_{3}$ or ${}_{3}^{5}Li_{2}$, will quickly (10^{-21} s) disintegrate into an alpha particle and a neutron or proton, and a nucleus with $A = 8$ such as ${}_{4}^{8}Be_{4}$ will quickly break apart into two alpha particles.

Unstable nuclei are transformed into other nuclear species by means of two different decay processes that change the Z and N of a nucleus. These two processes are *alpha decay* and *beta decay*. Excited states of nuclei can emit photons, called *gamma* rays, in making transitions that eventually lead to the ground state. However, no change of Z or N occurs. The three decay processes (alpha, beta, and gamma decay) are examples of the general subject of *radioactive decay*. In the remainder of this section, we establish some of the basic properties of radioactive decay, and in the following sections we treat alpha, beta, and gamma decay separately.

The rate at which radioactive nuclei decay in a sample of material is called the *activity* of the sample. The greater the activity, the more nuclear decays per second. (The activity has nothing to do with the *kind* of decays or of radiations emitted by the sample, or with the *energy* of the emitted radiations. The activity is determined only by the *number* of decays per second.)

Activity

The basic unit for measuring activity is the *curie*.* Originally, the curie was defined as the activity of one gram of radium; that definition has since

---

* The SI unit of activity is the becquerel (Bq), named for the discoverer of radioactivity. One becquerel equals one decay/s, so 1 Ci = 3.7 × 10^{10} Bq. The curie is the more widely used unit for activity.

Marie Curie (1867–1934, Poland–France). Her pioneering studies of the natural radioactivity of radium and other elements earned her two Nobel prizes, the physics prize in 1903 for the discovery of radioactivity (shared with Henri Becquerel and with her husband, Pierre) and the unshared chemistry prize in 1911 for the isolation of pure radium. She established the Institute of Radium at the University of Paris, where she continued to pursue research in the medical applications of radioactive materials. Her daughter Irene was awarded the 1935 Nobel prize in chemistry for the discovery of artificial radioactivity.

been replaced by a more convenient one:

$$1 \text{ curie (Ci)} = 3.7 \times 10^{10} \text{ decays/s}$$

One curie is quite a large activity, and so we work more often with units of millicurie (mCi), equal to 10^{-3} Ci, and microcurie (μCi), equal to 10^{-6} Ci.

Consider a sample with a mass of a few grams, containing the order of 10^{23} atoms. If the activity were as large as 1 Ci, about 10^{10} of the nuclei in the sample would decay every second. We could also say that for any one nucleus, the probability of decaying during each second is about $10^{10}/10^{23}$ or 10^{-13}. This quantity, the decay probability per nucleus per second, is called the *decay constant* (represented by the symbol λ). We assume that λ is a small number, and that it is constant in time for any particular material—the probability of any one nucleus decaying doesn't depend on the age of the sample. The activity $\mathcal{A}$ depends on the number N of radioactive nuclei in the sample and also on the probability λ for each nucleus to decay:

$$\mathcal{A} = \lambda N \tag{12.7}$$

Both $\mathcal{A}$ and N are functions of the time t. As our sample decays, N certainly decreases—there are fewer radioactive nuclei left. If N decreases and λ is constant, then $\mathcal{A}$ must also decrease with time so the number of decays per second becomes smaller with increasing time.

We can regard $\mathcal{A}$ as the change in the number of radioactive nuclei per unit time—the more nuclei decay per second, the larger is $\mathcal{A}$.

$$\mathcal{A} = -\frac{dN}{dt} \tag{12.8}$$

(We have included a minus sign because dN/dt is negative, since N is decreasing with time, and we want $\mathcal{A}$ to be a positive number.) From Equations 12.7 and 12.8 we have

$$\frac{dN}{dt} = -\lambda N \tag{12.9}$$

or

$$\frac{dN}{N} = -\lambda \, dt \tag{12.10}$$

This equation can be integrated directly to yield

Radioactive decay law

$$N = N_0 e^{-\lambda t} \tag{12.11}$$

where N_0 represents the number of radioactive nuclei originally present at $t = 0$. Equation 12.11 is the *exponential law of radioactive decay*, which tells us how the number of radioactive nuclei in a sample decreases with time. We can't really measure N, but we can put this equation in a more useful form by multiplying on both sides by λ, which gives

$$\mathcal{A} = \mathcal{A}_0 e^{-\lambda t} \tag{12.12}$$

where $\mathcal{A}_0$ is the original activity.

Suppose we count the number of decays of our sample in one second (by counting for one second the radiations resulting from the decays). We wait for a while, and then repeat the measurement. Continuing this process, we could then plot the activity $\mathcal{A}$ as a function of time, as shown in Figure 12.8. This plot shows the exponential dependence expected on the basis of Equation 12.12.

It is often more useful to plot $\mathcal{A}$ as a function of t on a semilogarithmic scale, as shown in Figure 12.9. On this kind of plot, Equation 12.12 gives a straight line; fitting a straight line to the data gives the value of λ.

The *half-life*, $t_{1/2}$, of the decay is the time that it takes for the activity to be reduced by half, as shown in Figure 12.8. That is, $\mathcal{A} = \mathcal{A}_0/2$ when $t = t_{1/2}$.

$$\frac{\mathcal{A}_0}{2} = \mathcal{A}_0 e^{-\lambda t_{1/2}}$$

from which we find

$$t_{1/2} = \frac{1}{\lambda}\ln 2$$

$$= \frac{0.693}{\lambda} \tag{12.13}$$

Another useful parameter is the *mean lifetime* τ (see Problem 22):

$$\tau = \frac{1}{\lambda} \tag{12.14}$$

When $t = \tau$, $\mathcal{A} = \mathcal{A}_0 e^{-1} = 0.37\mathcal{A}_0$.

EXAMPLE 12.5

The half-life of ^{198}Au is 2.70 days. (*a*) What is the decay constant of ^{198}Au? (*b*) What is the probability that any ^{198}Au nucleus will decay in one second? (*c*) Suppose we had a 1.00-μg sample of ^{198}Au. What is its activity? (*d*) How many decays per second occur when the sample is one week old?

SOLUTION

(*a*)

$$\lambda = \frac{0.693}{t_{1/2}} = \frac{0.693}{2.70\,\mathrm{d}} \cdot \frac{1\,\mathrm{d}}{24\,\mathrm{h}} \cdot \frac{1\,\mathrm{h}}{3600\,\mathrm{s}}$$

$$= 2.97 \times 10^{-6}\,\mathrm{s}^{-1}$$

(*b*) The decay probability per second is just the decay constant, so the probability of any ^{198}Au nucleus decaying in one second is 2.97×10^{-6}.

Half-life

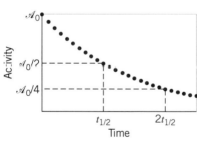

FIGURE 12.8 Activity of a radioactive sample as a function of time.

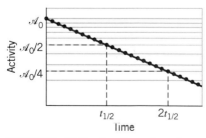

FIGURE 12.9 Semilog plot of activity versus time.

(*c*) The number of atoms in the sample is determined from the Avogadro constant N_A and the molar mass M:

$$N = \frac{mN_A}{M} = \frac{(1.00 \times 10^{-6} \text{ g})(6.02 \times 10^{23} \text{ atoms/mole})}{198 \text{ g/mole}}$$

$$= 3.04 \times 10^{15} \text{ atoms}$$

$$\mathcal{A} = \lambda N = (2.97 \times 10^{-6} \text{ s}^{-1})(3.04 \times 10^{15})$$

$$= 9.03 \times 10^9 \text{ decays per second}$$

$$= 0.244 \text{ Ci}$$

(*d*) The activity decays according to Equation 12.12

$$\mathcal{A} = \mathcal{A}_0 e^{-\lambda t}$$

$$= (9.03 \times 10^9 \text{ decays/s})e^{-(0.693/2.70 \text{ d})(7 \text{ d})}$$

$$= 1.50 \times 10^9 \text{ decays/s}$$

EXAMPLE 12.6

The half-life of ^{235}U is 7.04×10^8 y. A sample of rock, which solidified with the Earth 4.55×10^9 years ago, contains N atoms of ^{235}U. How many ^{235}U atoms did the same rock have at the time it solidified?

SOLUTION

The age of the rock corresponds to

$$\frac{4.55 \times 10^9 \text{ y}}{7.04 \times 10^8 \text{ y}} = 6.46 \text{ half-lives}$$

Since each half-life reduces N by a factor of 2, the overall reduction in N has been

$$2^{6.46} = 88.2$$

The original rock therefore contained $88.2N$ atoms of ^{235}U.

12.6 CONSERVATION LAWS IN RADIOACTIVE DECAYS

Our study of radioactive decays and nuclear reactions reveals that nature is not arbitrary in selecting the outcome of decays or reactions, but rather that certain laws limit the possible outcomes. We call these laws *conservation laws* and we believe these laws to give us important insight into the fundamental workings of nature. Several of these conservation laws are applied to radioactive decay processes.

1. *Conservation of energy.* Perhaps the most important of the conservation laws, conservation of energy tells us which decays are energetically possible and enables us to calculate rest energies or kinetic energies of decay products. A nucleus X will decay into a lighter nucleus X', with the emission of one or more particles we call collectively x ($X \rightarrow X' + x$), only if the rest energy of X is greater than the total rest energy of $X' + x$. The excess rest energy is known as the *Q value* of the decay:

$$m_N(X)c^2 = m_N(X')c^2 + m_N(x)c^2 + Q$$

$$Q = [m_N(X) - m_N(X') - m_N(x)]c^2 \qquad (12.15) \quad Q \ value$$

where m_N represents the *nuclear mass*. The decay is possible only if this Q value is positive. The excess energy Q appears as kinetic energy of the decay products (assuming X is initially at rest):

$$Q = K_{X'} + K_x \qquad (12.16)$$

2. *Conservation of linear momentum.* If the initially decaying nucleus is at rest, then the total momentum of all of the decay products must sum to zero

$$\mathbf{p}_{X'} + \mathbf{p}_x = 0 \qquad (12.17)$$

Usually the emitted particle or particles x are much less massive than the residual nucleus X', and the *recoil momentum* $p_{X'}$ yields a very small kinetic energy $K_{X'}$.

If there is only one emitted particle x, Equations 12.16 and 12.17 can be solved simultaneously for $K_{X'}$ and K_x. If x represents two or more particles, we have more unknowns than we have equations, and no unique solution is possible. In this case, a range of values from some minimum to some maximum is permitted for the decay products.

3. *Conservation of angular momentum.* The total spin angular momentum of the initial particle before the decay must equal the total angular momentum (spin plus orbital) of all of the product particles after the decay. For example, the decay of a neutron (spin angular momentum = $\frac{1}{2}$) into a proton plus an electron is forbidden by conservation of angular momentum, because the spins of the proton and electron, each equal to $\frac{1}{2}$, can be combined to give a total of either 0 or 1, neither of which is equal to the initial angular momentum of the neutron. Adding integer units of orbital angular momentum to the electron does not restore angular momentum conservation in this decay process.

4. *Conservation of electric charge.* This is such a fundamental part of all decay and reaction processes that it hardly needs elaborating. The total net electric charge before and after the decay must not change.

5. *Conservation of mass number.* In some decay processes, we can create particles (photons or electrons, for example) which did not exist before the decay occurred. (This of course must be done out of the available energy—that is, it takes 0.511 MeV of energy to create an electron.) However, nature does *not* permit us to create or destroy protons and neutrons, although in certain decay processes we can convert neutrons into protons or protons into neutrons. *The total mass number A does not change in decay*

or reaction processes. In some decay processes, A remains constant because both Z and N remain unchanged; in other processes Z and N both change in such a way as to keep their sum constant.

12.7 ALPHA DECAY

In alpha decay, an unstable nucleus disintegrates into a lighter nucleus and an alpha particle (a nucleus of ^{4}He), according to

$$\ _{Z}^{A}X_N \longrightarrow\ _{Z-2}^{A-4}X'_{N-2} +\ _{2}^{4}He_2 \tag{12.18}$$

where X and X' represent different nuclear species.

Decay processes of this sort liberate energy, since the decay products are more tightly bound than the initial nucleus. The liberated energy, which appears as the kinetic energy of the alpha particle and the "daughter" nucleus X', can be found from the masses of the nuclei involved according to Equation 12.15:

$$Q = [m(X) - m(X') - m(\ ^4He)]c^2 \tag{12.19}$$

As we did in our calculations of binding energy, we can show that the electron masses cancel in Equation 12.19, and so we can use *atomic masses*. This energy Q appears as kinetic energy of the decay products:

$$Q = K_{X'} + K_\alpha \tag{12.20}$$

assuming we choose a reference frame in which X is at rest. Linear momentum is also conserved in the decay process, as shown in Figure 12.10, so that

$$p_\alpha = p_{X'} \tag{12.21}$$

FIGURE 12.10 A nucleus X alpha decays, resulting in a nucleus X' and an alpha particle.

From Equations 12.20 and 12.21 we eliminate $p_{X'}$ and $K_{X'}$, since we normally don't observe the daughter nucleus in the laboratory (an alpha particle can travel relatively much further through matter than a heavy nucleus). Typical alpha decay energies are a few MeV; thus the kinetic energies of the alpha particle and the nucleus are much smaller than their corresponding rest energies, and so we can use nonrelativistic mechanics to find

Alpha particle kinetic energy

$$K_\alpha \cong \frac{A-4}{A}Q \tag{12.22}$$

EXAMPLE 12.7

Find the kinetic energy of the alpha particle emitted in the alpha decay of ^{226}Ra.

SOLUTION

From the list of isotopes in Appendix B we find the decay process to be:

$$^{226}_{88}\text{Ra}_{138} \longrightarrow {}^{222}_{86}\text{Rn}_{136} + \alpha$$

$$Q = [m(^{226}\text{Ra}) - m(^{222}\text{Rn}) - m(^{4}\text{He})]c^2$$

$$= (226.025403\ u - 222.017571\ u - 4.002603\ u)\ (931.5\ \text{MeV}/u)$$

$$= 4.871\ \text{MeV}$$

$$K_\alpha = \frac{A-4}{A}Q = \left(\frac{222}{226}\right) 4.871\ \text{MeV}$$

$$= 4.785\ \text{MeV}$$

Table 12.2 shows some sample alpha decays and their half-lives. You can see from the table that small changes in the decay energy (about a factor of 2) result in enormous changes in the half-life (24 orders of magnitude)! For example, for the isotopes ^{232}Th and ^{230}Th (which have the same Z and therefore the same Coulomb interaction between the alpha particle and the product nucleus) the kinetic energy changes by only 0.68 MeV (about 15%), while the half-life changes by about five orders of magnitude. Any successful calculation of the alpha decay probabilities must account for this sensitivity to the decay energy.

Alpha decay is an example of quantum-mechanical barrier penetration, as we discussed in Chapter 5. Let us imagine that two neutrons and two protons happen to come together inside a nucleus to form an alpha particle. (Picture the neutrons and protons swimming around within the nucleus, occasionally clumping together and breaking apart again.) That alpha particle is bound inside the nucleus by the nuclear force, which is represented in Figure 12.11 as a constant negative potential energy. Once it gets beyond the nuclear radius R, it feels the Coulomb repulsion of the daughter nucleus

TABLE 12.2 SOME ALPHA DECAY ENERGIES AND HALF-LIVES

Isotope	K_α (MeV)	$t_{1/2}$	λ (s^{-1})
^{232}Th	4.01	1.4×10^{10} y	1.6×10^{-18}
^{238}U	4.19	4.5×10^{9} y	4.9×10^{-18}
^{230}Th	4.69	8.0×10^{4} y	2.8×10^{-13}
^{241}Am	5.64	433 y	5.1×10^{-11}
^{230}U	5.89	20.8 d	3.9×10^{-7}
^{210}Rn	6.16	2.4 h	8.0×10^{-5}
^{220}Rn	6.29	56 s	1.2×10^{-2}
^{222}Ac	7.01	5 s	0.14
^{215}Po	7.53	1.8 ms	3.9×10^{2}
^{218}Th	9.85	0.11 μs	6.3×10^{6}

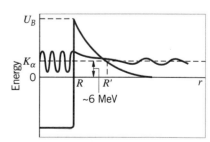

FIGURE 12.11 Barrier penetration by an alpha particle.

$(U \propto 1/r)$. The height of the barrier U_B is

$$U_B = \frac{1}{4\pi\varepsilon_0} \frac{2(Z-2)e^2}{R} \tag{12.23}$$

which gives 30 to 40 MeV for a typical heavy nucleus. (Here the factor of 2 in the numerator comes from the electric charge of the alpha particle, and the factor $Z - 2$ occurs because the daughter nucleus is responsible for the Coulomb force.) Typically, alpha particles have energies of 4 to 8 MeV, and so it is impossible for the alpha particle to surmount the barrier; the only way the alpha particle can escape is to "tunnel" through the barrier.

The probability per unit time λ for the alpha particle to appear in the laboratory is the probability of its penetrating the barrier multiplied by the number of times per second the alpha particle strikes the barrier in its attempt to escape. If the alpha particle is moving at speed v inside a nucleus of radius R, it will strike the barrier as it bounces back and forth inside the nucleus at time intervals of $2R/v$. In a heavy nucleus with $R \sim 6$ fm, the α particle strikes the "wall" of the nucleus about 10^{22} times per second!

The probability for the alpha particle to penetrate the barrier can be found by solving the Schrödinger equation for the potential energy shown in Figure 12.11. To simplify this calculation, we can replace the Coulomb barrier with a "flat" barrier, as shown in Figure 12.12. As we discussed in Chapter 5, the probability to penetrate a potential energy barrier is determined by the exponential factor e^{-2kL}, where L is the thickness of the barrier and where $k = \sqrt{(2m/\hbar^2)(U_0 - E)}$ for a barrier of height U_0 and a particle of energy E. The decay probability can then be estimated as

$$\lambda = \frac{v}{2R} e^{-2kL} \tag{12.24}$$

which includes both the rate at which the particle strikes the barrier and its probability to penetrate it. By making suitable rough estimates for the thickness and height of the barrier (see Problem 29), you should be able roughly to reproduce the range of values for the decay probabilities given in Table 12.2.

An exact calculation of the decay probability can be done by replacing the Coulomb barrier with a thin series of flat barriers that are chosen to fit the Coulomb barrier as closely as possible. This calculation was first done in 1928 and was one of the first successful applications of the quantum theory.

Some nuclei can be unstable to the emission of other particles or collections of particles. Nuclei that have an abundance of protons (those at the left-hand boundary of the light shaded region of Figure 12.7) have recently been discovered to emit protons in a rare process similar to alpha decay. In this way they reduce their proton excess and move closer to stability. An example of this process is $^{151}_{71}\text{Lu}_{80} \rightarrow \, ^{150}_{70}\text{Yb}_{80} + \text{p}$.

Other nuclei have recently been shown to emit clusters of particles such as ^{12}C, ^{14}C, or ^{20}Ne. The following example illustrates this process.

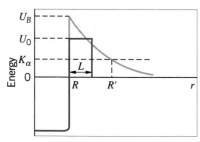

FIGURE 12.12 Replacing the Coulomb barrier for alpha decay with a flat barrier of height U_0.

EXAMPLE 12.8

The nucleus ^{226}Ra decays by alpha emission with a half-life of 1602 y. It also decays by emitting ^{14}C. Find the Q value for ^{14}C emission and compare with that for alpha emission (see Example 12.7).

SOLUTION

If ^{226}Ra emits ^{14}C, which contains 6 protons and 8 neutrons, the resulting nucleus is ^{212}Pb, so the decay process is

$$^{226}\text{Ra} \longrightarrow {}^{212}\text{Pb} + {}^{14}\text{C}$$

The Q value can be found from Equation 12.15, where we can again use atomic masses because the electron masses cancel.

$$Q = [m(^{226}\text{Ra}) - m(^{212}\text{Pb}) - m(^{14}\text{C})]c^2$$
$$= (226.025403 \text{ u} - 211.991871 \text{ u} - 14.003242 \text{ u})(931.5 \text{ MeV/u})$$
$$= 28.215 \text{ MeV}$$

Even though the Q value far exceeds the Q value for alpha decay (4.871 MeV), the Coulomb barrier for ^{14}C decay is roughly 3 times higher and thicker than it is for alpha decay [change the factor $2(Z - 2)$ to $6(Z - 6)$ in Equation 12.23]. As a result, the probability for ^{14}C decay turns out to be only about 10^{-9} of the probability for alpha decay; that is, ^{226}Ra emits one ^{14}C for every 10^9 alpha particles. See Problem 30 for a calculation of the relative decay probabilities.

12.8 BETA DECAY

In beta decay a neutron in the nucleus changes into a proton (or a proton into a neutron); Z and N each change by one unit, but A doesn't change. When this decay process was first studied, the emitted particles were called beta particles; later they were shown to be electrons. In the most basic beta decay process, a free neutron decays into a proton and an electron: n → p + e (plus a third particle, as we discuss later).

The emitted electron is *not* one of the orbital electrons of the atom. It also is not an electron that was previously present within the nucleus, for as we have seen (Example 4.7) the uncertainty principle forbids electrons of the observed energies to exist inside the nucleus. The electron is "manufactured" by the nucleus out of the available energy. If the rest energy *difference* between the nuclei is at least $m_e c^2$, this will be possible.

Early experiments with beta decay revealed two difficulties. First, the decay n → p + e$^-$ is forbidden because it appears to violate the law of

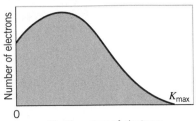

FIGURE 12.13 Spectrum of electrons emitted in beta decay.

conservation of angular momentum, as we discussed in Section 12.6. Another serious problem was revealed by measurements of the energy of the emitted electrons—the energy spectrum of the electrons was continuous, from zero up to some maximum value K_{max}, as shown in Figure 12.13. For example, in the neutron decay, the Q value is

$$Q = (m_n - m_p - m_e)c^2 \tag{12.25}$$

In calculating Q values for beta decay, we must exercise a bit of care, since the electron masses may *not* cancel out at they did in our previous calculations using atomic masses. This is because the initial and final nuclei have different Z values, and therefore different numbers of electrons. For neutron beta decay, we can group m_p and m_e together to give the atomic mass $m(^1_1H_0)$, and so

$$Q = [m_n - m(^1_1H_0)]c^2 \tag{12.26}$$

which we calculate to be 0.782 MeV.

Except for a very small correction, which accounts for the recoil energy of the proton, all of this energy should appear as kinetic energy of the electron, and all emitted electrons should have *exactly* this energy. We find by experiment that all the emitted electrons have less than this energy. In fact, they have a continuous range of energy, from 0 up to their maximum. Before 1930, physicists attacked the problem of this "missing" energy with considerable vigor and ingenuity, but without success.

Wolfgang Pauli found the solution to the apparent violations of conservation of angular momentum and energy in 1930; he suggested that there is a *third* particle emitted in beta decay. Since electric charge is already conserved by the proton and electron, this new particle cannot have electric charge. If it has spin $\frac{1}{2}$, it will satisfy conservation of angular momentum, since we can combine the spins of the three decay particles to give $\frac{1}{2}$. The "missing" energy is the energy carried away by this third particle, and the observed fact that the energy spectrum extends all the way to the value $Q = [m_n - m(^1_1H_0)]c^2$ suggests that this particle has a mass and a rest energy of 0, just like the photon. (We know this new particle can't be a photon, because a photon has a spin of 1.)

This new particle is called the *neutrino* ("little neutral one" in Italian) and has the symbol ν. As we discuss in Chapter 14, every particle has an *antiparticle*, and the antiparticle of the neutrino is the *antineutrino* $\bar{\nu}$. It is, in fact, the antineutrino that is emitted in neutron beta decay. The complete decay process is thus

$$n \longrightarrow p + e^- + \bar{\nu} \tag{12.27}$$

Neutron decay can also occur in a nucleus, in which a nucleus with Z protons and N neutrons decays to a nucleus with $Z + 1$ protons and $N - 1$ neutrons:

Negative beta decay Q value

$$^A_ZX_N \longrightarrow ^{A}_{Z+1}X'_{N-1} + e^- + \bar{\nu} \tag{12.28}$$

The Q value for this decay is

$$Q = [m(^A X) - m(^A X')]c^2 \qquad (12.29)$$

It can be shown (Problem 31) that the electron masses cancel in calculating Q, so it is *atomic masses* that appear in Equation 12.29. The neutrino does not appear in the calculation of the Q value because its mass is either zero or negligibly small.

The energy released in the decay (the Q value) appears as the energy E_ν of the antineutrino, the kinetic energy K_e of the electron, and a small (usually negligible) recoil kinetic energy of the nucleus X':

$$Q \cong E_\nu + K_e \qquad (12.30)$$

The electron (which must be treated relativistically, because its kinetic energy is *not* small compared with its rest energy) has its maximum kinetic energy when the antineutrino has an energy of approximately zero. Figure 12.13 shows the energy distribution of electrons emitted in a typical negative beta decay.

Another beta decay process is

$$p \longrightarrow n + e^+ + \nu \qquad (12.31)$$

in which a *positive electron*, or *positron*, is emitted. The positron is the antiparticle of the electron; it has the same mass as the electron but the opposite electric charge. This decay has a negative Q value, and so it is never observed in nature for free protons. (This is indeed fortunate—if the free proton were unstable to beta decay, stable hydrogen atoms, the basic material of the universe, could not exist!) Protons in nuclei can undergo such decay processes:

$$^A_Z X_N \longrightarrow {}_{Z-1}^A X'_{N+1} + e^+ + \nu \qquad (12.32)$$

The Q value for this process is (Problem 31)

$$Q = [m(^A X) - m(^A X') - 2m_e]c^2 \qquad (12.33)$$

Positive beta decay Q value

in which the masses are *atomic masses*. The positron and neutrino share this energy; the positron has its maximum energy when the neutrino has zero energy, according to Equation 12.30 and again neglecting the small recoil energy of the nucleus X'. Figure 12.14 shows the energy distribution of positrons emitted in a typical positive beta decay.

A nuclear decay process that competes with positron emission is *electron capture*; the basic electron capture process is

$$p + e^- \longrightarrow n + \nu \qquad (12.34)$$

in which a proton captures an electron from its orbit and converts into a neutron plus a neutrino. The electron necessary for this process is one of the inner orbital electrons in an atom, and we identify the capture process by the shell from which the captured electron comes: *K*-shell capture, *L*-shell capture, and so forth. (The electronic orbits that come closest to, or even penetrate, the nucleus have the higher probability to be captured.) The electron capture process does not occur for free protons, but in nuclei the process is

$$^A_Z X_N + e^- \longrightarrow {}_{Z-1}^A X'_{N+1} + \nu \qquad (12.35)$$

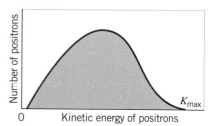

FIGURE 12.14 Spectrum of positrons emitted in beta decay.

TABLE 12.3 TYPICAL BETA DECAY PROCESSES

Decay	Type	Q (MeV)	$t_{1/2}$
$^{19}\text{O} \rightarrow {}^{19}\text{F} + e^- + \bar{\nu}$	β^-	4.82	27 s
$^{176}\text{Lu} \rightarrow {}^{176}\text{Hf} + e^- + \bar{\nu}$	β^-	1.19	3.6×10^{10} y
$^{25}\text{Al} \rightarrow {}^{25}\text{Mg} + e^+ + \nu$	β^+	3.26	7.2 s
$^{124}\text{I} \rightarrow {}^{124}\text{Te} + e^+ + \nu$	β^+	2.14	4.2 d
$^{15}\text{O} + e^- \rightarrow {}^{15}\text{N} + \nu$	EC	2.75	122 s
$^{170}\text{Tm} + e^- \rightarrow {}^{170}\text{Er} + \nu$	EC	0.31	129 d

and the Q value, using atomic masses, is

Electron capture decay Q value

$$Q = [m(^A X) - m(^A X')]c^2 \qquad (12.36)$$

In this case, neglecting the small initial kinetic energy of the electron and the recoil energy of the nucleus, the neutrino takes all of the available final energy:

$$E_\nu = Q \qquad (12.37)$$

In contrast to other beta-decay processes, a *monoenergetic* neutrino is emitted in electron capture.

Table 12.3 gives some typical beta decay processes, along with their Q values and half-lives.

EXAMPLE 12.9

^{23}Ne decays to ^{23}Na by negative beta emission. What is the maximum kinetic energy of the emitted electrons?

SOLUTION

This decay is of the form given by Equation 12.28:

$$^{23}_{10}\text{Ne}_{13} \longrightarrow {}^{23}_{11}\text{Na}_{12} + e^- + \bar{\nu}$$

and the Q value is found from Equation 12.29, using *atomic* masses:

$$Q = [m(^{23}\text{Ne}) - m(^{23}\text{Na})]c^2$$

$$= (22.994465 \text{ u} - 22.989768 \text{ u})(931.5 \text{ MeV/u})$$

$$= 4.375 \text{ MeV}$$

Except for a small correction for the kinetic energy of the recoiling nucleus, the maximum kinetic energy of the electrons is equal to this value. (This occurs when the neutrino has a negligible energy. Similarly, the maximum *neutrino* energy occurs when the electron has a negligibly small kinetic energy.)

EXAMPLE 12.10

^{40}K is an unusual isotope, in that it decays by negative beta emission, positive beta emission, and electron capture. Find the Q values for these decays.

SOLUTION

The process for negative beta decay is given by Equation 12.28:

$$^{40}_{19}\text{K}_{21} \longrightarrow \, ^{40}_{20}\text{Ca}_{20} + e^- + \bar{\nu}$$

and the Q value is found from Equation 12.29 using atomic masses:

$$Q_{\beta^-} = [m(^{40}\text{K}) - m(^{40}\text{Ca})]c^2$$

$$= (39.963999 \text{ u} - 39.962591 \text{ u})(931.5 \text{ MeV/u})$$

$$= 1.312 \text{ MeV}$$

Equation 12.32 gives the decay process for positive beta emission:

$$^{40}_{19}\text{K}_{21} \longrightarrow \, ^{40}_{18}\text{Ar}_{22} + e^+ + \nu$$

and the Q value is given by Equation 12.33:

$$Q_{\beta^+} = [m(^{40}\text{K}) - m(^{40}\text{Ar}) - 2m_e]c^2$$

$$= (39.963999 \text{ u} - 39.962384 \text{ u} - 2 \times 0.000549 \text{ u})(931.5 \text{ MeV/u})$$

$$= 0.482 \text{ MeV}$$

For electron capture,

$$^{40}_{19}\text{K}_{21} + e^- \longrightarrow \, ^{40}_{18}\text{Ar}_{22} + \nu$$

and from Equation 12.36:

$$Q_{\text{ec}} = [m(^{40}\text{K}) - m(^{40}\text{Ar})]c^2$$

$$= (39.963999 \text{ u} - 39.962384 \text{ u})(931.5 \text{ MeV/u})$$

$$= 1.504 \text{ MeV}$$

12.9 GAMMA DECAY

Following alpha or beta decay, the final nucleus may be left in an excited state. Just as an atom does, the nucleus will reach its ground state after emitting one or more photons, known as *nuclear gamma rays.* The energy of each photon is the energy difference between the initial and final nuclear states, less a negligibly small correction for the recoil kinetic energy of the

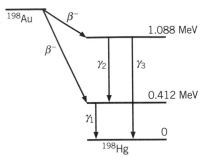

FIGURE 12.15   Some gamma rays emitted following beta decay.

nucleus. These energies are typically in the range of 100 keV to a few MeV. Nuclei can likewise be excited from the ground state to an excited state by absorbing a photon of the appropriate energy, in a process similar to the resonant absorption by atomic states.

Figure 12.15 shows a typical energy level diagram of excited nuclear states and some of the gamma-ray transitions that can be emitted. Typical values for the half-lives of the excited states are 10^{-9} to 10^{-12} s; the exact values of the half-lives (and the *selection rules* that allow certain transitions and forbid others) depend on rather sophisticated details of nuclear structure beyond the level of this book. Occasionally it happens that these details result in a half-life that is very long—hours or even days. Such states are known as *isomeric states* or *isomers*.

The study of nuclear gamma emission is an important tool of the nuclear physicist; the energies of the gamma rays can be measured with great precision, and they provide a powerful means of deducing the energies of the excited states of nuclei.

In calculating the energies of alpha and beta particles emitted in radioactive decays, we have assumed that no gamma rays are emitted. If there are gamma rays emitted, the available energy (Q value) must be shared between the other particles and the gamma ray, as the following example shows.

EXAMPLE 12.11

^{12}N beta decays to an excited state of ^{12}C, which subsequently decays to the ground state with the emission of a 4.43-MeV gamma ray. What is the maximum kinetic energy of the emitted beta particle?

SOLUTION

To determine the Q value for this decay, we first need to find the mass of the product nucleus ^{12}C *in its excited state*. In the ground state, ^{12}C has a mass of 12.000000 u, so its mass in the excited state is

$$12.000000 \text{ u} + \frac{4.43 \text{ MeV}}{931.5 \text{ MeV/u}} = 12.004756 \text{ u}$$

In this decay, a proton is converted to a neutron, so it must be an example of positron decay. The Q value is, according to Equation 12.33,

$$Q = (12.018613 \text{ u} - 12.004756 \text{ u} - 2 \times 0.000549 \text{ u})(931.5 \text{ MeV/u})$$

$$= 11.89 \text{ MeV}$$

(Notice that we could have just as easily found the Q value by first finding the Q value for decay to the *ground state*, 16.32 MeV, and then subtracting the excitation energy of 4.43 MeV, since the decay to the excited state has that much less available energy.)

Neglecting the small correction for the recoil kinetic energy of the ^{12}C nucleus, the maximum electron kinetic energy is 11.89 MeV.

12.10 NATURAL RADIOACTIVITY

All of the elements beyond the very lightest (hydrogen and helium) were produced by nuclear reactions in the interiors of stars. These reactions produce not only stable elements, but radioactive ones as well. Most of the radioactive elements have half-lives of the order of days or years, much smaller than the age of the Earth (about 4.5×10^9 y). Therefore, most of the radioactive elements that may have been present when the Earth was formed have decayed to stable elements. However, a few of the radioactive elements created long ago have half-lives that are of the same order as the age of the Earth, and so are still present and can still be observed to undergo radioactive decay. These elements are part of the background of *natural radioactivity* that surrounds us.

Radioactive decay processes either change the mass number A of a nucleus by four units (alpha decay) or don't change A at all (beta or gamma decay). A radioactive decay process can be part of a sequence or series of decays if a *radioactive* element of mass number A decays to another *radioactive* element of mass number A or $A - 4$. Such a series of processes will continue until a stable element is reached. A hypothetical such series is illustrated in Figure 12.16. Since gamma decays don't change Z or A, they are not shown; however, most of the alpha and beta decays are accompanied by gamma-ray emissions.

The A values of the members of such decay chains differ by a multiple of 4 (including zero as a possible multiple) and so we expect four possible decay chains, with A values that can be expressed as $4n$, $4n + 1$, $4n + 2$, and $4n + 3$, where n is an integer. One of the four naturally occurring radioactive series is illustrated in Figure 12.17. Each series begins with a relatively long-lived member, proceeds through many α and β decays, which may have very short half-lives, and finally ends with a stable isotope. Three of these series begin with isotopes having half-lives comparable to the age of the Earth, and so are still observed today. The neptunium series ($4n + 1$) begins with ^{237}Np which has a half-life of "only" 2.1×10^6 y, much less than the 4.5×10^9 y since the formation of the Earth. Thus all of the ^{237}Np that was originally present has long since decayed to ^{209}Bi.

EXAMPLE 12.12

Compute the Q value for the ^{238}U $\rightarrow$ ^{206}Pb decay chain, and find the rate of energy production per gram of uranium.

SOLUTION

Because A changes by 32, there must be 8 alpha decays in the chain. These 8 alpha decays would decrease Z by 16 units, from 92 to 76. However, the final Z must be 82, so there must also be 6 beta decays in the chain. We recall that for β^- decays, the electron masses combine with the nuclear masses in the computation of the Q value and we can therefore use atomic

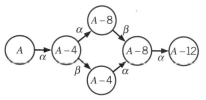

FIGURE 12.16 An example of a hypothetical radioactive decay chain.

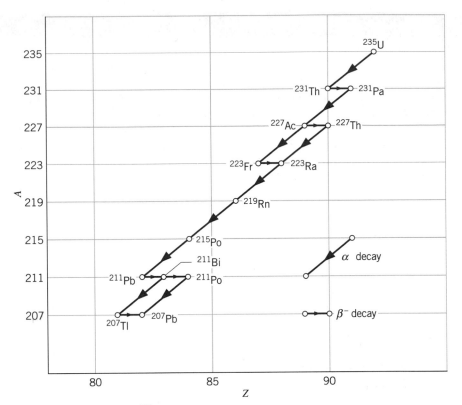

FIGURE 12.17 The ^{235}U decay chain.

masses. Thus for the entire decay chain

$$Q = [m(^{238}\text{U}) - m(^{206}\text{Pb}) - 8m(^4\text{He})]c^2$$

$$= (238.050786 \text{ u} - 205.974455 \text{ u} - 8 \times 4.002603 \text{ u})(931.5 \text{ MeV/u})$$

$$= 51.7 \text{ MeV}$$

One gram of ^{238}U is $\frac{1}{238}$ mole and therefore contains $\frac{1}{238} \times 6 \times 10^{23}$ atoms. The half-life of the decay is 4.5×10^9 y, so λ, the decay probability per atom, is

$$\lambda = \frac{0.693}{4.5 \times 10^9 \text{ y}} \times \frac{1 \text{ y}}{3.16 \times 10^7 \text{ s}} = 4.9 \times 10^{-18} \text{ s}^{-1}$$

Thus, on the average, the number of decays of the ^{238}U is

$$\left(\frac{1}{238} \times 6 \times 10^{23} \text{ atoms}\right) \times 4.9 \times 10^{-18} \frac{\text{decays}}{\text{atom·s}} = 12,000 \text{ decays/s}$$

Each decay liberates 51.7 MeV, and so the rate of energy liberation is

$$12,000 \frac{\text{decays}}{\text{s}} \times 51.7 \frac{\text{MeV}}{\text{decay}} \times 10^6 \frac{\text{eV}}{\text{MeV}} \times 1.6 \times 10^{-19} \frac{\text{J}}{\text{eV}} = 1.0 \times 10^{-7} \text{ W}$$

This may seem like a very small rate of energy release, but if the energy were to appear as thermal energy and were not dissipated by some means (radiation or conduction to other matter, for example) the 1-g sample of ^{238}U would increase in temperature by 25 C° per year and would be melted and vaporized in the order of one century! This calculation suggests that we can perhaps account for some of the internal heat of planets through natural radioactive processes.

Radioactive dating

If we examine a sample of uranium-bearing rock, we can find the ratio of ^{238}U atoms to ^{206}Pb atoms. If we assume that all of the ^{206}Pb was produced by the uranium decay and that none was present when the rock was originally formed (assumptions that must be examined with care both theoretically and experimentally), then this ratio can be used to find the age of the sample, as shown in the following example.

EXAMPLE 12.13

Three different rock samples have ratios of numbers of ^{238}U atoms to ^{206}Pb atoms of 0.5, 1.0, and 2.0. Compute the ages of the three rocks.

SOLUTION

Since all of the other members of the uranium series have half-lives that are much shorter than the half-life of ^{238}U (4.5×10^9 y), we ignore the intervening decays and consider only the ^{238}U decay. Let N_0 be the original number of ^{238}U atoms, so that $N_0 e^{-\lambda t}$ is the number that are still present today, and $N_0 - N_0 e^{-\lambda t}$ is the number that have decayed and are presently observed as ^{206}Pb. The ratio R of ^{238}U to ^{206}Pb is thus

$$R = \frac{\text{number of } ^{238}\text{U}}{\text{number of } ^{206}\text{Pb}} = \frac{N_0 e^{-\lambda t}}{N_0 - N_0 e^{-\lambda t}} = \frac{1}{e^{\lambda t} - 1}$$

Solving for t, we find:

$$t = \frac{1}{\lambda} \ln\left(\frac{1}{R} + 1\right) \tag{12.38}$$

and recalling that $\lambda = 0.693/t_{1/2}$ we can write this as

$$t = \frac{t_{1/2}}{0.693} \ln\left(\frac{1}{R} + 1\right) \tag{12.39}$$

We can then find the values of t corresponding to the three values of R,

$$R = 0.5 \qquad t = 7.1 \times 10^9 \text{ y}$$
$$R = 1.0 \qquad t = 4.5 \times 10^9 \text{ y}$$
$$R = 2.0 \qquad t = 2.6 \times 10^9 \text{ y}$$

The oldest rocks on Earth, dated by similar means, have ages of about 4.5×10^9 y. The age of the first rock analyzed above, 7.1×10^9 y, suggests either that the rock had an extraterrestrial origin, or else that our assumption of no initial ^{206}Pb was incorrect. The age of the third rock suggests that it solidified only 2.6×10^9 y ago; previous to that time it was molten and the decay product ^{206}Pb may have "boiled away" from the ^{238}U.

There are a number of other naturally occurring radioactive isotopes that are not part of the decay chain of the heavy elements. A partial list is given in Table 12.4; some of these can also be used for radioactive dating.

Other radioactive elements are being produced continuously in the Earth's atmosphere as a result of nuclear reactions between air molecules and the high-energy particles known as "cosmic rays." The most notable and useful of these is ^{14}C, which beta decays with a half-life of 5730 y. When a living plant absorbs CO_2 from the atmosphere, a small fraction (about 1 in 10^{12}) of the carbon atoms is ^{14}C, and the remainder is stable ^{12}C (99 percent) and ^{13}C (1 percent). When the plant dies, its intake of ^{14}C stops, and the ^{14}C decays. If we assume that the composition of the Earth's atmosphere and the flux of cosmic rays have not changed significantly in the last few thousand years, we can find the age of specimens of organic material by comparing their ^{14}C/^{12}C ratios to those of living plants. The following example shows how this *radiocarbon dating* technique is used.

Radiocarbon dating

EXAMPLE 12.14

(a) A sample of carbon dioxide gas from the atmosphere fills a vessel of volume 200.0 cm^3 to a pressure of 2.00×10^4 Pa (1 Pa = 1 N/m^2, about 10^{-5} atm) at a temperature of 295 K. Assuming that all of the ^{14}C beta decays were counted, how many counts would be accumulated in one week? (b) An old sample of wood is burned, and the resulting carbon dioxide is placed in an identical vessel at the same pressure and temperature. After one week, 1420 counts have been accumulated. What is the age of the sample?

SOLUTION

(a) We first find the number of moles present in the vessel, using the ideal gas law:

$$n = \frac{PV}{RT} = \frac{(2.00 \times 10^4 \, \text{N/m}^2)(2.00 \times 10^{-4} \, \text{m}^3)}{(8.314 \, \text{J/mole·K})(295 \, \text{K})}$$

$$= 1.63 \times 10^{-3} \, \text{mole}$$

Since each mole of CO_2 contains 6.02×10^{23} molecules, the number of molecules N is

$$N = (6.02 \times 10^{23} \, \text{molecules/mole})(1.63 \times 10^{-3} \, \text{mole})$$

$$= 9.82 \times 10^{20} \, \text{molecules}$$

Each molecule has one carbon atom, so N is also the number of carbon atoms in the sample. If the fraction of ^{14}C atoms is 10^{-12}, there are $9.82 \times$

TABLE 12.4   SOME NATURALLY OCCURRING RADIOACTIVE ISOTOPES

Isotope	$t_{1/2}$
^{40}K	1.28×10^9 y
^{87}Rb	4.8×10^{10} y
^{92}Nb	3.2×10^7 y
^{113}Cd	9×10^{15} y
^{115}In	5.1×10^{14} y
^{138}La	1.1×10^{11} y
^{176}Lu	3.6×10^{10} y
^{187}Re	4×10^{10} y

10^8 atoms of ^{14}C present. The activity is therefore

$$\mathcal{A} = \lambda N = \frac{0.693}{5730 \text{ y}} \cdot \frac{1 \text{ y}}{3.16 \times 10^7 \text{ s}} \cdot 9.82 \times 10^8$$

$$= 3.76 \times 10^{-3} \text{ decays/s}$$

In one week the number of decays is 2280.

(b) An identical sample which gives only 1420 counts must be old enough for only 1420/2280 of its original activity to remain.

$$\frac{1420}{2280} = e^{-\lambda t}$$

$$t = \frac{1}{\lambda} \ln \frac{2280}{1420}$$

$$= \frac{5730}{0.693} \ln \frac{2280}{1420} = 3920 \text{ y}$$

*12.11 THE MÖSSBAUER EFFECT

A typical atomic transition has a lifetime of about 10^{-8} s and an energy of a few electron-volts (for visible light). When an atom emits a photon of energy E_γ and momentum p_γ, the atom must recoil so that momentum is conserved. If the atom is initially at rest, its recoil momentum p_R is equal to the photon momentum (but in the opposite direction) and its recoil kinetic energy is $K = p_R^2/2M$, where M is the mass of the atom. (We assume that $K \ll Mc^2$, and so nonrelativistic kinematics is applicable.) Since $p_R = p_\gamma = E_\gamma/c$,

$$K = \frac{E_\gamma^2}{2Mc^2} \qquad (12.40)$$

With $E_\gamma \sim 1$ eV and $Mc^2 \sim 1000$ MeV $\times A$, even for the lightest atoms K is of the order of 10^{-10} eV. Thus if the energy available from the atomic transition is E (the energy difference of the electronic levels), conservation of energy demands that the photon energy be $E_\gamma = E - K$, where for the atomic case, $K \ll E$.

One productive way of studying atomic systems is to do *resonance* experiments. In such experiments, radiation from a collection of atoms in an excited state is incident on a collection of identical atoms in their ground state. The ground-state atoms can absorb the photons and jump to the corresponding excited state. However, as we have seen, the emitted photon energy is less than the transition energy by the recoil kinetic energy K; moreover, it is less than the photon energy required for resonance by $2K$,

---

* This is an optional section that may be skipped without loss of continuity.

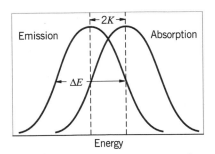

FIGURE 12.18   Representative emission and absorption energies in an atomic system.

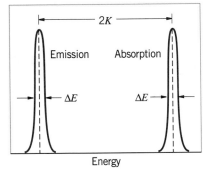

FIGURE 12.19   Representative emission and absorption energies in a nuclear system.

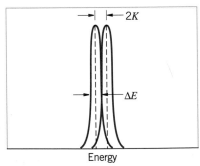

FIGURE 12.20   Emission and absorption energies for nuclei bound in a crystal lattice.

since the absorbing atom must recoil also. The absorption experiment is still possible, because the excited states don't have "exact" energies—a state with a mean lifetime τ has an energy uncertainty ΔE which is given by the uncertainty relationship: $\Delta E\,\tau \sim \hbar$. That is, the state lives on the average for a time τ, and during that time we can't determine its energy to an accuracy less than ΔE. For typical atomic states, $\tau \sim 10^{-8}$ s, so $\Delta E \sim 10^{-7}$ eV. Since K, which is of the order of 10^{-10} eV, is much less than the width ΔE, the "shift" caused by the recoil is not large, and the widths of the emitting and absorbing atomic states cause sufficient overlap for the absorption process to occur. Figure 12.18 illustrates this case.

The situation is different for nuclear gamma rays. A typical lifetime might be 10^{-10} s, and so the widths are the order of $\Delta E \sim 10^{-5}$ eV. The photon energies are typically 100 keV = 10^5 eV, and so K is of order 1 eV. This situation is depicted in Figure 12.19, and you can immediately see that since K is so much larger than the width ΔE, no overlap of emitter and absorber is possible.

In 1958, Rudolf Mössbauer made a discovery for which he was awarded the 1961 Nobel prize in physics. He discovered that he could eliminate most of the recoil by placing the radioactive nuclei and the absorbing atoms in crystals. Since the crystalline binding energies are large compared with K, the individual atoms are held tightly to their positions in the crystal lattice and are not free to recoil; if any recoil is to occur, it must be the whole crystal that recoils. This effect is to make the mass M that appears in Equation 12.40 not the mass of an atom, but the mass of the entire crystal, perhaps 10^{20} times larger than an atomic mass. (As an analogy, imagine the difference between striking a brick with a baseball bat, and striking a brick wall!) Once again the recoil kinetic energy is made small, and resonant absorption can occur (Figure 12.20).

In order to demonstrate the resonant absorption, the source of radiation can be slowly moved relative to the absorber, so that the gamma-ray energies are Doppler shifted in the vicinity of the resonance peak. We can estimate the speed necessary to achieve this effect. The frequency shift due to motion of the source toward the observer is given by (see Equation 2.22)

$$\nu' = \nu\left(1 + \frac{v}{c}\right) \tag{12.41}$$

where we ignore the $\sqrt{1 - v^2/c^2}$ term since $v \ll c$. Since the photon energy E is $h\nu$, we have

$$E' = E\left(1 + \frac{v}{c}\right) \tag{12.42}$$

If we take the width ΔE as a representative estimate of how far we would like to Doppler shift the photon energy, then $E' \cong E + \Delta E$, and so

$$E + \Delta E \cong E + E\frac{v}{c} \tag{12.43}$$

and, solving for v, we find

$$v \cong c\,\frac{\Delta E}{E} \tag{12.44}$$

We have estimated $\Delta E \sim 10^{-5}$ eV (the width of the state) and $E \sim 100$ keV (the energy of the photon). Thus

$$v \cong (3 \times 10^8 \text{ m/s}) \frac{10^{-5} \text{ eV}}{10^5 \text{ eV}} \cong 3 \text{ cm/s}$$

Such low speeds can easily and accurately be produced in the laboratory.

Figure 12.21 shows a diagram of the apparatus to measure the Mössbauer effect. The resonant absorption is observed by looking for decreases in the number of gamma rays that are transmitted through the absorber. At resonance, more gamma rays are absorbed and so the transmitted intensity decreases. Typical results are shown in Figure 12.22.

The Mössbauer effect is an extremely precise method for measuring small changes in the energies of photons. In one particular application, the Zeeman splitting of *nuclear* (not atomic) states can be observed. When a nucleus is placed in a magnetic field, the Zeeman effect causes an energy splitting of the nuclear *m* states, similar to the atomic case. However, nuclear magnetic moments are about 2000 times smaller than atomic magnetic moments, and a typical energy splitting would be about 10^{-6} eV. To observe such an effect directly we would need to measure photon energies to 1 part in 10^{11} (a photon energy of 10^5 eV is shifted by 10^{-6} eV), but using the Mössbauer effect, this is not difficult.

In Chapter 15 we discuss another application of this extremely precise technique, in which the energy gained when a photon "falls" through several meters of the Earth's gravitational field is measured in order to test one prediction of Einstein's general theory of relativity.

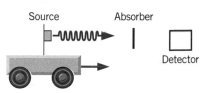

FIGURE 12.21 Mössbauer effect apparatus. A source of gamma rays is made movable, in order to Doppler shift the photon energies. The intensity of radiations transmitted through the absorber is measured as a function of the speed of the source.

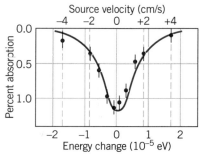

FIGURE 12.22 Typical results in a Mössbauer effect experiment. A velocity of 2 cm/s Doppler shifts the gamma rays enough to move the emission and absorption energies off resonance.

SUGGESTIONS FOR FURTHER READING

The following are some intermediate-level, comprehensive nuclear physics texts.

B. L. Cohen, *Concepts of Nuclear Physics* (New York, McGraw-Hill, 1971).

R. D. Evans, *The Atomic Nucleus* (New York, McGraw-Hill, 1955).

I. Kaplan, *Nuclear Physics* (Reading, Addison-Wesley, 1962).

K. S. Krane, *Introductory Nuclear Physics* (New York, Wiley, 1987).

A complete tabulation of nuclear masses, decay properties, isotopic abundances, excited states is:

C. M. Lederer and V. S. Shirley, editors, *Table of Isotopes*, 7th Edition (New York, Wiley, 1978).

QUESTIONS

1. The magnetic moment of a deuterium nucleus is about $\frac{1}{2000}$ Bohr magneton. What does this imply about the presence of an electron in the nucleus, as the proton-electron model requires?

2. Suppose we have a supply of 20 protons and 20 neutrons. Do we liberate more energy if we assemble them into a single ^{40}Ca nucleus or into two ^{20}Ne nuclei?

3. Atomic masses are usually given to a precision of about the sixth decimal place in atomic mass units (u). This is true both for stable and radioactive nuclei, even though the uncertainty principle requires that an atom with a lifetime Δt has a rest energy uncertain by $\hbar/\Delta t$. Based on the typical lifetimes given for nuclear decays, are we justified in expressing atomic masses to such precision? At what lifetimes would such precision not be justified?

4. Only two stable nuclei have $Z > N$. (a) What are these nuclei? (b) Why don't more nuclei have $Z > N$?

5. In a deuterium nucleus, the proton and neutron spins can be either parallel or antiparallel. What are the possible values of the total spin of the deuterium nucleus? (It is not necessary to consider any orbital angular momentum.) The magnetic moment of the deuterium nucleus is measured to be nonzero. Which of the possible spins is eliminated by this measured value?

6. Why is the binding energy per nucleon relatively constant? Why does it deviate from a constant value for low mass numbers? For high mass numbers?

7. A neutron, which has no electric charge, has a magnetic moment. How is this possible?

8. The electromagnetic interaction can be interpreted as an exchange force, in which photons are the exchanged particle. What does Equation 12.6 imply about the range of such a force? Is this consistent with the conventional interpretation of the electromagnetic force? What would you expect for the rest energy of the exchanged particle that carries the gravitational force?

9. What is meant by assuming that the decay constant λ is a constant, independent of time? Is this a requirement of theory, an axiom, or an experimental conclusion? Under what circumstances might λ change with time?

10. If we focus our attention on a specific nucleus in a radioactive sample, can we know exactly how long that nucleus will live before it decays? Can we predict which half of the nuclei in a sample will decay during one half-life? What part of quantum physics is responsible for this?

11. A certain radioactive sample is observed to undergo 10,000 decays in 10 s. Can we conclude that $\mathcal{A} = 1000$ decays/s if (a) $t_{1/2} \gg 10$ s; (b) $t_{1/2} \cong 10$ s; (c) $t_{1/2} \ll 10$ s?

12. Suppose we wish to do radioactive dating of a sample whose age we guess to be t. Should we choose an isotope whose half-life is (a) $\gg t$; (b) $\sim t$; (c) $\ll t$?

13. The alpha particle is a particularly tightly bound nucleus. Based on this fact, explain why heavy nuclei alpha decay and light nuclei don't.

14. Can you suggest a possible origin for the helium gas that is part of the Earth's atmosphere?

15. Estimate the recoil kinetic energy of the residual nucleus following alpha decay. (This energy is large enough to drive the residual nucleus out of certain radioactive sources; if the residual nucleus is itself radioactive, there is the chance of spread of radioactive material. A thin coating over the source is necessary to prevent this.)

16. Why does the electron energy spectrum (Figure 12.13) look different from the positron energy spectrum (Figure 12.14) at low energies?

17. Will electron capture always be energetically possible when positron beta decay is possible? Will positron beta decay always be energetically possible when electron capture is possible?

18. All three beta decay processes involve the emission of neutrinos (or antineutrinos). In which processes do the neutrinos have a continuous energy spectrum? In which is the neutrino monoenergetic?

19. Neutrinos always accompany electron capture decays. What other kind of radiation always accompanies electron capture? (*Hint*: It is not nuclear radiation.) What other kind of nonnuclear radiation might accompany β^- or β^+ decays in bulk samples?

20. The positron decay of ^{15}O goes directly to the ground state of ^{15}N; no excited states of ^{15}N are populated and no γ rays follow the beta decay. Yet a source of ^{15}O is found to emit γ rays of energy 0.51 MeV. Explain the origin of these γ rays.

21. Would ^{92}Nb be a convenient isotope to use for determining the age of the Earth by radioactive dating? (See Table 12.4.) What about ^{113}Cd?

22. The natural decay chain $^{238}_{92}$U $\rightarrow$ $^{206}_{82}$Pb consists of several alpha decays, which decrease A by 4 and Z by 2, and negative beta decays, which *increase* Z by 1. (a) How many alpha decays must occur in the chain? (b) For that number of alpha decays, how many beta decays must occur to make the final Z come out correctly? (c) As shown in Figure 12.17, sometimes a decay chain can proceed through different branches. Do your answers to (a) and (b) depend on this branching?

23. It has been observed that there is an increased level of radon gas ($Z = 86$) in the air just before an earthquake. Where does the radon come from? How is it produced? How is it released? How is it detected?

24. Which of the decay processes discussed in this chapter would you expect to be most sensitive to the chemical state of the radioactive sample?

25. In Figure 12.22, only 1 percent of the gamma intensity is absorbed, even at resonance. For complete resonance, we would expect 100 percent absorption. What factors might contribute to this small absorption?

PROBLEMS

1. Give the proper isotopic symbols for: (a) The isotope of fluorine with mass number 19. (b) An isotope of gold with 120 neutrons. (c) An isotope of mass number 107 with 60 neutrons.

2. Tin has more stable isotopes than any other element; they have mass numbers 114, 115, 116, 117, 118, 119, 120, 122, 124. Give the symbols for these isotopes.

3. (a) Compute the Coulomb repulsion energy between two nuclei of ^{16}O that just touch at their surfaces. (b) Do the same for two nuclei of ^{238}U.

4. What is the nuclear radius of (a) ^{197}Au; (b) ^{4}He; (c) ^{20}Ne?

5. Figure 12.2 suggests that the Rutherford scattering formula fails for 60° scattering when K is about 28 MeV. Use the results derived in Chapter 6 to find the

closest distance between alpha particle and nucleus for this case, and compare with the nuclear radius of ^{208}Pb. Suggest a possible reason for any discrepancy.

6. Assuming the nucleus to diffract like a circular disk, use the data shown in Figure 12.3 to find the nuclear radius for ^{12}C and ^{16}O. How does changing the electron energy from 360 MeV to 420 MeV affect the deduced radius for ^{16}O? (*Hint*: Use the extreme relativistic approximation from Chapter 2 to relate the electron's energy and momentum to find its de Broglie wavelength.)

7. Find the total binding energy, and the binding energy per nucleon, for (a) ^{208}Pb; (b) ^{133}Cs; (c) ^{90}Zr; (d) ^{59}Co.

8. Find the total binding energy, and the binding energy per nucleon, for (a) ^{4}He; (b) ^{20}Ne; (c) ^{40}Ca; (d) ^{55}Mn.

9. Calculate the total nuclear binding energy of ^{3}He and ^{3}H. Account for any difference by considering the Coulomb interaction of the extra proton of ^{3}He.

10. The *neutron separation energy* S_n is the amount of energy we must supply to remove a neutron from a nucleus. (a) Show that S_n can be found from the masses according to:

$$S_n = [m(\text{n}) + m(^{A-1}_{Z}X_{N-1}) - m(^{A}_{Z}X_N)]c^2$$

(b) Find the neutron separation energy of ^{17}O, ^{7}Li, and ^{57}Fe.

11. Find the *proton separation energy* (see Problem 10) of ^{4}He, ^{12}C, and ^{40}Ca.

12. The nuclear attractive force must turn into a repulsion at very small distances to keep the nucleons from crowding too close together. What is the mass of an exchanged particle that will contribute to the repulsion at separations of 0.25 fm?

13. The weak interaction (the force responsible for beta decay) is produced by an exchanged particle with a mass of roughly 80 GeV. What is the range of this force?

14. What fraction of the original number of nuclei present in a sample will remain after (a) two half-lives; (b) four half-lives; (c) 10 half-lives?

15. A certain sample of a radioactive material decays at a rate of 548 per second at $t = 0$. At $t = 48$ minutes, the counting rate has fallen to 213 per second. (a) What is the half-life of the sample? (b) What is its decay constant? (c) What will be the decay rate at $t = 125$ minutes?

16. What is the decay probability per second per nucleus of a substance with a half-life of 5.0 hours?

17. Tritium, the hydrogen isotope of mass 3, has a half-life of 12.3 y. What fraction of the tritium atoms remains in a sample after 50.0 y?

18. What is the activity of a container holding 125 cm^3 of tritium (^{3}H, $t_{1/2} =$ 12.3 y) at a pressure of 5.0×10^5 Pa (about 5 atm) at $T = 300$ K?

19. Suppose we have a sample containing 2.00 mCi of radioactive ^{131}I ($t_{1/2} =$ 8.04 d). (a) How many decays per second occur in the sample? (b) How many decays per second will occur in the sample after four weeks?

20. Ordinary potassium contains 0.012 percent of the naturally occurring radioactive isotope ^{40}K, which has a half-life of 1.3×10^9 y. (a) What is the activity of 1.0 kg of potassium? (b) What would have been the fraction of ^{40}K in natural potassium 4.5×10^9 y ago?

21. A radiation detector is in the form of a circular disc of diameter 3.0 cm. It is held 25 cm from a source of radiation, where it records 1250 counts per second. Assuming that the detector records every radiation incident upon it, find the activity of the sample (in curies).

22. With a radioactive sample originally of N_0 atoms, we could measure the mean, or average, lifetime τ of a nucleus by measuring the number N_1 that live for a time t_1 and then decay, the number N_2 that decay after t_2 and so on:

$$\tau = \frac{1}{N_0}(N_1 t_1 + N_2 t_2 + \cdots)$$

(a) Show that this is equivalent to $\tau = \lambda \int_0^\infty e^{-\lambda t} t \, dt$. (b) Show that $\tau = 1/\lambda$. (c) Is τ longer or shorter than $t_{1/2}$?

23. Complete the following decays:

(a) $^{27}\text{Si} \rightarrow {}^{27}\text{Al} +$

(b) $^{74}\text{As} \rightarrow {}^{74}\text{Se} +$

(c) $^{228}\text{U} \rightarrow \alpha +$

(d) $^{93}\text{Mo} + e^- \rightarrow$

(e) $^{131}\text{I} \rightarrow {}^{131}\text{Xe} +$

24. Derive Equation 12.22 from Equations 12.20 and 12.21.

25. For which of the following nuclei is alpha decay permitted? (a) ^{210}Bi (b) ^{203}Hg (c) ^{211}At

26. Find the kinetic energy of the alpha particle emitted in the decay of ^{234}U.

27. ^{239}Pu alpha decays with a half-life of 2.41×10^4 y. Compute the power output, in watts, which could be obtained from 1.00 gram of ^{239}Pu.

28. ^{228}Th alpha decays to an excited state of ^{224}Ra, which in turn decays to the ground state with the emission of a 217-keV photon. Find the kinetic energy of the alpha particle. The mass of ^{228}Th is 228.028731 u.

29. By replacing the Coulomb barrier in alpha decay with a flat barrier (see Figure 12.12) of thickness $L = \frac{1}{2}(R' - R)$, equal to half the thickness of the Coulomb barrier that the alpha particle must penetrate, and height $U_0 = \frac{1}{2}(U_B + K_\alpha)$, equal to half the height of the Coulomb barrier above the energy of the alpha particle, estimate the decay half-lives for ^{232}Th and ^{218}Th and compare with the measured values given in Table 12.2. (*Hint:* In calculating the speed of the alpha particle inside the nucleus, assume that the well depth is 30 MeV.) Although the results of this rough calculation do not agree well with the measured values, the calculation does indicate how barrier penetration is responsible for the enormous range of observed half-lives. How would you refine the calculation to obtain better agreement with the measured values?

30. (a) Using the same replacements described in Problem 29, estimate the decay probability of ^{226}Ra for alpha emission and for ^{14}C emission. (See Examples 12.7 and 12.8.) (b) Using the results of part (a), estimate the number of ^{14}C emitted relative to the number of alpha particles emitted by a source of ^{226}Ra.

31. Derive Equations 12.29, 12.33, and 12.36.

32. Compute the recoil proton kinetic energy in neutron beta decay (a) when the electron has its maximum energy; (b) when the neutrino has its maximum energy.

33. Find the maximum kinetic energy of the electrons emitted in the negative beta decay of ^{11}Be.

34. ^{15}O decays to ^{15}N by positron beta decay. (a) What is the Q value for this decay? (b) What is the maximum kinetic energy of the positrons?

35. ^{75}Se decays by electron capture to ^{75}As. Find the energy of the emitted neutrino.

36. In the beta decay of ^{24}Na, an electron is observed with a kinetic energy of 2.15 MeV. What is the energy of the accompanying neutrino?

37. The $4n$ radioactive decay series begins with $^{232}_{90}$Th and ends with $^{208}_{82}$Pb. (a) How many alpha decays are in the chain? (See Question 22.) (b) How many beta decays? (c) How much energy is released in the complete chain? (d) What is the radioactive power produced by 1.00 kg of ^{232}Th$(t_{1/2} = 1.40 \times 10^{10}$ y)?

38. A piece of wood from a recently cut tree shows 12.4 ^{14}C decays per minute. A sample of the same size from a tree cut thousands of years ago shows 3.5 decays per minute. What is the age of this sample?

39. The first excited state of ^{57}Fe decays to the ground state with the emission of a 14.4-keV photon in a mean lifetime of 141 ns. (a) What is the width ΔE of the state? (b) What is the recoil kinetic energy of an atom of ^{57}Fe that emits a 14.4-keV photon? (c) If the kinetic energy of recoil is made negligible by placing the atoms in a solid lattice, resonant absorptions will occur. What velocity is required to Doppler shift the emitted photon so that resonance does not occur?

40. CUPS Exercise 6.2.

41. CUPS Exercise 6.7.

42. CUPS Exercise 6.18.

C H A P T E R

13

Nuclear reactions can occur under slow and controlled conditions, as in a reactor, or under rapid and uncontrolled conditions, as in a nuclear ex- *plosion, in which the equivalent of approximately one cubic centimeter of matter is converted suddenly into energy.*

NUCLEAR
REACTIONS
AND
APPLICATIONS

The knowledge of the nucleus that we can obtain from studying radioactive decays is limited, because only certain radioactive processes occur in nature, only certain isotopes are made in those processes, and only certain excited states of nuclei (those that happen to follow radioactive decays) can be studied. Nuclear reactions, however, give us a controllable way to study *any* nuclear species, and to select any excited states of that species.

In this chapter we discuss some of the different nuclear reactions that can occur, and we study the properties of those reactions. Two nuclear reactions are of particular importance: fission and fusion; we pay special attention to those processes and we discuss how they are useful as sources of energy (or, more correctly, as *converters* of nuclear energy into thermal or electrical energy).

We conclude our study of nuclear physics with an introduction to some of the ways that methods of nuclear physics can be applied to problems in a variety of different areas.

13.1   TYPES OF NUCLEAR REACTIONS

In a typical nuclear reaction laboratory experiment, a beam of particles of type x is incident on a target containing nuclei of type X. After the reaction, an outgoing particle y is observed in the laboratory, leaving a residual nucleus Y. Symbolically, we write the reaction as

$$x + X \longrightarrow y + Y$$

For example,

$${}_{1}^{2}\text{H}_{1} + {}_{29}^{63}\text{Cu}_{34} \longrightarrow \text{n} + {}_{30}^{64}\text{Zn}_{34}$$

Like a chemical reaction, a nuclear reaction must be balanced—the number of protons and neutrons must be the same on both sides of the equation. (The forces responsible for nuclear beta decay can change neutrons into protons, but these forces act on a typical time scale of at least 10^{-10} s. The projectile and target nuclei are within the range of one another's nuclear forces for an interval of at most 10^{-20} s, so there is not enough time for this type of proton-neutron conversion to take place.)

Since a nuclear reaction takes place under the influence only of forces internal to the system of projectile and target, the reaction conserves energy, linear momentum, and angular momentum.

In most experiments, we observe only the outgoing light particle y; the heavy residual nucleus Y usually loses all its kinetic energy (by collisions with other atoms) and therefore stops within the target.

We assume that we produce the reaction by bombarding target nuclei X, initially at rest, with projectiles x of kinetic energy K_x. The product particles then share this kinetic energy, plus or minus any additional energy from the rest energy difference of the initial and final nuclei. (We consider the energetics of nuclear reactions in detail in the next section.)

The bombarding particles x can be either charged particles, supplied by a suitable nuclear accelerator, or neutrons, whose source may be a nuclear reactor. Accelerators for charged particles, illustrated in Figures 13.1 and 13.2, are of two basic types. In a cyclotron, a particle is held in a circular orbit by a magnetic field and receives a small "kick" by an electric field twice each time it travels around the circle; a particle may make perhaps 100 orbits before finally emerging with an energy of the order of 10 to 20 MeV per unit of electric charge. In the Van de Graaff accelerator, a particle is accelerated only once from a single high-voltage terminal, which may be at a potential of as much as 25 million volts; the energy of the particle is then about 25 MeV per unit of charge.

In nuclear reaction experiments, we usually measure two basic properties of the particle y: its energy, and its probability to emerge at a certain angle with a certain energy. We look briefly at these two types of measurements.

If neither the residual nucleus Y nor the outgoing particle y had excited states, then by using conservation of energy and momentum, we should be able to calculate exactly the energy of y when measured at a certain angle. If the nucleus Y is left in an excited state, then the kinetic energy of y is reduced by (approximately) the energy of the excited state above the ground state, since the two particles Y and y must still share the same amount of total energy. Each higher excited state of the nucleus Y corresponds to a certain reduced energy of the particle y, and a measurement of the different energies that the particle y can have tells us about the excited states of the nucleus Y. Figure 13.3 shows an example of a typical set of experimental results and the corresponding deduced excited states of the residual nucleus. Each peak in Figure 13.3 corresponds to a specific energy of y, and therefore to a specific excited state of Y; that is, when particles with

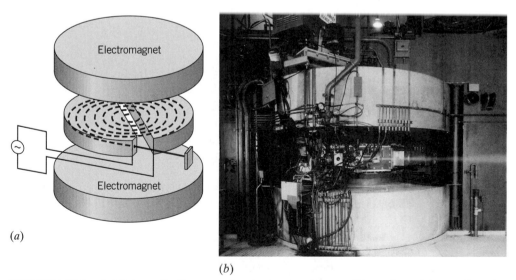

(a)

(b)

FIGURE 13.1 (a) Schematic diagram of cyclotron accelerator. Charged particles are bent in a circular path by the magnetic field and are accelerated by an electric field each time they cross the gap. (b) A cyclotron accelerator. The magnets are in the large cylinders at top and bottom. The beam is visible as it collides with air molecules after leaving the cyclotron.

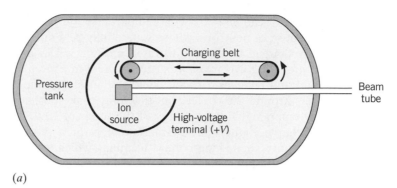

(a)

(b)

FIGURE 13.2   (a) Diagram of a Van de Graaff accelerator. Particles are accelerated from $+V$ to ground. (b) A typical Van de Graaff accelerator laboratory. The beam line and high-voltage terminal are inside the large pressure tank.

energy 9.0 MeV are observed, the nucleus Y is left in the excited state with energy 1.0 MeV.

Notice that the different peaks in Figure 13.3 have different heights. This feature of the results of our experiment tells us that it is more probable for the reaction to lead to one excited state than to another. This is an example of the reaction probability, the second of the properties of y that we can determine. For example, Figure 13.3 shows that the probability of leaving Y in its second excited state (1.0 MeV) is about twice the probability of leaving Y in its first excited state. If it were possible to solve the Schrödinger equation with the nuclear potential, we could calculate these reaction probabilities and compare them with experiment. Since we can't solve this many-body problem, we must work backward by measuring the reaction probabilities and then trying to infer some properties of the nuclear force.

Cross section of nuclear reactions   Reaction probabilities are usually expressed in terms of the *cross section*, which is a sort of effective area presented by the target nucleus to that projectile for a specific reaction, for all possible energies and directions of

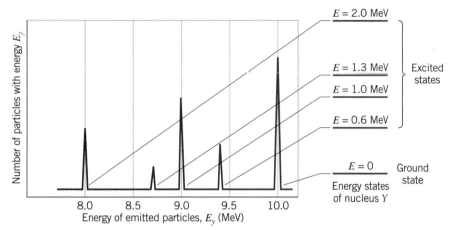

FIGURE 13.3 A sample spectrum of energies of the outgoing particle *y*, and the corresponding excited states of *Y*.

travel of the outgoing particle *y*. In general, the cross section depends on the energy of the incident particle, K_x.

The cross section σ is expressed in units of area, but the area is a very small one, of the order of 10^{-28} m². Nuclear physicists use this as a convenient unit of measure for cross sections, and it is known as one *barn* (b): 1 barn = 10^{-28} m². Notice that the area of the disc of a single nucleus of medium weight is about 1 barn; however, reaction cross sections can often be very much greater or less than one barn. For example, consider the cross section for these reactions involving certain isotopes of the neighboring elements iodine and xenon:

$$\text{I} + \text{n} \longrightarrow \text{I} + \text{n (inelastic scattering)} \qquad \sigma = 4\,\text{b}$$
$$\text{Xe} + \text{n} \longrightarrow \text{Xe} + \text{n (inelastic scattering)} \qquad \sigma = 4\,\text{b}$$
$$\text{I} + \text{n} \longrightarrow \text{I} + \gamma \text{ (neutron capture)} \qquad \sigma = 7\,\text{b}$$
$$\text{Xe} + \text{n} \longrightarrow \text{Xe} + \gamma \text{ (neutron capture)} \qquad \sigma = 10^6\,\text{b}$$

You can see that, although the neutron inelastic scattering cross sections of I and Xe are similar, the neutron capture cross sections are very different. These measurements are therefore telling us something interesting and unusual about the properties of the nucleus Xe.

Suppose a beam of particles is incident on a thin target of area *S*, which contains a total of *N* nuclei. The effective area of each nucleus is the cross section σ, and so the total effective area of all the nuclei in the target is (ignoring shadowing effects) σN. The fraction of the target area which this represents is $\sigma N/S$, and as long as this ratio is small, shadowing effects are negligible. This fraction is the probability for the reaction to occur.

Suppose the incident particles strike the target at a rate of I_0 particles per second, and suppose the outgoing particles *y* are emitted at a rate of *R* per second. (This is also the rate at which the product nucleus *Y* is formed.) Then the reaction probability can also be expressed as the probability to find *y* per incident particle *x*, or R/I_0. Combining the two expressions for the reaction probability, we obtain

$$R = \frac{\sigma N}{S} I_0 \qquad (13.1)$$

This gives a relationship between the reaction cross section and the rate of emission of y.

In a reactor, the intensity of neutrons is usually expressed in terms of the rate at which neutrons cross a unit area perpendicular to the beam, or *neutron flux* ϕ (neutrons/cm^2/s). The cross section is σ (square centimeter per nucleus per incident neutron). The rate R also depends on the number of target nuclei. Suppose the mass of the target is m; the number of target nuclei is then $(m/M)N_A$, where M is the molar mass, and N_A is Avogadro's constant (6.02×10^{23} atoms per mole). Thus, for neutron-induced reactions, using Equation 13.1 we obtain

$$R = \phi\sigma\frac{m}{M}N_A \qquad (13.2)$$

EXAMPLE 13.1

For a certain incident proton energy the reaction

$$p + {}^{56}Fe \longrightarrow n + {}^{56}Co$$

has a cross section of 0.60 b. If we bombard a target in the form of a 1.0-cm square, 1.0-μm thick iron foil, with a beam of protons equivalent to a current of 3.0 μA, and if the beam is spread uniformly over the entire surface of the target, at what rate are the neutrons produced?

SOLUTION

We first calculate the number of nuclei in the target. The volume of the target is

$$V = 1.0\,\text{cm} \times 1.0\,\text{cm} \times 1.0\,\mu\text{m} = 1.0 \times 10^{-4}\,\text{cm}^3$$

and its mass is (using the density of iron of 7.9 g/cm^3)

$$m = \rho V = (7.9\,\text{g/cm}^3)\,(1.0 \times 10^{-4}\,\text{cm}^3) = 7.9 \times 10^{-4}\,\text{g}$$

The number of atoms (or nuclei) is then

$$N = \frac{mN_A}{M} = \frac{(7.9 \times 10^{-4}\,\text{g})\,(6.02 \times 10^{23}\,\text{atoms/mole})}{56\,\text{g/mole}} = 8.5 \times 10^{18}$$

Next we need to find the number of particles per second in the incident beam. We are given that the current is 3.0×10^{-6} A = 3.0×10^{-6} C/s, and since each proton has a charge of 1.6×10^{-19} C, the beam intensity is

$$I_0 = \frac{3.0 \times 10^{-6}\,\text{C/s}}{1.6 \times 10^{-19}\,\text{C/particle}} = 1.9 \times 10^{13}\,\text{particle/s}$$

From Equation 13.1 we can now find R:

$$R = \frac{N\sigma I_0}{S}$$

$$= \frac{(8.5 \times 10^{18}\ \text{nuclei})\ (0.60 \times 10^{-24}\ \text{cm}^2/\text{nucleus})\ (1.9 \times 10^{13}\ \text{particle/s})}{1\ \text{cm}^2}$$

$$= 9.7 \times 10^7\ \text{particles/s}$$

About 10^8 neutrons per second are emitted from the target.

13.2 RADIOISOTOPE PRODUCTION IN NUCLEAR REACTIONS

Often we use nuclear reactions to produce radioactive isotopes. In this procedure, a stable (nonradioactive) isotope X is irradiated with the particle x to form the radioactive isotope Y; the outgoing particle y is of no interest and is not observed.

We would like now to calculate the activity of the isotope Y that is produced from a given exposure to a certain quantity of the particle x for a certain time t. Let R represent the constant rate at which Y is produced; this quantity is related to the cross section and to the intensity of the beam of x, as given in Equation 13.1. (In fact, you should convince yourself that the rate R at which Y is formed is identical to the rate R at which y is emitted.) In a time interval dt, the number of Y nuclei produced is $R\ dt$. Since the isotope Y is radioactive, the number of nuclei of Y that decay in the interval dt is $\lambda N\ dt$, where λ is the decay constant ($\lambda = 0.693/t_{1/2}$) and N is the number of Y nuclei present. The net change dN in the number of Y nuclei is

$$dN = R\ dt - \lambda N\ dt \tag{13.3}$$

or

$$\frac{dN}{dt} = R - \lambda N \tag{13.4}$$

The solution to this differential equation is

$$N(t) = \frac{R}{\lambda}(1 - e^{-\lambda t}) \tag{13.5}$$

and the activity is

$$\mathcal{A}(t) = \lambda N = R(1 - e^{-\lambda t}) \tag{13.6}$$

Notice that, as expected, at $t = 0$, $\mathcal{A} = 0$ (there are no nuclei of type Y present at the start). For large irradiation times $t \gg t_{1/2}$, this expression   *Activity produced in a nuclear reaction*

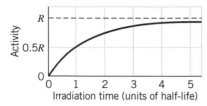

FIGURE 13.4 Formation of activity in a nuclear reaction.

approaches the constant value R. When t is small compared with the half-life $t_{1/2}$, the activity increases linearly with time:

$$\mathcal{A}(t) = R[1 - (1 - \lambda t + \cdots)]$$

$$\mathcal{A}(t) \cong R\lambda t \qquad (t \ll t_{1/2}) \tag{13.7}$$

Figure 13.4 shows the relationship between $\mathcal{A}(t)$ and t. As you can see, not much activity is gained by irradiating for more than about two half-lives.

EXAMPLE 13.2

Thirty milligrams of gold are exposed to a neutron flux of 3.0×10^{12} neutron/cm²/s for one minute. The neutron capture cross section of gold is 99 b. Find the resultant activity of ^{198}Au.

SOLUTION

From the table of isotopes in the appendix we find that the stable isotope of gold has a mass number of $A = 197$, and that radioactive ^{198}Au has a half-life of 2.7 d. Thus, using Equation 13.2,

$$R = \left(3.0 \times 10^{12} \frac{\text{neutrons}}{\text{cm}^2 \cdot \text{s}}\right)\left(99 \times 10^{-24} \frac{\text{cm}^2}{\text{neutron} \cdot \text{nucleus}}\right)\left(\frac{0.030 \text{ g}}{197 \text{ g/mole}}\right)$$

$$\times \,(6.02 \times 10^{23} \text{ atoms/mole})$$

$$= 2.7 \times 10^{10} \text{ s}^{-1}$$

Since $t \ll t_{1/2}$ in this case, we can use Equation 13.7:

$$\mathcal{A} = (2.7 \times 10^{10} \text{ s}^{-1})\left(\frac{0.693}{2.7 \text{ d}}\right)\left(1 \text{ min} \times \frac{1 \text{ h}}{60 \text{ min}} \times \frac{1 \text{ d}}{24 \text{ h}}\right)$$

$$= 4.8 \times 10^{6} \text{ s}^{-1}$$

$$= 130 \,\mu\text{Ci}$$

EXAMPLE 13.3

The radioactive isotope ^{61}Cu $(t_{1/2} = 3.41 \text{ h})$ is to be produced by alpha particle reactions on a target of ^{59}Co. A target of cobalt, measuring 1.5 cm × 1.5 cm in area and 2.5 μm in thickness, is placed in a 12.0-μA beam of alpha particles; the beam uniformly covers the target. For the alpha energy selected, the reaction has a cross section of 0.640 b. (*a*) At what rate is the ^{61}Cu produced? (*b*) What is the resulting activity of ^{61}Cu after 2.0 h of irradiation?

SOLUTION

(*a*) The reaction is ^{59}Co + ^{4}He → ^{61}Cu + 2n. The mass of the target is

$$m = \rho V = (8.9 \text{ g/cm}^3)(1.5 \text{ cm})^2 (2.5 \times 10^{-4} \text{ cm}) = 5.0 \times 10^{-3} \text{ g}$$

and the number of target atoms is then

$$N = \frac{mN_A}{M} = \frac{(5.0 \times 10^{-3}\text{ g}) (6.02 \times 10^{23}\text{ atoms/mole})}{58.9\text{ g/mole}}$$

$$= 5.12 \times 10^{19}\text{ atoms}$$

The rate at which the beam strikes the target is

$$I_0 = \frac{12.0 \times 10^{-6}\text{ A}}{2 \times 1.60 \times 10^{-19}\text{ C/particle}} = 3.75 \times 10^{13}\text{ particle/s}$$

The rate at which the ^{61}Cu is produced is, using Equation 13.1,

$$R = \frac{N\sigma I_0}{S} = \frac{(5.12 \times 10^{19}\text{ atoms}) (0.640 \times 10^{-24}\text{ cm}^2) (3.75 \times 10^{13}\text{ s}^{-1})}{(1.5\text{ cm})^2}$$

$$= 5.5 \times 10^8\text{ s}^{-1}$$

(*b*) The activity is determined from Equation 13.6:

$$\mathcal{A} = R(1 - e^{-\lambda t}) = (5.5 \times 10^8\text{ s}^{-1}) \left(1 - e^{-(0.693)(2.0\text{ h})/(3.41\text{ h})}\right)$$

$$= 1.8 \times 10^8\text{ s}^{-1} = 4.9\text{ mCi}$$

13.3 LOW-ENERGY REACTION KINEMATICS

We assume for this discussion that the velocities of the nuclear particles are sufficiently small that we can use nonrelativistic kinematics. We consider a projectile *x* moving with momentum $\mathbf{p}_x$ and kinetic energy K_x. The target is at rest, and the reaction products have momenta $\mathbf{p}_y$ and $\mathbf{p}_Y$, and kinetic energies K_y and K_Y. The particles *y* and *Y* are emitted at angles θ_y and θ_Y with respect to the direction of the incident beam. Figure 13.5 illustrates this reaction. We assume that the resultant nucleus *Y* is not observed in the laboratory (if it is a heavy nucleus, moving relatively slowly, it generally stops within the target).

As we did in the case of radioactive decay, we use energy conservation to compute the *Q* value for this reaction (assuming *X* is initially at rest):

$$\text{initial energy} = \text{final energy}$$

$$m_N(x)c^2 + K_x + m_N(X)c^2 = m_N(y)c^2 + K_y + m_N(Y)c^2 + K_Y \quad (13.8)$$

The *m*'s in Equation 13.8 represent the *nuclear* masses of the reacting particles. However, as we have discussed, the number of protons must be balanced in a nuclear reaction:

$$Z_x + Z_X = Z_y + Z_Y \quad (13.9)$$

We can therefore add equal numbers of electron masses to each side of Equation 13.8 and, neglecting as usual the electron binding energy, the

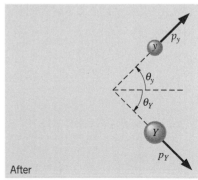

FIGURE 13.5 Momenta of reaction particles before (*x*, *X*) and after (*y*, *Y*) reaction.

nuclear masses become atomic masses with no additional corrections needed. Rewriting Equation 13.8, we obtain

$$[m(x) + m(X) - m(y) - m(Y)]c^2 = K_y + K_Y - K_x \qquad (13.10)$$

Reaction Q value The rest energy difference between the initial particles and final particles is defined to be the *Q value* of the reaction

$$Q = [m(x) + m(X) - m(y) - m(Y)]c^2 \qquad (13.11)$$

and, combining Equations 13.10 and 13.11, we see that the *Q* value is equal to the difference in kinetic energy between the final particles and initial particle:

$$Q = K_y + K_Y - K_x \qquad (13.12)$$

EXAMPLE 13.4

(*a*) Compute the *Q* value for the reaction

$$^2\text{H} + {}^{63}\text{Cu} \longrightarrow \text{n} + {}^{64}\text{Zn}$$

(*b*) Deuterons of energy 12.00 MeV are incident on a ^{63}Cu target, and neutrons are observed with 16.85 MeV of kinetic energy. Find the kinetic energy of the ^{64}Zn.

SOLUTION

(*a*) The atomic masses can be found in Appendix B:

^{2}H:	2.014102 u	n:	1.008665 u
^{63}Cu:	62.929599 u	^{64}Zn:	63.929145 u

The *Q* value can be found using Equation 13.11:

$$Q = (2.014102\,\text{u} + 62.929599\,\text{u} - 1.008665\,\text{u} - 63.929145\,\text{u})$$
$$\times (931.5\,\text{MeV/u})$$
$$= 5.487\,\text{MeV}$$

(*b*) From Equation 13.12, we find

$$K_Y = Q + K_x - K_y$$
$$= 5.487\,\text{MeV} + 12.00\,\text{MeV} - 16.85\,\text{MeV}$$
$$= 0.64\,\text{MeV}$$

Reactions for which $Q > 0$ convert nuclear energy to kinetic energy of *y* and *Y* and are called *exothermic* or *exoergic* reactions. Reactions with $Q < 0$ require energy input, in the form of the kinetic energy of *x*, to be converted into nuclear binding energy. These are known as *endothermic* or *endoergic* reactions.

In an endoergic reaction, we must supply at least enough kinetic energy to provide the additional rest energy of the reaction products. There is thus some minimum, or *threshold*, kinetic energy of x, below which the reaction will not take place. This threshold kinetic energy not only must supply the additional rest energy of the products, but also must supply some kinetic energy of the products; even at the minimum energy, the products cannot be at rest, for that would violate conservation of linear momentum—the momentum p_x before the collision would not be equal to the momentum of the final products after the collision.

This problem is most easily analyzed in the center-of-mass reference frame. In the lab frame, before the reaction, the center of mass moves with speed $v = m(x)v_x/[m(x) + m(X)]$. If we travel with that speed and observe the reaction, we would see x moving with speed $v_x - v$ and X moving with speed $-v$, as shown in Figure 13.6. If x is moving with the threshold kinetic energy, in this reference frame the reaction products y and Y appear to be at rest.

We must conserve total relativistic energy $K + mc^2$ in the reaction, and we restrict our discussion to small velocities $v \ll c$ so that the classical expression for the kinetic energy can be used. Energy conservation in the center-of-mass frame gives:

$$\tfrac{1}{2}m(x)(v_x - v)^2 + \tfrac{1}{2}m(X)(-v)^2 + m(x)c^2 + m(X)c^2 = m(y)c^2 + m(Y)c^2 \quad (13.13)$$

where v_x represents the threshold speed in the lab frame. Substituting the value of v and doing a bit of algebra, we can find the threshold kinetic energy (in the laboratory reference frame):

$$K_{\text{th}} = -Q\left(1 + \frac{m(x)}{m(X)}\right) \quad (13.14)$$

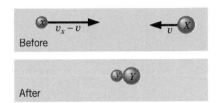

FIGURE 13.6   Reaction at threshold in center-of-mass reference frame.

Threshold kinetic energy

EXAMPLE 13.5

Calculate the threshold kinetic energy for the reaction

$$p + {}_1^3H_2 \longrightarrow {}_1^2H_1 + {}_1^2H_1$$

(*a*) If protons are incident on ^3H at rest. (*b*) If ^3H (tritons) are incident on protons at rest.

SOLUTION

The atomic masses are:

p: 1.007825 u ^3H: 3.016049 u ^2H: 2.014102 u

The Q value is thus

$$Q = (1.007825 \text{ u} + 3.016049 \text{ u} - 2 \times 2.014102 \text{ u})(931.5 \text{ MeV/u})$$

$$= -4.033 \text{ MeV}$$

(*a*) When protons are incident on ^{3}H, the identification is $x = \,^1$H, $X = \,^3$H, so

$$K_{\text{th}} = (4.033 \text{ MeV}) \left(1 + \frac{1.007825 \text{ u}}{3.016049 \text{ u}} \right)$$

$$= 5.381 \text{ MeV}$$

(*b*) When ^{3}H is incident on protons, the identification of x and X is reversed, so

$$K_{\text{th}} = (4.033 \text{ MeV}) \left(1 + \frac{3.016049 \text{ u}}{1.007825 \text{ u}} \right)$$

$$= 16.10 \text{ MeV}$$

This calculation illustrates a general result: Less energy is required for a nuclear reaction if a light particle is incident on a heavy target than if a heavy particle is incident on a light target.

13.4 FISSION

In the process of fission, a heavy nucleus such as uranium splits into two lighter nuclei. Since the lighter nuclei are about 1 MeV per nucleon more tightly bound than the heavy nucleus (see Figure 12.4), there is an energy conversion of about 200 MeV (200 nucleons × 1 MeV per nucleon) in each fission process. Compare this with typical electronic processes in atoms that involve a few electron-volts of energy—the energy converted per atom is roughly 10^8 times greater in nuclear reactions than in chemical reactions!

In a nucleus, there is a competition between the nuclear force, which holds the nucleus together, and the electrostatic repulsion of the nuclear protons, which tries to tear the nucleus apart. For most nuclei, the nuclear force dominates this competition, but for heavy nuclei there is a delicate balance between the nuclear and electric forces, a balance that is easily upset.

We can imagine a stable heavy nucleus as a sort of liquid drop, with a slightly elongated equilibrium shape, as shown in Figure 13.7. When that nucleus is disturbed, such as by absorbing a neutron or a high-energy photon, the drop begins to vibrate and quiver. The shape of the nucleus changes rapidly back and forth from more elongated to rather spherical. When the nucleus is stretched to highly elongated shapes, the Coulomb repulsion energy is not changed very much, but the nuclear force, being of short range, is reduced significantly. With sufficient stretching the center becomes "pinched off" somewhat, and the nucleus readily splits into two pieces, with the Coulomb repulsion driving the two pieces apart, a process illustrated in Figure 13.8. The energy necessary to cause fission is typically about 6 MeV (see Problem 21).

The sizes of the two fragments may vary; Figure 13.9 shows the mass distribution of the fragments from the fission of ^{235}U, and we can see that

FIGURE 13.7 The elongated shape of a heavy nucleus.

FIGURE 13.8   Sequence of nuclear shapes in fission.

it is most likely that one fragment will have a mass number of about 90 and the other about 140.

Most of the energy released in fission goes into kinetic energy of the two fission fragments. We can understand that statement with a rough calculation of the Coulomb potential energy of two electric charges of $Z_1 \cong$ 35 and $Z_2 \cong$ 55 (as expected for $A_1 \cong$ 90 and $A_2 \cong$ 140) separated by a distance of $R = R_1 + R_2$, where R_1 and R_2 are the radii of the two fragments (which are assumed to be just touching at their surfaces). The potential energy $Z_1 Z_2 e^2 / 4\pi\varepsilon_0 R$ is easily calculated to be about 200 MeV. The two fragments separate rapidly, with the potential energy converted into about 200 MeV of kinetic energy. Although this is a very rough calculation, it does show that we might expect most of the fission energy to go to the fragments; in fact about 80 percent of the energy released in fission does appear as the kinetic energy of the fragments, and the remaining 20 percent appears as decay products (beta and gamma decays) and kinetic energy of neutrons emitted in the fission process. The neutrons typically have energies of one to several MeV.

A typical fission nuclear reaction might be

$$^{235}_{92}\text{U}_{143} + \text{n} \longrightarrow {}^{93}_{37}\text{Rb}_{56} + {}^{141}_{55}\text{Cs}_{86} + 2\text{n}$$

Of course, many different fission reactions are possible, with many different final products; the mass distribution of fragments was illustrated in Figure 13.9. The number of neutrons produced in the fission process can likewise vary, but the average is about 2.5. Each neutron can then initiate another fission process, resulting in the emission of still more neutrons, followed by more fissions, and so forth. This avalanche or *chain reaction* of fission events, each with the release of about 200 MeV of energy, can either occur under very rapid and uncontrolled conditions, as in a nuclear weapon, or else under slower and carefully controlled conditions, as in a nuclear reactor.

There are three features of the fission reaction that make it useful as a means to generate electrical energy:

1. **Energy dissipation.**   Most of the energy is released as kinetic energy of the fission fragments. These relatively heavy fragments do not travel very far through the reactor fuel element before they dissipate most of their kinetic energy in collisions with the atoms of the fuel element. The energy can be extracted as heat and used to boil water; the resulting steam can then be used in a conventional way to drive a turbine to generate electricity.

2. **Neutron multiplicity.**   The second feature of the fission reaction that makes it useful is that the average number of neutrons produced is greater than one, making possible the chain reaction. How much greater than one it must be, in order to achieve a chain reaction, depends on the construction of the reactor.

3. **Delayed neutrons.**   The third advantage of the fission process is the one that enables an operator or mechanical system to control the reaction

Lise Meitner (1878–1968, Germany-Sweden). Known for her research into radioactivity, Meitner discovered the radioactive element protactinium ($Z = 91$), and was among the first to investigate the properties of beta decay. Her most important work was the explanation of the puzzling results obtained when uranium was bombarded by neutrons; she suggested that the uranium nucleus could split into lighter fragments, and she proposed the name "fission" for this process.

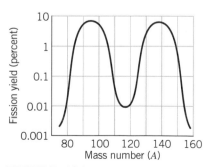

FIGURE 13.9   Mass distribution of fragments from fission of ^{235}U.

and keep it from proceeding too rapidly. The two neutrons emitted in the fission process

$$^{235}_{92}U_{143} + n \longrightarrow \,^{93}_{37}Rb_{56} + \,^{141}_{55}Cs_{86} + 2n$$

Delayed neutrons

are *prompt neutrons*—they are emitted essentially at the instant of fission. About 1 percent of the neutrons in the fission process are *delayed neutrons* emitted following the decays of the heavy fragments. For example, the fission products in the above reaction are unstable, and decay according to the following sequences:

$$^{93}_{37}Rb_{56} \xrightarrow[\beta,\gamma]{6\,s} \,^{93}_{38}Sr_{55} \xrightarrow[\beta,\gamma]{7\,min} \,^{93}_{39}Y_{54} \xrightarrow[\beta,\gamma]{10\,h} \,^{93}_{40}Zr_{53} \xrightarrow[\beta,\gamma]{10^6\,y} \,^{93}_{41}Nb_{52}$$

$$n(1.4\%) \searrow$$

$$^{92}_{38}Sr_{54} \xrightarrow[\beta,\gamma]{3\,h} \,^{92}_{39}Y_{53} \xrightarrow[\beta,\gamma]{4\,h} \,^{92}_{40}Zr_{52}$$

$$^{141}_{55}Cs_{86} \xrightarrow[\beta,\gamma]{25\,s} \,^{141}_{56}Ba_{85} \xrightarrow[\beta,\gamma]{18\,min} \,^{141}_{57}La_{84} \xrightarrow[\beta,\gamma]{4\,h} \,^{141}_{58}Ce_{83} \xrightarrow[\beta,\gamma]{33\,d} \,^{141}_{59}Pr_{82}$$

$$n(0.03\%) \searrow$$

$$^{140}_{56}Ba_{84} \xrightarrow[\beta,\gamma]{13\,d} \,^{140}_{57}La_{83} \xrightarrow[\beta,\gamma]{40\,h} \,^{140}_{58}Ce_{82}$$

The fission product ^{93}Rb decays to ^{93}Sr with a half-life of 6 s. In some of the decays, ^{93}Sr is left in an excited state that may decay by emitting a neutron: ^{93}Sr $\rightarrow \,^{92}$Sr $+$ n. This emission process is very rapid, but the neutron appears to be delayed by the original half-life of the ^{93}Rb decay. Only 1.4 percent of the decays of ^{93}Rb result in the emission of these delayed neutrons. Similarly, delayed neutrons are emitted following 0.03 percent of the decays of ^{141}Cs to an excited state of ^{141}Ba, which decays according to ^{141}Ba $\rightarrow \,^{140}$Ba $+$ n.

Were it not for these delayed neutrons, mechanical control of the reaction rate would not be possible. In practice this control is accomplished by inserting into the core of the reactor a rod of material, such as cadmium, which has a high absorption cross section for neutrons. With the control rod fully inserted, enough neutrons are absorbed so that the average number of neutrons available to cause new fissions is less than one per fission reaction; as the rod is slowly withdrawn, the average number of available neutrons climbs until it is just equal to one per reaction, at which time the reactor is said to be *critical*. During operation, the position of the control rod can be continually adjusted, so that the chain reaction rate and the power level can be held constant. No mechanical system can respond rapidly enough to control the fluctuations in the reaction rate caused by prompt neutrons, but if the reactor is carefully designed to be less than critical for prompt neutrons, and critical for prompt plus delayed neutrons, mechanical control is possible.

Several technological problems required solutions before the nuclear reactor became a useful power generator:

1. **Enrichment.**   The only naturally occurring material with a reasonably large fission cross section is the isotope ^{235}U. Naturally occurring uranium is only 0.7 percent ^{235}U; the other 99.3 percent is ^{238}U, which is

for all practical purposes not fissionable. In order to build a fission reactor or a fission weapon, the concentration of ^{235}U must be substantially increased. This process is known as *enrichment*. Since ^{235}U and ^{238}U are chemically identical, the only means of enrichment is to take advantage of their small mass difference. This is a relatively difficult process, but one that can be accomplished with large quantities of uranium. For example, the gaseous diffusion method is based on the less massive ^{235}U diffusing through porous materials more easily than ^{238}U. Another easily fissionable material is ^{239}Pu. This substance does not occur in nature but can be produced through neutron capture by the nonfissionable ^{238}U; the resulting ^{239}U beta decays to ^{239}Np, which in turn decays to ^{239}Pu:

$$^{238}_{92}U_{146} + n \longrightarrow {}^{239}_{92}U_{147} \longrightarrow {}^{239}_{93}Np_{146} + e^- + \bar{\nu}$$

$$^{239}_{93}Np_{146} \longrightarrow {}^{239}_{94}Pu_{145} + e^- + \bar{\nu}$$

The plutonium can then be separated by chemical means from the uranium. This process of plutonium fuel production from uranium is known as *breeding* and a reactor that is designed to produce plutonium fuel is known as a *breeder*.

2. **Moderation.** A second difficulty in trying to produce a chain reaction is the energy of the neutrons emitted in the fission process. Typically, the kinetic energies of these neutrons are a few MeV; such energetic neutrons have a relatively low probability of inducing new fissions, since the fission cross section generally decreases rapidly with increasing neutron energy. We therefore must slow down, or *moderate*, these neutrons in order to increase their chances of initiating fission events. The fissionable material is surrounded by a *moderator*, and the neutrons lose energy in collisions with the atoms of the moderator. When a neutron is scattered from a heavy nucleus like uranium, the energy of the neutron is changed hardly at all, but in a collision with a very light nucleus, the neutron can lose substantial energy. The most effective moderator is one whose atoms have about the same mass as a neutron; hydrogen is therefore the first choice. Ordinary water is frequently used as a moderator, since collisions with the protons are very effective in slowing the neutrons; however, neutrons have a relatively high probability of being absorbed by the water according to the reaction $p + n \rightarrow {}^{2}_{1}H_1 + \gamma$. So-called "heavy water," in which the hydrogen is replaced by deuterium, is more useful as a moderator, since it has virtually no neutron absorption cross section. A heavy-water reactor, since it has more available neutrons, can use ordinary (nonenriched) uranium as fuel; a reactor using ordinary water as moderator has fewer neutrons available to produce fission, and must therefore have more ^{235}U in its core.

Carbon is a light material that is solid, stable, and abundant, and that has a relatively small neutron absorption cross section. The first nuclear reactor was constructed by Enrico Fermi and his co-workers in 1942 at the University of Chicago; this reactor used carbon, in the form of graphite blocks, as moderator.

Neutron moderators

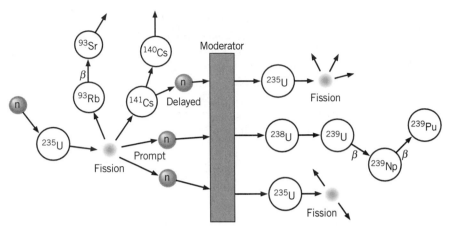

FIGURE 13.10   A typical sequence of processes in fission. A ^{235}U nucleus absorbs a neutron and fissions; two prompt neutrons and one delayed neutron are emitted. Following moderation, two neutrons cause new fissions and the third is captured by ^{238}U resulting finally in ^{239}Pu.

3. **Loss of neutrons.**   There is also a problem in reactor design associated with neutrons that do not produce fission reactions. If every neutron produced a fission reaction, then a self-sustaining chain reaction could occur if the average number of neutrons produced per fission were exactly 1. There are, however, many ways that neutrons can be "lost" and be unable to produce fission reactions: (1) escape through the surface; (2) absorption in the moderator; (3) absorption by ^{238}U. Escape through the surface is minimized by making the core of the reactor large enough so that the surface-to-volume ratio is small, and absorption by the moderator can be eliminated by use of a heavy-water moderator.

Figure 13.10 summarizes some of the processes that can occur in fission. A nucleus of ^{235}U captures a neutron and fissions into two heavy fragments and two prompt neutrons; one of the fragments emits a delayed neutron. The three neutrons are slowed by passage through the moderator. Two of the neutrons cause new fissions, and the third is captured by ^{238}U, eventually to form fissionable ^{239}Pu, which can be recovered from the fuel by chemical means. Not shown in this diagram are other processes that can occur: escape of neutrons through the surface of the reactor, capture in the moderator, and fission of ^{238}U by fast (unmoderated) neutrons.

The heat produced in the reactor fuel must be extracted to generate electrical power. (It must also be extracted for reasons of safety, since enough heat is produced to melt the core and cause a serious accident. For this reason, reactors contain an emergency core cooling system that is designed to prevent the core from overheating if the heat extraction system should fail.) Extracting the fission energy from the reactor core can be accomplished through several different techniques. Reactor designs using these techniques include:

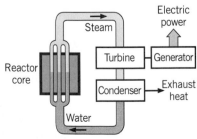

FIGURE 13.11   A   boiling-water reactor.

1. **Boiling-water reactors.**   As indicated in Figure 13.11, a stream of water circulates through the core. The heat turns the water to steam, which is then used to generate electricity. The disadvantage of this system is

that the water can become radioactive, and a rupture of the pipes near the turbines could result in a serious accident, with the spread of radioactive materials.

2. **Pressurized-water reactors.** In this case, as shown in Figure 13.12, the heat is extracted in a two-step process. Water circulates through the core under great pressure, to prevent its turning to steam. This hot water then in turn heats a second water system, which actually delivers steam to the turbine. Since the steam never enters the reactor core, it is not radioactive, and so there is no radioactive material in the vicinity of the turbine.

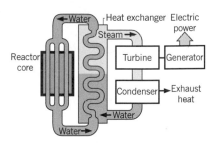

FIGURE 13.12 A pressurized-water reactor.

The power reactors in the United States are mostly the pressurized-water type using enriched uranium as fuel and ordinary water as moderator. Canada also uses pressurized water reactors, but heavy water and natural uranium are used. In another variation on this design, the pressurized water is replaced with a liquid metal such as sodium, which has the advantages of remaining liquid at much higher temperatures than water and of having a larger thermal conductivity than water. Yet another design uses gas flow through the core to extract the heat; the hot gas is then used to produce steam. Reactors in Great Britain are gas-cooled and graphite-moderated.

There are yet other technological problems associated with nuclear power that are the subjects of active debate and investigation. Some of the radioactive isotopes among the fission fragments have very long half-lives, of the order of many years. The radioactive waste from reactors must be stored in a manner that prevents leakage of radioactive material into the biological environment. Many people are concerned about the safety of nuclear reactors, not only regarding proper design and operation, but about their resistance to such external forces as earthquakes and acts of terrorism or sabotage.

In 1986, a graphite-moderated power reactor at Chernobyl in the former U.S.S.R. suffered a serious accident due to the disabling of the core cooling system, which is designed to extract the intense heat generated in the reactor core. The resulting temperature rise ignited the graphite moderator and caused an explosion of the reactor containment vessel, releasing radioactive fission products and exposing the inhabitants of the region to life-threatening radiation doses. Crops and livestock throughout western Europe were contaminated by the spread of radioactive materials. This accident emphasizes the need for redundant safety systems in reactor designs. The water-moderated power reactors used in the United States cannot suffer this kind of accident.

Finally, as in all heat engines, the disposal of the exhaust or waste heat (primarily from the steam recondensing to water) generates considerable thermal pollution. Nuclear power plants are generally less efficient at converting fuel to electrical power compared with plants that burn fossil fuels, because nuclear plants operate at lower temperatures; while fossil-fuel plants can have efficiencies as large as 40 percent, nuclear plants are generally in the range of 30–35 percent. A plant operating at 30 percent efficiency produces 50 percent more thermal pollution than one that generates the same amount of power at 40 percent efficiency.

Natural fission reactor

We conclude this section with a fascinating example of nature at work—the first sustained nuclear fission reactor on Earth was *not* the one constructed by Fermi in Chicago in 1942, but a *natural* fission reactor in Africa, which is believed to have operated two billion years ago for a period of perhaps several hundred thousand years. This reactor of course used naturally occurring uranium as a fuel, and naturally occurring water as a moderator. As we have discussed, it would not be possible to build such a reactor today, because the capture of neutrons by the protons in water results in too few neutrons remaining to sustain a chain reaction in uranium with only 0.7 percent of ^{235}U. However, two billion years ago, naturally occurring uranium contained a much larger fraction of ^{235}U than does present-day uranium. Both ^{235}U and ^{238}U are radioactive, but the half-life of ^{235}U is only about one-sixth as great as the half-life of ^{238}U. If we go back in time about 2×10^9 y, which is half of one half-life of ^{238}U, there was about 40 percent more ^{238}U than there is today, but there was $2^3 = 8$ times as much ^{235}U. Naturally occurring uranium was then about 3 percent ^{235}U, and, at such enrichments, ordinary water can serve as an effective moderator. A deposit of such uranium, in a large enough mass and with ground water present to act as moderator, could have "gone critical" and begun to react. The reaction could have been controlled by the boiling of the water—when enough heat had been generated to evaporate some of the water, the reaction would slow down and perhaps stop, because of the lack of a moderator. When the uranium had cooled sufficiently to allow more liquid water to collect, the reactor would have started up again. This cycle could in principle have continued indefinitely, until enough ^{235}U were used up or until geological changes resulted in the removal of the water.

The discovery of this reactor followed the observation of a French researcher that the uranium that was being mined from that region in Africa contained too little ^{235}U. The discrepancy was a very small one—the samples contained 0.7171 percent ^{235}U, compared with the usual 0.7202 percent—but it was enough to stimulate the curiosity of the French workers. They guessed that the only mechanism that could result in the consumption of ^{235}U was the nuclear fission process, and this guess was tested by searching in the ore for stable isotopes that result from the radioactive decay of fission products. When such isotopes were found, and in particular when they were found in abundances very different from what would be expected from "natural" mineral deposits, the existence of the natural reactor was confirmed. An interesting account of this reactor is given in the reference listed at the end of this chapter.

13.5   FUSION

Energy may also be released in nuclear reactions in the process of fusion, in which two light nuclei combine to form a heavier nucleus. The energy released in this process is the excess binding energy of the heavy nucleus compared with the lighter nuclei; from Figure 12.4, we see that this process can release energy as long as the final nucleus is no heavier than about $A = 60$.

For example, consider the reaction

$$^2_1H_1 + {}^2_1H_1 \longrightarrow {}^3_1H_2 + {}^1_1H_0$$

The Q value is 4.0 MeV, and so this nuclear reaction liberates about 1 MeV per nucleon, roughly the same as the fission reaction. This reaction can be performed in the laboratory, by accelerating a beam of deuterons on to a deuterium target. In order to observe the reaction, we must get the incident and target deuterons close enough that the nuclear force can produce the reaction; that is, we must overcome the mutual Coulomb repulsion of the two particles. We can estimate this Coulomb repulsion by calculating the electrostatic repulsion of two deuterons when they are just touching. The radius of a deuteron is about 1.5 fm, and the electrostatic potential energy of the two charges separated by about 3 fm is about 0.5 MeV. A deuteron with 0.5 MeV of kinetic energy can overcome the Coulomb repulsion and initiate a reaction in which 4.5 MeV of energy (0.5 MeV of incident kinetic energy plus the 4-MeV Q value) is released.

We can produce such a beam of deuterons in many of the accelerators available in nuclear physics laboratories. The beam currents of such accelerators are typically of the order of microamperes. If every particle in the beam produced a reaction (hardly a reasonable assumption!), the total power produced would be 4 W. An output of 4 W (assuming we could extract every bit of the energy liberated in the reaction, which appears as the kinetic energies of the products ^1H and ^3H) hardly makes this device a useful power source!

A more promising approach consists of heating deuterium gas to a high enough temperature so that each atom of deuterium has about 0.25 MeV of thermal kinetic energy (hence the name *thermo*nuclear fusion). Then in a collision between two deuterium atoms, the total of 0.5 MeV of kinetic energy would be sufficient to overcome the Coulomb repulsion. If we could use this method to extract the fusion energy from the deuterium in a cup of "heavy water" (D_2O), we would have an energy of about 5×10^{12} J available; even if the conversion were done over the time of one day, the power output would be about 50 MW! Ordinary water contains about 0.015 percent D_2O; the fusion energy from the deuterium in a liter of ordinary water is equivalent to the chemical energy obtained from burning about 300 liters of gasoline.

The difficulty with this approach is in heating the deuterium gas to a sufficient temperature; from the expression $\frac{3}{2}kT$ for the thermal kinetic energy of a gas molecule, we can calculate that an energy of 0.25 MeV corresponds to a temperature of the order of 10^9 K. Even assuming that barrier penetration (Section 5.7) would allow a reasonable probability to penetrate the Coulomb barrier at lower kinetic energies (perhaps corresponding to one-tenth of the calculated temperature), it is hard to imagine any conditions under which such temperatures can be created. However, such conditions do exist in the interiors of stars, which produce their energy through fusion reactions. Fusion processes thus support all life on Earth. Many scientists and engineers are seeking to develop fusion processes for electrical power generation; they face the challenge of duplicating, for a brief instant of time and on a much smaller scale, the conditions in the interior of stars.

Proton-proton cycle   Let us begin with a brief look at the fusion processes inside the Sun. In the basic fusion process, which can occur through several different paths, four protons combine to make one ^{4}He. Since the Sun is composed of ordinary hydrogen, rather than deuterium, it is first necessary to convert the hydrogen to deuterium. This is done according to the reaction

$$^1_1H_0 + {}^1_1H_0 \longrightarrow {}^2_1H_1 + e^+ + \nu$$

This process involves converting a proton to a neutron and is analogous to the beta-decay processes discussed in Chapter 12. Once we have obtained ^{2}H (deuterium), the next reaction that can occur is

$$^2_1H_1 + {}^1_1H_0 \longrightarrow {}^3_2He_1 + \gamma$$

followed by

$$^3_2He_1 + {}^3_2He_1 \longrightarrow {}^4_2He_2 + 2\,{}^1_1H_0$$

Note that the first two reactions must occur *twice* in order to produce the two ^{3}He we need for the third reaction; see the schematic diagram of Figure 13.13. We can write the net process as

$$4^1_1H_0 \longrightarrow {}^4_2He_2 + 2e^+ + 2\nu + 2\gamma$$

For the calculation of the Q value in terms of *atomic* masses, four electrons must be added to the left side to make four neutral hydrogen atoms. To balance the reaction we must also add four electrons to the right side; two of these are associated with the ^{4}He atom, and the other two can be combined with the two positrons according to the reaction $e^+ + e^- \rightarrow 2\gamma$, so that the additional gamma rays are available as energy from the reaction. Since the two positrons disappear in this process, the only masses remaining are four hydrogen *atoms* and the one helium *atom*, and so

$$Q = (m_i - m_f)c^2$$
$$= (4 \times 1.007825 \text{ u} - 4.002603 \text{ u}) (931.5 \text{ MeV/u})$$
$$= 26.7 \text{ MeV}$$

Each fusion reaction liberates about 26.7 MeV of energy. Let us try now to calculate the rate at which these reactions occur in the Sun. About 1.4×10^3 W of solar power is incident on each square meter of the Earth's surface. At our distance of about 1.5×10^{11} m from the Sun, its energy is spread over a sphere of area $4\pi r^2 = 28 \times 10^{22}$ m^2, and thus the power output from the Sun is about 4×10^{26} W, which corresponds to about 2×10^{39} MeV/s. Each fusion reaction liberates about 26 MeV, and thus there must be about 10^{38} fusion reactions per second, consuming about 4×10^{38} protons per second. (Don't worry about running out of protons—the Sun's mass is about 2×10^{30} kg, which corresponds to about 10^{57} protons, enough to burn for the next few billion years.)

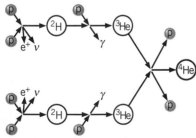

FIGURE 13.13   Schematic diagram of processes in fusion of protons to form helium.

The sequence of reactions described above is called the *proton-proton* cycle and probably represents the source of the Sun's energy. However, it is probably *not* the primary source of fusion energy in many stars, because the first reaction (in which two protons combine to form a deuteron), which is a sort of beta decay, takes place only on a very long time scale (as we discuss in the next chapter), and is therefore very unlikely to occur. A more likely sequence of reactions is the *carbon cycle*:

$$^{12}C + {}^{1}H \longrightarrow {}^{13}N + \gamma$$

$$^{13}N \longrightarrow {}^{13}C + e^{+} + \nu$$

$$^{13}C + {}^{1}H \longrightarrow {}^{14}N + \gamma$$

$$^{14}N + {}^{1}H \longrightarrow {}^{15}O + \gamma$$

$$^{15}O \longrightarrow {}^{15}N + e^{+} + \nu$$

$$^{15}N + {}^{1}H \longrightarrow {}^{12}C + {}^{4}He$$

Carbon cycle

A symbolic diagram of the process is shown in Figure 13.14. Notice that the ^{12}C plays the role of catalyst; we neither produce nor consume any ^{12}C in these reactions, but the presence of the carbon permits this sequence of reactions to take place at a much greater rate than the previously discussed proton-proton cycle. The net process is still described by $4{}^{1}H \rightarrow {}^{4}He$, and of course the Q value is the same. Since the Coulomb repulsion between H and C is larger than the Coulomb repulsion between two H nuclei, more thermal energy and a correspondingly higher temperature are needed for the carbon cycle. The carbon cycle probably becomes important at a temperature of about 20×10^{6} K, while the Sun's interior temperature is "only" 15×10^{6} K.

When all of the hydrogen has been converted to helium, the Sun will contract and its temperature will increase until helium burning occurs, by processes such as

$$3{}^{4}He \longrightarrow {}^{12}C$$

Two He nuclei have a larger mutual Coulomb repulsion than two H nuclei, so helium fusion needs more thermal energy than hydrogen fusion.

When the helium is used up, a still higher temperature will allow carbon fusion to make even heavier elements, for example, ^{24}Mg. Such processes will continue until ^{56}Fe is reached; beyond this point (see Figure 12.4) no further energy is gained by fusion. The production of elements in fusion processes is discussed in greater detail in Chapter 15.

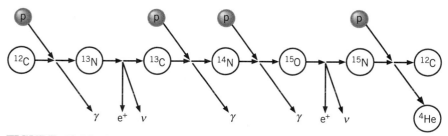

FIGURE 13.14 Sequence of events in carbon cycle.

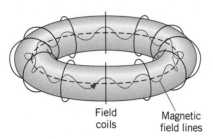

Field coils   Magnetic field lines

FIGURE 13.15 The toroidal geometry of plasma confinement. The ionized atoms circulate around the ring, trapped by the magnetic field lines. The field coils produce a magnetic field along the axis of the toroid (dashed line). Another field component is produced by a current along the axis in the plasma. The two components of the field combine to give the helical field lines shown.

Fusion Reactors

For a controlled thermonuclear reactor, several reactions could be used, such as

$$^{2}\text{H} + {}^{2}\text{H} \longrightarrow {}^{3}\text{H} + {}^{1}\text{H} \qquad Q = 4.0 \,\text{MeV}$$

$$^{2}\text{H} + {}^{2}\text{H} \longrightarrow {}^{3}\text{He} + \text{n} \qquad Q = 3.3 \,\text{MeV}$$

$$^{2}\text{H} + {}^{3}\text{H} \longrightarrow {}^{4}\text{He} + \text{n} \qquad Q = 17.6 \,\text{MeV}$$

The third reaction, known as the D-T (deuterium-tritium) reaction, has the largest energy release and is perhaps the best candidate for a fusion reactor.

When deuterium gas (or a deuterium-tritium mixture) is heated to a high temperature, the atoms become ionized; the resulting gas of hot, ionized particles is called a *plasma*. To increase the probability of collisions between the ions that would result in fusion, there are three requirements for the plasma: (1) *a high density n*, so that the particles have a high probability of collision; (2) *a high temperature T*, in the range of 10^8 K, which increases the probability for the particles to penetrate their mutual Coulomb barrier; and (3) *a long confinement time τ*, during which the high temperature and density must be maintained. The first and third of these parameters can be combined using some fairly general considerations based on the power needed to heat the plasma (which is proportional to the density n) and the power derived from fusions in the plasma (which is proportional to $n^2\tau$). For the fusion power to exceed the input power, the product $n\tau$ must exceed a certain minimum value; this condition is

$$n\tau \geq 10^{20} \,\text{s} \cdot \text{m}^{-3} \tag{13.15}$$

which is known as *Lawson's criterion*. The capability of a plasma to produce energy through fusion can be characterized by the value of its Lawson's parameter $n\tau$ and its temperature T.

The electrical repulsion of the ionized particles in a plasma tends to force the ions away from one another and toward the walls of their container, where they would lose energy in collisions with the cooler atoms of the walls. To maintain the density and temperature, two techniques are under development. In *magnetic confinement*, intense magnetic fields are used to trap the motion of the particles, and in *inertial confinement*, the plasma is heated and compressed so quickly that fusion occurs before the fuel can expand and cool.

Magnetic confinement   A magnetic field can confine a plasma because the charged particles spiral around the magnetic field lines. Figure 13.15 shows a toroidal magnetic confinement geometry. There are two contributions to the magnetic field: One is along the toroid axis and another is around the axis. The combination of these two fields gives a helical field along the toroid axis, and the charged particles are confined as they spiral about the field lines. This type of device is a called a *tokamak* (from the Russian acronym for "toroidal magnetic chamber"). A current passed through the plasma serves both to heat the plasma and to create one of the magnetic field components. Color Plate 11 shows the Tokamak Fusion Test Reactor at Princeton University, which has achieved a fusion power level of 6 MW for a time of 1 s. This device

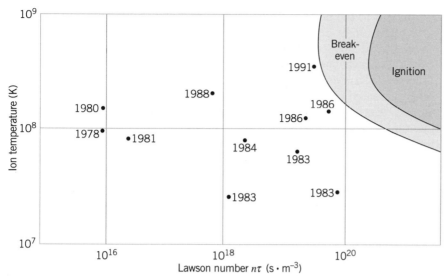

FIGURE 13.16 The approach to breakeven and ignition in controlled fusion reactors, shown as a plot of Lawson number against temperature.

satisfies Lawson's criterion with a plasma density of $n = 10^{20}$ particles/m^3 (five orders of magnitude smaller than an ordinary gas) and a confinement time of $\tau = 1$ s.

The development of magnetic confinement devices has produced a steady march toward achieving a self-sustaining fusion reactor by increasing the values of both Lawson's parameter and the temperature, as illustrated in Figure 13.16. Devices are currently approaching "breakeven," where the power produced by fusion reactions equals the power necessary to heat the plasma. A true self-sustaining reactor requires the attainment of "ignition," where the power produced by fusion reactions can maintain the reactor with no external source of energy.

Inertial confinement takes the opposite approach by compressing the fuel to high densities for very short confinement times. In one method, a small pellet of fuel (0.1–1 mm in diameter) containing deuterium and tritium is struck simultaneously from many directions by intense laser beams that first vaporize the pellet and convert it to a plasma, and then heat and

Inertial confinement

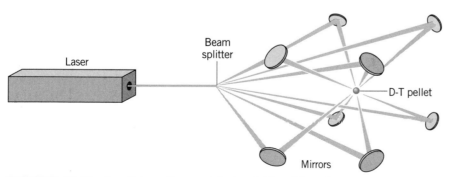

FIGURE 13.17 Inertial confinement fusion initiated by a laser.

compress it to the point at which fusion can occur (Figure 13.17). A typical laser pulse might deliver 10^5 J in a time of 10^{-9} s, for an instantaneous power of 10^{14} W (which exceeds the electrical generating power of the United States by two orders of magnitude!). For a confinement time of 10^{-9} s, the density must exceed 10^{29} particles/m³ (characteristic of ordinary solids) to satisfy Lawson's criterion, but because of inefficiencies of the lasers and other losses a self-sustaining laser fusion reactor must exceed this minimum by perhaps 2–3 orders of magnitude. Color Plate 12 shows the NOVA laser fusion test facility at the Lawrence Livermore National Laboratory, where 10 laser beams converge to deliver simultaneous pulses to a target.

In the D-T fusion reaction, most of the energy is carried by the neutrons (recall in the fission reaction only a small fraction of the energy went to the neutrons). This presents some difficult problems for the recovery of the energy and its conversion into electrical power. One possibility for a fusion reactor design is shown in Figure 13.18. The reaction area is surrounded by lithium, which captures neutrons by the reaction

$$\ce{^6_3Li_3} + n \longrightarrow \ce{^4_2He_2} + \ce{^3_1H_2}$$

The kinetic energies of the reaction products are rapidly dissipated as heat, and the thermal energy of the liquid lithium can be used to convert water to steam in order to generate electricity. This reaction has the added advantage of producing tritium (^{3}H), which is needed as a fuel for the fusion reactor.

One difficulty with the D-T fusion reaction is the large quantity of neutrons released in the reaction. Although fusion reactors will not produce the radioactive wastes that fission reactors do, the neutrons are sure to make radioactive the immediate area surrounding the reactor, and the structural damage to materials resulting from exposure to large fluxes of

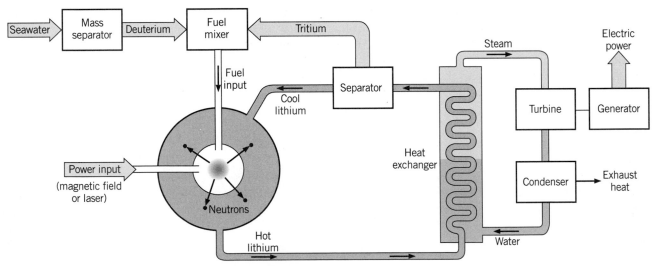

FIGURE 13.18   Proposed design of a fusion reactor.

neutrons may weaken critical parts of the reactor vessel. Here once again the lithium is helpful, since a 1-m thickness of lithium should be sufficient to stop essentially all of the neutrons.

Fusion energy is the subject of vigorous research in many laboratories in the United States and around the world; the technological problems are being attacked with a variety of methods, and researchers are hopeful that solutions can be found during the next 20 years so that fusion can help to supply our electrical power needs.

13.6 APPLICATIONS OF NUCLEAR PHYSICS

In this chapter, we have discussed how fission and fusion reactions can be used to generate electrical power, and in the last chapter we discussed how the radioactive decay of various isotopes can be used to date the historical origin of material containing those isotopes. These are but a few of the many ways that nuclear decays and reactions can be applied to the solution of practical problems. In this section we discuss briefly some other applications of the techniques of nuclear physics.

Neutron Activation Analysis

Nearly every radioactive isotope emits characteristic gamma rays, and many chemical elements can be identified by their gamma ray spectra. For example, when ^{59}Co (the only stable isotope of cobalt) is placed in a flux of neutrons (such as is found near the core of a reactor), neutron absorption results in the production of the radioactive isotope ^{60}Co, which beta decays with a half-life of 5.27 years. Following the beta decay, ^{60}Ni emits two gamma rays of energies 1.17 MeV and 1.33 MeV and of equal intensity. If we place in a flux of neutrons a material of unknown composition, and if we observe, following the neutron bombardment, two gamma rays of equal intensity and energies 1.17 MeV and 1.33 MeV, it is a safe bet that the unknown sample contained cobalt. In fact, from the rate of gamma emission we could deduce exactly how much cobalt the material contains, assuming that we know the neutron flux and the neutron capture cross section of ^{59}Co. This technique is known as *neutron activation analysis*, and has been used in many applications in which the elements are present in such small quantities that chemical identification is not practical. Typically, neutron activation analysis can be used to identify elements in quantities of the order of 10^{-9} g, and sensitivity down to 10^{-12} g is often possible.

Such a sensitive and precise technique finds application in a variety of areas, in which the chemical composition must be determined for samples which are available only in microscopic quantities or which must be analyzed in a nondestructive manner. For example, the chemical composition of

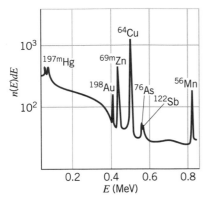

FIGURE 13.19 Gamma-ray spectrum following neutron activation of a sample of human hair. Trace elements revealed include mercury, gold, zinc, copper, arsenic, antimony, and manganese. From D. DeSoete et al., *Neutron Activation Analysis* (New York: Wiley-Interscience, 1972).

various types of pottery can help us trace the geographical origin of the clay from which they were made; such analyses of pottery shards can trace the trading routes of prehistoric people. Art forgeries can be detected by a knowledge of the chemical composition of paints, since techniques for producing pigments have changed over the last four centuries with corresponding changes in the level of impurities in paints. The chemical analysis of minute quantities of material such as paint, gunshot residues, soil, or hair can provide important evidence in criminal investigations. As an extreme example, neutron activation analysis of samples of the hair of such historical figures as Napoleon or Newton has revealed the chemicals to which they were exposed centuries ago. An example of a neutron activation analysis study of a sample of hair is shown in Figure 13.19.

Medical Radiation Physics

One of the most important applications of nuclear physics has been in medicine, both for diagnostic and therapeutic purposes. The use of X rays for producing images for medical diagnosis is well known, but X rays are of limited value. They show distinct and detailed images of bones, but they are generally less useful in making images of soft tissue. Radioactive isotopes can be introduced into the body in chemical forms that have an affinity for certain organs, such as bone or the thyroid gland. A sensitive detector (called a "gamma-ray camera") can observe the radiations from the isotopes that are concentrated in the organ and can produce an image that shows how the activity is distributed in the patient. These detectors are capable of determining where each gamma-ray photon originates in the patient. Figure 13.20 shows an image of the brain, taken after the patient was injected with the radioactive isotope ^{99}Tc ($t_{1/2} = 6$ h). The images clearly show an area of the brain where the activity has concentrated. Ordinarily the brain does not absorb impurities from the blood, so such concentrations often indicate a tumor or other abnormality.

Another technique that reveals a wealth of information is *positron emission tomography* (PET), in which the patient is injected with a positron-emitting isotope that is readily absorbed by the body. Examples of isotopes used are ^{15}O ($t_{1/2} = 2$ min), ^{13}N ($t_{1/2} = 10$ min), ^{11}C ($t_{1/2} = 20$ min), and ^{18}F ($t_{1/2} = 110$ min). These isotopes are produced with a cyclotron, and because of the short half-lives the cyclotron must be present at the site of the diagnostic facility. When a positron emitter decays, the positron quickly annihilates with an electron and produces two 511-keV gamma rays that travel in opposite directions. By surrounding the patient with a ring of detectors, it is possible to determine exactly where the decay occurred, and from a large number of such events, the physician can produce an image that reconstructs the distribution of the radioisotope in the patient. One advantage of the PET scan over X-ray techniques such as the CAT (computerized axial tomography) scan, is that it can produce a dynamic image—changes in the patient during the measuring time can be observed. Color Plate 13 shows a brain scan of a patient who was injected with glucose labeled with ^{18}F. Active areas of the brain metabolize glucose more rapidly, and so they become more concentrated with ^{18}F. The figure shows which areas of the brain are more active for language or music.

Radiation therapy takes advantage of the effect of radiations in destroying unwanted tissue in the body, such as a cancerous growth or an overactive thyroid gland. The effect of the passage of radiation through matter is often to ionize the atoms. The ionized atoms can then participate in chemical reactions that lead to their incorporation into molecules and subsequent alteration of their biological function, possibly the destruction of a cell or the modification of its genetic material. For example, an overactive thyroid gland is often treated by giving the patient radioactive ^{131}I, which collects in the thyroid. The beta emissions from this isotope damage the thyroid cells and ultimately lead to their destruction. Certain cancers are treated by implanting needles or wires containing radium or other radioactive substances. The decays of these radioisotopes cause localized damage to the cancerous cells.

Other cancers can be treated using beams of particles that cause nuclear reactions within the body at the location of the tumor. Pions and neutrons are used for this purpose. The absorption of a pion or a neutron by a nucleus causes a nuclear reaction, and the subsequent emission of particles or decays by the reaction products again causes local damage that is concentrated at the site of the tumor, inflicting maximum damage to the tumor and minimum damage to the surrounding healthy tissue.

Alpha Decay Applications

Radioactive sources emitting alpha particles have been used in a variety of applications. Most of these take advantage of the persistence of radioactive decay—the decays can be depended upon to occur at a fixed rate in any location.

Alpha particles from radioactive decay can be absorbed and their energy converted into another form, such as electrical power obtained through thermoelectric conversion. The power levels are not large (of the order of 1 W per gram of material; see Problem 33), but they are sufficient to power many devices, from cardiac pacemakers to the Voyager spacecraft, which photographed Jupiter, Saturn, and Uranus.

Scattering of alpha particles emitted by a radioactive source is the basis of operation of ionization-type smoke detectors; alpha particles from the decay of ^{241}Am are scattered by the ionized atoms that result from combustion. When the smoke detector senses a decrease in the rate at which alphas are counted (due to some of them being scattered away from the detector), the alarm is triggered.

Other applications of alpha-particle scattering are used for materials analysis. In *Rutherford backscattering*, the analysis uses the reduction in energy of an alpha particle that is scattered through an angle of 180°. Although our discussion of Rutherford scattering in Chapter 6 assumed that the target nucleus was infinitely heavy and thus acquired no energy in the scattering, in practice a small amount of energy is given even to a heavy nucleus. By allowing the target nucleus to recoil, we can find the

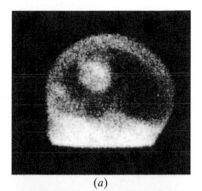

(a)

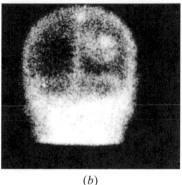

(b)

FIGURE 13.20 Scintillation camera image of the brain, following intravenous injection of 20 mCi of ^{99m}Tc. Photo (*a*) shows a side view, with the patient's face to the left; photo (*b*) is the back view. The bright circular spot shows concentration of blood in a lesion, possibly a tumor. Other bright areas show the scalp and the major veins.

Rosalind Yalow (1921–, United States). After receiving her Ph.D. in nuclear physics, she researched the medical applications of radioactive isotopes. She developed the technique of radioimmunoassay, which uses radioactive tracers to measure small amounts of substances in the blood or other fluids. She first applied this technique to study insulin in the blood of diabetics. Her development of this technique was recognized with the award of the Nobel prize in medicine in 1977.

loss in energy ΔK of an alpha particle of kinetic energy K that scatters through 180° (see Problem 31):

$$\Delta K = K \left[\frac{4m/M}{(1 + m/M)^2} \right] \qquad (13.16)$$

where m is the mass of the alpha particle and M is the mass of the target nucleus. For a heavy nucleus ($m/M = 0.02$), the loss in energy is of order 0.5 MeV, which is easily measurable. Figure 13.21 shows a sample of the spectrum of alpha particles backscattered from a thin foil containing copper, silver, and gold. Note the Z^2 dependence of the scattering probability that characterizes Rutherford scattering (see Equation 6.18), and also note the sensitivity of the technique even to the two naturally occurring isotopes of copper. (However, the two isotopes of silver cannot be resolved.) The Surveyor spacecraft that landed on the Moon and the Viking landers on Mars carried Rutherford backscattering experiments to analyze the chemical composition of the surface of those bodies.

Synthetic Elements

The elements up through Fe were made by means of fusion reactions in stars. As we have seen, it is not energetically favorable for fusion reactions to continue beyond ^{56}Fe. The remaining elements were made by neutron

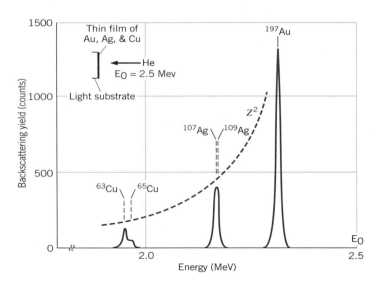

FIGURE 13.21 Backscattering spectrum of 2.5-MeV α particles from a thin film of copper, silver, and gold. The dashed line shows the Z^2 behavior of the cross section expected from the Rutherford formula. Note the appearance of the two isotopes of copper. From M.-A. Nicolet, J. W. Mayer, and I. V. Mitchell, *Science* **177**, 841 (1972). Copyright © 1972 by the AAAS.

capture, producing isotopes with an excess of neutrons, followed by beta decay in which a neutron converts to a proton. Thus,

$$^{56}_{26}\text{Fe}_{30} + \text{n} \longrightarrow {}^{57}_{26}\text{Fe}_{31} \quad \text{(stable)}$$

$$^{57}_{26}\text{Fe}_{31} + \text{n} \longrightarrow {}^{58}_{26}\text{Fe}_{32} \quad \text{(stable)}$$

$$^{58}_{26}\text{Fe}_{32} + \text{n} \longrightarrow {}^{59}_{26}\text{Fe}_{33} \xrightarrow{\beta} {}^{59}_{27}\text{Co}_{32} \quad \text{(stable)}$$

The stable isotope ^{59}Co is produced following the radioactive decay of ^{59}Fe, which is produced from ^{56}Fe following the capture of three neutrons. This process continues as follows:

$$^{59}_{27}\text{Co}_{32} + \text{n} \longrightarrow {}^{60}_{27}\text{Co}_{33} \xrightarrow{\beta} {}^{60}_{28}\text{Ni}_{32} \quad \text{(stable)}$$

Notice that each beta decay *increases* the atomic number by one unit. In principle this process of neutron capture and β decay would go on indefinitely, producing atoms of larger and larger Z. However, the known atoms beyond uranium ($Z = 92$) are *all* radioactive, with half-lives short compared with the age of the Earth. They are therefore not present in terrestrial matter, but they can be produced in the laboratory. The production process for the series of elements beginning with neptunium ($Z = 93$), called *transuranic* elements, follows the same process outlined above: neutron capture followed by beta decay. Using similar techniques researchers have produced elements all the way up to $Z = 109$. Many of the elements in this series have half-lives of only minutes or seconds, and thus the production and identification of these elements requires painstaking experimental efforts—the isotopes are often produced in quantities of a few atoms! Although most of these elements have not been produced in sufficient quantity to study their chemical properties, it is expected that their place in the periodic table will be as shown in Figure 13.22, up to the inert gas $Z = 118$.

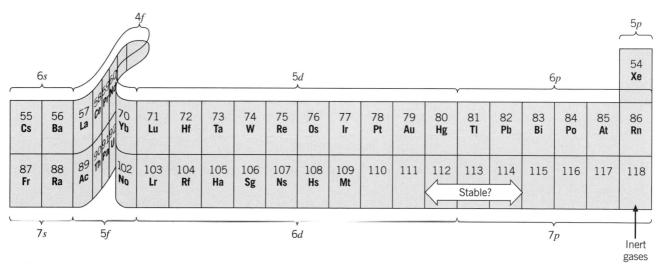

FIGURE 13.22 The location of possible new elements in the periodic table.

The extreme instability of these transuranic elements results from the increased Coulomb repulsion of the nuclear protons as Z increases; these elements decay by alpha decay or by spontaneous fission (without first having to absorb a neutron). However, there is strong theoretical evidence which suggests that, for complicated reasons based on the nuclear structure, elements around $Z = 114$, $N = 184$ should be stable against alpha decay, beta decay, and spontaneous fission. Owing to the short half-lives and microscopic quantities of the elements between $Z = 100$ and $Z = 114$, it is unlikely that the neutron-capture, beta-decay process could be used to produce and observe element 114. Elements 110 and 111 have been observed through reactions with heavy ions ($_{28}$Ni + $_{82}$Pb $\rightarrow$ $_{110}$X and $_{28}$Ni + $_{83}$Bi $\rightarrow$ $_{111}$X), and it is possible that similar methods will be necessary to produce element 114. It might also be possible to bring together two heavy particles, such as ^{238}U + ^{238}U, in the hope that an intermediate state would be formed that would then decay (fission, perhaps) to the stable isotope at $Z = 114$.

Although the production and observation of such *superheavy* nuclei are not likely to have any immediate applications, their study would be of great interest to test our understanding of the ordering of the periodic table, and the comparison of their chemical and physical properties with those of the $5d$ and $6p$ elements would be a test of our ordering of the elements. Many of the artificially produced transuranic elements have already found applications in research and technology. The alpha decays of ^{238}Pu and ^{239}Pu have been used as power sources for spacecraft, and ^{241}Am serves as an alpha source for smoke detectors. The radioisotope ^{252}Cf decays by spontaneous fission; the neutrons released in the decay have many applications, including medical treatment and materials analysis.

SUGGESTIONS FOR FURTHER READING

See the references listed in Chapter 12 for more detail on nuclear reactions. Other references at about the same level as this chapter are:

H. Semat and J. R. Albright, *Introduction to Atomic and Nuclear Physics* (New York, Holt, Rinehart and Winston, 1972).

M. R. Wehr, J. A. Richards, and T. W. Adair, *Physics of the Atom* (Reading, Addison-Wesley, 1984). Particularly good discussion of fission reactors.

The natural fission reactor is discussed in:

G. A. Cowan, "A Natural Fission Reactor," *Scientific American* **235**, 36 (July 1976).

Fusion power has frequently been treated in general articles, including the following:

B. Coppi and J. Rem, "The Tokamak Approach in Fusion Research," *Scientific American* **227**, 65 (July 1972).

J. L. Emmett, J. Nuckolls, and L. Wood, "Fusion Power by Laser Implosion," *Scientific American* **230**, 24 (June 1974).

W. C. Gough and B. J. Eastland, "The Prospects of Fusion Power," *Scientific American* **224**, 50 (February 1971).

M. J. Lubin and A. P. Fraas, "Fusion by Laser," *Scientific American* **224**, 21 (June 1971).

H. Furth, "Progress toward a Tukamak Fusion Reactor," *Scientific American* **241**, 50 (August 1979).

R. W. Conn, V. A. Chuyanov, N. Inoue, and D. R. Sweetman, "The International Thermonuclear Experimental Reactor," *Scientific American* **266**, 103 (April 1992).

For a comprehensive and nontechnical history of fusion, see:

R. Herman, *Fusion: The Search for Endless Energy* (Cambridge, Cambridge University Press, 1990).

A detailed history of the development of fission is:

R. Rhodes, *The Making of the Atomic Bomb* (New York, Simon and Schuster, 1986).

For more information on medical applications of nuclear physics, see:

N. A. Dyson, *An Introduction to Nuclear Physics with Applications in Medicine and Biology* (New York, Halsted Press, 1981).

More detail on the effects of radiation on living organisms can be found in:

S. C. Bushong, "Radiation Exposure in our Daily Lives," *The Physics Teacher,* March 1977, p. 135.

For more information on the possible stable elements beyond uranium, see:

G. T. Seaborg and J. L. Bloom, "The Synthetic Elements: IV," *Scientific American* **220**, 56 (April 1969).

QUESTIONS

1. The cross sections for reactions induced by protons generally increase as the kinetic energy of the proton increases, while cross sections for neutron-induced reactions generally decrease with increasing neutron energy. Explain this behavior.

2. Cross sections for reactions induced by thermal neutrons ($K \sim kT$, where T is room temperature) are often several orders of magnitude larger than cross sections for the same reaction induced by fast neutrons ($K \sim$ MeV). Justify this difference by comparing the time spent in the vicinity of a target nucleus by a thermal neutron and a fast neutron.

3. When two nuclei approach one another in a nuclear reaction, there is a Coulomb repulsion between them. Does this potential energy affect the kinematics of the reaction? Does it affect the cross section?

4.   The most abundant component of our atmosphere is ^{14}N. Assuming that cosmic rays supply sufficient high-energy protons and neutrons, explain how the radioactive isotopes ^{14}C and ^{3}H can be formed.

5.   Consider the photodisintegration reaction $A + \gamma \rightarrow B + C$. In terms of the binding energies of A, B, and C, what are the requirements for this reaction? Would you expect to observe photodisintegration more readily for light nuclei ($A < 56$) or heavy nuclei ($A > 56$)?

6.   Would you expect to observe the radiative capture of an alpha particle $X + \alpha \rightarrow X' + \gamma$ for heavy nuclei?

7.   In what sense is photodisintegration (see Question 5) the inverse of radiative capture (see Question 6)? How are the photon energies related?

8.   Comment on the following statement: The fission reaction is useful for energy production because of the large kinetic energies given to the neutrons emitted following fission.

9. ^{238}U is fissionable, but only with neutrons in the MeV range. Explain why ^{238}U is not a suitable reactor fuel.

10.   What is the difference between a slow neutron and a delayed neutron? Between a fast neutron and a prompt neutron?

11.   In a typical fission reaction, which fragment (heavier or lighter) has the larger kinetic energy? the larger momentum? the larger speed?

12.   When charged particles travel in a medium faster than light travels in that medium, Cerenkov radiation is emitted. This is the origin of the blue glow of the water which surrounds a reactor core. What might be the identity of these charged particles from a reactor? What velocities and kinetic energies do they need to produce Cerenkov radiation? (The index of refraction of water is 1.33.)

13.   In general, would you expect fission fragments to decay by positive or negative beta decay? Why?

14.   Among the fission products that build up in reactor fuel elements is xenon, which has an extremely large neutron capture cross section (see Section 13.1). What effect does this buildup have on the operation of the reactor?

15.   Helium has virtually no neutron absorption cross section. Would helium be a better reactor moderator than carbon, which has a small but nonzero cross section?

16.   Estimate the number of fissions per second that must occur in a 1000-MW power plant, assuming a 30 percent efficiency of energy conversion.

17.   The fission cross section for ^{235}U for slow neutrons is about 10^6 times the fission cross section of ^{238}U for slow neutrons, yet for fast neutrons the fission cross sections of ^{235}U and ^{238}U are roughly the same. Explain this effect.

18.   Consider two fragments of uranium fission with atomic numbers Z and $92 - Z$. Estimating their mass numbers as 2.5 times their atomic numbers, find an expression for the Coulomb potential energy of the two fragments when they are just touching, and show that this expression is maximized when the two fragments are identical. Why then is the fission fragment mass distribution, Figure 13.9, not a maximum at $A \cong 118$?

19.   Assume that ^{235}U splits into two fragments with mass numbers of 90 and 145, with each fragment having roughly the same ratio of Z/A as ^{235}U. On this basis, explain why neutrons are emitted in fission.

20. Why is it necessary to convert a proton to a neutron in the first step of the proton-proton fusion cycle? Why can't two protons fuse directly?

21. Explain why a fusion reactor requires a high particle density, a high temperature, and a long confinement time.

22. In the argument leading to Lawson's criterion, Equation 13.15, it was mentioned that the power necessary to heat the plasma is proportional to the particle density n, while the power obtained from fusion is proportional to n^2. Explain these two proportionalities.

23. Why do radioactive decay power sources use alpha emitters rather than beta emitters?

PROBLEMS

1. Fill in the missing particle in these reactions:
   (a) $^4\text{He} + {}^{14}\text{N} \rightarrow {}^{17}\text{O} +$
   (b) $^9\text{Be} + {}^4\text{He} \rightarrow {}^{12}\text{C} +$
   (c) $^{27}\text{Al} + {}^4\text{He} \rightarrow \text{n} +$
   (d) $^{12}\text{C} + \quad \rightarrow {}^{13}\text{N} + \text{n}$

2. In a certain nuclear reaction, outgoing protons are observed with energies 16.2 MeV, 14.8 MeV, 11.6 MeV, 8.9 MeV, and 6.7 MeV. No energies higher than 16.2 MeV are observed. Construct a level scheme of the product nucleus.

3. A radioactive isotope of half-life $t_{1/2}$ is produced in a nuclear reaction. What fraction of the maximum possible activity is produced in a time of (a) $t_{1/2}$; (b) $2t_{1/2}$; (c) $4t_{1/2}$?

4. A beam of alpha particles is incident on a target of ^{63}Cu, resulting in the reaction $\alpha + {}^{63}\text{Cu} \rightarrow {}^{66}\text{Ga} + \text{n}$. Assume the cross section for the particular alpha energy to be 1.25 b. The target is in the form of a foil, 2.5 μm thick. The beam has a circular cross section of diameter 0.50 cm and a current of 7.5 μA. Find the rate of neutron emission.

5. A beam of 20.0 μA of protons is incident on 2.0 cm^2 of a target of ^{107}Ag of thickness 4.5 μm producing the reaction $\text{p} + {}^{107}\text{Ag} \rightarrow {}^{105}\text{Cd} + 3\text{n}$. Neutrons are observed at a rate of 8.5×10^6 per second. What is the cross section for this reaction at this proton energy?

6. Show that Equation 13.5 is a solution to Equation 13.4.

7. Neutron capture in sodium occurs with a cross section of 0.53 b and leads to radioactive ^{24}Na ($t_{1/2} = 15$ h). What is the activity that results when 1.0 μg of Na is placed in a neutron flux of 2.5×10^{13} neutrons/cm^2/s for 4.0 h?

8. A small sample of paint is placed in a neutron flux of 3.0×10^{12} neutrons/cm^2/s for a period of 2.5 min. At the end of that period the activity of the sample is found to include 105 decays/s of ^{51}Ti ($t_{1/2} = 5.8$ min) and 12 decays/s ^{60}Co ($t_{1/2} = 5.27$ y). Find the amount, in grams, of titanium and cobalt in the original sample. Use the following information: Cobalt is pure ^{59}Co, which has a cross section of 19 b; titanium is 5.25 percent ^{50}Ti, which has a cross section of 0.14 b.

9. A 2.0-mg sample of copper (69% ^{63}Cu, 31% ^{65}Cu) is placed in a reactor where it is exposed to a neutron flux of 5.0×10^{12} neutrons/cm^2/s. After 10.0 min

the resulting activities are 72 μCi of ^{64}Cu ($t_{1/2} = 12.7$ h) and 1.30 mCi of ^{66}Cu ($t_{1/2} = 5.1$ min). Find the cross sections of ^{63}Cu and ^{65}Cu.

10. A beam of neutrons of intensity I is incident on a thin slab of material of area A, thickness dx, density ρ, and atomic weight M. The neutron absorption cross section is σ. (a) What is the loss in intensity dI of this beam in passing through the material? (b) A beam of original intensity I_0 passes through a thickness x of the material. Show that the intensity of the emerging beam is $I = I_0 e^{-n\sigma x}$, where n is the number of absorber nuclei per unit volume.

11. Derive Equation 13.14 from Equation 13.13.

12. Assume that the total cross section for neutrons incident on copper is 5.0 b. What fraction of the intensity of a neutron beam is lost after traveling through copper of thickness (a) 1.0 mm; (b) 1.0 cm; (c) 1.0 m? (See Problem 10.)

13. Find the Q value of the reactions:
    (a) $p + {}^{55}Mn \rightarrow {}^{54}Fe + 2n$
    (b) $^3He + {}^{40}Ar \rightarrow {}^{41}K + {}^2H$

14. Find the Q values of the reactions:
    (a) $^6Li + n \rightarrow {}^3H + {}^4He$
    (b) $p + {}^2H \rightarrow 2p + n$
    (c) $^7Li + {}^2H \rightarrow {}^8Be + n$

15. In the reaction $^2H + {}^3He \rightarrow p + {}^4He$ deuterons of energy 5.000 MeV are incident on ^3He at rest. Both the proton and the alpha particle are observed to travel along the same direction as the incident deuteron. Find the kinetic energies of the proton and the alpha particle.

16. (a) What is the Q value of the reaction $p + {}^4He \rightarrow {}^2H + {}^3He$? (b) What is the threshold energy for protons incident on ^4He at rest? (c) What is the threshold energy if ^4He are incident on protons at rest?

17. A reaction in which two particles join to form a single excited nucleus, which then decays to its ground state by photon emission, is known as *radiative capture*. Find the energy of the gamma ray emitted in the radiative capture of an alpha particle by ^7Li. Assume alpha particles of very small kinetic energy are incident on ^7Li at rest.

18. How much energy is required (in the form of gamma-ray photons) to break up ^7Li into $^3H + {}^4He$? This reaction is known as *photodisintegration*.

19. The nucleus ^{113}Cd captures a thermal neutron ($K = 0.025$ eV), producing ^{114}Cd in an excited state; the excited state of ^{114}Cd decays to the ground state by emitting a photon. Find the energy of the photon.

20. When a neutron collides head-on with an atom at rest, the loss in its kinetic energy is given by Equation 13.16. (a) What fraction of its energy will a neutron lose in a head-on collision with an atom of hydrogen, deuterium, or carbon? (b) Consider a neutron with an initial energy of 2.0 MeV. How many head-on collisions must it make with carbon atoms for its energy to be reduced to the thermal range (0.025 eV)? (c) Is the result of part (b) an underestimate or an overestimate of the actual number of collisions necessary to "thermalize" the neutrons? Explain.

21. We can understand why ^{235}U is readily fissionable, and ^{238}U is not, with the following calculation. (a) Find the energy difference between $^{235}U + n$ and ^{236}U. We can regard this as the "excitation energy" of ^{236}U. (b) Repeat for $^{238}U + n$ and ^{239}U. (c) Comparing your results for (a) and (b) explain why ^{235}U

will fission with very low energy neutrons, while ^{238}U requires fast neutrons of 1 to 2 MeV of energy to fission. (d) From a similar calculation, predict whether ^{239}Pu requires low-energy or higher-energy neutrons to fission.

22. Find the energy released in the fission of 1.00 kg of uranium that has been enriched to 3.0 percent in the isotope ^{235}U.

23. Find the Q value (and therefore the energy released) in the fission reaction ^{235}U + n → ^{93}Rb + ^{141}Cs + 2n. Use $m(^{93}$Rb$) = 92.92195$ u and $m(^{141}$Cs$) = 140.92005$ u.

24. To what temperature must helium gas be heated before the Coulomb barrier is overcome and fusion reactions begin?

25. (a) Calculate the Q value for the six reactions or decays of the carbon cycle of fusion. (b) By accounting for the electron masses, show that the total Q value for the carbon cycle is identical with that of the proton–proton cycle.

26. Show that the D-T fusion reaction releases 17.6 MeV of energy.

27. In the D-T fusion reaction, the kinetic energies of ^{2}H and ^{3}H are small, compared with typical nuclear binding energies. (Why?) Find the kinetic energy of the emitted neutron.

28. Suppose we have 100.0 cm^3 of water, which is 0.015 percent D$_2$O. (a) Compute the energy that could be obtained if all the deuterium were consumed in the ^{2}H + ^{2}H → ^{3}H + p reaction. (b) As an alternative, compute the energy released if two-thirds of the deuterium were fused to form ^{3}H, which is then combined with the remaining one-third in the D-T reaction.

29. Find the energy released when three alpha particles combine to form ^{12}C.

30. (a) If a tokamak fusion reactor were able to achieve a confinement time of 0.60 s, what minimum particle density is required? (b) If the reactor were able to achieve 10 times the density found in part (a), what is the minimum plasma temperature required for ignition of a self-sustaining fusion reaction?

31. An alpha particle of mass m makes an elastic head-on collision with an atom of mass M at rest. Show that the loss in kinetic energy of the alpha particle is given by Equation 13.16.

32. (a) Calculate the energy loss of a 2.50-MeV alpha particle after backscattering from an atom of copper, silver, and gold. Compare your calculated values with the peak energies in Figure 13.21. (b) Calculate the expected energy difference between the peaks for the two isotopes of copper and also for the two isotopes of silver. Explain why the silver peaks are closer together than the copper peaks. Can you estimate the relative abundances of the two isotopes of copper from the figure?

33. A radioactive source is to be used to produce electrical power from the alpha decay of ^{238}Pu ($t_{1/2} = 88$ y). (a) What is the Q value for the decay? (b) Assuming 100% conversion efficiency, how much power could be obtained from the decay of 1.0 g of ^{238}Pu?

34. CUPS Exercise 6.17.

35. CUPS Exercise 6.30.

36. CUPS Exercise 6.32.

The fundamental nature of matter is revealed in experiments in which particles collide at high energy to produce other kinds of particles. Photographs of the tracks left by the particles are analyzed by elementary particle physicists to determine the identity of the particles and their energy and momentum. From such studies it is possible to determine whether the particles are elementary or composite and through which forces they interact.

ELEMENTARY PARTICLES

The search for the basic building blocks of nature has occupied the thoughts of scientific investigators since the Greeks introduced the idea of atomism 2500 years ago. As we look carefully at complex structures, we find underlying symmetries and regularities, which help us to understand the laws that determine how they are put together. The regularities of crystal structure, for example, suggest to us that the atoms of which the crystal is composed must follow certain rules for arranging themselves and joining together. As we look more deeply, we find that although nature has constructed all material objects out of roughly 100 different kinds of atoms, we can understand these atoms in terms of only three fundamental particles: the electron, proton, and neutron. Our attempts to look further within the electron have been unsuccessful—the electron seems to be a fundamental particle, with no internal structure. However, when nucleons collide at high energy, the result is more complexity rather than simplicity; hundreds of new particles can emerge as products of these reactions. If there are hundreds of basic building blocks, it seems unlikely that we could ever uncover any fundamental dynamic laws of their behavior. However, in recent years, experiments have suggested a new, underlying regularity that seems to be consistent with an analysis in terms of a small number of fundamental particles called *quarks*.

In this chapter, we examine the properties of many of the particles of physics, the laws that govern their behavior, and the classifications of these particles. We also show how the quark model helps us to understand some properties of the particles.

In the early days of the study of atomic physics, before the insights of the Bohr model helped us to understand atomic structure, all we had available were such guidelines as the Balmer formula, the Ritz combination principle, and so forth. These were rules based not on a fundamental understanding of atomic structure, but rather on the regularities that were revealed in experimental results. The study of particle physics was for many years in much the same state as atomic physics before Bohr. Just as the regularities and classifications of atomic properties led to the Bohr model and to the new mechanics of quantum physics, the regularities and classifications of particle properties have led to the quark model and to a new mechanical system called *quantum chromodynamics*. The mathematics associated with this new mechanics is beyond the level of this book, but the fundamental details of the model and the properties of the quarks can be appreciated without the mathematics.

14.1   THE FOUR BASIC FORCES

All of the known forces in the universe can be grouped into four basic types. In order of increasing strength, these are: *gravitation*, the *weak interaction*, *electromagnetism*, and the *strong interaction*.

1. *The gravitational interaction.* Gravity is of course exceedingly important in our daily lives, but on the scale of fundamental interactions between particles in the subatomic realm, it is of no importance at all. To give

a relative figure, the gravitational force between two protons just touching at their surfaces is about 10^{-38} of the strong force between them. The principal difference between gravitation and the other interactions is that, on the practical scale, gravity is cumulative and infinite in range. The strong and weak interactions, as we shall see, have no effect on a distance scale larger than the size of a nucleon (10^{-15} m = 1 fm), and although the electromagnetic interaction also has infinite range, the shielding of electromagnetic effects (such as that of the nuclear charge by the electrons) prevents their cumulative build-up. No such shielding occurs for gravity, which allows its tiny effects to build up cumulatively for large numbers of particles (such as all the atoms of the Earth).

2. *The weak interaction.* The weak interaction is responsible for nuclear beta decay (see Section 12.8) and other similar decay processes involving fundamental particles. It does not play a major role in the binding of nuclei. The weak force between two neighboring protons is about 10^{-7} of the strong force between them, and the range of the weak force is smaller than 0.001 fm. Nevertheless, the weak force is important in understanding the behavior of fundamental particles, and it is critical in understanding the evolution of the universe.

3. *The electromagnetic interaction.* Electromagnetism is important in the structure and the interactions of the fundamental particles. For example, some particles interact or decay primarily through this mechanism. Electromagnetic forces are of infinite range, but the shielding effect generally diminishes their effect for ordinary objects. Many common macroscopic forces (such as friction, air resistance, drag, and tension) are ultimately due to electromagnetic forces. Within the atom, electromagnetic forces dominate. The electromagnetic force between neighboring protons in a nucleus is about 10^{-2} of the strong force, but within the nucleus the electromagnetic forces can act cumulatively because there is no shielding. As a result, the electromagnetic force can compete with the strong force in determining the stability and the structure of nuclei.

4. *The strong force.* The strong force, which is responsible for the binding of nuclei, is the dominant one in the reactions and decays of most of the fundamental particles. However, as we shall see, some particles (such as the electron) do not feel this force at all. It has a relatively short range, on the order of 1 fm.

The relative strength of a force determines the time scale over which it acts. If we bring two particles close enough together for any of these forces to act, then a longer time is required for the weak force to cause a decay or reaction than for the strong force. As we shall see, the mean lifetime of a decay process is often a signal of the type of interaction responsible for the process, with strong forces being at the shortest end of the time scale (often down to 10^{-23} s). Table 14.1 summarizes the four forces and some of their properties.

Particles can interact with one another in decays and reactions through any of the basic forces. Table 14.1 indicates which particles can interact through each of the four forces. All particles can interact through the gravitational and weak forces. A subset of those can interact through the electromagnetic force (for example, the neutrinos are excluded from this

TABLE 14.1 THE FOUR BASIC FORCES

Type	Range	Relative Strength	Characteristic Time	Typical Particles
Strong	1 fm	1	$<10^{-22}$ s	π, K, n, p
Electromagnetic	∞	10^{-2}	$10^{-14} - 10^{-20}$ s	e, μ, π, K, n, p
Weak	10^{-3} fm	10^{-7}	$10^{-8} - 10^{-13}$ s	All
Gravitational	∞	10^{-38}	Years	All

category), and a still smaller subset can interact through the strong force. When two strongly interacting particles are within the range of each other's strong force, we can often neglect the effects of the weak and electromagnetic forces in decay and reaction processes; because their relative strengths are so much smaller than that of the strong force, their effects are much smaller than those of the strong force. (However, these forces are not always negligible—the weak interaction between protons is responsible for a critical step in one of the fusion processes that occurs in stars.)

Even though the proton is a strongly interacting particle, a proton and an electron will *never* interact through the strong force. The electron is able to ignore the strong force of the proton and respond only to its weak or electromagnetic force.

Each of the four forces can be represented in terms of the emission or absorption of particles that carry the interaction, just as we represent the force between nucleons in the nucleus in terms of the exchange of pions (see Section 12.4). For example, the electromagnetic force between particles can be represented in terms of the emission and absorption of photons. Associated with each type of force is a field that is carried by its characteristic field particle, as shown in Table 14.2. The *weak bosons* $W^{\pm}$ and Z^0 are responsible for processes such as nuclear beta decay. For example, the beta decay of the neutron (a weak interaction) can be represented as

$$n \longrightarrow p + W^-$$

followed by

$$W^- \longrightarrow e^- + \bar{\nu}_e$$

Because the decay $n \rightarrow p + W^-$ would violate energy conservation, the existence of the W^- is restricted by the uncertainty principle, and its range

TABLE 14.2 THE FIELD PARTICLES

Force	Field Particle	Symbol	Charge (e)	Spin ($\hbar$)	Rest Energy (GeV)
Strong	Gluon	g	0	1	0
Electromagnetic	Photon	γ	0	1	0
Weak	Weak boson	W^+, W^-	± 1	1	80.2
	Weak boson	Z^0	0	1	91.2
Gravitation	Graviton		0	2	0

can be determined in a manner similar to that of the pion (see Equation 12.6). The strong force between quarks is carried by particles called *gluons*, which have been observed through indirect techniques. The *graviton*, which is expected to exist based on theories of gravitation, has not yet been observed.

14.2 PARTICLES AND ANTIPARTICLES

One way of studying the elementary particles is to classify them into different categories based on certain behaviors or properties and then to look for similarities or common characteristics among the classifications. We have already classified some particles in Table 14.1 according to the types of forces through which they interact. Another way of classifying them might be according to their masses. In the early days of particle physics, it was observed that the lightest particles (including electrons, muons, and neutrinos) showed one type of behavior, the heaviest group (including protons and neutrons) showed a different behavior, and a middle group (such as pions and kaons) showed a still different behavior. The names originally given to these groups are based on the Greek words for light, middle, and heavy: *leptons* for the light particles, *mesons* for the middle group, and *baryons* for the heavier particles. Even though the classification by mass is now obsolete (leptons and mesons have been discovered that are more massive than protons or neutrons), we keep the original names, which now describe instead a group or *family* of particles with similar properties. When we compare our first two ways of classifying particles, we find an interesting result: The leptons do not interact through the strong force, but the mesons and baryons do.

We can also classify particles by their intrinsic spins. Every particle has such an intrinsic spin; you will recall that the electron has a spin of $\frac{1}{2}$, as do the proton and neutron. We find that the leptons all have spins of $\frac{1}{2}$, the mesons all have integral spins $(0, 1, 2, \ldots)$, and the baryons all have half-integral spins $(\frac{1}{2}, \frac{3}{2}, \frac{5}{2}, \ldots)$.

One additional property that is used to classify a particle is the nature of its *antiparticle*.* Every particle has an antiparticle, which is identical to the particle in such properties as mass and lifetime, but differs from the particle in the sign of its electric charge (and in the sign of certain other properties, as we discuss later). The antiparticle to the electron is the positron e^+, which was discovered in the 1930s through reactions initiated by cosmic rays. The positron has a charge of $+e$ (opposite to that of the electron) and a rest energy of 0.511 MeV (identical to that of the electron). The antiproton $\overline{p}$ was discovered in 1956 (see Example 2.17); it has a charge

---

* We use two systems to indicate antiparticles. Sometimes the symbol for the particle will be written along with the electric charge to indicate particle or antiparticle, as, for example, e^+ and e^-, or μ^+ and μ^-. Other times the antiparticle will be written with a bar over the symbol, for example, ν and $\overline{\nu}$ or p and $\overline{p}$.

of $-e$ and a rest energy of 938 MeV. A stable atom of antihydrogen could be constructed from a positron and an antiproton; the properties of this atom would be identical to those of ordinary hydrogen.

Antiparticles of stable particles (such as the positron and the antiproton) are themselves stable. However, when a particle and its antiparticle meet, the *annihilation reaction* can occur—the particle and antiparticle both vanish, and instead two or more photons can be produced. Conservation of energy and momentum require that, neglecting the kinetic energies of the particles, when two photons are emitted each must have an energy equal to the rest energy of the particle. Examples of annihilation reactions are:

$$e^- + e^+ \longrightarrow \gamma_1 + \gamma_2 \qquad (E_{\gamma_1} = E_{\gamma_2} = 0.511 \text{ MeV})$$
$$p + \overline{p} \longrightarrow \gamma_1 + \gamma_2 \qquad (E_{\gamma_1} = E_{\gamma_2} = 938 \text{ MeV})$$

We call the kind of stuff of which we are made *matter* and the other kind of stuff *antimatter*. There may indeed be galaxies composed of antimatter, but we cannot tell by the ordinary techniques of astronomy, because *light and antilight are identical*! To put it another way, the photon and antiphoton are the same particle. We cannot tell by looking at the light (or other electromagnetic radiation) that reaches us from distant galaxies whether they are made from matter or antimatter. The only way to tell is by sending a chunk of our matter to the distant galaxy and seeing whether or not it is destroyed with the corresponding emission of a burst of photons. (It is indeed possible, but *highly unlikely*, that the first astronaut to travel to another galaxy may suffer such a fate! The first intergalactic handshake would indeed be quite an event!)

In our classification scheme it is usually easy to distinguish particles from antiparticles. We begin by defining *particles* to be the stuff of which ordinary matter is made—electrons, protons, and neutrons. Ordinary matter is not composed of neutrinos, so we have no basis for distinguishing a neutrino from an antineutrino, but the conservation laws in the beta decay process can be understood most easily if we define the *antineutrino* to be the particle that accompanies negative beta decay and the *neutrino* to be the particle that accompanies positron decay and electron capture. For a heavy baryon, such as the Λ, we take advantage of its radioactive decay, which leads eventually to ordinary protons and neutrons; that is, the Λ is the particle that decays to n, and the $\overline{\Lambda}$ ("anti-lambda") therefore decays to $\overline{n}$. Similarly, in the case of the leptons, the μ^- and the μ^+ are antiparticles of one another; since μ^- decays to ordinary e^- (and has many properties in common with the electron) it is the *particle*, while μ^+ is the antiparticle. The case of the pions is not so easy to resolve. The π's come with three different electric charges: π^+, π^0, π^-. The π^0 seems to be, like the photon, its own antiparticle and the π^+ and π^- seem to be antiparticles of one another, but which (π^+ or π^-) is the particle and which the antiparticle? Ordinary matter is not composed of π mesons, so we get no clue there. The decays of the π^+ and π^- always give one lepton and one antilepton so again we get no clue there either. There seems in fact to be no way (*or no need*) to distinguish particles

TABLE 14.3 FAMILIES OF PARTICLES

Family	Structure	Interactions	Spin	Examples
Leptons	Fundamental	Weak, electromagnetic	Half integral	e, ν
Mesons	Composite	Weak, electromagnetic, strong	Integral	π, K
Baryons	Composite	Weak, electromagnetic, strong	Half integral	p, n
Field quanta	Fundamental	Weak, electromagnetic, strong	Integral	γ, W, Z

from antiparticles for the π mesons, and so we regard the set of π's as three *particles* π^+, π^0, π^- whose antiparticles are, respectively, π^-, π^0, π^+.

14.3 FAMILIES OF PARTICLES

Table 14.3 summarizes the three families of material particles and the field particles that are responsible for the interactions between them.

The leptons interact only through the weak or electromagnetic interactions. No experiment has yet been able to reveal any internal structure for the leptons; they appear to be truly fundamental particles that cannot be split into still smaller particles. This lack of structure is consistent with current theories, which regard the leptons and the quarks as dimensionless point particles. All known leptons have spin $\frac{1}{2}$.

Table 14.4 shows the six known leptons, grouped as three pairs of particles. Each pair includes a charged particle (e^-, μ^-, τ^-) and an uncharged neutrino (ν_e, ν_μ, ν_τ). Each lepton has a corresponding antiparticle. We have already discussed the electron neutrino and antineutrino in connection with beta decay (Section 12.8), and the decay of cosmic-ray muons was discussed as confirming the time dilation effect in special relativity (Section 2.4).

Mesons are strongly interacting particles having integral spin. A partial list of some mesons is given in Table 14.5. Mesons can be produced in

TABLE 14.4 THE LEPTON FAMILY

Particle	Antiparticle	Particle Charge (e)	Spin ($\hbar$)	Rest Energy (MeV)	Mean Life (s)	Typical Decay Products
e^-	e^+	-1	$\frac{1}{2}$	0.511	∞	—
ν_e	$\bar{\nu}_e$	0	$\frac{1}{2}$	<10 eV	∞	—
μ^-	μ^+	-1	$\frac{1}{2}$	105.7	2.2×10^{-6}	$e^- + \bar{\nu}_e + \nu_\mu$
ν_μ	$\bar{\nu}_\mu$	0	$\frac{1}{2}$	<0.3	∞	—
τ^-	τ^+	-1	$\frac{1}{2}$	1777	3.0×10^{-13}	$\mu^- + \bar{\nu}_\mu + \nu_\tau$
ν_τ	$\bar{\nu}_\tau$	0	$\frac{1}{2}$	<40	∞	—

TABLE 14.5 SOME SELECTED MESONS

Particle	Antiparticle	Charge[a] (e)	Spin ($\hbar$)	Strangeness[a]	Rest Energy (MeV)	Mean Life (s)	Typical Decay Products
π^+	π^-	+1	0	0	140	2.6×10^{-8}	$\mu^+ + \nu_\mu$
π^0	π^0	0	0	0	135	8.4×10^{-17}	$\gamma + \gamma$
K^+	K^-	+1	0	+1	494	1.2×10^{-8}	$\mu^+ + \nu_\mu$
K^0	$\overline{K}^0$	0	0	+1	498	0.9×10^{-10}	$\pi^+ + \pi^-$
η	η	0	0	0	547	8.0×10^{-19}	$\gamma + \gamma$
ρ^+	ρ^-	+1	1	0	769	4.5×10^{-24}	$\pi^+ + \pi^0$
η'	η'	0	0	0	958	2.2×10^{-21}	$\eta + \pi^+ + \pi^-$
D^+	D^-	+1	0	0	1869	1.1×10^{-12}	$K^- + \pi^+ + \pi^+$
ψ	ψ	0	1	0	3097	1.0×10^{-20}	$e^+ + e^-$
B^+	B^-	+1	0	0	5278	1.5×10^{-12}	$D^- + \pi^+ + \pi^+$
Y	Y	0	1	0	9460	1.3×10^{-20}	$e^+ + e^-$

[a] The charge and strangeness are those of the particle. Values for the antiparticle have the opposite sign. The spin, rest energy, and mean life are the same for a particle and its antiparticle.

reactions through the strong interaction; they decay to other mesons or leptons through the strong, electromagnetic, or weak interactions. For example, pions can be produced in reaction of nucleons, such as

$$p + n \longrightarrow p + p + \pi^- \quad \text{or} \quad p + n \longrightarrow p + n + \pi^0$$

and the pions can decay according to

$$\pi^- \longrightarrow \mu^- + \overline{\nu}_\mu \quad \text{(mean life} = 2.6 \times 10^{-8}\,\text{s)}$$

$$\pi^0 \longrightarrow \gamma + \gamma \quad \text{(mean life} = 8.4 \times 10^{-17}\,\text{s)}$$

The first decay is caused by the weak interaction (indicated by the lifetime and by the presence of a neutrino among the decay products) and the second is caused by the electromagnetic interaction (indicated by the lifetime and the photons).

Because mesons are not observed in ordinary matter, the classification into particles and antiparticles is somewhat arbitrary. For the charged mesons, we often choose the positive member as the particle and the negative partner as the antiparticle (for example, π^+ and π^-). For some uncharged mesons (such as π^0 and η) the particle and antiparticle are identical, while for others (such as K^0 and $\overline{K}^0$) they may be distinct.

The baryons are strongly interacting particles with half-integral spins ($\frac{1}{2}$, $\frac{3}{2}$, ...). A partial listing of some baryons is given in Table 14.6. Like the leptons, the baryons have distinct antiparticles. Like the mesons, the baryons can be produced in reactions with nucleons through the strong interaction; for example, the Λ^0 baryon can be produced in the following reaction:

$$p + p \longrightarrow p + \Lambda^0 + K^+$$

The Λ^0 then decays through the weak interaction according to

$$\Lambda^0 \longrightarrow p + \pi^- \quad \text{(mean life} = 2.6 \times 10^{-10}\,\text{s)}$$

TABLE 14.6   SOME SELECTED BARYONS

Particle	Antiparticle	Charge^a (e)	Spin ($\hbar$)	Strangeness^a	Rest Energy (MeV)	Mean Life (s)	Typical Decay Products
p	$\bar{\text{p}}$	+1	$\frac{1}{2}$	0	938	∞	
n	$\bar{\text{n}}$	0	$\frac{1}{2}$	0	940	889	$\text{p} + \text{e}^- + \bar{\nu}_e$
Λ^0	$\bar{\Lambda}^0$	0	$\frac{1}{2}$	−1	1116	2.6×10^{-10}	$\text{p} + \pi^-$
Σ^+	$\bar{\Sigma}^+$	+1	$\frac{1}{2}$	−1	1189	0.8×10^{-10}	$\text{p} + \pi^0$
Σ^0	$\bar{\Sigma}^0$	0	$\frac{1}{2}$	−1	1192	7.4×10^{-20}	$\Lambda^0 + \gamma$
Σ^-	$\bar{\Sigma}^-$	−1	$\frac{1}{2}$	−1	1197	1.5×10^{-10}	$\text{n} + \pi^-$
Ξ^0	$\bar{\Xi}^0$	0	$\frac{1}{2}$	−2	1315	2.9×10^{-10}	$\Lambda^0 + \pi^0$
Ξ^-	$\bar{\Xi}^-$	−1	$\frac{1}{2}$	−2	1321	1.6×10^{-10}	$\Lambda^0 + \pi^-$
Δ^*	$\bar{\Delta}^*$	+2, +1, 0, −1	$\frac{3}{2}$	0	1232	6×10^{-24}	$\text{p} + \pi$
Σ^*	$\bar{\Sigma}^*$	+1, 0, −1	$\frac{3}{2}$	−1	1385	2×10^{-23}	$\Lambda^0 + \pi$
Ξ^*	$\bar{\Xi}^*$	−1, 0	$\frac{3}{2}$	−2	1530	6×10^{-23}	$\Xi + \pi$
Ω^-	$\bar{\Omega}^-$	−1	$\frac{3}{2}$	−3	1672	8.2×10^{-11}	$\Lambda^0 + \text{K}^-$

^{a} The charge and strangeness are those of the particle. Values for the antiparticle have the opposite sign. The spin, rest energy, and mean life are the same for a particle and its antiparticle.

Even though neutrinos are not produced in the decay process, the lifetime indicates that the decay proceeds through the weak interaction. Other baryons can be identified in Table 14.6 that decay through the strong, electromagnetic, or weak interactions.

14.4   CONSERVATION LAWS

We frequently use the conservation of energy, linear momentum, and angular momentum in our analysis of physical phenomena. These conservation laws are closely connected with the fundamental properties of space and time; we believe those laws to be absolute and inviolable.

We also use other kinds of conservation laws in analyzing various processes. For example, when we combine two elements in a chemical reaction, such as hydrogen + oxygen → water, we must balance the reaction in the following way:

$$2H_2 + O_2 \longrightarrow 2H_2O$$

The process of balancing a reaction can also be regarded as a way of accounting for the electrons that participate in the process: A molecule of water contains 10 electrons, and so the atoms that combine to make up the molecule must likewise include 10 electrons.

In nuclear processes, we are concerned not with electrons but with protons and neutrons. In the alpha decay of a nucleus, such as

$$^{235}_{92}\text{U}_{143} \longrightarrow ^{231}_{90}\text{Th}_{141} + ^{4}_{2}\text{He}_2$$

Emmy Noether (1882–1935, Germany-United States). Known both as a mathematician and as a theoretical physicist, she explored the role of conservation laws in physics. In an important result now known as Noether's theorem, she discovered that each symmetry of the mathematical equations describing a phenomenon gives a conserved quantity. For example, conservation of energy results from the equations being invariant to translations in time, and conservation of linear momentum from invariance to translations in space.

or in a reaction such as

$$p + {}^{63}_{29}Cu_{34} \longrightarrow {}^{63}_{30}Zn_{33} + n$$

we balance the number of protons and also the number of neutrons. We might be tempted to conclude that nuclear processes conserve both proton number and neutron number, but the separate conservation laws are not satisfied in beta decays, for example

$$n \longrightarrow p + e^- + \bar{\nu}_e$$

which does not conserve either neutron number or proton number. However, it does conserve the total neutron number plus proton number, which is equal to one on both sides of the decay. (This conservation law of total nucleon number includes the separate laws of conservation of proton number and neutron number as a special case.)

In the decays and reactions of elementary particles, conservation laws provide us with a means to understand why some processes occur and others are not observed, even though they are expected on the basis of other considerations. In negative beta decay, for example, we always find an antineutrino emitted, never a neutrino. Conversely, in positron decay, it is the neutrino that is always emitted. We account for these processes by assigning each particle a *lepton number L*. The electron and neutrino are assigned lepton numbers of +1, and the positron and antineutrino are assigned lepton numbers of −1; all mesons and baryons are assigned lepton numbers of zero. Lepton number conservation in positive and negative beta decay then works as follows:

$$n \longrightarrow p + e^- + \bar{\nu}_e$$
$$L = 0 \longrightarrow 0 + 1 + (-1)$$

$$p \longrightarrow n + e^+ + \nu_e$$
$$L = 0 \longrightarrow 0 + (-1) + 1$$

You can see that the total lepton number remains zero on both sides of these decays, which accounts for the appearance of the antineutrino in negative beta decay and the neutrino in positron decay.

According to the lepton conservation law, these processes are forbidden:

$$e^- + p \longrightarrow n + \bar{\nu}_e$$
$$L = 1 + 0 \longrightarrow 0 + (-1)$$

$$p \longrightarrow e^+ + \gamma$$
$$L = 0 \longrightarrow -1 + 0$$

In keeping track of leptons, we must count each type of lepton (e, μ, τ) separately. Evidence for this comes from a variety of experiments. The distinction between electron-type and muon-type leptons can be revealed in an experiment in which a beam of muon-type antineutrinos is incident on a target of protons:

$$\bar{\nu}_\mu + p \longrightarrow n + \mu^+$$

If there were no difference between electron-type and muon-type leptons, the following reaction would be possible:

$$\bar{\nu}_\mu + p \longrightarrow n + e^+$$

The failure to observe this second reaction indicates the fundamental difference between the types of leptons and the need to account separately for the different types.

Another example of the difference between the types of leptons comes from the failure to observe the decay

$$\mu^- \longrightarrow e^- + \gamma$$

If there were only one type of lepton number, this decay would be possible. The failure to observe this decay (in comparison with the commonly observed decay $\mu^- \to e^- + \bar{\nu}_e + \nu_\mu$, which conserves both muon-type and electron-type lepton number) suggests the need for the different kinds of lepton numbers. We call these lepton numbers L_e, L_μ, and L_τ, and we have the following conservation law for leptons:

In any process, the lepton number for electron-type leptons, muon-type leptons, and tau-type leptons must each remain constant.   *Conservation of lepton number*

The following examples illustrate the conservation of these lepton numbers.

$$\begin{array}{ccccccc} & \bar{\nu}_e & + & p & \longrightarrow & e^+ & + & n \\ L_e = & -1 & + & 0 & \longrightarrow & -1 & + & 0 \end{array}$$

$$\begin{array}{ccccccc} & \nu_\mu & + & n & \longrightarrow & \mu^- & + & p \\ L_\mu = & 1 & + & 0 & \longrightarrow & 1 & + & 0 \end{array}$$

$$\begin{array}{ccccccccc} & \mu^- & \longrightarrow & e^- & + & \bar{\nu}_e & + & \nu_\mu \\ L_e = & 0 & \longrightarrow & 1 & + & (-1) & + & 0 \\ L_\mu = & 1 & \longrightarrow & 0 & + & 0 & + & 1 \end{array}$$

$$\begin{array}{ccccccc} & \pi^- & \longrightarrow & \mu^- & + & \bar{\nu}_\mu \\ L_\mu = & 0 & \longrightarrow & 1 & + & (-1) \end{array}$$

Studying these examples, we can understand why sometimes neutrinos appear and sometimes antineutrinos appear.

Baryons are subject to a similar conservation law. All baryons are assigned a baryon number $B = +1$, and all antibaryons are assigned $B = -1$. All nonbaryons (mesons and leptons) have $B = 0$. We then have the law of conservation of baryon number:

In any process, the total baryon number must remain constant.   *Conservation of baryon number*

(The conservation of nucleon number A is a special case of conservation of baryon number, in which all the baryons are nucleons. In particle physics, it is customary to use B instead of A to represent all baryons, including the nucleons.) No violation of the law of baryon conservation has ever been observed, although the Grand Unified Theories (see Section 14.9) suggest that the proton can decay in a way that would violate conservation of baryon number.

As an example of conservation of baryon number, consider the reaction that was responsible for the discovery of the antiproton:

$$p + p \longrightarrow p + p + p + \bar{p}$$

On the left side, the total baryon number is $B = +2$. On the right side, we have three baryons with $B = +1$ and one antibaryon with $B = -1$, so the total baryon number is $B = +2$ on the right side also. On the other hand, the process

$$p + p \longrightarrow p + p + \bar{n}$$

violates baryon number conservation and is therefore forbidden.

The number of mesons that can be created or destroyed in decays or reactions is not subject to a conservation law like the number of leptons or baryons. For example, the following reactions can be used to produce pions:

$$p + p \longrightarrow p + n + \pi^+$$
$$p + p \longrightarrow p + p + \pi^0$$
$$p + p \longrightarrow p + n + \pi^+ + \pi^0$$
$$p + p \longrightarrow p + p + \pi^+ + \pi^-$$

As long as enough energy is available, any number of pions can be produced in these reactions. Note the conservation of baryon number (nucleon number) in these reactions. The pion can decay to a lepton and an antilepton (for example, $\pi^+ \to e^+ + \nu_e$) without violating any conservation law, despite the disappearance of a meson in this process.

Even though the meson number itself is not conseved, the unusual behavior of another type of meson (the K mesons, or kaons) leads to a new and different type of conservation law. The uncharged η and π^0 mesons decay very rapidly ($10^{-16} - 10^{-18}$ s) into two photons; on the basis of the systematic behavior of mesons, we would expect the K^0 to decay similarly to two photons in a comparable time. The obseved decay of the K^0 takes place much more slowly (10^{-10} s); moreover, the decay products are not photons, but π mesons and leptons. On the basic premise that anything that is not observed must be forbidden by some rule of nature, we suspect that there is a reason for the observed decay modes of the K^0 (just as the failure to observe the decay $\mu^- \to e^- + \gamma$ leads us to the rule for conserving both electron-type and muon-type lepton numbers). We therefore guess that there is some "number" associated with the K^0 meson that forbids its decay into two photons. As another example, the heavy charged mesons are all strongly interacting particles, and we expect such particles as the ρ meson to decay into lighter strongly interacting particles (π mesons, for example) in times of the order of 10^{-23} s. But the decay $K^+ \to \pi^+ + \pi^0$ occurs very slowly, in a time of the order of 10^{-8} s, and in fact the different decay mode $K^+ \to \mu^+ + \nu_\mu$ is more probable.

This unusual behavior is explained by the introduction of a new quantum number. This number is called the *strangeness S*, and we can use it to explain the properties of the K-meson decays. The K^0 and K^+ are assigned strangeness of $S = +1$; the π mesons and leptons are nonstrange particles ($S = 0$). The decay $K^0 \to \gamma + \gamma$, which is an electromagnetic decay (as indicated by the photons), is forbidden because the electromagnetic inter-action conserves strangeness ($S = +1$ on the left, $S = 0$ on the right). The decay $K^+ \to \pi^+ + \pi^0$ does not occur in the typical strong interaction time of 10^{-23} s because the strong interaction cannot change strangeness. It occurs in 10^{-8} s (and the corresponding decay $K^+ \to \mu^+ + \nu_\mu$ occurs) because the weak interaction *does not* conserve strangeness; decays that are caused by the weak interaction can change the strangeness by one unit. (Recall our discussion in the first section of this chapter that a decay time of the order of 10^{-8} s is typical of the weak interaction.)

We can summarize these results in the law of conservation of strangeness:

In processes governed by the strong or electromagnetic interactions, the total strangeness must remain constant. In processes governed by the weak interaction, the strangeness either remains constant or changes by one unit.

The strangeness quantum numbers of the mesons and baryons are given in Tables 14.5 and 14.6. The strangeness of an antiparticle has the opposite sign to that of the corresponding particle.

Strangeness conservation also helps to explain another curious aspect of the behavior of K mesons. Pi mesons are produced in nuclear collisions, for example $p + p \to p + p + \pi$ or $p + p \to p + p + \pi + \pi$. Since there is no meson conservation law, any number of π mesons can be produced. But K mesons and all other "strange" particles are always produced in pairs: $p + p \to p + p + K + K$ or $p + p \to p + \Lambda^0 + K^+$. If one of the K mesons in the first reaction is a K^+ or K^0 with $S = +1$ and the other is a K^- or $\overline{K}^0$ with $S = -1$, conservation of strangeness explains this phenome-non of *associated production*. Similarly if the K^+ has strangeness $+1$ and the Λ^0 has strangeness -1, strangeness is conserved in the second reaction.

The baryons also come in strange and nonstrange varieties. Looking at the lifetimes in Table 14.6, we see that the Λ^0 decays into $p + \pi^-$ with a lifetime of about 10^{-10} s, while we would expect a strongly interacting particle to decay to other strongly interacting particles with a lifetime of about 10^{-23} s. If the strangeness of the Λ^0 is assigned as -1, these decays change S and are forbidden to go by the strong interaction, and so must be due to the weak interaction, with the expected 10^{-10}-s lifetime. The strangeness violation also tells us why the electromagnetic decay $\Lambda^0 \to n + \gamma$ does not occur (while the decay $\Sigma^0 \to \Lambda^0 + \gamma$ does occur, with a typical electromagnetic lifetime of 10^{-19} s). Also, the weak decay can change the strangeness by at most *one* unit, and so $\Xi^0 \to n + \pi^0$ ($S = -2 \to S = 0$) is forbidden.

Strangeness

The lepton number, baryon number, and strangeness are useful concepts for describing the occurrence and nonoccurrence of various decays and reactions. We don't understand why lepton number or baryon number are conserved, or what strangeness really represents, but the hope is that one day a complete theory of the structure of particles and their interactions will provide that understanding.

EXAMPLE 14.1

The Ω^- baryon has $S = -3$. (*a*) It is desired to produce the Ω^- using a beam of K^- incident on protons. What other particles are produced in this reaction? (*b*) How might the Ω^- decay?

SOLUTION

(*a*) Reactions usually proceed only through the strong interaction, which conserves strangeness. We consider the reaction

$$K^- + p \longrightarrow \Omega^- + ?$$

On the left side, we have $S = -1$, $B = +1$, and electric charge $Q = 0$. On the right side, we have $S = -3$, $B = +1$, and $Q = -1$. We must therefore add to the right side particles with $S = +2$, $B = 0$, and $Q = +1$. Scanning through the tables of mesons and baryons, we find that we can satisfy these criteria with K^+ and K^0, so the reaction is

$$K^- + p \longrightarrow \Omega^- + K^+ + K^0$$

(*b*) The Ω^- cannot decay by the strong interaction, because no $S = -3$ final states are available. It must therefore decay to particles having $S = -2$ through the weak interaction, which can change S by one unit. One of the product particles must be a baryon in order to conserve baryon number. Two possibilities are

$$\Omega^- \longrightarrow \Lambda^0 + K^- \quad \text{and} \quad \Omega^- \longrightarrow \Xi^0 + \pi^-$$

14.5 PARTICLE INTERACTIONS AND DECAYS

In this section we briefly summarize the properties of the elementary particles and how they are measured.

Atoms and molecules can be taken apart relatively easily and nonviolently, enabling us to study their structure. However, the elementary particles, most of which are unstable and do not exist in nature, must be created in violent collisions. (The particle theorist Richard Feynman once compared this process with studying fine Swiss watches by smashing them together and looking at the pieces that emerge from the collision.) For this purpose we need a high-energy beam of particles and a suitable target of elementary particles. The only strongly interacting, stable elementary particle is the proton, and thus a hydrogen target is a logical choice. To get a reasonable density of target atoms, it is necessary to use liquid, rather than gaseous, hydrogen.

For a suitable beam, we must be able to accelerate a particle to very high energies (so high that the energy of the particle may be hundreds of times its rest energy mc^2). A stable charged particle is the logical choice for the beam; stability is required because of the relatively long time necessary to accelerate the particle to such an energy, and a charged particle is required so that electromagnetic fields may be used to accelerate the particle. Once again the proton is the logical choice, and so many particle physics reactions are initiated by accelerating protons on to a proton target, which gives

$$p + p \longrightarrow \text{product particles}$$

Among the product particles may be a variety of mesons or even heavier particles of the baryon family, of which the nucleons are the lightest members. The study of the nature and properties of these particles is the goal of particle physics.

For example, the Fermi National Accelerator Laboratory (Fermilab) near Chicago was originally designed to produce a beam of 500-GeV protons ($v/c = 0.999998$) circulating around a track of radius 1000 m (Color Plate 14). Since it began operation, the beam energy has been doubled to 1000 GeV.

In many cases, conservation laws restrict the nature of the product particles, and it would be desirable to have other types of beams available. One possibility is indicated in Figure 14.1. A proton beam is incident on a target—the nature of the target is not important. Like Feynman's Swiss watch parts, many different particles emerge. By suitable focusing and selection of the momentum, we can extract a beam of the *secondary* particles created in the reactions. The particle must live long enough to be delivered to a second target, which might be tens of meters away; even if the particle were traveling at the speed of light, it would need about 10^{-7} s to make its journey. Although this is a very short time interval by ordinary standards, on the time scale of elementary particles, it is a very long time—in fact, none of the unstable mesons or baryons (except the neutron) lives that long. Although our efforts to make a secondary beam would seem to be in vain, we have forgotten one very important detail. The lifetime of the particle is measured in its rest frame, while we are observing its flight in the laboratory frame, in which the particle is moving at speeds extremely close to the speed of light. The *time dilation* factor results in a lifetime, observed in our frame of reference, which might be hundreds of times longer than the *proper lifetime*. This factor extends the range of available secondary beams to those particles with lifetimes as short as 10^{-10} s, and makes it possible to obtain secondary beams to study such reactions as

$$\pi + p \longrightarrow \text{particles}$$

Richard P. Feynman (1918–1988, United States). Seldom is one person known for both exceptional insights into theoretical physics and exceptional methods of teaching first-year physics. He received the Nobel prize for his work on the theory that couples quantum mechanics to electromagnetism, and his text and film *Lectures on Physics* give unusual perspectives to many areas of basic physics and are enjoyed by undergraduate students, graduate students, and instructors.

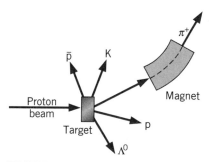

FIGURE 14.1 The production of secondary particle beams. The magnet helps to select the mass and momentum of the desired particle.

and

$$K + p \longrightarrow \text{particles}$$

even though the proper lifetimes of the π and K are in the range of 10^{-10} to 10^{-8} s.

Observing the products of these reactions, which may involve dozens of high-energy charged and uncharged particles, poses a great technological problem for the experimenter. The detector must completely surround the reaction area, so that particles are recorded no matter what direction they travel after the reaction. The particles must produce visible tracks in the detector, so that their identity and direction of travel can be determined. It must provide sufficient mass to stop the particles and measure their energy. A magnetic field must be present, so that the resulting curved trajectory of a charged particle can be used to determine its momentum and the sign of its charge. Figure 14.2 shows tracks left in a *bubble chamber*, a large tank filled with liquid hydrogen in which the passage of a charged particle causes microscopic bubbles resulting from the ionization of the hydrogen atoms. The bubbles can be illuminated and photographed to reveal the tracks. Color Plate 15 shows a large detector system that is used both to display the tracks of particles and to measure their energies; Color Plate 16 shows a sample of the results that can be obtained with this type of detector.

From a careful analysis of the paths of particles, such as revealed in bubble chamber photographs, we can deduce the desired quantities of mass, linear momentum, and energy. The other important property we would like to know is the lifetime of the decay of the product particles, since

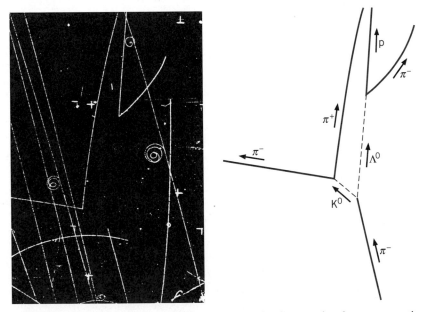

FIGURE 14.2 A bubble chamber photograph of a reaction between particles. At right is shown a diagram indicating the particles that participate in the reaction (courtesy Lawrence Berkeley Laboratory).

many of the products are often unstable. If we know the speed of a particle, we can find its lifetime by simply observing the length of its track in a bubble chamber photograph. (Even for uncharged particles, which leave no tracks, we can use this method to deduce the lifetime, since the subsequent decay of the uncharged particle into two charged particles defines the length of its path rather clearly, as shown in Figure 14.2.)

This method works well if the lifetime is of the order of 10^{-10} s or so, such that the particle leaves a track long enough to be measured (millimeters to centimeters). With careful experimental technique and clever data analysis, this can be extended to track lengths of the order of 10^{-6} m, and so lifetimes down to about 10^{-16} s can be measured in this way (with a little help from the time dilation factor). But many of our particles have lifetimes of only 10^{-23} s, and a particle moving at even the speed of light travels only the diameter of a nucleus in that time! How can we measure such a lifetime? Furthermore, how do we even know such a particle exists at all? Consider the reaction

$$\pi + p \longrightarrow \pi + p + x$$

where x is an unknown particle with a lifetime of about 10^{-23} s, which decays into two π mesons according to $x \rightarrow \pi + \pi$. How do we distinguish the above reaction from the reaction

$$\pi + p \longrightarrow \pi + p + \pi + \pi$$

which leads to the same particles as actually observed in the laboratory?

Experimental evidence suggests that the two π mesons in such reactions may combine for an instant (10^{-23} s) to form an entity with all of the usual properties of a particle—a definite mass, charge, spin, lifetime, etc. Such states are known as *resonance particles*, and we now look at the indirect evidence from which we infer their existence.

Suppose you receive a package in the mail from a friend. When you open it, you find it contains many small, irregular pieces of broken glass. How do you learn whether your friend sent you a beautiful glass vase that was broken in shipment or a package of broken glass as a practical joke? You try to put the pieces together! If the pieces fit together, it is a good assumption that the vase was once whole, although the mere fact that they fit together doesn't *prove* that it was once whole. It's just the simplest possible assumption *consistent with our experience*. (An alternative assumption that the pieces were manufactured separately and just happen by chance to fit together is highly improbable.)

How then do we detect a "particle" which lives for only 10^{-23} s? We look at its decay products (which live long enough to be seen in the laboratory), and putting the pieces back together, we infer that they once may have been a whole particle.

For example, suppose in the laboratory we observe two π mesons emitted as shown in Figure 14.3. We measure the direction of travel and the linear momentum of the π mesons as shown. A second and a third event each produces two π mesons as also shown in the figure. Are these three events consistent with the existence of the same resonance particle?

Resonance particles

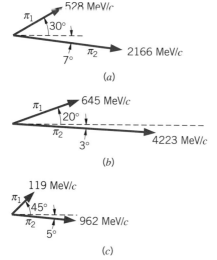

FIGURE 14.3   Three possible decays of an unknown particle into two pi mesons. The direction and momentum of each pi meson are indicated.

Let us assume that in each case, a particle moving at an unknown speed decayed into the two particles as shown. Since each decay must conserve energy and momentum, we can use the decay information to work backward and find the energy and momentum of the decaying particle, and therefore we can find its rest energy according to $mc^2 = \sqrt{(E_1 + E_2)^2 - c^2(\mathbf{p}_1 + \mathbf{p}_2)^2}$. Carrying out the calculation, we find that, for the decay shown in part (a) of Figure 14.3, $mc^2 = 764$ MeV, while for part (b), $mc^2 = 775$ MeV. It is therefore possible that these two events result from the decays of identical particles. Part (c) of the figure gives $mc^2 = 498$ MeV, which differs considerably from parts (a) and (b).

Of course, these two events are not sufficient to identify conclusively the existence of a resonance particle with a rest energy in the range of 770 MeV. It could be a mere accident, just like the chance fitting together of two pieces of broken glass. What is needed is a large (statistically significant) number of events, in which we can combine the momenta of the two emitted π mesons in such a way that the deduced mass of the resonance particle is always the same. Figure 14.4 is an example of such a result. There is a background of events with a continuous distribution of energies, like beta decay electrons; these come from events like part (c) of Figure 14.3. There is also present a very prominent peak at 770 MeV. We identify this energy as the rest energy of the resonance particle, which is known as the ρ (rho) meson. (How do we know it is a meson? It must be a strongly interacting particle, since it decays so rapidly. The only possibilities are therefore mesons, with integral spin, or baryons, with half-integral spin. Since π mesons have integral spin, and since two integral spins can combine to give only another integral spin, it must be a meson.)

We can also infer the lifetime of the particle from Figure 14.4. The particle lives only for about 10^{-23} s, and so if we are to measure its rest

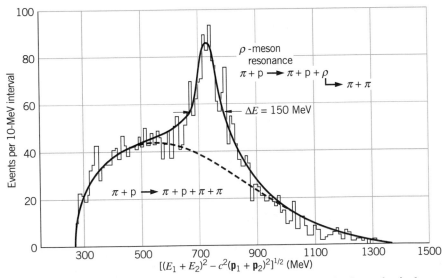

FIGURE 14.4 The resonance identified as the ρ meson. The horizontal axis shows the energy and momenta of the two decay pi mesons, combined to be equivalent to the mass of the resonance particle.

energy we have only 10^{-23} s in which to do it. But the uncertainty principle requircs that an energy measurement made in a time interval Δt be uncertain by an amount roughly $\Delta E \cong \hbar/\Delta t$. This energy uncertainty ΔE is observed as the *width* of the peak in Figure 14.4. We don't always deduce the same value 770 MeV for the rest energy of the ρ meson; sometimes our value is a bit larger and sometimes a bit smaller. *The width of the resonance peak tells us the lifetime of the particle.* (The width is not really precisely defined, but physicists usually take as the width the interval between the two points where the height of the resonance is one-half its maximum value above the background, as shown in Figure 14.4.) The width of $\Delta E = 150$ MeV leads to a value of $\Delta t = \hbar/\Delta E = 4.4 \times 10^{-24}$ s for the lifetime of the ρ meson.

14.6 ENERGETICS OF PARTICLE DECAYS

In analyzing the decays and reactions of elementary particles, we apply many of the same laws that we used for nuclear decays and reactions: energy, linear momentum, and total angular momentum must be conserved, and the total value of the quantum numbers associated with electric charge, lepton number, and baryon number (which we previously called nucleon number) must be the same before and after the decay or reaction. In reactions of elementary particles, we are often concerned with the production of new varieties of particles. The energy necessary to manufacture these particles comes from the kinetic energy of the reaction constituents (usually the incident particle), and since this energy is usually quite large (hence the name *high-energy physics* for this type of research), *relativistic equations* must be used for energy and momentum.

The decays of elementary particles can be analyzed in a way similar to the decays of nuclei, following the same two basic rules:

1. The energy available for the decay (assuming the decaying particle is at rest) is the difference in rest energy between the initial decaying particle and the particles that are produced in the decay. By analogy with our study of nuclear decays, we call this the Q value:

$$Q = (m_i - m_f)c^2 \qquad (14.1) \qquad \textit{Q value of particle decay}$$

where $m_i c^2$ is the rest energy of the initial particle and $m_f c^2$ is the total rest energy of all the final product particles. (Of course, the decay will occur only if Q is positive.)

2. The available energy Q is shared as kinetic energy of the decay products in such a way as to conserve linear momentum. As in the case of nuclear decays, for a decay of a particle at rest into two final particles, the particles have equal and opposite momenta, and we can find unique values for the energies of the two final particles. For decays into three or more particles, each particle has a spectrum or distribution of energies from zero up to some maximum value (as was the case with nuclear beta decay).

EXAMPLE 14.2

Compute the energies of the proton and π meson that result from the decay of the Λ^0.

SOLUTION

The decay process is

$$\Lambda^0 \longrightarrow p + \pi^-$$

Using the rest energies from Tables 14.5 and 14.6, we have:

$$Q = (m_{\Lambda^0} - m_p - m_{\pi^-})c^2$$

$$= 1116\,\text{MeV} - 938\,\text{MeV} - 140\,\text{MeV}$$

$$= 38\,\text{MeV}$$

and so the total kinetic energy of the decay products must be:

$$K_p + K_\pi = 38\,\text{MeV}$$

Using the relativistic formula for kinetic energy, we can write this as

$$K_p + K_\pi = (\sqrt{c^2 p_p^2 + m_p^2 c^4} - m_p c^2) + (\sqrt{c^2 p_\pi^2 + m_\pi^2 c^4} - m_\pi c^2) = 38\,\text{MeV}$$

Conservation of momentum requires $p_p = p_\pi$. Substituting for one of the unknown momenta in the above equation and solving algebraically for the other, we obtain

$$p_\pi = p_p = 101\,\text{MeV}/c$$

The kinetic energies can be found by substituting these momenta into the relativistic formula:

$$K_\pi = 33\,\text{MeV}$$

$$K_p = 5\,\text{MeV}$$

EXAMPLE 14.3

What is the maximum kinetic energy of the electron emitted in the decay $\mu^- \rightarrow e^- + \bar{\nu}_e + \nu_\mu$?

SOLUTION

The Q value for this decay is $Q = m_\mu c^2 - m_e c^2 = 105.2$ MeV, since the neutrinos have negligible or zero rest energy. If the μ^- is at rest, this energy is shared by the electron and the neutrinos: $Q = K_e + E_{\bar{\nu}_e} + E_{\nu_\mu}$. When the electron has its maximum kinetic energy, the two neutrinos carry away the minimum energy. This minimum cannot be zero, because that would violate momentum conservation: the electron would be carrying momentum that would not be balanced by the neutrino momenta to give a net of zero (since we assumed the μ^- to be at rest, $p_{\text{initial}} = p_{\text{final}} = 0$). We assume that the electron has its maximum energy when the neutrinos are emitted in exactly the opposite direction to the electron; otherwise some of the decay energy is "wasted" by providing transverse momentum components for the neutrinos, and not as much energy will be available for the electron. Since it does not matter which of the neutrinos carry the energy and momentum (they may even share it in any proportion), we let E_ν and p_ν be the total neutrino energy and momentum; these are of course related by $E_\nu = cp_\nu$, since neutrinos are presumed to be massless and to travel at the speed of light. If we let E_e and p_e represent the energy and momentum of the electron, then linear momentum conservation gives

$$p_e - p_\nu = 0$$

For the electron, $E_e = \sqrt{c^2 p_e^2 + m_e^2 c^4}$. Together, these equations give:

$$Q = E_e - m_e c^2 + cp_\nu$$
$$= E_e - m_e c^2 + cp_e$$
$$= E_e - m_e c^2 + \sqrt{E_e^2 - m_e^2 c^4}$$

Solving, we find:

$$E_e = \frac{Q^2}{2m_\mu c^2} + m_e c^2$$

$$K_e = E_e - m_e c^2 = Q^2/2m_\mu c^2 = 52.3 \text{ MeV}$$

The original rest energy of the μ^- is shared essentially equally by the electron and the two neutrinos in this case: $(K_e)_{\text{max}} \cong (E_\nu)_{\text{max}} \cong Q/2$. Note how different this is from the case of the beta decay of the neutron, where the heavy proton resulting from the decay could absorb considerable recoil momentum at a cost of very little energy, so nearly all of the available energy could be given to the electron: $(K_e)_{\text{max}} \cong Q$.

EXAMPLE 14.4

Find the maximum energy of the positrons and of the π mesons produced in the decay $K^+ \rightarrow \pi^0 + e^+ + \nu_e$

SOLUTION

The Q value for this decay is

$$Q = (m_K - m_\pi - m_e)c^2$$
$$= 494 \text{ MeV} - 135 \text{ MeV} - 0.5 \text{ MeV}$$
$$= 358.5 \text{ MeV}$$

This energy must be shared among the three products:

$$Q = K_\pi + K_e + E_\nu$$

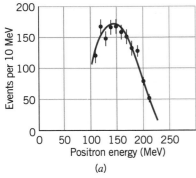

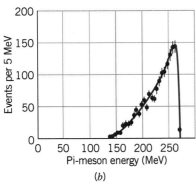

FIGURE 14.5 The spectrum of positrons and pi mesons from the decay of the K^+ meson.

The electron and π meson have their maximum energies when the neutrino has negligible energy:

$$Q = K_\pi + K_e$$

and conservation of momentum in this case (if the neutrino has negligible momentum) requires $p_\pi = p_e$. Using the relativistic equation for kinetic energy, we have

$$Q = \sqrt{(pc)^2 + (m_\pi c^2)^2} - m_\pi c^2 + \sqrt{(pc)^2 + (m_e c^2)^2} - m_e c^2$$

where $p = p_e = p_\pi$. Inserting the numbers, we obtain

$$494\,\text{MeV} = \sqrt{(pc)^2 + (135\,\text{MeV})^2} + \sqrt{(pc)^2 + (0.5\,\text{MeV})^2}$$

Clearing the two radicals involves quite a bit of algebra, but we can simplify the problem if we inspect this expression and notice that the solution must have a large value of pc, certainly greater than 100 MeV. (Otherwise the two terms could not sum to nearly 500 MeV.) Thus $(pc)^2 \gg (0.5\,\text{MeV})^2$, and we can neglect the electron rest energy term in the second radical, which simplifies the equation somewhat:

$$494\,\text{MeV} = \sqrt{(pc)^2 + (135\,\text{MeV})^2} + pc$$

Solving, we find $pc = 229$ MeV, which gives $(E_e)_{max} = 229$ MeV and $(E_\pi)_{max} = 266$ MeV. Figure 14.5 shows the observed energy spectra of e^+ and π^0 from the K^+ decay, and the energy maxima are in agreement with the calculated values. (The shapes of the energy distributions are determined by statistical factors, as in the case of nuclear beta decay. The statistical factors are different for e^+ and π^0, since the π^0 also has its maximum energy when the e^+ appears at rest and the ν carries the recoil momentum.)

You should repeat this calculation and convince yourself that (1) the π^0 has its maximum energy also when $K_e = 0$ ($E_e = m_e c^2$) and (2) the e^+ does *not* have its maximum energy when $K_\pi = 0$.

14.7 ENERGETICS OF PARTICLE REACTIONS

The basic experimental technique of particle physics consists of studying the product particles that result from a collision between an incident particle (accelerated to high energies) and a target particle (usually at rest). The kinematics of the reaction process must be analyzed using relativistic formulas, since the kinetic energies of the particles are usually comparable to or greater than their rest energies. In this section we derive some of the relationships that are needed to analyze these reactions, using the formulas for relativistic kinematics we obtained in Chapter 2. Since an important purpose of these reactions is the production of new varieties of particles, we concentrate on calculating the threshold energy needed to produce these particles. (You might find it helpful to review the discussion in Chapter 13 on *nonrelativistic* reaction thresholds.)

Consider the following reaction:

$$m_1 + m_2 \longrightarrow m_3 + m_4 + m_5 + \cdots$$

where the m's represent both the particles and their masses. Any number of particles can be produced in the final state. Here m_1 is the incident particle, which has total energy E_1, kinetic energy $K_1 = E_1 - m_1 c^2$, and momentum $p_1 = \sqrt{E_1^2 - m_1^2 c^4}$ in the *laboratory* frame of reference. The target particle m_2 is at rest in the laboratory. Figure 14.6 illustrates this reaction in the laboratory frame of reference.

Just as we did for nuclear reactions, we define the Q value to be

$$Q = [m_1 + m_2 - (m_3 + m_4 + m_5 + \cdots)]c^2 \qquad (14.2)$$

Q value of particle reaction

If Q is positive, energy is "liberated" (actually, rest energy is turned into kinetic energy, so that the product particles m_3, m_4, m_5, ... have more combined kinetic energy than the initial particles m_1 and m_2). If Q is negative, some of the initial kinetic energy of m_1 is turned into rest energy.

EXAMPLE 14.5

Compute the Q values for the reactions

$$\pi^- + p \longrightarrow K^0 + \Lambda^0$$

$$K^- + p \longrightarrow \Lambda^0 + \pi^0$$

SOLUTION

For the first reaction we have, using rest energies from Tables 14.5 and 14.6,

$$Q = [m_{\pi^-} + m_p - (m_{K^0} + m_{\Lambda^0})]c^2$$

$$= 140\,\text{MeV} + 938\,\text{MeV} - 498\,\text{MeV} - 1116\,\text{MeV}$$

$$= -536\,\text{MeV}$$

This reaction has a negative Q value, and energy must be supplied in the form of initial kinetic energy to produce the additional rest energy of the products. For the second reaction we have:

$$Q = [m_{K^-} + m_p - (m_{\Lambda^0} + m_{\pi^0})]c^2$$

$$= 494\,\text{MeV} + 938\,\text{MeV} - 1116\,\text{MeV} - 135\,\text{MeV}$$

$$= 181\,\text{MeV}$$

A positive Q value indicates that there is enough rest energy in the initial particles to produce the final particles; in fact there is 181 MeV of energy (plus the kinetic energy of the incident particle) left over for kinetic energy of the Λ^0 and π^0.

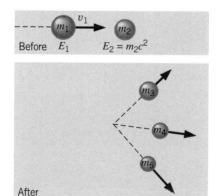

FIGURE 14.6 A reaction between particles in the laboratory reference frame.

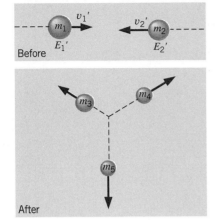

FIGURE 14.7   The same reaction as Figure 14.6, but viewed in the CM reference frame.

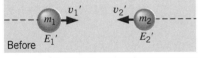

FIGURE 14.8   A reaction in the CM reference frame when m_1 has the threshold kinetic energy.

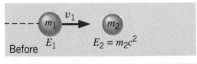

FIGURE 14.9   The reaction of Figure 14.8 in the laboratory reference frame.

Only for negative Q values do we have a *threshold energy* K_{th}, a minimum kinetic energy that m_1 must have in order to initiate the reaction. As in the nuclear physics case, the threshold kinetic energy K_{th} is larger than the magnitude of Q. We must not only create the additional particles, but the product particles must be given sufficient kinetic energy so that linear momentum is conserved in the reaction.

This calculation, like the nuclear physics case, is easiest to do if we switch to the center-of-mass (or center-of-momentum) reference frame, in which the total momentum is zero. Above threshold, the reaction viewed in this frame (which we call the CM frame) would look like the illustration in Figure 14.7; the total momentum is zero, both before and after the collision. If we supply only the threshold energy required to create the product particles, the reaction in the CM frame would look like Figure 14.8; this same reaction viewed in the laboratory would be as illustrated in Figure 14.9. In the CM frame, the product particles are at rest (this is possible because the total momentum is zero in this frame). In the laboratory frame, they move together with a common speed which is the same as the relative velocity between the laboratory and CM frames.

We begin by deriving a *relativistic* expression for this velocity. In the laboratory frame, m_1 moves with speed v_1 and therefore has momentum $p_1 = m_1 v_1 / \sqrt{1 - v_1^2/c^2}$. In the CM frame, m_1 moves with speed v_1' and momentum p_1', and m_2 moves with speed v_2' and momentum p_2'. Let v be the velocity of the CM frame relative to the laboratory frame; we wish to find v in terms of m_1, m_2, and v_1.

From Chapter 2 we take the relativistic expression for velocity transformation:

$$v_x' = \frac{v_x - v}{1 - v_x v/c^2} \tag{14.3}$$

and so

$$v_1' = \frac{v_1 - v}{1 - v_1 v/c^2} \tag{14.4}$$

and

$$v_2' = -v \tag{14.5}$$

The last result follows directly from $v_2 = 0$. Since $p_1' = p_2'$ in this frame, we must have:

$$\frac{m_1 v_1'}{\sqrt{1 - v_1'^2/c^2}} = \frac{m_2 v_2'}{\sqrt{1 - v_2'^2/c^2}} \tag{14.6}$$

Substituting Equations 14.4 and 14.5 into 14.6, we can derive the result

$$v = \frac{m_1 v_1}{m_1 + m_2 \sqrt{1 - v_1^2/c^2}} \tag{14.7}$$

That is, if we were to travel at speed v (in the direction of v_1), then the reaction shown in Figure 14.6 would appear to us as the illustration in Figure 14.7.

When m_1 is given just the threshold kinetic energy, the final products remain at rest in the CM frame (Figure 14.8). That is, the total energy of the reaction products m_3, m_4, m_5, ... is their total rest energy $m_3c^2 + m_4c^2 + m_5c^2 + \ldots$. Since energy is conserved in all frames of reference, the total energy of m_1 and m_2 before the collision must be equal to the total energy after the collision. Letting E_1' and E_2' be the total energies of m_1 and m_2 in the CM frame, the *threshold condition* is

$$E_1' + E_2' = m_3c^2 + m_4c^2 + m_5c^2 + \cdots \tag{14.8}$$

where

$$E_1' = \frac{m_1c^2}{\sqrt{1 - v_1'^2/c^2}} \tag{14.9}$$

$$E_2' = \frac{m_2c^2}{\sqrt{1 - v_2'^2/c^2}} \tag{14.10}$$

The total energy in the CM frame, $E_1' + E_2'$, can be found by adding Equations 14.9 and 14.10 and substituting our previous expressions for v_1' and v_2', Equations 14.4 and 14.5. Using our deduced value for v, after considerable algebraic manipulation we find:

$$E_1' + E_2' = \sqrt{m_1^2c^4 + m_2^2c^4 + 2E_1m_2c^2} \tag{14.11}$$

This is a general expression for the total CM energy when m_1 has *laboratory* total energy E_1; this expression is always valid, not only at threshold.

We now apply the threshold condition, Equation 14.8:

$$\sqrt{m_1^2c^4 + m_2^2c^4 + 2E_1m_2c^2} = m_3c^2 + m_4c^2 + m_5c^2 + \cdots \tag{14.12}$$

This expression can be solved for E_1, and the threshold kinetic energy K_{th} is then $E_1 - m_1c^2$; after a bit of algebra the result is as follows:

$$K_{\text{th}} = (-Q)\frac{m_1 + m_2 + m_3 + m_4 + m_5 + \cdots}{2m_2} \tag{14.13}$$

Reaction threshold kinetic energy

This can also be written as

$$K_{\text{th}} = (-Q)\frac{\text{total mass of all particles involved in reaction}}{2 \times \text{mass of target particle}} \tag{14.14}$$

In the limit of low speeds, the relativistic threshold formula reduces to the nonrelativistic formula for nuclear reactions derived in Chapter 13 (see Problem 17).

EXAMPLE 14.6

Calculate the threshold kinetic energy to produce π mesons from the reaction $\text{p} + \text{p} \rightarrow \text{p} + \text{p} + \pi^0$.

SOLUTION

The Q value is

$$Q = m_p c^2 + m_p c^2 - (m_p c^2 + m_p c^2 + m_\pi c^2)$$

$$= -m_\pi c^2 = -135 \text{ MeV}$$

Using Equation 14.13 we can find the threshold kinetic energy

$$K_{th} = (-Q) \frac{4m_p + m_\pi}{2m_p}$$

$$= (135 \text{ MeV}) \frac{4 \times 938 \text{ MeV} + 135 \text{ MeV}}{2 \times 938 \text{ MeV}}$$

$$= 280 \text{ MeV}$$

Such energetic protons are produced at many accelerators throughout the world, and as a result the properties of the π mesons can be carefully investigated.

EXAMPLE 14.7

In 1956 an experiment was performed at Berkeley to search for the antiproton, which could be produced in the reaction

$$p + p \longrightarrow p + p + p + \bar{p}$$

What is the threshold energy for this reaction?

SOLUTION

Since the rest energy of the antiproton is identical to the rest energy of the proton (938 MeV), the Q value is

$$Q = m_p c^2 + m_p c^2 - (4 \times m_p c^2)$$

$$= -2m_p c^2$$

Thus

$$K_{th} = (2m_p c^2) \frac{6m_p c^2}{2m_p c^2}$$

$$= 6m_p c^2 = 5628 \text{ MeV} = 5.628 \text{ GeV}$$

For the discovery of the antiproton produced in this reaction, Owen Chamberlain and Emilio Segrè were awarded the Nobel prize in physics in 1959.

It is interesting to compute the "efficiency" of these reactions; that is, how much of the initial kinetic energy we supply actually goes into producing the final particles, and how much is "wasted" in the laboratory kinetic energies of the reaction products. In the first example, we supply 280 MeV of kinetic energy to produce 135 MeV of rest energy, for an efficiency of about 50 percent. In the second example, $6m_pc^2$ of kinetic energy produces only $2m_pc^2$ of rest energy, for an efficiency of only 33 percent. As the rest energies of the product particles become larger, the efficiency decreases, and relatively more energy must be supplied. For example, to produce a particle with a rest energy of 50 GeV in a proton-proton collision, we need to supply about 1250 GeV of initial kinetic energy. Only 4 percent of the energy supplied actually goes into producing the new particles; the remaining 96 percent must go to kinetic energy of the products in order to balance the large initial momentum of the incident particle. To produce a 100-GeV particle requires not twice as much energy, but four times as much.

This is obviously not a pleasant situation for particle physicists, who must build increasingly more powerful accelerators to accomplish their goals of producing more massive particles. One way out of this difficulty would be to do an experiment in the CM frame, where at threshold the production of new particles is 100 percent efficient—*none* of the initial kinetic energy goes into kinetic energy of the products, which are at rest in the CM frame. Thus a 50-GeV particle could be produced by a head-on collision between two protons with as little as 25 GeV of kinetic energy. Of course, this great gain in efficiency is at a cost of the technological difficulty of making such collisions occur. There are now *colliding beam* accelerators in operation, in which beams of particles (such as electrons or protons) can occasionally be made to collide. For example, in the Fermilab accelerator (Color Plate 14), beams of protons and antiprotons (each of energy 1 TeV = 1000 GeV) circulate around the ring in opposite directions and collide once during each revolution. Other colliding beam accelerators bring together electrons and positrons at energies of 50 to 100 GeV. In each case, all of the available energy can go into the production of new particles.

14.8 THE QUARK MODEL

Although the classes and properties of the elementary particles seem like a complicated and disordered collection, there is an underlying order that suggests that a scheme of remarkable simplicity is at work. We can illustrate this order if we plot a diagram that has strangeness along the y axis and electric charge along the x axis. If the families of particles are placed in their proper locations on the graphs, regular geometrical patterns begin to emerge. Figures 14.10 to 14.12 show such plots for the lower mass spin-0 mesons, the spin-$\frac{1}{2}$ baryons, and the spin-$\frac{3}{2}$ baryons. In 1964, Murray Gell-Mann and George Zweig independently and simultaneously recognized that such regular patterns are evidence of an underlying structure in the particles. They showed that they could duplicate these patterns if the mesons and baryons were composed of three fundamental particles, which soon became known as *quarks*. The three quarks, known as up (u), down (d), and strange (s), have the properties listed in Table 14.7. We will shortly

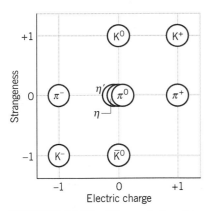

FIGURE 14.10 The relationship between electric charge and strangeness for the spin-0 mesons.

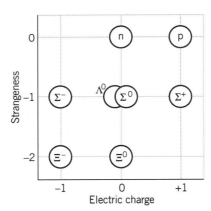

FIGURE 14.11 The relationship between electric charge and strangeness for the spin-$\frac{1}{2}$ baryons.

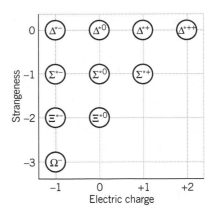

FIGURE 14.12 The relationship between electric charge and strangeness for the spin-$\frac{3}{2}$ baryons.

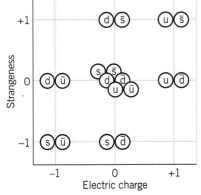

FIGURE 14.13 Spin-0 quark-antiquark combinations; compare with Figure 14.10.

see that we now believe that six quarks are necessary to account for all known mesons and baryons.

Let us see how the quark model works in the case of the mesons. Since the quarks have spin $\frac{1}{2}$ and the mesons have spin zero, the simplest scheme would be to combine two quarks, with their spins directed oppositely, to make a meson. However the mesons have baryon number $B = 0$, while a combination of two quarks would have $B = \frac{1}{3} + \frac{1}{3} = +\frac{2}{3}$. A combination of a quark and an antiquark, on the other hand, would have $B = 0$, since the antiquark has $B = -\frac{1}{3}$. For example, suppose we combine a u quark with a $\overline{d}$ ("antidown") quark, obtaining the combination u$\overline{d}$. This combination has spin zero and electric charge $\frac{2}{3}e + \frac{1}{3}e = +e$. (A d quark has charge $-\frac{1}{3}e$, so $\overline{d}$ has charge $+\frac{1}{3}e$.) The properties of this combination are identical with the π^+ meson, and so we identify the π^+ with the combination u$\overline{d}$. Continuing in this way, we find nine possible combinations of one of the three original quarks from Table 14.7 with an antiquark, as listed in Table 14.8, and plotting those nine combinations on a graph of strangeness against electric charge, we obtain Figure 14.13, which looks identical to Figure 14.10.

The baryons have $B = +1$ and spin $\frac{1}{2}$ or $\frac{3}{2}$, which suggests immediately that three quarks make a baryon. The 10 possible combinations of the three original quarks are listed in Table 14.9, and we can arrange them into two

TABLE 14.7   PROPERTIES OF THE THREE ORIGINAL QUARKS

Name	Symbol	Charge (e)	Spin ($\hbar$)	Baryon Number	Strangeness	Antiquark
Up	u	$+\frac{2}{3}$	$\frac{1}{2}$	$+\frac{1}{3}$	0	$\overline{u}$
Down	d	$-\frac{1}{3}$	$\frac{1}{2}$	$+\frac{1}{3}$	0	$\overline{d}$
Strange	s	$-\frac{1}{3}$	$\frac{1}{2}$	$+\frac{1}{3}$	-1	$\overline{s}$

TABLE 14.8 POSSIBLE QUARK-ANTIQUARK COMBINATIONS

Combination	Charge (e)	Spin (ℏ)	Baryon Number	Strangeness
$u\bar{u}$	0	0, 1	0	0
$u\bar{d}$	+1	0, 1	0	0
$u\bar{s}$	+1	0, 1	0	+1
$d\bar{u}$	−1	0, 1	0	0
$d\bar{d}$	0	0, 1	0	0
$d\bar{s}$	0	0, 1	0	+1
$s\bar{u}$	−1	0, 1	0	−1
$s\bar{d}$	0	0, 1	0	−1
$s\bar{s}$	0	0, 1	0	0

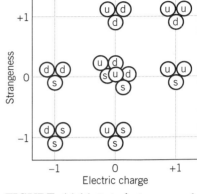

FIGURE 14.14 Spin-$\frac{1}{2}$ three-quark combinations; compare with Figure 14.11.

patterns as shown in Figures 14.14 and 14.15, which are identical to those for the spin-$\frac{1}{2}$ and spin-$\frac{3}{2}$ baryons.

Using the quark model, we can analyze the decays and reactions of the elementary particles, based on two rules:

1. Quark-antiquark pairs can be created from energy quanta, and conversely can annihilate into energy quanta. For example,

$$d + \bar{d} \longrightarrow \text{energy} \qquad \text{or} \qquad \text{energy} \longrightarrow u + \bar{u}$$

This energy can be in the form of gamma rays (as in electron-positron annihilation), or else it can be transferred to or from other particles in the decay or reaction.

2. The weak interaction can change one type of quark into another through emission or absorption of a W^+ or W^-, for example $s \rightarrow u + W^-$. The W then decays by the weak interaction, such as $W^- \rightarrow \mu^- + \bar{\nu}_\mu$. The strong and electromagnetic interactions cannot change one type of quark into another.

TABLE 14.9 POSSIBLE THREE-QUARK COMBINATIONS

Combination	Charge (e)	Spin (ℏ)	Baryon Number	Strangeness
uuu	+2	$\frac{3}{2}$	+1	0
uud	+1	$\frac{1}{2}, \frac{3}{2}$	+1	0
udd	0	$\frac{1}{2}, \frac{3}{2}$	+1	0
uus	+1	$\frac{1}{2}, \frac{3}{2}$	+1	−1
uss	0	$\frac{1}{2}, \frac{3}{2}$	+1	−2
uds	0	$\frac{1}{2}, \frac{3}{2}$	+1	−1
ddd	−1	$\frac{3}{2}$	+1	0
dds	−1	$\frac{1}{2}, \frac{3}{2}$	+1	−1
dss	−1	$\frac{1}{2}, \frac{3}{2}$	+1	−2
sss	−1	$\frac{3}{2}$	+1	−3

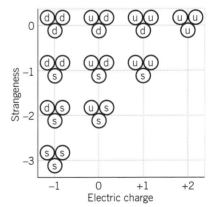

FIGURE 14.15 Spin-$\frac{3}{2}$ three-quark combinations; compare with Figure 14.12.

EXAMPLE 14.8

Analyze (*a*) the reaction $\pi^- + p \rightarrow \Lambda^0 + K^0$ and (*b*) the decay $\pi^+ \rightarrow \mu^+ + \nu_\mu$ in terms of the constituent quarks.

SOLUTION

(*a*) The reaction can be rewritten as follows:

$$d\bar{u} + uud \longrightarrow uds + d\bar{s}$$

Each side contains one u quark and two d quarks, which don't change in the reaction. The remaining transformation is:

$$\bar{u} + u \longrightarrow s + \bar{s}$$

The u and $\bar{u}$ annihilate, and from the resulting energy s and $\bar{s}$ are created.

(*b*) The π^+ has the quark composition $u\bar{d}$. Since there are no quarks in the final state $(\mu^+ + \nu_\mu)$, we must find a way to get rid of the quarks. One possible way is to change the u quark into a d quark: $u \rightarrow d + W^+$. The remaining processes are then

$$d + \bar{d} \longrightarrow \text{energy} \qquad \text{and} \qquad W^+ \longrightarrow \mu^+ + \nu_\mu$$

You may have noticed that some of the heavier mesons listed in Table 14.5 were not included in Figure 14.10, and they cannot be accounted for among the quark-antiquark combinations listed in Table 14.8. Where do these particles fit in our scheme?

In 1974, a new meson ψ (psi) was discovered at a rest energy of 3.1 GeV. This new meson was expected to decay to lighter mesons in a characteristic strong interaction time of around 10^{-23} s. Instead, its lifetime was stretched by 3 orders of magnitude to about 10^{-20} s, and its decay products were e^+ and e^-, which are more characteristic of an electromagnetic process. Why is the rapid, strong interaction decay path blocked for this particle? This was soon explained by assuming the ψ to be composed of a new quark c, called the *charm* quark, and its antiquark $\bar{c}$. The existence of the c quark had been predicted 4 years earlier as a way to explain the failure to observe the decay $K^0 \rightarrow \mu^+ + \mu^-$, which violates no previously known law but is nevertheless not observed.

The c quark, which carries a charge of $+\frac{2}{3}e$, has another property, charm, which operates somewhat like strangeness. We assign a charm quantum number $C = +1$ to the c quark (and assign $C = -1$ to its antiquark $\bar{c}$). All other quarks are assigned $C = 0$. We can now construct a new set of mesons by combining the c quark with the $\bar{u}$, $\bar{d}$, and $\bar{s}$ antiquarks and by combining the $\bar{c}$ antiquark with the u, d, and s quarks. Instead of nine spin-0 mesons, there are now 16, and the two-dimensional graphs of Figures 14.10 and 14.13 must be extended to a third dimension to show the C axis (Figure 14.16). All of these new mesons, called D, have been observed in high-energy collision experiments.

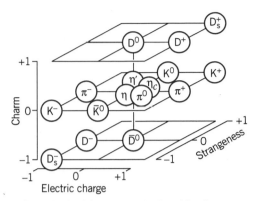

FIGURE 14.16 The relationship between electric charge, strangeness, and charm for the spin-0 mesons.

In 1977, the same sequence of events was repeated with another meson, Υ (upsilon). The rest energy was determined to be about 9.5 GeV, and again its decay was slowed to about 10^{-20} s and occurred into $e^+ + e^-$ rather than into mesons. Once again, a new quark was postulated: the b (bottom) quark with a new quantum "bottomness" number $B = -1$ and a charge of $-\frac{1}{3}e$. The Υ is assigned as the combination $b\bar{b}$, and new particles such as the B meson ($b\bar{u}$) were discovered.

The sixth, and presumably last, quark was discovered in 1994 in proton-antiproton collisions at Fermilab. These collisions created this new quark and its antiquark, both of which decayed into a shower of secondary particles. By measuring the energy and momentum of the secondary particles, the experimenters were able to determine the mass of the new quark to be 180 GeV (roughly the mass of a gold atom). This new quark is known as t (top) and has a new associated property of "topness" with a quantum number $T = +1$.

It may now seem that we are losing sight of our goal to achieve simplicity (to add the "bottomness" axis to Figure 14.16 we would need to depict a four-dimensional space!) and that we are moving toward replacing a complicated array of particles with an equivalently complicated array of quarks. In the next section, we see that we are indeed on the path to a simple explanation of the fundamental particles.

The quark model does a great deal more than allow us to make geometrical arrangements of particles such as Figure 14.16. It can be used to explain many observed properties of the particles, such as their masses and magnetic moments, and to account for their decay lifetimes and reaction probabilities. Nevertheless, a free quark has never been observed, despite heroic experiments to search for them. How can we be sure that they exist? In experiments that scatter high-energy electrons from protons, we observe more particles scattered at large angles than we would expect if the electric charge of the proton were uniformly distributed throughout its volume, and from the analysis of the distribution of the scattered electrons we conclude that inside the proton are 3 point-like objects that are responsible for the scattering. This experiment is exactly analogous to Rutherford scattering, in which the presence of the nucleus as a compact object inside the atom was revealed by the distribution of scattered alpha particles at angles larger than expected. Like Rutherford's experiment, the observed cross section depends on the electric charge of the object doing the scattering, and from these experiments we can deduce charges of magnitude $\frac{1}{3}e$ and $\frac{2}{3}e$ for these point-like objects. These experiments give clear evidence for the presence of quarks inside the proton.

We don't yet know why free quarks don't exist. Perhaps they are so massive that no accelerator yet built has enough energy to liberate one. Perhaps the force between quarks increases with distance (in contrast with electromagnetism or gravitation, which *decrease* with separation distance), so that an infinite amount of energy would be required to separate a quark from a nucleon. Or (as is now widely believed) perhaps the basic theory of quark structure forbids the existence of free quarks.

14.9 THE STANDARD MODEL

Ordinary matter is composed of protons and neutrons, which are in turn composed of u and d quarks. Ordinary matter is also composed of electrons. In the radioactive decay of ordinary matter, electron-type neutrinos are emitted. Our entire world can thus be regarded as composed of four spin-$\frac{1}{2}$ particles (and their antiparticles), which can be grouped into a pair of leptons and a pair of quarks:

$$(e, \nu_e) \qquad \text{and} \qquad (u, d)$$

Within each pair, the charges of the two particles differ by one unit: -1 and 0, $+\frac{2}{3}$ and $-\frac{1}{3}$.

When we do experiments with high-energy accelerators, we find new types of particles: muons and muon neutrinos, plus mesons and baryons with the new properties of strangeness and charm. We can account for the structure of these particles with another pair of leptons and another pair of quarks:

$$(\mu, \nu_\mu) \qquad \text{and} \qquad (c, s)$$

Once again, the particles come in pairs differing by one unit of charge.

At even higher energies, we find a new generation of particles consisting of another pair of leptons (tau and its neutrino) and a new pair of quarks (top and bottom), which permits us to continue the symmetric arrangement of the fundamental particles in pairs:

$$(\tau, \nu_\tau) \qquad \text{and} \qquad (t, b)$$

Table 14.10 shows the six quarks and their properties. The masses of the quarks cannot be directly determined, because a free quark has yet to be observed. The rest energies shown in Table 14.10 are estimates based

TABLE 14.10 PROPERTIES OF THE QUARKS

Type	Symbol	Antiparticle	Charge (e)	Spin (ℏ)	Baryon Number	Rest Energy[a] (MeV)	C	S	T	B
Up	u	$\overline{u}$	$+\frac{2}{3}$	$\frac{1}{2}$	$+\frac{1}{3}$	300	0	0	0	0
Down	d	$\overline{d}$	$-\frac{1}{3}$	$\frac{1}{2}$	$+\frac{1}{3}$	300	0	0	0	0
Charm	c	$\overline{c}$	$+\frac{2}{3}$	$\frac{1}{2}$	$+\frac{1}{3}$	1500	+1	0	0	0
Strange	s	$\overline{s}$	$-\frac{1}{3}$	$\frac{1}{2}$	$+\frac{1}{3}$	500	0	-1	0	0
Top	t	$\overline{t}$	$+\frac{2}{3}$	$\frac{1}{2}$	$+\frac{1}{3}$	180,000	0	0	+1	0
Bottom	b	$\overline{b}$	$-\frac{1}{3}$	$\frac{1}{2}$	$+\frac{1}{3}$	4700	0	0	0	-1

[a] The rest energies are those of *constituent* quarks, which are bound in particles. The rest energies of free quarks are unknown.

on the "apparent" masses that quarks have when bound in various particles. For example, the observed rest energy of the proton is the sum of the rest energies of its three quark constituents less the binding energy of the quarks. Since we don't know the binding energy, we can't determine the rest energy of a free quark. The rest energies shown in Table 14.10 are often called those of *constituent* quarks.

Is it possible that there are more pairs of leptons and quarks that have not yet been discovered? At this point, we believe the answer to be "No." Every particle so far discovered can be fit into this scheme of 6 leptons and 6 quarks. Furthermore, the number of lepton generations can be determined by the decay rates of the heaviest particles, and a limit of 3 emerges from these experiments. Finally, according to present theories the evolution of the universe itself would have proceeded differently if there had been more than 3 types of neutrinos. For these reasons, it is generally believed that there are no more than 3 generations of particles.

The strong force between quarks is carried by an exchanged particle, called the *gluon*, which provides the "glue" that binds quarks together in mesons and baryons. (There are actually eight different gluons in the theory.) A theory known as *quantum chromodynamics* describes the interactions of quarks and the exchange of gluons. In this theory, the internal structure of the proton consists of three quarks "swimming in a sea" of exchanged gluons. Like the quarks, the gluons cannot be observed directly, but there is indirect evidence of their existence from a variety of experiments.

The theory of the structure of the elementary particles we have described so far is known as the *Standard Model*. It consists of 6 leptons and 6 quarks (and their antiparticles), plus the field particles (photon, 3 weak bosons, 8 gluons) that carry the various forces. It is remarkably successful in accounting for the properties of the fundamental particles, but it lacks the unified treatment of forces we would expect from a complete theory.

The first step toward unification was taken in 1967 with the development of the *electroweak* theory by Stephen Weinberg and Abdus Salam. In this theory, the weak and electromagnetic interactions are regarded as separate aspects of the same basic force (the electroweak force), just as electric and magnetic forces are distinct but part of a single phenomenon, electromagnetism. The theory predicted the existence of the W and Z particles; their discovery in 1983 provided a dramatic confirmation of the theory.

The next higher level of unification would be to combine the strong and electroweak forces into a single interaction. Theories that attempt to do this are called *Grand Unified Theories* (GUTs). By incorporating leptons and quarks into a single theory, the GUTs explain many observed phenomena: the fractional electric charge of the quarks and the difference of one unit of charge between the members of the quark and lepton pairs within each generation. The GUTs also predict new phenomena, such as the conversion of quarks into leptons, which would permit the proton (which we have so far assumed to be an absolutely stable particle) to decay into lighter particles with a lifetime of at least 10^{31} y. Searches for photon decay (by looking for evidence of decays in a large volume of matter; see Figure 14.17) have so far been unsuccessful and have placed lower limits on the proton lifetime of at least 10^{32} y.

Another consequence of the GUTs is that the neutrinos are no longer massless. The experimental limits on the masses of the neutrinos (see Table 14.4) are not very restrictive, and masses of a few eV would be consistent with experiments. Such a mass would have enormous consequences for cosmology and for our theories of the evolution and future of the universe (see Chapter 16).

There is so far no conclusive verification for any of the GUTs, nor is there a successful theory that incorporates the remaining force, gravity, into a unified theory. The quest for unification and its experimental tests remains an active area of research in particle physics.

FIGURE 14.17   An underground chamber, lined with plastic, in the Morton salt mine near Cleveland. This chamber is filled with 10,000 tons of water in which are suspended 2048 detectors that respond to the tiny flashes of light that would be emitted in the decay of one of the protons in the water.

SUGGESTIONS FOR FURTHER READING

Advanced books on particle physics tend to be mathematically difficult, full of field theory and relativistic quantum mechanics. Fortunately there are many popular-level books and articles that can be read for background material. These are generally descriptive and nonmathematical. For example, see

G. Feinberg, *What is the World Made Of?* (Garden City, Anchor Press, 1977).
J. C. Polkinghorne, *The Particle Play* (Oxford, W. H. Freeman, 1981).

Other general books that are now somewhat out of date but are still interesting for their background material are:

D. H. Frisch and A. M. Thorndike, *Elementary Particles* (Princeton, Van Nostrand, 1964).
R. Gourian, *Particles and Accelerators* (New York, McGraw-Hill, 1967).
C. N. Yang, *Elementary Particles* (Princeton, Princeton University Press, 1961).

A little more challenging, but still containing lots of general introductory material:

G. D. Coughlan and J. E. Dodd, *The Ideas of Particle Physics*, 2nd ed. (Cambridge, Cambridge University Press, 1991).

For histories of recent discoveries in particle physics, see:

A. Pickering, *Constructing Quarks* (Chicago, University of Chicago Press, 1984).
R. P. Crease and C. C. Mann, *The Second Creation*, (New York, MacMillan, 1986).

A historical and lavishly illustrated introduction is:

F. Close, M. Marten, and C. Sutton, *The Particle Explosion*, (New York, Oxford University Press, 1987).

For speculations about unification by one of the developers of the electroweak theory, see:

S. Weinberg, *Dreams of a Final Theory* (New York, Pantheon, 1992).

Developments occur so rapidly in particle physics that textbooks are often two years outdated when they are published. A better source of current, popular level information on particle physics is the magazine *Scientific American*, which usually includes a major article on current developments in particle physics every other month or so. A good background survey article is:

V. F. Weisskopf, "The Three Spectroscopies," *Scientific American* **218**, 15 (May 1968).

Some excellent summaries of the quark model are:

S. L. Glashow, "Quarks with Color and Flavor," *Scientific American* **233**, 38 (October 1975).
Y. Nambu, "The Confinement of Quarks," *Scientific American* **235**, 48 (November 1976).

C. Quigg, "Elementary Particles and Forces," *Scientific American* **255**, 84 (April 1985).

For information on experiments on the decay of the proton, see:

J. M. LoSecco, F. Reines, and D. Sinclair, "The Search for Proton Decay," *Scientific American* **252**, 54 (1985).

Many articles on particle physics from *Scientific American* are collected in:

Particles and Forces: At the Heart of the Matter, edited by R. A. Carrigan, Jr. and W. P. Trower (New York, Freeman, 1990).

QUESTIONS

1. Some conservation laws are based on fundamental properties of nature, while others are based on systematics of decays and reactions and have as yet no fundamental basis. Give the basis for the following conservation laws: energy, linear momentum, angular momentum, electric charge, baryon number, lepton number, strangeness.

2. Does the presence of neutrinos among the decay products of a particle always indicate that the weak interaction is responsible for the decay? Do all weak interaction decays have neutrinos among the decay products? Which decay product indicates an electromagnetic decay?

3. Do all strongly interacting particles also feel the weak interaction?

4. In what ways would physics be different if there were another member of the lepton family less massive than the electron? What if there were another lepton more massive than the tau?

5. Suppose a proton is moving with high speed, so that $E \gg mc^2$. Is it possible for the proton to decay, such as into $n + \pi^+$ or $p + \pi^0$?

6. On planet anti-Earth, antineutrons beta decay into antiprotons. Is a neutrino or an antineutrino emitted in this decay?

7. List some experiments that might distinguish antineutrons from neutrons. Among others, you might consider (a) neutron capture by a nucleus; (b) beta decay; (c) the effect of a magnetic field on a beam of neutrons.

8. The Σ^0 can decay to Λ^0 without changing strangeness, so it goes by the electromagnetic interaction; the charged $\Sigma^\pm$ decay to p or n by the weak interaction in characteristic lifetimes of 10^{-10} s. Why can't $\Sigma^\pm$ decay to Λ^0 by the strong interaction in a much shorter time?

9. The Ω^- particle decays to $\Lambda^0 + K^-$. Why doesn't it also decay to $\Lambda^0 + \pi^-$?

10. Explain why we do not account for the number of mesons in decays or reactions with a "meson number" in analogy with lepton number or baryon number.

11. Consider that leptons and baryons both obey conservation laws and are both fermions; mesons do not obey a conservation law and are bosons. Can you

think of another particle (other than a meson) that has integral spin and can be emitted or absorbed in unlimited numbers?

12. Can antibaryons be produced in reactions between baryons and mesons?

13. List some similarities and differences between the properties of photons and neutrinos.

14. Is it reasonable to describe a resonance as a definite particle, when its mass is uncertain (and therefore variable) by 20 percent?

15. Why are most particle physics reactions endothermic ($Q < 0$)?

16. Although doubly charged baryons have been found, no doubly charged mesons have yet been found. What would be the effect on the quark model if a meson with charge $+2e$ were found? How could such a meson be interpreted within the quark model?

17. All direct quark transformations must involve a change of charge; for example, $u \rightarrow d$ is allowed (accompanied by the emission of a W^-), but $s \rightarrow d$ is not. Can you suggest a two-step process that might permit the transformation of an s quark into a d?

18. The decay $K^+ \rightarrow \pi^+ + e^+ + e^-$ is at least five orders of magnitude less probable than the decay $K^+ \rightarrow \pi^0 + e^+ + \nu_e$. Based on Question 17, can you explain why?

19. The D mesons decay to π and K mesons with a lifetime of 10^{-13} s. (a) Why is the lifetime so much slower than a typical strong interaction lifetime? Is a quantum number not conserved in the decay? (b) What interaction is responsible for the decay?

20. The Δ^* baryons are found with electric charges $+2, +1, 0$, and -1. Based on the quark model, why do we expect no Δ^* with charge -2?

21. Although we cannot observe quarks directly, indirect evidence for quarks in nucleons comes from the scattering of high-energy particles, such as electrons. When the de Broglie wavelength of the electrons is small compared with the size of a nucleon (~ 1 fm), the electrons appear to be scattered from massive, compact objects much smaller than a nucleon. To which phenomenon discussed previously in this text is this similar? Can the scattering be used to deduce the mass of the struck object? How does the scattering depend on the electric charge of the struck object? What would be the difference between scattering from a particle of charge e and one of charge $\frac{2}{3}e$?

PROBLEMS

1. Identify the interaction responsible for the following decays:
   (a) $\Delta^* \rightarrow p + \pi$ (c) $K^+ \rightarrow \mu^+ + \nu_\mu$ (e) $\eta' \rightarrow \eta + 2\pi$
   (b) $\eta \rightarrow \gamma + \gamma$ (d) $\Lambda^0 \rightarrow p + \pi^-$ (f) $K^0 \rightarrow \pi^+ + \pi^-$

2. Name the conservation law that would be violated in each of the following decays:
   (a) $\pi^+ \rightarrow e^+ + \gamma$ (d) $\Lambda^0 \rightarrow \pi^- + \pi^+$ (g) $\Xi^0 \rightarrow \Sigma^0 + \pi^0$
   (b) $\Lambda^0 \rightarrow p + K^-$ (e) $\Lambda^0 \rightarrow n + \gamma$ (h) $\mu^- \rightarrow e^- + \gamma$
   (c) $\Omega^- \rightarrow \Sigma^- + \pi^0$ (f) $\Omega^- \rightarrow \Xi^0 + K^-$

3. Each of the following reactions violates one (or more) of the conservation laws. Name the conservation law violated in each case:
   (a) $\nu_e + p \rightarrow n + e^+$
   (b) $p + p \rightarrow p + n + K^+$
   (c) $p + p \rightarrow p + p + \Lambda^0 + K^0$
   (d) $\pi^- + n \rightarrow K^- + \Lambda^0$
   (e) $K^- + p \rightarrow n + \Lambda^0$

4. Table 14.5 lists the most likely decay mode of the K^+ meson; Example 14.4 gives another possible decay. List four other possible decays that are allowed by the conservation laws.

5. Supply the missing particle in each of the following decays:
   (a) $K^- \rightarrow \pi^0 + e^- +$
   (b) $K^0 \rightarrow \pi^0 + \pi^0 +$
   (c) $\eta \rightarrow \pi^+ + \pi^- +$

6. List one possible decay mode of the following antiparticles: (a) $\overline{\Lambda^0}$; (b) $\overline{\Omega^-}$; (c) $\overline{K^0}$; (d) $\overline{n}$.

7. Carry out the calculations of mc^2 for the three decays of Figure 14.3.

8. Repeat the calculation of Example 14.4 for the case in which the π meson has zero kinetic energy, and show that the electron energy in this case is less than the maximum value.

9. It is desired to form a beam of Λ^0 particles to use for the study of reactions with protons. The Λ^0 are produced by reactions at one target and must be transported to another target 2.0 m away so that at least half of the original Λ^0 remain in the beam. Find the speed and the kinetic energy of the Λ^0 for this to occur.

10. A Σ^- baryon is produced in a certain reaction with a kinetic energy of 3642 MeV. If the particle decays after one mean lifetime, what is the longest possible track this particle could leave in a detector?

11. Determine the energy uncertainty or width of (a) η; (b) η'; (c) Σ^0; (d) Δ^*.

12. Find the kinetic energies of each of the two product particles in the following decays (assume the decaying particle is at rest):
    (a) $\Omega^- \rightarrow \Lambda^0 + K^-$
    (b) $\pi^+ \rightarrow \mu^+ + \nu_\mu$
    (c) $K^0 \rightarrow \pi^+ + \pi^-$

13. Find the Q values of the following decays:
    (a) $\pi^- \rightarrow \mu^- + \nu_\mu$
    (b) $\pi^0 \rightarrow \gamma + \gamma$
    (c) $K^0 \rightarrow \pi^+ + \pi^-$
    (d) $\Sigma^+ \rightarrow p + \pi^0$
    (e) $\Sigma^0 \rightarrow \Lambda^0 + \gamma$

14. Find a decay mode, other than that listed in Table 14.6, for (a) Ω^-; (b) Λ^0; (c) Σ^+, that satisfies the applicable conservation laws.

15. Each of the reactions below is missing a single particle. Supply the missing particle in each case.
    (a) $p + p \rightarrow p + \Lambda^0 +$
    (b) $p + \overline{p} \rightarrow n +$
    (c) $\pi^- + p \rightarrow \Xi^0 + K^0 +$
    (d) $K^- + n \rightarrow \Lambda^0 +$
    (e) $\overline{\nu}_\mu + p \rightarrow n +$
    (f) $K^- + p \rightarrow K^+ +$

16. In this problem you will derive Equation 14.13 using a slightly different procedure. See Figure 14.9, and let M be the total mass of the product particles, which move together and can therefore be considered as a single particle. Let p_1 be the momentum of m_1. (a) What is the momentum of M? (b) Write an expression for conservation of total relativistic energy. (c) Combine (a) and (b) and solve for the kinetic energy.

17. Show Equation 14.13 reduces to Equation 13.14 in the nonrelativistic limit.

18. Determine the Q values of the following reactions:

    (a) $K^- + p \rightarrow \Lambda^0 + \pi^0$ (d) $p + p \rightarrow p + \pi^+ + \Lambda^0 + K^0$

    (b) $\pi^+ + p \rightarrow \Sigma^+ + K^+$ (e) $\gamma + n \rightarrow \pi^- + p$

    (c) $K^- + p \rightarrow \Omega^- + K^+ + K^0$

19. Find the threshold kinetic energy for the following reactions. In each case the first particle is in motion and the second is at rest.

    (a) $p + p \rightarrow n + \Sigma^+ + K^0 + \pi^+$ (c) $p + n \rightarrow p + \Sigma^- + K^+$

    (b) $\pi^- + p \rightarrow \Sigma^0 + K^0$ (d) $\pi^+ + p \rightarrow p + p + \bar{n}$

20. A K^0 with a kinetic energy of 276 MeV decays in flight into π^+ and π^-, which move off at equal angles with the original direction of the K^0. Find the energies and directions of motion of the π^+ and π^-.

21. A Σ^- with a kinetic energy of 0.250 GeV decays into $\pi^- + n$. The π^- moves at $90°$ to the original direction of travel of the Σ^-. Find the kinetic energies of π^- and n and the direction of travel of n.

22. Analyze the following reactions in terms of the quark content of the particles:

    (a) $K^- + p \rightarrow \Lambda^0 + \pi^0$ (d) $p + p \rightarrow p + \pi^+ + \Lambda^0 + K^0$

    (b) $\pi^+ + p \rightarrow \Sigma^+ + K^+$ (e) $\gamma + n \rightarrow \pi^- + p$

    (c) $K^- + p \rightarrow \Omega^- + K^+ + K^0$

23. Analyze the following decays in terms of the quark content of the particles:

    (a) $\Omega^- \rightarrow \Lambda^0 + K^-$ (d) $K^0 \rightarrow \pi^+ + \pi^-$

    (b) $n \rightarrow p + e^- + \bar{\nu}_e$ (e) $\Delta^{*++} \rightarrow p + \pi^+$

    (c) $\pi^0 \rightarrow \gamma + \gamma$ (f) $\Sigma^- \rightarrow n + \pi^-$

24. Based on Figure 14.16, give the quark content of the six D mesons.

25. The D_s^+ meson (rest energy = 1969 MeV, $S = +1$, $C = +1$; see Figure 14.16) has a lifetime of 0.5×10^{-12} s. (a) Which interaction is responsible for the decay? (b) Among the possible decay modes are $\phi + \pi^+$, $\mu^+ + \nu_\mu$, and $K^+ + K^0$. How do the S and C quantum numbers change in these three decays? (The ϕ meson has a spin of 1, a rest energy of 1020 MeV, and a quark content of $s\bar{s}$.) (c) Analyze the three decay modes according to the quark content of the initial and final particles. (d) Why is the decay into $K^+ + \pi^+ + \pi^-$ allowed, while the decay into $K^- + \pi^+ + \pi^+$ is forbidden?

Although observations of visible light with optical telescopes gave us our first clues about the nature of the universe, we now make observations across the entire spectrum of electromagnetic radiation. The Very Large Array near Socorro, New Mexico is an array of 27 radiotelescopes, each 25 m in diameter, that can be moved to various positions along several miles of tracks in a Y-shaped configuration. By operating the telescopes as an interferometer, detailed maps of radio emissions can be produced that reveal the structure of galaxies.

ASTROPHYSICS AND GENERAL RELATIVITY

We now discuss what is perhaps the grandest of all human endeavors: the study of our universe on the large scale, and of the stars and galaxies of which it is made. Some aspects of this subject, such as the motion of the planets, can be well understood on the basis of Newtonian gravitation. However, Newton's theory is insufficient to explain a number of observations of the motion of celestial objects. For this purpose we need a different theory, the *general theory of relativity*, which was proposed by Albert Einstein in 1916. Although the mathematics of this theory are beyond the level of this text, we can summarize some of its features and discuss its experimental predictions and their verification. Like the special theory, the general theory of relativity offers us a new way of thinking about space and time.

For our study of the universe and its evolution, we require many details of modern physics that we have previously discussed: special relativity, quantum theory, atomic structure, statistical mechanics, and nuclear and particle physics. In fact, we can infer details about the properties of matter at its most fundamental level—quarks and neutrinos—from observations of the structure of the universe. In this respect, physics comes full circle, as the very large and the very small converge.

In this chapter, we discuss the general theory of relativity and some observational properties of our universe. In the following chapter, we consider the more general properties of the origin, age, and composition of the universe.

15.1   THE PRINCIPLE OF EQUIVALENCE

The special theory of relativity arose from a thought experiment of Einstein's in which he imagined trying to catch up with a light beam. The general theory also arose from a thought experiment. Here are Einstein's words:

> *I was sitting in a chair in the patent office at Bern when all of a sudden a thought occurred to me: If a person falls freely he will not feel his own weight. This simple thought made a deep impression on me. It impelled me toward a theory of gravitation.*

Figure 15.1 illustrates a freely falling person in two situations: in the Earth's gravity and in interstellar space where the gravitational field is negligibly weak. In both cases the person is in an isolated chamber and therefore unable to use outside objects to deduce the motion of the chamber. From within the chamber the two cases look exactly equivalent; no measuring instrument operating entirely within the chamber can distinguish between the two cases. An acceleration $\mathbf{a} = \mathbf{g}$ in a gravitational field $\mathbf{g}$ is equivalent to an acceleration of 0 in a vanishing gravitational field.

It appears that an acceleration is able to "cancel out" the effects of a gravitational field. Let us go one step further and ask whether an acceleration can *produce* the effects of a gravitational field. Consider the situations illustrated in Figure 15.2. In one case the observer is at rest near the Earth, where the gravitational field is **g**. In the other case, the observer is in empty space where the gravitational field is negligibly small, but the rocket engines are firing so that the chamber has an acceleration **a** = −**g**. There are various experiments in the chamber: a scale displays the weight of the observer (actually, the normal force between the observer and the scale), a ball drops to the floor, a mass stretches a spring, and a pendulum oscillates. All of these experiments give identical results in the two chambers. Once again, there is no experiment that can be done within the chamber to distinguish the two cases.

This leads us to the *principle of equivalence*:

There is no local experiment that can be done to distinguish between the effects of a uniform gravitational field in a nonaccelerating inertial frame and the effects of a uniformly accelerating (noninertial) reference frame.

(a)

By "local" we mean that the experiments must be done within the chamber, and also that the chamber must be sufficiently small that the gravitational field is uniform. Near the surface of the Earth, for example, not all **g** vectors inside the chamber would be parallel; they point toward the center of the Earth, so there would be a slight angle between the **g** vectors on opposite sides of the chamber. If we make the chamber small, this effect is negligible and the **g** vectors everywhere in the chamber, like the **a** vectors in the accelerating chamber, are parallel to one another.

The principle of equivalence appears in a slightly different (and weaker) form in introductory physics, where it is stated in terms of the equivalence of *inertial* and *gravitational* mass. That is, the mass m that appears in the expression $F = ma$ (inertial mass) is identical to the mass m that appears in the expression $F = GMm/r^2$ (gravitational mass). It follows from this form of the principle of equivalence that all objects, regardless of their masses, fall with the same acceleration in the Earth's gravity. This was first tested by Galileo in the famous (and perhaps apocryphal) experiment in which he dropped two different masses from the top of the leaning tower of Pisa and observed them to fall at the same rate. In recent years other more precise experiments have established the equivalence of gravitational and inertial mass to about 1 part in 10^{11}.

Einstein realized that the principle of equivalence applied not only to mechanical experiments but to all experiments, even ones based on electromagnetic radiation. Consider the arrangement shown in Figure 15.3. At the top of the chamber is a light source that emits a wave of frequency ν. At the bottom of the chamber and a distance H away is a detector that observes the wave and measures its frequency. When the light wave is emitted in the accelerating chamber, the source has speed u, which we assume to be small compared with the speed of light c. When it is detected, after a time of flight $t \approx H/c$, the floor is moving with a speed $u + at$. In effect there is a relative speed $\Delta u = at$ between the source and the detector,

(b)

FIGURE 15.1 The effects of freely falling in (*a*) the Earth's gravity and (*b*) interstellar space appear identical from within the chamber.

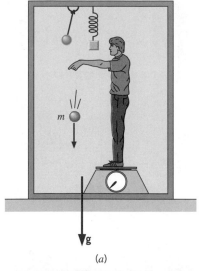

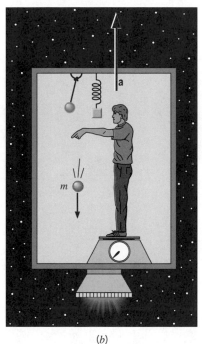

FIGURE 15.2 The effects of (a) resting in a uniform gravitational field and (b) accelerating in interstellar space appear identical from within the chamber.

so there is a Doppler shift in the frequency given by Equation 2.22:

$$\nu' = \nu \sqrt{\frac{1 + \Delta u/c}{1 - \Delta u/c}} \approx \nu(1 + \Delta u/c) \qquad (15.1)$$

or, in terms of the frequency difference $\Delta \nu = \nu' - \nu$,

$$\frac{\Delta \nu}{\nu} = \frac{\Delta u}{c} = \frac{at}{c} = \frac{aH}{c^2} \qquad (15.2)$$

Now let us compare the result of this experiment with a similar one done in a chamber at rest in a uniform gravitational field g. If the results of the two experiments are to be identical (as required by the principle of equivalence), there must be a frequency shift given by Equation 15.2 with $a = g$:

$$\frac{\Delta \nu}{\nu} = \frac{gH}{c^2} \qquad (15.3)$$

The principle of equivalence thus predicts a change in frequency of a light wave falling in the Earth's gravity.

In 1959, R. V. Pound and G. A. Rebka allowed 14.4-keV photons from the radioactive decay of ^{57}Co to fall down the Harvard tower, a distance of 22.6 m. The expected fractional change in frequency, $\Delta \nu/\nu = gH/c^2$, was 2.46×10^{-15}; that is, to detect the effect, they had to measure the frequency or energy of the photon at the bottom of the tower to a precision of about 1 part in 10^{15}! The Mössbauer effect (Section 12.11) makes it possible to achieve such a level of precision, and the measured result was $\Delta \nu/\nu = (2.57 \pm 0.26) \times 10^{-15}$, consistent with the equivalence principle. Similar experiments based on comparisons between the frequency of radiation emitted by satellites and received by ground stations have confirmed the predictions of the principle of equivalence to a precision of about 1 part in 10^4.

Note that in Equation 15.3 the frequency shift depends on the difference in gravitational potential (potential energy per unit mass) between the source and the detector:

$$\Delta V = \frac{\Delta U}{m} = \frac{(mgH - 0)}{m} = gH \qquad (15.4)$$

For the satellite, even though the gravitational field through which the radiation travels is not uniform, the same conclusion holds: the frequency shift depends on the difference in gravitational potential between the source and the observer. Consider, for example, light leaving the surface of a star of mass M and radius R. The gravitational potential at the surface is $V = -GM/R$. If the light is observed on the Earth, where the gravitational potential is negligible compared with that of the star, the frequency shift is

$$\frac{\Delta \nu}{\nu} = \frac{\Delta V}{c^2} = -\frac{GM}{Rc^2} \qquad (15.5)$$

Photons climbing out of a star's gravitational field lose energy and are therefore shifted to smaller frequencies or longer wavelengths (red shifted).

This effect is difficult to observe for two reasons: (1) the motion of the star causes a Doppler shift which is generally greater than the gravitational shift (see Section 16.1), and (2) the spectral lines are Doppler broadened by the thermal motion of the atoms near the surface of a star (see Section 10.3). Nevertheless, the effect has been confirmed for a few stars including the Sun.

EXAMPLE 15.1

The Lyman α line in the hydrogen spectrum has a wavelength of 121.5 nm. Find the change in wavelength of this line in the solar spectrum due to the gravitational shift and compare it to the thermal Doppler broadening of the line.

SOLUTION

From Equation 15.5, we have

$$\frac{\Delta\lambda}{\lambda} = -\frac{\Delta\nu}{\nu} = \frac{GM}{Rc^2} = \frac{(6.67 \times 10^{-11}\,\text{N} \cdot \text{m}^2/\text{kg}^2)(1.99 \times 10^{30}\,\text{kg})}{(6.96 \times 10^8\,\text{m})(3.00 \times 10^8\,\text{m/s})^2}$$
$$= 2.12 \times 10^{-6}$$

The shift in wavelength is

$$\Delta\lambda = (2.12 \times 10^{-6})(121.5\,\text{nm}) = 0.257\,\text{pm}$$

The width of the spectral line, due to the thermal Doppler broadening, can be found from the result of Example 10.2, $\Delta\lambda/\lambda = 5.5 \times 10^{-5}$, so

$$\Delta\lambda = (5.5 \times 10^{-5})(121.5\,\text{nm}) = 6.7\,\text{pm}$$

It is difficult to measure such a small change in wavelength for this relatively broad spectral line. How would you expect the Sun's rotation to affect this measurement?

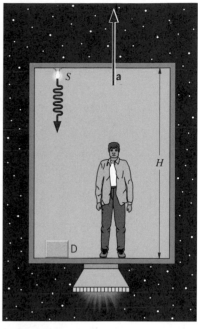

FIGURE 15.3 A source *S* emits a light wave that is recorded by a detector *D*. The chamber is accelerating upward.

15.2 THE GENERAL THEORY OF RELATIVITY

From *special* relativity we learn that the laws of physics must be the same in all inertial frames, and that there is no preferred inertial frame relative to which it is possible to determine the absolute velocity of an observer. From this point of view, it seems that an accelerated (noninertial) frame is a preferred frame, because it is possible to determine the absolute acceleration. In the *general* theory, Einstein sought to remove this restriction, so that motion would be relative for *all* observers, even accelerated ones. The principle of equivalence, which tells us that we can't distinguish between acceleration and a gravitational field, removes acceleration from its privileged role.

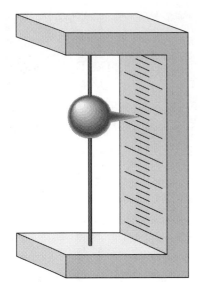

FIGURE 15.4   A device for measuring acceleration or gravity.

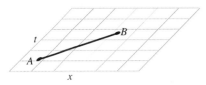

FIGURE 15.5   The path through flat spacetime of a particle moving at constant velocity.

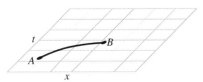

FIGURE 15.6   The path through flat spacetime of an accelerating particle.

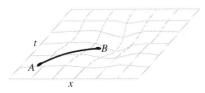

FIGURE 15.7   The path of a particle through spacetime curved by matter. Note the similarity of the path to that of Figure 15.6.

Ultimately, general relativity is a theory of geometry. The motion of a particle is determined by the properties of the space and time coordinates through which it moves. Because space and time are intimately coupled in relativity (see, for example, the Lorentz transformation in Section 2.5), we regard them as components of a combined coordinate system called *spacetime*.

The equivalence between accelerated motion and gravity suggests a relationship between spacetime coordinates and gravity. In the classical description, we would say that the presence of matter sets up a gravitational field, which then determines how objects move in response to that field. According to general relativity, we say that the presence of matter (and energy) causes spacetime to warp or curve; the motion of particles is then determined by the shape of the coordinate system. It is sometimes said that "Geometry tells matter how to move, and matter tells geometry how to curve." From the configuration of mass and energy, general relativity gives us a procedure for calculating the curvature of spacetime, and the motion of a particle or a light beam then follows directly.

Consider the simple device shown in Figure 15.4, which consists of a frame that holds a wire on which a small bead can slide without friction. A scale indicates the position of the bead. The spatial coordinate system of the bead has only one dimension, and we can plot the motion of the bead on an *xt* coordinate system, which is the two-dimensional spacetime of the bead. For example, if the bead moves with constant velocity along the wire from position x_A at time t_A to position x_B at time t_B, the motion in spacetime is represented by the straight line shown in Figure 15.5.

Now suppose the apparatus stands upright on a table in our accelerating chamber. To an observer in the chamber, the bead will appear to accelerate downward as the frame is accelerated upward along with the chamber. The path in spacetime is now curved, as suggested by Figure 15.6.

If the acceleration is replaced by the equivalent gravitational field, the motion of the bead, as observed from inside the chamber, is exactly the same—the bead appears to accelerate downward. General relativity describes this situation as a change in the shape of spacetime; the presence of matter (which classically we would describe as the source of the gravitational field) distorts the *xt* spacetime as indicated in Figure 15.7. If we imagine the spacetime coordinate system as a grid laid out on a rubber sheet, the gravitating matter stretches the sheet, and the particle moves from A to B along the most direct path in the curved spacetime.

It is convenient to define the *spacetime interval ds*, in effect the separation between two events (such as the particle passing through successive points) in two-dimensional spacetime, as

$$(ds)^2 = (c\,dt)^2 - (dx)^2 \tag{15.6}$$

This quantity is invariant under the Lorentz transformation, as you can prove by substituting dx' and dt' from Equation 2.23. The trajectory of the particle in spacetime can be regarded as a collection of infinitesimal intervals. Since the particle is merely following the contour of spacetime, the interval serves both to define the trajectory and to represent the shape of spacetime.

The interval of Equation 15.6 and its four-dimensional extension (three space dimensions, one time dimension) are characteristic of our familiar Euclidian space, which we call "flat." Figure 15.8 summarizes some of the characteristics of that space: a straight line is the shortest distance between two points, the sum of the angles of a triangle is 180°, parallel lines never meet, the ratio between the circumference and the diameter of a circle is π, and so forth. To characterize a "curved" four-dimensional spacetime, we might write the interval as

$$(ds)^2 = g_0(c\,dt)^2 - g_1(dx)^2 - g_2(dy)^2 - g_3(dz)^2 \qquad (15.7)$$

where the four coefficients g_i describe the curvature of the spacetime and its deviation from a Euclidian nature (for which all $g_i = 1$).

Figure 15.9 shows a curved non-Euclidian geometry, the surface of a sphere. Here the shortest distance between points is an arc of a great circle, the sum of the angles of a triangle is greater than 180°, parallel lines can meet, and the ratio between the circumference and the diameter of a circle is less than π. The saddle-shaped geometry shown in Figure 15.10 has a different kind of curvature, in which the ratio between the circumference and the diameter of a circle is greater than π.

To appreciate the importance of curved spacetime, consider the experiment illustrated in Figure 15.11 on page 495. A light beam is emitted by a source in the chamber and travels across the chamber to the opposite wall. If the chamber is in an inertial frame and free from gravitational fields, the beam travels horizontally across the chamber and strikes the opposite wall at the same height above the floor as the source. (This holds even if the chamber moves at constant velocity, as you can prove using special relativity.) Observers inside and outside the chamber agree on this conclusion.

If the chamber accelerates, the situation is different. Suppose the light is emitted when the chamber is at rest, relative to a particular inertial frame. The light beam then has no transverse component of velocity in this frame and moves horizontally. As the chamber accelerates, the light beam acquires no transverse velocity component, but the chamber's velocity increases. To the inertial observer, the beam travels along a horizontal straight line, but the chamber accelerates forward so that the beam strikes the opposite wall at a lower height than the source. To an observer in the chamber, the beam appears to follow the curved path shown in Figure 15.11a and strikes the wall at a lower height than the source.

According to the principle of equivalence, the observer in the chamber should record the same outcome if the chamber is at rest in a uniform gravitational field (Figure 15.11b). General relativity explains this observation through the curvature of spacetime in the vicinity of the mass that is responsible for the gravitational field. The light beam is merely seeking out the shortest possible path in the curved spacetime, just like an ant crawling along a line on the spherical surface of Figure 15.9. All paths in the curved spacetime (as in Figure 15.7, for example) are curved.

It is tempting to seek an alternative explanation for the outcome shown in Figure 15.11b. For instance, we can assume each photon in the light beam to have an effective mass $m = E/c^2$ and then calculate its trajectory in the gravitational field as we would that of any classical particle of mass

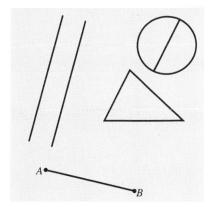

FIGURE 15.8 A flat space and its Euclidian geometrical properties.

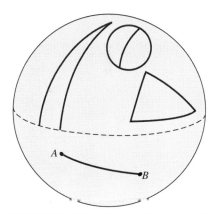

FIGURE 15.9 A curved space and its non-Euclidian geometrical properties.

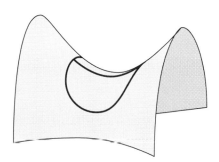

FIGURE 15.10 Another curved space.

m. However, as we discuss in the next section, this method gives results that do not agree with observations for the path of photons in a gravitational field. The curvature of spacetime, which provides the correct explanation, is an inescapable consequence of the principle of equivalence.

General relativity gives a relationship between curvature and the density of mass and energy in space, which can be written symbolically as

$$\text{curvature of space} = \frac{8\pi G}{c^4} (\text{mass-energy density}) \tag{15.8}$$

Note that this expression incorporates gravitation (Newton's constant G) and special relativity (the speed of light c). If no matter or energy is present, the right-hand side is zero; as a result, the curvature is zero and space is flat. In the limit of classical kinematics ($c \to \infty$) and in the limit of weak gravitational fields ($G \to 0$), space is nearly flat and we can safely use the Newtonian gravitational theory. This is equivalent to saying that if we take a small enough region of the sphere of Figure 15.9, or if we increase its radius to a sufficiently large value, the geometry is approximately Euclidian.

Just as classical kinematics can be regarded as the limiting case of *special* relativity (for low speeds), so can classical gravitation be regarded as the limiting case of *general* relativity (for weak fields). In calculating the orbit of an Earth satellite or the trajectory of a space probe to Mars, Newton's theory gives entirely satisfactory results. Close to the Sun and to compact or massive stars, the curving of space can lead to observable effects, as we discuss in the next section.

15.3 TESTS OF GENERAL RELATIVITY

Newtonian gravitation and Einstein's general relativity each give predictions that can be tested against experiment, but under most circumstances the differences between the two predictions are extremely small. At the surface of the Earth, space is curved by only about 1 part in 10^8; that is, the ratio of the curvature of space caused by the Earth's mass to the curvature of the Earth's surface itself is about 10^{-8}, and deviations from Euclidian geometry would occur only at about that level. Even at the surface of the Sun, the curvature is only about 1 part in 10^6.

Nevertheless, there are experiments we can do that are precise enough to detect the difference between flat spacetime and curved spacetime. In this section we discuss several of these experiments.

Deflection of Starlight

When a beam of light from a star passes close to the Sun, it is deflected from its original direction, as shown in Figure 15.12. The star appears to be displaced from its truc position by an angle θ.

It is possible to analyze this situation using Newtonian gravitation and special relativity by assigning the photons in the beam an effective mass $m = E/c^2$ and assuming them to be deflected by the Newtonian gravitational force. The experiment then looks very much like Rutherford scattering, and by analogy with the Rutherford scattering formula (see Problem 7) it is possible to calculate the deflection angle:

$$\theta = \frac{2GM}{Rc^2} \tag{15.9}$$

where M is the mass of the Sun and R is its radius. Substituting the numbers gives $\theta = 0.87''$ as the prediction of special relativity and Newtonian gravitation.

General relativity gives a different view. Spacetime in the vicinity of the Sun is curved, and the light beam is simply following the most direct path along the curved spacetime (Figure 15.13 is a two-dimensional representation of this effect). According to general relativity, the expected deflection is $1.74''$, exactly twice the value predicted by the Newtonian formula.

Measuring this effect requires the observation of a beam of light, such as from a star, which passes near the edge of the Sun. Starlight near the Sun can be observed only during a total solar eclipse. In 1919, just a few years after Einstein completed his general theory, two expeditions of British astronomers traveled to Africa and to South America to observe the solar eclipse and to measure the apparent changes in positions of stars whose light grazed the Sun. Their results for the deflection angles, $1.98'' \pm 0.18''$ and $1.69'' \pm 0.45''$, gave strong support for the new general theory. In the years since those early results, this experiment has been repeated at nearly each total solar eclipse, and the overall agreement with general relativity is within 10 percent. Radio emission from quasars has also been used to confirm this effect, and here the agreement with general relativity is within 2 percent.

These experimental results give a clear distinction between Newtonian gravity (even with special relativity included) and general relativity.

Delay of Radar Echoes

When a line joining Earth and another planet (Venus, for example) passes through the Sun, the situation is known as "superior conjunction" and is illustrated in Figure 15.14. Based on the orbits of Earth and Venus, we can calculate how long it takes a radar signal sent from Earth to be reflected from Venus and return to Earth (about 20 minutes). Near superior conjunction, the signal passes close to the Sun, and therefore, according to general relativity, it does not travel in a Euclidian straight line, but instead follows a path through curved spacetime (Figure 15.15). It therefore takes the signal a bit longer than the expected time to make the round trip. This time delay is expected to be about 10^{-4} s, and it has been confirmed to within a few percent (the limit on precision being uncertainties in the surface of Venus, since we don't know if the signal is being reflected from a mountain or a valley). More precise experiments were done in the late

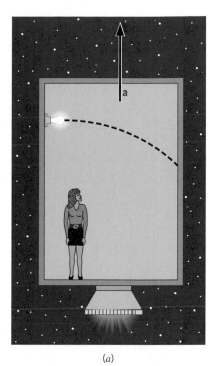

(a)

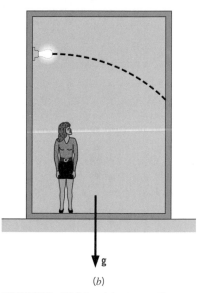

(b)

FIGURE 15.11 (a) According to an observer in an accelerating chamber, the light beam follows a curved path. (b) An observer in a uniform gravitational field finds the same outcome.

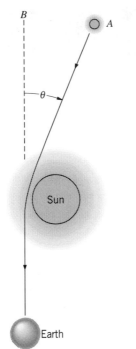

FIGURE 15.12 A light beam passing near the Sun is deflected. To an observer on Earth, the star at *A* appears to be at *B*.

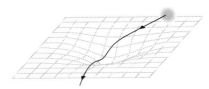

FIGURE 15.13   The path of a light beam from a star through curved spacetime.

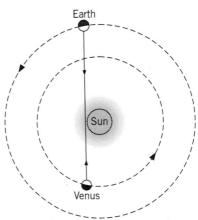

FIGURE 15.14   Superior conjunction of Earth and Venus.

1970s using a signal sent between Earth and the Viking landers on Mars. In this case, the result was consistent with general relativity to within about 0.1%.

Precession of Perihelion of Mercury

Consider a simple solar system, shown in Figure 15.16, consisting of a single planet in orbit about a star of mass M such as the Sun. According to Newtonian gravitation, the orbit is a perfect ellipse with the star at one focus. The equation of the ellipse is

$$r = r_{min} \frac{1 + e}{1 + e \cos \phi} \tag{15.10}$$

where r_{min} is the minimum distance between planet and star and e is the *eccentricity* of the orbit (the degree to which the ellipse is noncircular; $e = 0$ for a circle). When $r = r_{min}$, the planet is said to be at *perihelion*; this occurs regularly, at exactly the same point in space, whenever $\phi = 0, 2\pi, 4\pi, \ldots$. According to general relativity, the orbit is not quite a closed ellipse; the effect of the curved spacetime near the star is to cause the perihelion direction to *precess* somewhat, as shown in Figure 15.17. After completing one orbit, the planet returns to r_{min}, but at a slightly different ϕ; the difference $\Delta\phi$ can be computed from general relativity, according to which the orbit is

$$r = r_{min} \frac{1 + e}{1 + e \cos(\phi - \Delta\phi)} \tag{15.11}$$

where

$$\Delta\phi = \frac{6\pi GM}{c^2 r_{min}(1 + e)} \tag{15.12}$$

For the Sun, $6\pi GM/c^2 = 27.80$ km, and thus even for the smallest value of r_{min} (for Mercury, 46×10^6 km) $\Delta\phi$ is of order 10^{-6} rad, an extremely small quantity. However, this effect is *cumulative*; that is, it builds up orbit after orbit, and after N orbits, the perihelion has advanced by $N \Delta\phi$. We usually express this precession in terms of the total precession per century (per 100 Earth years), and some representative values are shown in Table 15.1.

The expected precessions are very small, of the order of seconds of arc per century, but nevertheless have been measured with great accuracy; for the three planets closest to the Sun, and for the asteroid Icarus, the measured values are in agreement with the predictions of general relativity. In the best case, the agreement is within about 1 percent.

These experiments are very difficult to do because (except for Mercury and Icarus) the eccentricities are small and locating the perihelion is difficult. A more serious problem is that other effects, not associated with general relativity, also cause an apparent precession of the perihelion. In the case of Mercury, the observed precession is actually about 5601″ per century;

TABLE 15.1   PRECESSION OF PERIHELIA

Planet	N (Orbits per Century)	e	r_{min} (10⁶ km)	$N \Delta\phi$ (arc seconds per century) General Relativity	Observed
Mercury	415.2	0.206	46.0	43.0	43.1 ± 0.5
Venus	162.5	0.0068	107.5	8.6	8.4 ± 4.8
Earth	100.0	0.017	147.1	3.8	5.0 ± 1.2
Mars	53.2	0.093	206.7	1.4	
Jupiter	8.43	0.048	740.9	0.06	
Icarus	89.3	0.827	27.9	10.0	9.8 ± 0.8

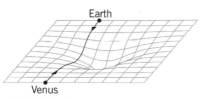

FIGURE 15.15 Path of signal between Earth and Venus in curved spacetime.

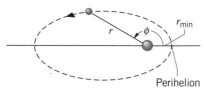

FIGURE 15.16 The elliptical orbit of a planet about a star.

of that, 5026″ are due to the precession of the Earth's equinox (a classical Newtonian effect of the spinning Earth) and 532″ are due to the gravitational pull of the other planets on Mercury (also a classical Newtonian effect). Only the difference of 43″ is due to general relativity.

Gravitational Radiation

Just as an accelerated charge emits electromagnetic radiation that travels with the speed of light, an accelerated mass emits gravitational radiation that also travels with the speed of light. In effect, gravity waves are ripples that travel through spacetime. Waves produced by such motions as the planets around the Sun are exceedingly weak and beyond any hope of detection. Cataclysmic events in the universe, such as supernova explosions, and highly accelerated systems, such as compact binary objects, may produce observable gravity waves. Detection of gravity waves would provide another important confirmation of the general relativity theory.

In analogy with the effect of a passing electromagnetic wave on a charge, a passing gravity wave could be detected by its effect on matter. Several antennas have been built to search for gravity waves, but no conclusive experimental evidence has yet been obtained. Indirect evidence has been obtained from the observations of the change in the orbital period of a binary pulsar (see Section 15.7). Interferometric techniques are being used to build new detectors to search for gravity waves (see Color Plate 17).

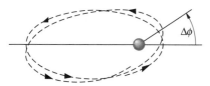

FIGURE 15.17 Precession of the perihelion (greatly exaggerated). After each orbit, the perihelion advances by an angle $\Delta\phi$.

15.4   STELLAR EVOLUTION

The problem of understanding the structure and evolution of stars is exceedingly complex. We have identified nuclear fusion as the basic source of energy production in stars, but our knowledge of the cross sections for many of the nuclear reactions of the fusion cycle is not sufficiently complete for us to be confident of calculations of the rates at which these reactions occur. (We usually think of the center of the Sun as a very hot place, full of energetic particles, but its temperature of about 10^7 K corresponds to particle kinetic energies of only about 1 keV. We routinely study nuclear reactions in our Earthbound laboratories at MeV energies. It is ironic that our understanding of the energy production in stars is limited, not because we can't reach *high* enough energies in our laboratories, but rather because it is very difficult to study these reactions at *low* energies!)

Even if our knowledge of nuclear physics were more complete, we would still have the difficulties of understanding the physical structure of a star. For example, we know the masses and radii of many stars, so we can find their *average* density, but this does not tell us how this density varies inside the star. The rate at which nuclear reactions occur depends on the higher density at the center of the star, not on its average density. The temperature of the surface of a star can be determined from the radiation emitted by the star, but the surface temperature (10^3 to 10^4 K) is much lower than the interior temperature (10^7 K), and we can't determine how the temperature varies over the interior of the star. The brightness of a star depends on how the nuclear energy (mostly gamma rays) produced in nuclear reactions near the center gets absorbed and reradiated as visible light, a process that will also vary over the interior of a star.

In spite of these difficulties, we can construct a reasonably good theoretical model of the evolution and structure of stars. We explore some of the details of our model in this section and look at the fate of stars in the following sections.

We begin with a large, cold, diffuse cloud of hydrogen gas, the end product of the early evolution of the universe. (As we learn in the next chapter, the cloud contains about 25 percent helium, but that is not important for this calculation.) The atoms are electrically neutral, and their nuclei are far apart, so the only force that we need to consider is the gravitational force between the atoms. Under its influence the cloud contracts. As the average distance between the atoms decreases, the gravitational potential energy $U = -Gm_1m_2/r$ decreases, and to balance the total energy, the kinetic energy of the atoms must increase, with a corresponding increase in temperature. As the cloud contracts, more atoms are attracted toward the center, and so the density and temperature near the center increase more rapidly than near the outer regions.

As the temperature slowly rises, the gas begins to radiate energy, much like a blackbody: the higher the temperature, the more radiation is emitted. Each decrease in gravitational potential energy is accompanied by an increase in the kinetic energy K and in the radiant energy E_R:

$$-\Delta U = \Delta K + \Delta E_R \qquad (15.13)$$

Just like an ordinary object when we heat it, the cloud begins to glow with a dull red color. Its size is still much larger than the final star, perhaps 10 times as large, and its temperature might be of the order of 1000 K.

As the cloud continues to contract, the temperature rises especially rapidly near the center. Eventually, temperatures in the range of 10^7 K are reached, nuclear fusion reactions begin, and a star is born. Perhaps 10^6 y have passed since the contraction began.

The star now enters a stable period, in which further contraction is halted by the outward pressure due to the radiation (photons) traveling from the core to the surface. It continues to generate energy in fusion reactions at a rate determined by its mass; for a star such as the Sun, this period may last for 10^{10} y, while a much heavier star (10 to 100 solar masses) may burn in this phase for only 10^7 y.

The basic fusion reactions discussed in Section 13.5 provide the energy radiated by the star, as four protons combine to make one helium nucleus, with the release of 26.7 MeV. Stars like our Sun produce most of their energy from the proton-proton cycle; such stars are either not hot enough for the carbon cycle, or else have no carbon present. (Recall from Section 13.5 that the carbon acts only as a catalyst.)

Depending on the availability of ^{4}He, other proton-proton cycles may occur:

$$^3\text{He} + {}^3\text{He} \longrightarrow {}^4\text{He} + 2\,{}^1\text{H}$$

$$\begin{aligned} ^3\text{He} + {}^4\text{He} &\longrightarrow {}^7\text{Be} + \gamma \\ ^7\text{Be} + e^- &\longrightarrow {}^7\text{Li} + \nu \\ ^7\text{Li} + {}^1\text{H} &\longrightarrow 2\,{}^4\text{He} \end{aligned}$$

$$\begin{aligned} ^1\text{H} + {}^1\text{H} &\longrightarrow {}^2\text{H} + e^+ + \nu \\ ^2\text{H} + {}^1\text{H} &\longrightarrow {}^3\text{He} + \gamma \end{aligned}$$

$$\begin{aligned} ^3\text{He} + {}^4\text{He} &\longrightarrow {}^7\text{Be} + \gamma \\ ^7\text{Be} + {}^1\text{H} &\longrightarrow {}^8\text{B} + \gamma \\ ^8\text{B} &\longrightarrow {}^8\text{Be} + e^+ + \nu \\ ^8\text{Be} &\longrightarrow 2\,{}^4\text{He} \end{aligned}$$

Proton-proton cycles

Our Sun probably produces most of its energy by the second of these possible sets of reactions.

The helium which was present in the initial gas cloud, along with that produced in fusion, forms a core at the center of the star, since helium is heavier than hydrogen (see Figure 15.18). Surrounding the helium is a layer in which the hydrogen fusion reactions take place. The outer radius of this shell is less than 10 percent of the radius of the star. The fusion energy is liberated mostly in the form of gamma rays (95 percent) and neutrinos (5 percent), which make their way to the surface to be radiated into space. The gamma rays undergo repeated Compton scatterings in the dense solar material, and may take 10^5 to 10^6 y to arrive at the surface—the light that we see today from the Sun comes from energy generated in nuclear reactions perhaps a million years ago! Moreover, the gamma rays lose energy until they become photons of visible light, so what we see is more characteristic of the solar surface than of its interior. The neutrinos, on the other hand, have a very small cross section for interaction, and they go directly through the solar material at the speed of light. Observing the solar neutrinos is therefore a way of studying the Sun's core and its nuclear processes.

Several experiments have been done to observe solar neutrinos. One experiment has been operating since 1970 in a gold mine 1 mile underground

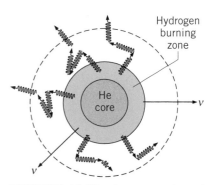

FIGURE 15.18 A cross section of the Sun. The fusion energy is generated in a region near the center. Emitted gamma rays are scattered many times in the outer zones, but neutrinos penetrate easily.

FIGURE 15.19 The solar neutrino experiment. The large tank, one mile underground, contains 100,000 gallons of C_2Cl_4.

(to shield against cosmic rays). The neutrino detector consists of a 100,000-gallon tank of tetrachloroethylene cleaning fluid (C_2Cl_4); see Figure 15.19. When a neutrino enters the detector, it can be captured by a chlorine nucleus:

$$\nu_e + {}^{37}Cl \longrightarrow {}^{37}Ar + e^-$$

The Ar gas is inert, and it bubbles up through the liquid in the tank where it can be collected. It is also radioactive; the activity of the Ar gas is a measure of the rate at which solar neutrinos have been captured in the tank.

The neutrinos produced by the p-p reaction (the first reaction of the proton-proton cycle) are expected to be the most intense component of the solar neutrinos. They have a maximum energy of 0.42 MeV, but the Q value for the ${}^{37}Cl$ capture reaction is -0.814 MeV. The p-p neutrinos thus do not have enough energy to produce ${}^{37}Ar$, but the neutrinos from the ${}^7Be$ and ${}^8B$ decays of the proton-proton cycle are more energetic and are able to produce the reaction.

For the same reason that neutrinos pass easily through the solar material, their chance of being captured by a Cl nucleus is very low—even though the tank contains about 10^{30} atoms of Cl, the production rate of ${}^{37}Ar$ is less than *1 atom per day*! After nearly 25 years of measurement, the experiment has determined a neutrino rate that is only about $\frac{1}{3}$ of the value predicted on the basis of our current understanding of fusion processes in the Sun and of the properties of neutrinos.

Other experiments use a different reaction to signal the presence of neutrinos:

$$\nu_e + {}^{71}Ga \longrightarrow {}^{71}Ge + e^-$$

The Q value for this reaction is -0.23 MeV, so it is sensitive to the more abundant neutrinos from the p-p reaction. Two gallium experiments are currently operating; one uses 60 tons of liquid gallium metal, and the other uses 30 tons of gallium in the form of a solution. The ${}^{71}Ge$ is radioactive, and by removing it from the gallium liquid, the experimenters can determine the rate of neutrino capture in the detectors. Both gallium experiments began operating (also deep underground) about 1990, and their initial results indicate a slightly greater neutrino rate than the C_2Cl_4 experiment, but their results are still only about $\frac{1}{2}$ the expected value.

These puzzling results have not yet been satisfactorily explained. They may indicate that perhaps we do not fully understand the fusion processes that occur deep in the Sun's interior, or that perhaps our knowledge of the properties of neutrinos is incomplete. For example, one theory explains the shortage by assuming that the neutrinos are not massless but have a small mass (consistent with experimental upper limits given in Table 14.4), which would allow one type of neutrino to transform into another, such as electron neutrinos into muon neutrinos. The shortage of solar neutrinos, according to this explanation, then occurs because most of the electron neutrinos are transformed on their journey from the Sun into muon or tau neutrinos; since the experiments are sensitive only to electron neutrinos, they will miss any neutrinos that have been changed into another kind.

Although this theory has not yet been confirmed (it violates the conservation law for each type of lepton number), assigning a small but nonzero mass to the neutrinos would be consistent with cosmological theories that favor neutrinos with mass as an explanation for the "missing mass" in the universe (see Chapter 16).

EXAMPLE 15.2

Compute the maximum energy of the neutrino emitted in the first reaction of the proton-proton cycle.

SOLUTION

The reaction is

$$p + p \longrightarrow {}^2H + e^+ + \nu_e$$

Written in this way, the reacting particles are nuclei, not neutral atoms. The Q value is then, using nuclear mass energies,

$$Q = m_i c^2 - m_f c^2 = 2(938.280\,\text{MeV}) - 1875.628\,\text{MeV} - 0.511\,\text{MeV}$$
$$= 0.421\,\text{MeV}$$

Because there are three particles in the final state, the energies are not determined uniquely. As in beta decay, the three particles share the available energy. The neutrino has its maximum energy when the positron has zero energy. This maximum equals the Q value, except for the recoil energy carried by the ${}^2H$ in order to conserve momentum. We can show that this energy is negligible. When the neutrino has its maximum momentum of $0.421\,\text{MeV}/c$, the recoil energy of the ${}^2H$ is

$$K - \frac{p^2}{2m} - \frac{(pc)^2}{2mc^2} = \frac{(0.421\,\text{MeV})^2}{2(1875.6\,\text{MeV})} = 4.7 \times 10^{-5}\,\text{MeV}$$

which is a negligible amount. The maximum neutrino energy is therefore

$$(E_\nu)_{max} = 0.421\,\text{MeV}$$

15.5 NUCLEOSYNTHESIS

The fate of a star that has used up its hydrogen fuel depends on the mass of the star. What is true in general is that once the hydrogen fusion stops, gravitational contraction sets in again, raising the temperature and permitting fusion reactions with helium and heavier elements. The star may in the process settle into periods of stability while these other processes occur, but one thing is clear: the star is entering its old age and will soon use up all its available fusion energy and die, either by slowly dissipating its remaining energy as a white dwarf star, by contracting to an unimaginably large density as a neutron star or black hole, or perhaps by throwing out all its remaining energy (and a good deal of its matter) in one explosion as a supernova, and for a brief few days outshining an entire galaxy.

The physical processes that occur in the collapse of a star are discussed in detail in the next few sections of this chapter. In this section we look at the nuclear processes that accompany the collapse.

After a star's hydrogen has been converted to helium through fusion reactions, gravitational collapse begins again, and the core of the star heats up from about 10^7 K to about 10^8 K. At this point there is enough thermal kinetic energy to overcome the Coulomb repulsion of the helium nuclei, and helium fusion can begin. In this process three ^{4}He are converted into ^{12}C by the two-step process.

$$^4\text{He} + {}^4\text{He} \longrightarrow {}^8\text{Be}$$

$$^8\text{Be} + {}^4\text{He} \longrightarrow {}^{12}\text{C}$$

The first reaction is endothermic, with a Q value of 92 keV. The nucleus ^{8}Be is unstable and decays back into two alpha particles in a time of the order of 10^{-16} s. Even so, the Boltzmann factor $e^{-\Delta E/kT}$ suggests that at 10^8 K there will be a small concentration of ^{8}Be. The second reaction proceeds through a resonance and therefore has a particularly large cross section; in spite of the rapid breakup of ^{8}Be, there is still a good chance to form ^{12}C. The net Q value for the process is 7.3 MeV, or about 0.6 MeV per nucleon, much less than the 6.7 MeV per nucleon produced by hydrogen burning.

Once enough ^{12}C has formed in the core, other alpha particle reactions become possible, such as

$$^{12}\text{C} + {}^4\text{He} \longrightarrow {}^{16}\text{O}$$

$$^{16}\text{O} + {}^4\text{He} \longrightarrow {}^{20}\text{Ne}$$

$$^{20}\text{Ne} + {}^4\text{He} \longrightarrow {}^{24}\text{Mg}$$

Each of these reactions is exothermic, releasing a few MeV of energy and contributing to the star's energy production. At still higher temperatures (10^9 K) carbon burning and oxygen burning begin:

$$^{12}\text{C} + {}^{12}\text{C} \longrightarrow {}^{20}\text{Ne} + {}^4\text{He}$$

$$^{16}\text{O} + {}^{16}\text{O} \longrightarrow {}^{28}\text{Si} + {}^4\text{He}$$

Eventually ^{56}Fe is reached, at which point no further energy is gained by fusion (Figure 12.4).

If this explanation of the formation of elements is correct, we expect the abundances of the elements to have the following properties:

1. Large relative abundances of the light, even-Z elements; small relative abundances of odd-Z elements.

2. Little or none of the elements between He and C (Li, Be, B), which are not produced in these reactions.

3. Large relative abundance of Fe, the end product of the fusion cycle.

Figure 15.20 shows the relative abundances of the light elements in the solar system, and they are in agreement with all of the three above expectations. Each even-Z element is 10 to 100 times more abundant than its odd-Z neighbors, there is a prominent peak at Fe, the heavy elements with $Z > 30$ *combined* are less abundant than every element but one in the range C to Zn, and the three elements Li, Be, B are far less abundant than the elements in the range C to Zn.

The light odd-Z elements can be produced by alternative reactions among the fusion products, for example:

$$^{12}C + {}^{12}C \longrightarrow {}^{23}Na + {}^{1}H$$

$$^{16}O + {}^{16}O \longrightarrow {}^{31}P + {}^{1}H$$

The abundance of nitrogen is nearly equal to that of its neighbors C and O, which are the most abundant of elements beyond H and He; nitrogen has a greater abundance than any other odd-Z element shown, and greater than all even-Z elements with $Z > 8$. The formation of nitrogen must therefore be a relatively common process in stars. Since the element B is so rare, alpha particle reactions are of no help in forming nitrogen. The most likely sources of N are

$$^{12}C + {}^{1}H \longrightarrow {}^{13}N + \gamma$$

$$^{13}N \longrightarrow {}^{13}C + e^{+} + \nu$$

$$^{13}C + {}^{1}H \longrightarrow {}^{14}N + \gamma$$

and

$$^{16}O + {}^{1}H \longrightarrow {}^{17}F + \gamma$$

$$^{17}F \longrightarrow {}^{17}O + e^{+} + \nu$$

$$^{17}O + {}^{1}H \longrightarrow {}^{14}N + {}^{4}He$$

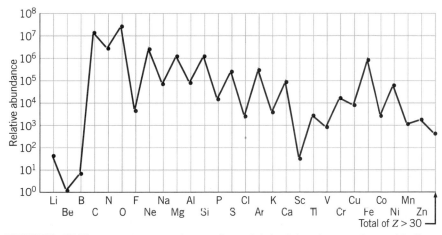

FIGURE 15.20   Relative abundances (by weight) of the elements beyond helium in the solar system.

The stable isotopes ^{13}C and ^{17}O are found in natural carbon and oxygen with abundances of 1.1 percent and 0.04 percent, which suggests that reactions of this sort can indeed take place.

The production of the elements beyond iron requires the presence of neutrons, which are not produced in the reactions we have studied so far, because neutrons are likely to be emitted only in reactions with nuclei that have an excess of neutrons. If enough of the heavier isotopes, such as ^{13}C, ^{17}O, or ^{21}Ne, are formed, the following reactions can produce neutrons:

$$^{13}\text{C} + {}^{4}\text{He} \longrightarrow {}^{16}\text{O} + \text{n}$$

$$^{17}\text{O} + {}^{4}\text{He} \longrightarrow {}^{20}\text{Ne} + \text{n}$$

$$^{21}\text{Ne} + {}^{4}\text{He} \longrightarrow {}^{24}\text{Mg} + \text{n}$$

How are the heavy elements built up by neutron capture? Let us consider the effect of neutron capture on ^{56}Fe:

$$^{56}\text{Fe} + \text{n} \longrightarrow {}^{57}\text{Fe} \quad \text{(stable)}$$

$$^{57}\text{Fe} + \text{n} \longrightarrow {}^{58}\text{Fe} \quad \text{(stable)}$$

$$^{58}\text{Fe} + \text{n} \longrightarrow {}^{59}\text{Fe} \quad (t_{1/2} = 45 \text{ d})$$

What happens next depends on the number of available neutrons. If that number is small, the chances of ^{59}Fe encountering a neutron *before* it decays to ^{59}Co are small, and the process might continue as follows:

$$^{59}\text{Fe} \longrightarrow {}^{59}\text{Co} + \text{e}^{-} + \bar{\nu}$$

$$^{59}\text{Co} + \text{n} \longrightarrow {}^{60}\text{Co} \quad (t_{1/2} = 5 \text{ y})$$

$$^{60}\text{Co} \longrightarrow {}^{60}\text{Ni} + \text{e}^{-} + \bar{\nu}$$

On the other hand, if the number of neutrons is very large, a different sequence might result:

$$^{59}\text{Fe} + \text{n} \longrightarrow {}^{60}\text{Fe} \quad (t_{1/2} = 3 \times 10^{5} \text{ y})$$

$$^{60}\text{Fe} + \text{n} \longrightarrow {}^{61}\text{Fe} \quad (t_{1/2} = 6 \text{ m})$$

$$^{61}\text{Fe} \longrightarrow {}^{61}\text{Co} + \text{e}^{-} + \bar{\nu}$$

$$^{61}\text{Co} \longrightarrow {}^{61}\text{Ni} + \text{e}^{-} + \bar{\nu}$$

If the density of neutrons is so low that the chance of encountering a neutron is, on the average, less than once every 45 days, the first process ought to dominate, with the production of ^{60}Ni. If the chance of encountering a neutron is more like once every few minutes, the second process should dominate, and no ^{60}Ni is produced.

The first type of process, which occurs *slowly* and allows the nuclei time to beta decay, is known as the *s process* (s for slow); the second process occurs very *rapidly* and is known as the *r process* (r for rapid).

Figure 15.21 illustrates how the *r* and *s* processes can proceed from ^{56}Fe. The *s* process never strays very far from the region of the stable nuclei, while the *r* process can produce many nuclei that have a large excess of

s process and r process

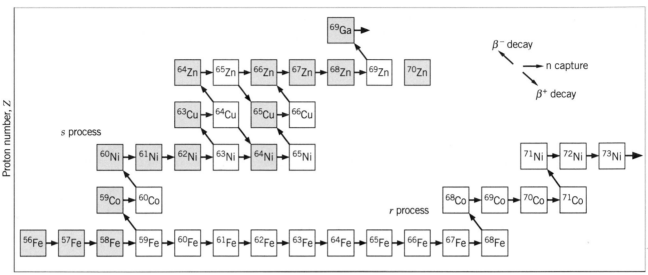

FIGURE 15.21   A section of the chart of the nuclides (Figure 12.4), showing the *s*- and *r*-process paths from ^{56}Fe. Shaded squares represent stable nuclei, and unshaded squares represent radioactive nuclei. Many *r*-process paths are possible, as the short-lived nuclei beta decay; only one such path is shown. All the nuclei in the *r*-process path are unstable and may beta decay toward the stable nuclei.

neutrons. The larger the excess of neutrons, the shorter is the half-life of these nuclei. Eventually the half-life becomes so short that no neutron is captured before the beta decay occurs to the next higher Z. All nuclei produced in the *r* process will eventually decay toward the stable nuclei, generally moving by beta decays along the diagonal line of constant mass number A.

Some stable nuclei are produced only through the *s* process, others are produced only through the *r* process, and some may be produced through both processes. Often the natural abundance of the isotopes of an element can suggest the relative roles of these two processes. In Figure 15.21, you can see that the stable isotope ^{70}Zn cannot be produced in the *s* process, because the half-life of ^{69}Zn is too short (56 min). Other isotopes for which the *r* process is important are ^{76}Ge, ^{82}Se, ^{86}Kr, ^{96}Zr, and ^{122}Sn. The isotope ^{64}Ni can be produced either through the *s* process (as shown in Figure 15.21) or through the *r* process (such as through beta decays beginning with ^{64}Fe). On the other hand, ^{64}Zn (the most abundant isotope of zinc) is produced *only* through the *s* process, because *r* process beta decays proceeding along the $A = 64$ line are stopped at stable ^{64}Ni and cannot reach ^{64}Zn.

The heaviest element that can be built up out of *s*-process neutron captures is ^{209}Bi; the half-lives of the isotopes beyond ^{209}Bi are too short to allow the *s* process to continue. The presence in nature of heavier elements such as thorium or uranium suggests that the *r* process must operate in this region as well.

The *r* process most likely occurs during supernova explosions, following the breakdown and implosion of a star that has used up its fusion reserves. In a very short time, lasting of the order of seconds, the star implodes, produces an enormous flux of neutrons (perhaps 10^{32} n/cm²/s), and builds up all elements to about $A = 260$. When the final explosion occurs, these elements are hurled out into space, to become part of new star systems. The heavy atoms of which the Earth is made may have been produced in such an explosion.

15.6 WHITE DWARF STARS

When a star uses up its hydrogen fuel and heats up to allow helium burning to begin, there is an increased radiation pressure that pushes the outer layers of a star outward until its radius becomes hundreds or thousands of times as large. The total energy production in the star must be spread over a much larger surface area and so the energy per unit area decreases—the surface of the star seems to reach a lower temperature or to become redder. This is the "red giant" stage; when it becomes a red giant, our Sun will become large enough to swallow up the orbits of Mercury and Venus.

The star now continues to burn helium, and perhaps heavier elements, in the red giant stage until the fusion cycle ends with the formation of ^{56}Fe.

At this point there is no further fusion energy available to prevent the gravitational collapse of the entire star, and so gravity takes over again. The star eventually contracts to a *white dwarf*, with a density of perhaps 10^6 g/cm³ and a surface temperature of the order of 10^4 K. Eventually it cools down as its energy is radiated away and becomes a black dwarf, a dark, cold chunk of the ash left over from the original star.

What keeps further gravitational collapse from occurring? At such enormous densities, the atoms are crowded so close together that their electron wave functions begin to overlap. Just as in the case of the formation of molecules, there is a repulsive force which tends to oppose that overlap, because the Pauli exclusion principle prevents the electrons from having identical quantum numbers. One way to picture this situation is as a solid, in which the energies of the electrons are governed by the Fermi-Dirac distribution; the electrons occupy very closely spaced, nearly continuous, energy levels from 0 up to the Fermi energy E_F, where

$$E_F = \frac{\hbar^2}{2m_e}\left(\frac{3\pi^2 N_e}{V}\right)^{2/3} \tag{15.14}$$

The electron density is N_e/V in this expression. The average electron energy is $\frac{3}{5}E_F$, and therefore the total electron energy is $\frac{3}{5}N_e E_F$. The total gravitational potential energy can be shown (see Problem 13) to be $-\frac{3}{5}GM^2/R$, where M is the total mass and R is the radius, and thus the total energy of the white dwarf, which we assume to be spherical, of uniform density and at a constant temperature, is

$$E = \frac{3}{5}N_e \frac{\hbar^2}{2m_e}\left(\frac{3\pi^2 N_e}{V}\right)^{2/3} - \frac{3}{5}\frac{GM^2}{R} + \frac{3}{2}N_a kT + E_{rad} \tag{15.15}$$

The first term represents the motion of the electrons, the second term is the gravitational potential energy, the third term takes into account the thermal motions of the atoms (N_a = number of atoms), and the fourth term gives the energy radiated by the star. For the moment we neglect the last two terms, and we assume the star to be composed of N nucleons (in the form of fusion products such as iron and lighter elements) and approximately $\frac{1}{2}N$ electrons. To find the equilibrium radius we set $dE/dR = 0$ and find

$$R = \frac{3^{4/3}\pi^{2/3}}{8}\frac{\hbar^2}{Gm_e m_n^2}N^{-1/3} \qquad (15.16)$$

Radius of white dwarf star

For a star of the mass of the Sun ($M = 2.0 \times 10^{30}$ kg, so $N = 1.2 \times 10^{57}$) we find $R = 7.1 \times 10^3$ km, about the same as the Earth's radius, and this corresponds to a density of 1.1×10^6 g/cm^3. Going back to Equation 15.14 we find $E_F = 0.194$ MeV, so the total contribution of the electrons is

$$E_{\text{electrons}} = \tfrac{3}{5}N_e E_F = 7 \times 10^{55} \text{ MeV} \qquad (15.17)$$

We now justify neglecting the last two terms of Equation 15.15. The thermal energy can be easily computed at $T \cong 10^8$ K (the central temperature is $\sim 10^9$ K, while the surface temperature is only 10^4 K):

$$E_{\text{thermal}} = \tfrac{3}{2}N_a kT$$

$$= \frac{3}{2}\left(\frac{1.2 \times 10^{57}}{30}\right)\left(8.6 \times 10^{-5}\frac{\text{eV}}{\text{K}}\right)10^8 \text{ K}$$

$$= 5 \times 10^{53} \text{ MeV} \qquad (15.18)$$

(where we take the average atom to contain 30 nucleons) and so $E_{\text{thermal}} \ll E_{\text{electrons}}$, as we assumed. The radiant energy can be estimated from Stefan's law as σT^4, which gives the *power per unit area*; the power is radiant energy per unit time, so

$$E_{\text{rad}} = (\sigma T^4)(4\pi R^2)t \qquad (15.19)$$

where t is the time over which the star has been radiating. The observed surface temperatures of white dwarfs are of the order of 10^4 K, and so we estimate

$$E_{\text{rad}} = (2.3 \times 10^{36} \text{ MeV/s})t \qquad (15.20)$$

Even if the star had been radiating uniformly for the entire age of the universe (15×10^9 y $\cong 5 \times 10^{17}$ s) the total power radiated would be only about 10^{54} MeV, and so $E_{\text{rad}} \ll E_{\text{electrons}}$.

The white dwarf star represents an extraordinary state of matter; its average density is about 10^6 g/cm^3, and its central density may approach 10^8 g/cm^3. A cubic centimeter of such matter would, on the Earth, weigh 100 tons! We have based our analysis of this unusual condition on expressions that we derived for ordinary matter, and it is surprising, perhaps comforting, to learn that we can understand this state of matter from these expressions. Quantum physics works well even in the interiors of white dwarf stars!

15.7 NEUTRON STARS

J. Robert Oppenheimer (1904–1967, United States). A theoretical physicist with a philosophical nature, it is ironic that he was the director of the laboratory at Los Alamos that developed the first nuclear weapons. His early theoretical work speculated on the existence of black holes.

A white dwarf star is prevented from collapse by the Pauli principle, which prevents the electron wave functions from being squeezed too close together. The larger the mass of the star, the greater is the gravitational force trying to collapse the star. Will the repulsion of the electron wave functions due to the Pauli principle be able to prevent the collapse of any star, no matter how massive?

There is a mass limit of about 1.4 solar masses, called the *Chandrasekhar mass*, beyond which the Pauli principle applied to electrons cannot prevent further gravitational collapse. Let us look at how this might occur. Equation 15.14 gives* $E_F = 0.304$ MeV for a white dwarf having the Chandrasekhar mass, but remember that this is only the point at which the Fermi-Dirac distribution has the value 0.5. There is a high-energy tail of the Fermi-Dirac distribution, and some of the electrons in that tail have enough energy to produce the inverse beta decay reaction

$$e^- + p \longrightarrow n + \nu_e$$

for which the Q value is 0.782 MeV, not too far above E_F. This reaction removes some electrons from the star, reducing the effects of Pauli repulsion, and allowing the star to collapse a bit (R depends on $N_e^{5/3}$). The Fermi energy *increases*, pushing more electrons above the 0.782-MeV Q value, resulting in more electrons being lost, and so on, until all (or very nearly all) of the electrons vanish. The star is now composed of neutrons, instead of protons and electrons. The "pressure" of the electrons no longer can oppose gravitational collapse, and so the star contracts until the Pauli principle applied to the *neutrons* (which also obey Fermi-Dirac statistics) prevents further collapse. The situation is now described by expressions identical with Equations 15.14 to 15.16 but applied to neutrons; in particular

$$E_F = \frac{\hbar^2}{2m_n} \left(\frac{3\pi^2 N}{V} \right)^{2/3} \tag{15.21}$$

Radius of neutron star

$$R = \frac{3^{4/3}\pi^{2/3}}{2^{4/3}} \frac{\hbar^2}{Gm_n^3} N^{-1/3} \tag{15.22}$$

For a star of 1.5 solar masses, for example, $R = 11$ km and its density is about 4×10^{14} g/cm³. This is about the same density as the interior of atomic nuclei, so in this sense the star has become a giant nucleus, 20 km across, with a mass number of about 10^{57}! Such objects are known as *neutron stars*.

Are these neutron stars merely figments of the physicist's imagination or do they really exist? In 1967, radio astronomers at Cambridge University discovered an unusual signal among their observations—a regular pulsa-

---

* The expression for E_F derived in Chapter 10 was based on the classical expression $p^2/2m$ for the electron's kinetic energy. In the previous section, we found $E_F = 0.194$ MeV, which is not sufficiently smaller than the electron's rest energy (0.511 MeV) to justify using the classical expression. However, the error introduced by using the classical expression is not large, and it is acceptable for this discussion.

tion, such as is shown in Figure 15.22, with a period of 1.34 s. No previously known astronomical object could produce such sharp and regular pulses, and at first the Cambridge group suspected that they might have discovered signals from an extraterrestrial intelligent civilization. (The object emitting the pulses was at first called LGM-1; LGM stands for "Little Green Men.") This notion was later discarded (unfortunately) and the object became known as a *pulsar*. Since 1967, hundreds of other pulsars have been discovered; all have extremely regular periods typically in the range 0.01–1 s.

Pulsars

The connection between pulsars and neutron stars was made soon after their discovery. The collapse of a rotating star to a neutron star ought to cause the neutron star to rotate much more rapidly. (Since angular momentum is conserved in the collapse, a decrease in the rotational inertia, as the star contracts, must be balanced by an increase in the angular speed.) The intense magnetic field of such a rapidly rotating object traps any emitted charged particles and accelerates them to high speeds, especially near the magnetic poles, where they give off radiation (Figure 15.23). As the star rotates, this beam of emitted radiation sweeps around like a searchlight or a lighthouse, and we see a pulse of radiation whenever the beam sweeps through the Earth. The observed interval between the pulses is, according to this interpretation, the rotational period of the neutron star.

If this explanation of a pulsar as a rotating neutron star is correct, we ought to see the pulsars slowing down somewhat, as the radiated energy is compensated by a decrease in the neutron star's rotational kinetic energy. This effect has been seen for nearly all pulsars, and amounts to about 1 part in 10^9 per day.

Pulsars have now been observed at many different wavelengths (optical, X ray, γ ray, radio) and with such great precision of timing that the slowing down of 10^{-9} per day is easily observable.

Although the exact mechanism of the collapse of a star to a neutron star is not yet understood, we suspect that the violent explosions known as *supernovas* leave a neutron star as a remnant. In 1054 Chinese astronomers observed a supernova explosion (which they called a "guest star") that was visible in the daytime over many days. Today we see the expanding shell of that explosion as the Crab Nebula (Figure 15.24). At the center of the Crab Nebula is a pulsar, rotating with a period of 30 Hz. It is remarkable that none of the many photographs of the Crab that were taken before 1967 revealed this pulsar blinking on and off every 0.033 s; all of these photographs were taken over long exposure times, and so the pulsations were not observable. When careful measures are taken, however, the blinking effect can be seen quite clearly (Figure 15.25). This suggests that, at least in this instance, pulsars may be identified as supernova remnants.

A Binary Pulsar

In 1974, an unusual pulsar was discovered by Joseph Taylor and Russell Hulse. The period of the pulsar was measured to be 59 ms, making it among the fastest observed up to that time. More surprising, the pulse rate appeared to be slowing down by about 0.1 percent per hour, but later was

Jocelyn Bell Burnell (1943–, England). As a graduate student at Cambridge University, she helped to build a radio telescope designed to search for cosmic radio sources. Using the telescope, she discovered the periodic sources of radio waves now known as pulsars, and she showed them to be extraterrestrial in origin.

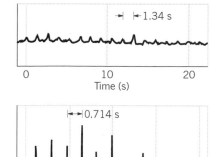

FIGURE 15.22   The radio signals from two different pulsars. The top signal is the record of the first pulsar discovered.

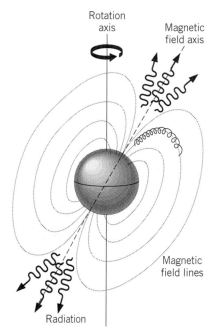

FIGURE 15.23   Charged particles trapped by the magnetic field lines of a neutron star are given large accelerations near the magnetic poles, from which a directional radiation beam emerges. If this radiation beam intercepts the Earth as the neutron star rotates, we see it as a pulse of radiation.

observed to be *increasing* by about the same amount. It was quickly realized that the decrease and increase of the pulse rate could be explained as a Doppler shift if the neutron star were moving first away from and later toward the Earth. To move in this way, the pulsar must be in orbit around an unseen companion. Thus we have a pulsar as part of a binary star system, or a "binary pulsar."

The orbital period of the binary system was determined to be about 8 hours. This is an extremely short period; for example, it is more than 250 times faster than that of Mercury, the fastest moving planet in our solar system. To have such a short orbital period, the pulsar must be orbiting very close to its companion (which is believed to be another neutron star). At such close distances, the curvature of spacetime is large and the effects of general relativity should be measurable. In effect, the binary pulsar provides us with a "general relativity laboratory."

Among the general relativity effects that have been observed in the binary pulsar system is the delay of the pulses due to the curvature of spacetime. This situation is similar to Figures 15.14 and 15.15, except that the pulsar (instead of Venus) is the origin of the signals and the curvature is due to the companion star (instead of the Sun). An effect analogous to the precession of Mercury's perihelion has also been observed (except that, in the case of a star, the point of closest approach is called *periastron* rather than perihelion). The periastron of the binary pulsar changes by 4.23° per year, about 35,000 times more rapidly than that of Mercury; the change of the periastron is known to an accuracy of about 1 part in 10^5, three orders of magnitude more precisely than Mercury's.

The most remarkable observation from the binary pulsar is the slowing of its orbital period, which general relativity explains as caused by the emission of gravitational radiation. Because of its large centripetal acceleration (its orbital speed is about $0.001c$), the binary pulsar should emit gravity

FIGURE 15.24   The Crab Nebula, remnant of a supernova observed in the year 1054.

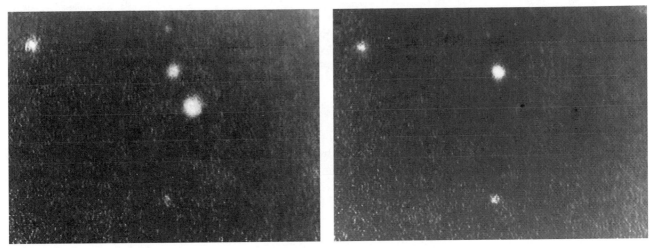

FIGURE 15.25   The visible pulsar of the Crab Nebula. The two exposures show the pulsar blinking on and off relative to the other stars in the photograph.

waves, and the energy radiated away is compensated by a loss in the orbital energy. This loss amounts of 76 μs per year or 67 ns per orbit, and the measured change in the orbital rate confirmed the prediction of general relativity to within 1 percent. In the absence of direct observation, this is the strongest evidence yet obtained for the existence of gravitational radiation.

For the discovery of the binary pulsar and its contributions to the study of gravitation, Taylor and Hulse were awarded the 1993 Nobel prize in physics.

15.8   BLACK HOLES

A neutron star is not the ultimate fate of the collapse of massive stars. Stars with masses less than two or three solar masses probably do end up as white dwarf stars or neutron stars. For more massive stars, the gravitational force is strong enough to overcome even the Pauli principle applied to the neutrons, and there is nothing to prevent the material in the star from suffering complete collapse down to a single point in space. To understand gravitational collapse, we must turn again to general relativity.

Within a year after Einstein's 1916 publication of the general theory, Karl Schwarzschild worked out the solutions to the equations for the curvature of spacetime near a spherically symmetric mass M. In spherical coordinates (r, θ, ϕ), the spacetime interval for this solution is

$$(ds)^2 = c^2 \left(1 - \frac{2GM}{c^2 r} \right) (dt)^2 - \frac{(dr)^2}{\left(1 - \frac{2GM}{c^2 r} \right)}$$

$$- r^2 (d\theta)^2 - r^2 \sin^2 \theta \, (d\phi)^2$$

(15.23)   *Schwarzschild solution*

The radial part of this solution (the *dr* term) has what appears to be a serious problem: the factor in the denominator can become zero for a particular *r*, causing that term in the equation to "blow up." This occurs when *r* has the value

$$r_S = \frac{2GM}{c^2} \tag{15.24}$$

which is known as the *Schwarzschild radius*. None of the physical coordinates actually "blows up" at $r = r_S$, and an object falling toward *M* would notice no change in its motion as it crossed the Schwarzschild radius.

For external observers watching the falling object, the situation is very different. As the object falls, general relativity predicts that its clocks would appear to run ever more slow, stopping completely when the object reaches r_S. The object appears to be frozen forever at that location! While the object is falling, the light it emits becomes increasingly red shifted, and the red shift becomes infinite at $r = r_S$, so the object disappears from view! The outside observer can obtain no information about the object once it passes through the Schwarzschild radius. For that reason, the Schwarzschild radius is often called the *event horizon*; no external observer can see beyond that horizon.

The falling object does encounter one crisis at r_S. At any time before crossing r_S, the object could reverse its fall and escape from the gravitational pull of *M*, for example, by firing its rockets. Once it passes r_S, no escape is possible. Inside r_S, the escape speed exceeds the speed of light, and nothing (*not even light*) can escape. No travel or communication is permitted from inside r_S to the outside world. However, the object inside r_S can continue to exert a gravitational force on external objects or, in the language of general relativity, to curve spacetime beyond r_S.

Black hole   An object whose mass *M* lies totally within the corresponding radius r_S is said to be a *black hole*. To form such an object requires that matter be compressed to exceptional densities. Table 15.2 shows some values of r_S for representative objects. For the Earth to become a black hole, we would need to compress it into a sphere of radius less than 1 cm, and the Sun would become a black hole only if compressed to a 3-km radius! Nevertheless, it has been speculated that black holes are the end products of the collapse of massive stars, and that tiny (atom-sized) black holes may have been formed by the extreme densities and pressures in the early universe (see

TABLE 15.2   BLACK HOLE EVENT HORIZONS

Object	Mass (kg)	Ordinary Radius (m)	r_S (m)
^{238}U nucleus	4×10^{-25}	7×10^{-15}	6×10^{-52}
Physics book	1	0.1	1.5×10^{-27}
Earth	6×10^{24}	6×10^6	8.9×10^{-3}
Sun	2×10^{30}	7×10^8	3×10^3
Galaxy	$\sim 2 \times 10^{41}$	$\sim 10^{20}$	3×10^{14}
Universe	$\sim 10^{51}$	$10^{26}(?)$	$\sim 10^{24}$

Section 16.5). It is also speculated that galaxies, including the Milky Way, have massive black holes at their center.

Far from a black hole, the gravitational field is Newtonian in character; the effects of curved spacetime are small, and the black hole cannot be distinguished from any other gravitating object. Close to a black hole (or close to any massive, compact object), the effects of curved spacetime can become significant. The discovery of a massive black hole could therefore provide another "laboratory" for testing the predictions of general relativity.

Although there are several good candidates, the existence of a black hole has yet to be unambiguously confirmed. The leading candidates are members of binary systems that are intense X-ray sources in which one object, usually a visible star, orbits about an unseen massive companion. The masses of the stars can be estimated from the luminosity of the visible star and the orbital period of the pair, and the possible candidates are those in which the invisible star has a mass great enough to have evolved into a black hole (at least three solar masses). The X rays originate from material attracted by the black hole from the visible star; the material heats up and emits X rays as it falls toward the black hole.

It has long been suspected that massive black holes lie at the center of galaxies. In 1994, the repaired Hubble Space Telescope provided evidence for a black hole at the center of the giant elliptical galaxy M87, which is about 50 million light years away. The telescope revealed a rotating disk of gas at the center of the galaxy. As the disk rotates, one side is moving toward the Earth and the other away from the Earth; by comparing the Doppler shifts of the light coming from the two sides of the disk, it is possible to determine the rotational speed. At a distance of 60 light years from the center of the disk, the rotational speed was found to be about 500 km/s. At locations closer to the center, the speed was found to increase. This suggests that a concentrated mass at the center is holding the material of the disk in rotational motion (just as the Sun holds the planetary disk of our solar system). We can use Kepler's laws to estimate the mass of the central object that provides the gravitational force; the mass must be more than 10^9 times the mass of the Sun! No known phenomenon other than a black hole permits so much mass to be concentrated in so small a region. Similar evidence derived from rotational motions suggests that even our near neighbor, the Andromeda galaxy, has a black hole of a few million solar masses at its center. Radio emissions from the center of our own Milky Way galaxy also suggest a black hole of a few million solar masses.

A surprising development in black hole theory occurred in 1974, when Steven Hawking showed that black holes could be sources of particle emission. According to quantum mechanics, particle-antiparticle pairs can spontaneously appear, as long as they exist for a short enough time that the uncertainty principle is not violated. That is, the particles can "borrow" an energy of $2mc^2$ as long as the loan is paid back (the particles vanish) within a time of at most $\Delta t \sim \hbar/2mc^2$. If a particle-antiparticle pair arises outside the event horizon of a black hole, its gravitational field can provide the energy necessary to pay back the loan so that the particle and antiparticle can become real. Usually the particle and antiparticle fall back into the black hole, restoring the energy balance. However, one of the members of the pair may have enough energy to escape into the outside world. The black hole thus appears to be emitting particles. In the process, the black hole loses mass. The rate of mass loss is inversely proportional to the mass of the black hole; massive black holes that result from the collapse of stars emit particles at too low a rate to be observed. However, tiny black holes of atomic or nuclear size, which may have been formed in the early evolution of the universe, could be very bright sources of radiation.

Black holes provide both a fertile area for theoretical speculation and a challenge for the skill of experimenters. It has been suggested that material that falls into a black hole reappears somewhere else in the universe, or perhaps in another universe. Thus a black hole, if this speculation is correct, could be used for travel between different universes. Other proposals suggest harnessing a black hole as an energy source. It has been estimated that, if black holes are indeed the end products of the evolution of massive stars, there could be as many as 10^9 massive black holes in our galaxy, which makes it likely that many black holes are within our observational reach. Or, perhaps we will someday observe a minihole ending its existence with a burst of radiation. As we continue to refine our ability to study the skies at visible, X-ray, and γ-ray wavelengths, black holes will figure prominently in our investigations.

SUGGESTIONS FOR FURTHER READING

There are many introductory, advanced, and popular-level books and articles that cover in more detail the subjects touched only briefly in this chapter. To pursue any of these subjects, you may first need some background material on astronomy and astrophysics:

F. Golden, *Quasars, Pulsars, and Black Holes* (New York, Scribner's, 1976). A popular-level, nonmathematical introduction.

M. Harwit, *Astrophysical Concepts* (New York, Wiley, 1973). Nearly everything you would want to know about astrophysics, using undergraduate math.

L. Motz, *Astrophysics and Stellar Structure* (Waltham, Ginn & Co., 1970). An advanced presentation of astrophysics, complete but mathematically sophisticated.

L. Motz and A. Duveen, *Essentials of Astronomy* (New York, Columbia University Press, 1977). A complete introductory undergraduate astronomy text.

H. L. Shipman, *Black Holes, Quasars, and the Universe* (Boston, Houghton Mifflin, 1976). An excellent popular-level survey; see especially Chapter 4 on clock rates and red shifts during the descent into a black hole.

Some additional detail on general relativity, without the advanced math:

P. G. Bergmann, *The Riddle of Gravitation* (New York, Scribner's, 1968). Very descriptive and successful presentation of the philosophy of general relativity by a student of Einstein's.

M. Berry, *Principles of Cosmology and Gravitation* (Cambridge, Cambridge University Press, 1976). Reduces the mathematics to undergraduate level; see especially Chapter 5 on the tests of general relativity.

P. C. W. Davies, *Space and Time in the Modern Universe* (Cambridge, Cambridge University Press, 1977). Excellent nonmathematical introduction to spacetime and general relativity.

R. P. Feynman, R. B. Leighton, and M. Sands, *The Feynman Lectures on Physics* (Reading, Addison-Wesley, 1964). Chapter 42 contains an interesting elementary discussion on the properties of curved spacetimes.

S. W. Hawking, *A Brief History of Time* (New York, Bantam, 1988). General relativity and black holes for the general public, written by a brilliant scientist.

W. J. Kaufmann, III, *The Cosmic Frontiers of General Relativity* (Boston, Little, Brown and Co., 1977); also by Kaufmann, *Black Holes and Warped Spacetime* (San Francisco, W. H. Freeman and Co., 1979). Both are excellent as elementary introductions to the properties of black holes.

W. Rindler, *Essential Relativity* (New York, Van Nostrand, 1969). Uses a bit of tensor math; a good beginning if you really want to get into the complete theory.

R. v. B. Rucker, *Geometry, Relativity, and the Fourth Dimension* (New York, Dover, 1977). A good nonmathematical introduction to higher dimensions; includes a particularly comprehensive list of references, including Abbott's classic work *Flatland*.

R. M. Wald, *Space, Time, and Gravity,* 2nd ed. (Chicago, University of Chicago Press, 1992). Brief but thorough introduction using a bit of calculus but no tensors; particularly good on black holes.

If you want to see the complete theory, in all its mathematical glory:

S. W. Hawking and G. F. R. Ellis, *The Large-Scale Structure of Space-Time* (Cambridge, Cambridge University Press, 1973).

C. W. Misner, K. S. Thorne, and J. A. Wheeler, *Gravitation* (San Francisco, Freeman, 1973). Figure 33.2 shows how to use a black hole as a combined energy source and garbage dump.

S. Weinberg, *Gravitation and Cosmology* (New York, Wiley, 1972).

Each of these is a difficult but standard work, and each contains a little that you should understand.

As with many of the subjects covered in this book, the best sources for current, popular-level articles are journals such as *Scientific American*. From time to time, articles appear covering techniques and results in current astronomy and astrophysics. Some articles that relate directly to this chapter are:

J. N. Bahcall, "The Solar Neutrino Problem," *Scientific American* **262**, 54 (May 1990).

T. Folger, "The Ultimate Vanishing," *Discover* **14**, 98 (October 1993).

S. W. Hawking, "The Quantum Mechanics of Black Holes," *Scientific American* **236**, 34 (January 1977).

M. J. Rees, "Black Holes in Galactic Centers," *Scientific American* **263**, 56 (November 1990).

K. S. Thorne, "The Search for Black Holes," *Scientific American* **231**, 32 (December 1974).

QUESTIONS

1. If we were to measure the equivalence of gravitational and inertial mass, would we show that $m_{\text{inertial}} = m_{\text{gravitational}}$ or merely that $m_{\text{inertial}} \propto m_{\text{gravitational}}$?

2. Do tidal effects distinguish between Newtonian gravity and curved spacetime? What would be the shape of a drop of liquid following a path in a curved spacetime? Can such a drop distinguish between a uniform gravitational field and a uniform acceleration?

3. Suppose that the first measurement of deflection of starlight during a solar eclipse had been done after 1905, when the special theory of relativity was introduced, but before 1916, when the general theory was introduced. What would have been the effect of this measurement on the special theory?

4. If we could make a precise comparison of light from the Sun with light from the Moon, would the moonlight be red shifted, blue shifted, or unshifted relative to sunlight?

5. What difficulties might arise in the Pound and Rebka experiment on the gravitational red shift if the temperature of the source or the absorber varied?

6. Why are the abundances of Li, Be, and B so small?

7. Based on the model for nucleosynthesis of light nuclei, we expect greater abundance of nuclei with even numbers of protons or neutrons. (Why?) Figure 15.20 shows that this is true for protons. Is it also true for neutrons? (See the abundances in Appendix B.)

8. The isotope ^{64}Zn is "shielded" from the r process by ^{64}Ni, which is a stable isotope with the same A but smaller Z. Explain how this causes the shielding. What isotopes are responsible for shielding ^{76}Ge, ^{82}Se, ^{86}Kr, ^{96}Zr, and ^{122}Sn? Can you find other examples of isotopes that are shielded from the r process?

9. Because ^{69}Zn has too short a half-life, ^{70}Zn is not likely to be produced in the s process. Can you find other examples of stable isotopes that are similarly unlikely to be produced in the s process?

10. How "sharp" is the Fermi distribution for the electrons in a white dwarf star? That is, how does E_F compare with kT? Are we justified in using the $T = 0$ expressions for E_F and E_m?

11. Why does the radius of a white dwarf or neutron star depend inversely on the number of nucleons N? Shouldn't a star with more matter have a larger radius?

12. Do the neutrons in a typical neutron star have enough energy to create pi mesons in collisions such as $n + n \rightarrow n + n + \pi$? As a neutron star collapses toward a black hole, the neutrons become increasingly energetic, and greater numbers of pi mesons are produced. Why doesn't the Pauli principle applied to the pi mesons prevent collapse?

PROBLEMS

1. In Example 15.1 we calculated the change in wavelength of the Lyman α line due to the general relativistic red shift. Compare this value with the special

relativistic Doppler shift due to the rotation of the Sun. The Sun's rotational period at the equator is 26 days.

2. Verify that the gravitational potential at the Earth is negligible in Equation 15.5.

3. Show that the spacetime interval given by Equation 15.6 is invariant with respect to the Lorentz transformation. That is, show that $(ds)^2 = (ds')^2$, where $(ds')^2 = (c\,dt')^2 - (dx')^2$.

4. A satellite is in orbit at an altitude of 150 km. We wish to communicate with it using a radio signal of frequency 10^9 Hz. What is the gravitational change in frequency between a ground station and the satellite? (Assume g doesn't change appreciably.)

5. According to the uncertainty principle, what is the minimum time interval necessary to measure a change in frequency of the magnitude observed in the Pound and Rebka experiment?

6. A certain star is 80.0 light-years from Earth. Directly along the line of sight from Earth to the star, and at a distance of 20.0 light-years from the star, is a white dwarf star, with a mass equal to the Sun's mass and a radius of 7.0×10^3 km. The deflection of light from the star by the white dwarf causes us to observe two images of the star, as shown in Figure 15.26. Find the angle α between the two images. (Such an effect, called a gravitational lens, has been observed for galaxies; see *Scientific American*, November 1980.)

FIGURE 15.26 Problem 6.

7. By drawing analogies between the Coulomb force law and the gravitational force law, use Equation 6.13 for the deflection in Rutherford scattering to obtain Equation 15.9 for the deflection of photons. Assume the photon behaves as if it has a mass $m = E/c^2$. (*Hint*: Write Equation 6.13 in terms of the velocity of the particle instead of kinetic energy.)

8. (a) Show that the reaction ^{7}Be $+ e^- \to {}^7$Li $+ \nu_e$, which occurs in the proton-proton cycle, results in a neutrino of energy 0.862 MeV. (b) Compute the Q value of the reaction $\nu + {}^{37}$Cl $\to {}^{37}$Ar $+ e^-$, which is used to detect the neutrinos, and show that the neutrino energy found in part (a) is sufficient to produce the reaction.

9. Calculate the Q value for neutrino capture by ^{71}Ga in the reaction used in the gallium solar neutrino detectors.

10. (a) Find the Q value of the reaction ^{4}He $+ {}^4$He $\to {}^8$Be. (b) In a gas of ^{4}He at a temperature of 10^8 K, estimate the relative amount of ^{8}Be present.

11. Trace the path of the s process from the stable isotope ^{63}Cu to the stable isotope ^{75}As, showing the neutron capture and beta decay processes.

12. Show how the s process proceeds from stable ^{81}Br to stable ^{95}Mo.

13. Consider a spherical distribution of mass of uniform density ρ and radius R. (a) An element of mass dm is at a radius $r < R$. Assuming there is a "Gauss' law" for gravitation, what is the quantity of mass that attracts dm? (b) What is the gravitational potential energy dU of the mass element dm? (c) Express dm in terms of the volume element in spherical coordinates. (d) Integrate over the volume of the sphere to obtain the total gravitational potential energy that was used in Equation 15.15.

14. Neglecting the last two terms of Equation 15.15, show that $dE/dR = 0$ at the value given by Equation 15.16.

15. Show that Equation 15.16 can be written

$$R = (7145 \text{ km}) \left(\frac{M}{M_\odot} \right)^{-1/3}$$

where $M_\odot$ is the mass of the Sun.

16. (a) For a white dwarf star of the mass of the Sun, compute the de Broglie wavelength of electrons at the Fermi energy. (Use nonrelativistic kinematics.) (b) Assuming the star to be made of iron atoms and to be of uniform density, compute the distance between atoms and compare with the electron de Broglie wavelength. What do you conclude from this comparison? Do the electrons "see" the lattice of iron atoms? Are they easily scattered by the atoms?

17. In deriving Equation 15.16, it was assumed that $N_e = \frac{1}{2}N$, as would be true for a star composed of lightweight elements with equal numbers of protons and neutrons. When electrons begin to be consumed in the development of a neutron star, this is not true. Repeat the derivation of Equation 15.16 for the case $N_e = fN$, where f is the fraction of electrons. Show that $R \propto N_e^{5/3}$, as claimed in Section 15.7.

18. Estimate the angular speed of a neutron star in the following way. Consider a star 1.5 times as massive as the Sun, rotating on its axis about once per year. (This is quite a slow rate of rotation—our Sun rotates about once per month.) Assume the star to have about the same radius as the Sun (7×10^5 km) and to be relatively uniform in density. If angular momentum is conserved in the collapse, what will be the final angular velocity? Is this of the right order of magnitude for a neutron star? What errors are made in this estimate, and how do they affect the final result?

19. (a) Compute the value of Gm_n^2 in units of J·m. (b) Compute the value of $\hbar^2/m_n$ in MeV·m^2. (Multiply top and bottom by c^2). (c) Combine the results of (a) and (b) to find $\hbar^2/Gm_n^3$ in meters. (d) Combine the results of (a) and (b) to find $G^2m_n^5/\hbar^2$ in MeV.

20. Use the results of the previous problem to rewrite Equations 15.21 and 15.22 as:

$$E_F = (56.27 \text{ MeV}) \left(\frac{M}{M_\odot} \right)^{4/3}$$

and

$$R = (12.34 \text{ km}) \left(\frac{M}{M_\odot} \right)^{-1/3}$$

where $M_\odot$ is the mass of the Sun.

21. (a) Compute the neutron Fermi energy in a neutron star of 2.0 solar masses. (b) What is the average neutron energy? (See Equation 10.41.) (c) What is the total energy of the neutrons? (d) Calculate the total gravitational potential energy of the neutron star. (e) The gravitational potential energy resulted from the collapse of the original star (or of the gas cloud that formed the star). Only part of that energy went into the neutron kinetic energies. What fraction of the gravitational energy ended with the neutrons? Where did the rest go?

22. Find the Fermi energy of the neutrons in a neutron star of 1.5 solar masses. Are we justified in using nonrelativistic kinematics to describe their motion?

23. Find the de Broglie wavelength of the neutrons at the Fermi energy in a neutron star of mass $1.80 M_\odot$. Compare the de Broglie wavelength with the distance between neutrons. Do the neutrons "see" one another's deBroglie waves? Are they scattered by the lattice of neutrons?

24. A neutron star of 2.00 solar masses is rotating at a rate of 1.00 revolutions per second. (a) What is the radius of the neutron star? (b) Assuming its density is uniform, find its rotational inertia and its angular momentum. (c) Find its rotational kinetic energy. (d) If its rotational speed slows by 1 part in 10^9 per day, find the loss in rotational kinetic energy per day. (e) Assuming that the entire energy loss goes into radiation, find the radiative power.

25. Assume that the neutron star of Problem 24 is 10^4 light-years from Earth. (a) If its radiative power were distributed uniformly in space, what power would be delivered to 1.0 m^2 of antenna? (b) A typical pulsar, 10^4 light-years from Earth, has an observed average power level of 5.0×10^{-20} W/m^2. What does the comparison of this power with the value deduced in part (a) tell us about the radiative power?

Today we scan the skies at all wavelengths, from the very long (radioastronomy) to the very short (X-ray and gamma-ray astronomy): many new and unexpected phenomena have been discovered at these wavelengths. Quasars, pulsars, novas, supernovas, black holes—all these suggest that the universe is not at all static and eternal but instead is active and evolving and teeming with radiation. Van Gogh's painting The Starry Night *suggests exactly that view, even though it was painted in 1889, long before any of these discoveries were made.*

COSMOLOGY: THE ORIGIN AND FATE OF THE UNIVERSE

Edwin Hubble (1889–1953, United States). His observational work with large telescopes revealed the existence of galaxies, and he was the first to measure their size and distance. His discovery of the recessional motion of the galaxies was one of the most exciting and important in the history of astronomy.

In the short time of a few hundred years, developments in astronomy have taken us from a situation in which the Earth and its human population were believed to be quite literally the center of the universe, to a situation in which we shrink into insignificance. Before the sixteenth century, it was widely believed that the planets, Sun, Moon, and stars revolved about a central Earth. In the sixteenth and seventeenth centuries, as a result of discoveries by Copernicus, Tycho, and Kepler, the Earth was removed from its privileged place and replaced by the Sun. The use of the telescope in the seventeenth century showed that the Sun is merely one rather ordinary star among a galaxy of others, although it was assumed again that the Sun was at the center of our galaxy. By the early twentieth century, astronomers had discovered that the Sun is far from the center of our galaxy, and that the universe contains a vast number of other galaxies. We occupy one star out of perhaps 10^{11} in our Milky Way galaxy, and through the vastness of space are perhaps 10^{11} more galaxies, each with as many stars as our own.

Far from being preferred observers, we are viewing the universe from a vantage point of no special significance. In fact, we have no reason to believe that any other observer on a distant plant in another galaxy would observe the universe to be any different than we do. This is the fundamental underlying assumption of our speculations, the *cosmological principle*—at any given time the universe looks the same from all vantage points (homogeneity) and in all directions (isotropy).

In this chapter, we briefly survey the field of *cosmology*, the study of the universe on the large scale, including its origin, evolution, and future. For this study we must rely not only on relativity (special and general) and quantum theory, but also on fundamental results from atomic and molecular physics, statistical physics, thermodynamics, nuclear physics, and particle physics.

We begin with three discoveries that fundamentally altered our concept of the universe: it is expanding, it is filled with electromagnetic radiation, and most of its mass is mysteriously hidden from our view. We show how these discoveries have been incorporated into a theory of the origin of the universe known as the Big Bang theory. We then consider other measurements that support this theory, and we conclude with some speculations on the future of the universe.

16.1 THE EXPANSION OF THE UNIVERSE

The evidence for the expansion of the universe comes from the Doppler shift of light from distant galaxies. In Chapter 2, we obtained the expression for the relativistic Doppler shift:

$$\nu' = \nu \sqrt{\frac{1 - v/c}{1 + v/c}} \tag{16.1}$$

where now v represents the relative speed and we have assumed that the source and observer are moving away from one another. We can rewrite

Equation 16.1 in terms of wavelength

$$\lambda' = \lambda \sqrt{\frac{1 + v/c}{1 - v/c}}$$

$$= \frac{\lambda(1 + v/c)}{\sqrt{1 - v^2/c^2}} \tag{16.2}$$

Here λ' is the wavelength we measure on Earth and λ is the wavelength emitted by the moving star or galaxy in its own rest frame.

The light emitted by a star such as the Sun has a continuous spectrum. As it passes through the star's atmosphere, some of the light is absorbed by the gases in the atmosphere, so the continuous *emission* spectrum has a few dark *absorption* lines superimposed (see Figure 6.17 and Color Plate 4). Comparison between the known wavelengths of these lines (measured in laboratories at rest on the Earth) and the Doppler-shifted wavelengths allows the speed of the star to be deduced from Equation 16.2.

Of the stars in our galaxy, some are found to be moving toward us, with their light shifted toward the shorter wavelengths (blue), and others are moving away from us, with their light shifted toward the longer wavelengths (red). The average speed of these stars relative to us is about 30 km/s ($10^{-4}c$). The change in wavelength for these stars is very small. Light from nearby galaxies, those of our "local" group, again shows either small blue shifts or small red shifts.

However, when we look at the light from distant galaxies, we find it to be systematically red shifted. Some examples of these measurements are shown in Figure 16.1. We do *not* see a comparable number of red and blue shifts, as we would expect if the galaxies were in random motion. All of the galaxies beyond our local group seem to be moving away from us.*

If we accept the cosmological principle that the universe must look the same from any vantage point, we must conclude that any other observer in the universe would draw the same conclusion: *The galaxies are observed to recede from every point in the universe.*

In the 1920s, astronomer Edwin Hubble was using the 100-inch telescope on Mount Wilson in California to study the wispy nebulae. By resolving individual stars in the nebulae, Hubble was able to show that they are galaxies like the Milky Way, composed of hundreds of billions of stars. When Hubble measured the Doppler shifts of the galaxies and deduced their speeds, he made two remarkable conclusions: the galaxies are moving away from us, and *the farther away a galaxy is from us, the faster it is moving.* This proportionality between the speed of the galaxy and its distance d is known as *Hubble's law:*

$$v = Hd \tag{16.3}$$ *Hubble's law*

The proportionality constant H is known as the *Hubble parameter.*

---

* The observed red shifts are not the gravitational red shifts of general relativity discussed in Section 15.1. As we showed in Example 15.1, these red shifts are small, of the order of 10^{-6} for a typical star. The motional Doppler shifts are much larger, occasionally doubling or tripling the wavelength. The distant galaxies are not sufficiently compact to produce such large gravitational red shifts.

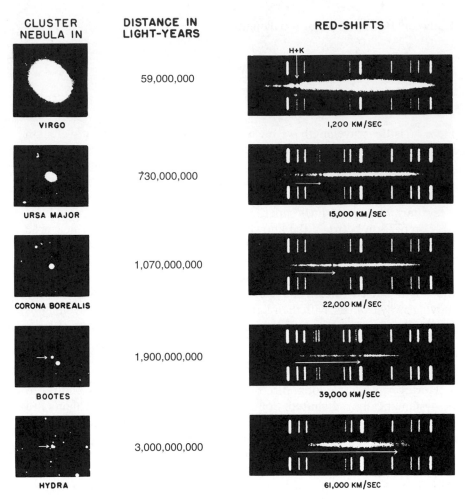

CLUSTER NEBULA IN	DISTANCE IN LIGHT-YEARS	RED-SHIFTS
VIRGO	59,000,000	1,200 KM/SEC
URSA MAJOR	730,000,000	15,000 KM/SEC
CORONA BOREALIS	1,070,000,000	22,000 KM/SEC
BOOTES	1,900,000,000	39,000 KM/SEC
HYDRA	3,000,000,000	61,000 KM/SEC

FIGURE 16.1 Red shifts for several galaxies. The spectra from the galaxies are absorption line spectra, continuous emission spectra containing dark absorption lines. For these galaxies, the two dark lines indicated by the arrows are from calcium. The emission lines above and below each absorption spectrum are for calibration. Below each spectrum is the velocity determined from the red shift. (Courtesy Hale Observatories.)

Figure 16.2 shows a plot of the deduced speeds against the distance. Although the points scatter quite a bit (due primarily to uncertainties in the distance measurements), there is a definite indication of a linear relationship. The slope of the line gives us the best value of the Hubble parameter.**

$$H = 67 \frac{\text{km/s}}{\text{Mpc}}$$

** A parsec, pc, is a measure of distance on the cosmic scale; it is the distance that corresponds to one angular second of parallax. Since parallax is due to the Earth's motion around the Sun, the parallax angle 2α is the diameter $2R$ of the Earth's orbit divided by the distance d to the star or galaxy. Thus $\alpha = R/d$ radians, which gives 1 pc = 3.26 light-years = 3.084×10^{13} km. One megaparsec, Mpc, is 10^6 pc.

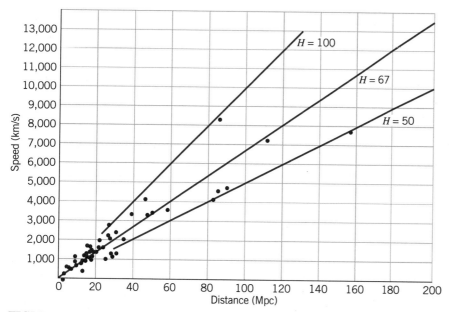

FIGURE 16.2 The velocity-distance relationship for groups and clusters of galaxies. The straight lines show the Hubble law for various values of the Hubble parameter H (in km/s·Mpc).

There is a fairly large uncertainty in the Hubble parameter—values in the range 50–100 km/s·Mpc would also fit the data.

The Hubble parameter has units of inverse time. As we show later, H^{-1} is a rough measure of the age of the universe. The best value of H gives an age of 15×10^9 y, and the uncertainty in the value of H permits ages in the range $10–19 \times 10^9$ y. If the speed of recession has been changing, the true age can be less than H^{-1}.

How does the Hubble law show that the universe is expanding? Consider the unusual universe represented by the three-dimensional coordinate system shown in Figure 16.3a, where each point represents a galaxy. With the Earth at the origin, we can determine the distance d to each galaxy. If this universe were to expand, with all the points becoming further apart, as in Figure 16.3b, the distance to each galaxy would be increased to d'. Suppose the expansion were such that every dimension increased by a constant ratio k in a time t; that is, $x' = kx$, and so forth. Then $d' = kd$, and a given galaxy moves away from us by a distance $d' - d$ in a time t, so its apparent recessional speed is

$$v = \frac{d' - d}{t}$$

$$= d\frac{k - 1}{t} \qquad (16.4)$$

If we compare two galaxies 1 and 2,

$$\frac{v_1}{v_2} = \frac{d_1}{d_2} \qquad (16.5)$$

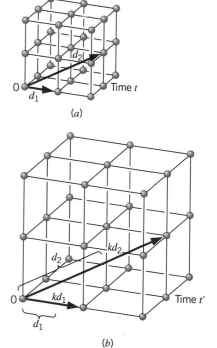

FIGURE 16.3 The expansion of a coordinate space, showing that the apparent speed of recession depends on the distance; d_2 is greater than d_1, and d_2 increases faster than d_1.

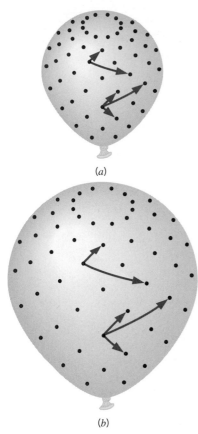

(a)

(b)

FIGURE 16.4 As a balloon is inflated, every observer on the surface observes a velocity-distance relationship of the form of the Hubble law.

a relationship identical with Hubble's law, Equation 16.3. Thus, in an expanding universe, it is perfectly natural that, the further away from us a galaxy might be, the faster we observe it to be receding.

Notice also from Figure 16.3 that this is true no matter which point we happen to choose as our origin. From *any* point in the "universe" of Figure 16.3, the other points would be observed to satisfy Equation 16.5 and thus also Hubble's law. We can further demonstrate this with two analogies. If we paint some spots on a balloon (Figure 16.4) and then inflate it, *every* spot observes every other spot to be moving away from it, and the farther away a spot is, the faster its separation grows. For a three-dimensional analogy, consider a loaf of raisin bread shown in Figure 16.5 rising in an oven. As the bread rises, every raisin observes all the others to be moving away from it, and the speed of recession varies with the separation.

The information presented here for the expansion of the universe, most of which was gathered in the 1920s and 1930s, shows that the galaxies are moving away from one another as the universe expands. We must therefore regard this expansion as experimentally established, and indeed it has been accepted by the scientific community since its introduction. There are, however, two interpretations of this expansion. (1) If the galaxies are moving apart, long ago they must have been closer together. The universe was much denser in its past history, and if we look back far enough we find a single point of infinite density. This is the "Big Bang" hypothesis, first developed by George Gamow and his colleagues. (2) The universe has always had about the same density it does now. As the galaxies move apart, additional matter is continuously created in the empty space between the galaxies, to keep the density more or less constant. This is the "Steady State" hypothesis, of astronomer Fred Hoyle and others. New galaxies created from this new matter would make the universe look the same not only from all vantage points, but also *at all times* in the present and future. (To keep the density constant, the rate of creation need be only about one hydrogen atom per cubic meter every billion years.)

Both hypotheses had their supporters, and during the 1940s and 1950s the experimental evidence did not seem to favor either one over the other. (Note that both hypotheses violate conservation of energy, the first at a single time by an infinite amount, the other at all times by a large number of very small amounts.) In the 1960s, the new field of radio astronomy revealed the presence of a universal background radiation in the microwave region, which is believed to be the remnant radiation from the Big Bang. This single observation has propelled the Big Bang theory to the forefront of cosmological models.

16.2 THE COSMIC MICROWAVE BACKGROUND RADIATION

When a gas expands adiabatically, it cools. The same is true for the universe: the expansion is accompanied by cooling. As we go back in time, we find a hotter, denser universe. Far enough back in time, the universe must have a hotter, denser universe. Far enough back in time, the universe must have been too hot for stable matter to form. Its composition was then a "gas" of particles and photons. The unstable particles eventually decayed to stable ones, and the stable particles eventually clumped together to form matter. The photons that filled the universe remained, but their wavelengths were stretched by the continuing expansion. Today those photons have a much lower temperature, but they still uniformly fill the universe.

The wavelength spectrum of those photons must have been that of a blackbody at whatever temperature T characterized the universe at that time. Although the wavelengths changed as the universe expanded, the spectrum should still be that of a blackbody, but at a much lower temperature. In the 1940s, the Big Bang cosmologists (Gamow and others) predicted that this "fireball" would today be at a temperature of the order of 5 to 10 K; such photons would have a typical energy kT of the order of 10^{-3} eV or a wavelength of order 1 mm, in the microwave region of the spectrum.

Let us first review some of the properties of blackbody radiation we discussed in Chapter 3. The wavelength spectrum is given by the Planck distribution, Equation 10.30:

$$u(\lambda)\,d\lambda = \frac{8\pi hc}{\lambda^5}\frac{1}{e^{hc/\lambda kT}-1}\,d\lambda \tag{16.6}$$

where $u(\lambda)\,d\lambda$ is the energy density (energy per unit volume) of the radiation emitted between the wavelengths λ and $\lambda + d\lambda$. The wavelength distribution for a specific temperature has a peak at λ_{max}, determined by Wien's displacement law, Equation 3.29:

$$\lambda_{max}T = 2.898 \times 10^{-3}\,\text{m·K} \tag{16.7}$$

The total radiant energy density U at all wavelengths is found from Stefan's law, Equation 3.28:

$$U = \int u(\lambda)\,d\lambda$$

$$= \frac{\sigma}{c/4}T^4 \tag{16.8}$$

where σ is the Stefan-Boltzmann constant, with a value of 5.67×10^{-8} W/m^2·K^4.

As our universe expands, the radiation from the Big Bang is stretched to longer wavelengths. If every wavelength component of the spectrum increases by a factor of f (from λ to $f\lambda$), then you can see from Equation 16.6 that the present energy spectrum remains a blackbody spectrum, but it is characterized by the lower temperature T/f. The energy density is decreased by the factor $1/f^4$.

Before discussing the experimental results, let us use Equations 16.6 to 16.8 to derive some other basic results. First we calculate the *energy* spectrum of the photons from Equation 16.6 by replacing λ with hc/E

$$u(E)\,dE = \frac{8\pi E^3}{(hc)^3}\frac{1}{e^{E/kT}-1}\,dE \tag{16.9}$$

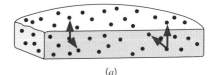
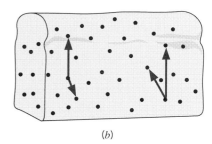

FIGURE 16.5 Another system in which the Hubble law is valid.

George Gamow (1904–1969, United States). His significant contributions to nuclear physics (theories of alpha decay, beta decay, and nuclear structure), astrophysics (nucleosynthesis, stellar structure), and cosmology (the Big Bang theory) would alone be enough to put him among the first rank of scientists, but he was also one of the most successful writers of popular science, to which he brought unusual and amusing perspectives.

Since this gives the *energy density*, we can divide by the energy E to find the *number* of photons of energy E per unit volume, $n(E)$:

$$n(E)\,dE = \frac{u(E)\,dE}{E}$$

$$= \frac{8\pi E^2}{(hc)^3} \frac{1}{e^{E/kT} - 1}\,dE \tag{16.10}$$

To find the *total* number of photons of all energies per unit volume, N, we integrate Equation 16.10 over energy:

Total number density of blackbody photons

$$N = \int_0^\infty n(E)\,dE$$

$$= \frac{8\pi}{(hc)^3} \int_0^\infty \frac{E^2\,dE}{e^{E/kT} - 1}$$

$$= \frac{8\pi}{(hc)^3}(kT)^3 \int_0^\infty \frac{x^2\,dx}{e^x - 1} \tag{16.11}$$

where we have substituted $x = E/kT$. The definite integral is a standard form and is approximately equal to 2.404. Equation 16.11 shows that the total number of photons per unit volume is proportional to the cube of the temperature, and evaluating the constants we find

$$N = (2.03 \times 10^7\,\text{photons/m}^3 \cdot \text{K}^3)T^3 \tag{16.12}$$

We can write Equation 16.8 in the same form by evaluating the constants:

$$U = (4.73 \times 10^3\,\text{eV/m}^3 \cdot \text{K}^4)T^4 \tag{16.13}$$

and the mean (average) energy per photon at temperature T is

$$E_\text{m} = \frac{U}{N} = (2.33 \times 10^{-4}\,\text{eV/K})T \tag{16.14}$$

We turn now to the experimental evidence for the existence of this microwave radiation and the determination of its temperature. From Equation 16.6 we see that the measurement of the blackbody radiant energy density at *any* wavelength is enough for a determination of the temperature T, although to demonstrate that the radiation actually has a blackbody spectrum requires measurement over a range of wavelengths.

In their 1965 experiment, Arno Penzias and Robert Wilson used a microwave antenna tuned to a wavelength of 7.35 cm. At this wavelength they recorded an annoying "hiss" from their antenna that could not be eliminated, no matter how much care they took in refining the measurement. After painstaking efforts to eliminate the "noise," they concluded that it was coming from no identifiable source and was striking their antenna from all directions, day and night, summer and winter. (The story of this discovery and of other recent discoveries in this field can be found in the references listed at the end of this chapter.) From the radiant energy at that wavelength they deduced a temperature of 3.1 ± 1.0 K, and it was later concluded that the radiation was the present remnant of the Big Bang "fireball."

Since that original experiment there have been many additional studies, at various wavelengths in the range 0.05 to 100 cm, all giving about the same temperature. The most recent measurements were made with the Cosmic Background Explorer (COBE) satellite, which was launched in 1989. Previously, no precise data were available below a wavelength of 1 cm due to atmospheric absorption. The COBE satellite was able to obtain very precise data on the intensity of the background radiation in the wavelength range between 0.05 cm and 1 cm.

The experiments are summarized in Figure 16.6. The data points fall precisely on the solid line, which is calculated from Equation 16.6 for a temperature $T = 2.735$ K. Other experiments show that the radiation has a uniform intensity in all directions. It comes from no particular source, but instead fills the universe today as it did in the early times just after the Big Bang.

Using the deduced temperature of 2.7 K, we can calculate from Equations 16.12 to 16.14 that there are about 4.0×10^8 of these photons in every cubic meter of space (there are about 10^8 of them passing through your body right now!), that they contribute to the universe an energy density of about 2.5×10^5 eV/m^3 (about half the rest energy of an electron), and that each photon has an average energy of about 0.00063 eV. The number of photons is particularly important, since for nearly all of the last 15×10^9 y the *ratio* of the number of nucleons (protons and neutrons) to photons has been almost constant. This has important consequences for the Big Bang cosmology.

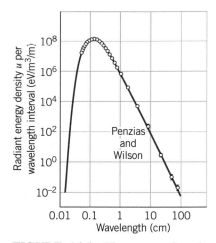

FIGURE 16.6 The wavelength spectrum of the cosmic microwave background radiation. The solid line is the Planck blackbody spectrum for $T = 2.735$ K.

EXAMPLE 16.1

The molecule CN (cyanogen) is observed in interstellar space by its absorption spectrum. From its ground state, it absorbs at a wavelength of 387.461 nm, and from the first excited rotational state of the molecule (which is 4.70×10^{-4} eV above the ground state), it absorbs at a wavelength of 387.400 nm. Measurements of the intensity of the absorption lines show that interstellar CN has about 25 percent of its molecules in the first excited state and 75 percent in the ground state. Calculate the resulting temperature of interstellar space.

SOLUTION

Figure 16.7 represents the excited states of CN and the absorption transitions. If the CN were at $T = 0$, we would expect all of the molecules to be in the ground state. At temperature T, the population of the excited state is determined by the Boltzmann factor $e^{-E/kT}$ and by the statistical weights of the levels, as in Equation 9.20. The ratio between the population N_2 of the excited state and N_1 of the ground state is then

$$\frac{N_2}{N_1} = \frac{2L_2 + 1}{2L_1 + 1} e^{-(E_2 - E_1)/kT} \tag{16.15}$$

where L_2 and L_1 are respectively the rotational quantum numbers of the excited state and the ground state. Substituting the known quantities, we

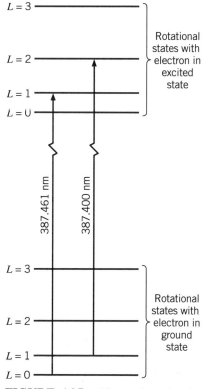

FIGURE 16.7 Absorption of radiation by cyanogen (CN).

obtain

$$\frac{0.25}{0.75} = \frac{2 \times 1 + 1}{2 \times 0 + 1} e^{-(4.70 \times 10^{-4} \text{eV})/kT}$$

Solving, we find

$$kT = 2.13 \times 10^{-4} \text{ eV}$$

or

$$T = 2.5 \text{ K}$$

This measurement clearly shows us that, although we might expect interstellar space to be at a temperature of 0 K, something is "warming" these molecules to 2.5 K. The molecules are immersed in a bath of background radiation which raises their temperature.

16.3 DARK MATTER

Figure 16.8 shows spiral galaxies that are similar to our Milky Way galaxy, in which about 10^{11} stars are bound together by the gravitational force. The diameter of a typical galaxy might be 10–50 kpc (0.3–1.5×10^{15} km). Many galaxies have this spiral structure, with a bright central region (containing most of the galaxy's mass) and several spiral arms in a flat disk.

The entire structure rotates about an axis perpendicular to the plane of the disk. The Sun, which is in one of the spiral arms of our galaxy at a distance of 8.5 kpc from the center (about $\frac{2}{3}$ of the radius of the disk), has a tangential velocity of 220 km/s. At this speed, it takes about 240 million years for a complete rotation; during the lifetime of the solar system of about 4.5 billion years, the Sun has made about 20 revolutions.

FIGURE 16.8 Spiral galaxies similar to the Milky Way, viewed from two different perspectives, one normal to the plane and one along the plane.

Because the stars in the galaxy are bound by the gravitational force, we can use Kepler's laws to analyze the motion. We assume that the gravitational force on the Sun is due primarily to the dense region at the center of the galaxy; the total mass of the other stars in the spiral arm is much smaller than the central mass, so they make a negligible contribution to the force on the Sun. Kepler's third law relates the period T of the orbit to the radius:

$$T^2 = \left(\frac{4\pi^2}{GM}\right) r^3 \qquad (16.16)$$

With $T = 2\pi r/v$, where v is the tangential velocity, we obtain

$$v = \sqrt{\frac{GM}{r}} \qquad (16.17)$$

Here M refers to the mass contained within the region of radius r. The Sun's tangential velocity suggests that a mass equivalent to 10^{11} solar masses lies within the Sun's orbit.

According to this model, we expect stars beyond the Sun to have tangential velocities that decrease with increasing radius like $r^{-1/2}$. (The planets in the solar system follow this expectation to very high precision.) However, we observe that for the rotation of the galaxy v is constant or perhaps increases slightly for stars beyond the Sun (Figure 16.9).

Other spiral galaxies show the same effect. The tangential speeds of stars in distant galaxies can be measured by the Doppler shift of their light. If we are viewing a galaxy along the plane of the disk, then one side will always be moving toward us and the other will always be moving away from us. The *difference* between the Doppler shifts of the light from the two sides of the galaxy then tells us about its rotational speed, independently of the net motion of the entire galaxy. From this measurement, we can determine how the tangential velocity of the galaxy depends on the distance from its center. A typical set of results is shown in Figure 16.10. Once again, the velocity fails to follow the expected relationship and instead remains constant throughout the visible part of the galaxy.

These results are not consistent with Kepler's law, which is based on a large central mass attracting each star toward the center of the galaxy. On the contrary, to explain a velocity that is constant as a function of radius we must have a mass M that increases linearly with r (see Equation 16.17). However, this explanation is inconsistent with visual observations of the galaxies, which clearly show that most of the light, and therefore presumably most of the mass, is concentrated in the central region.

We conclude that there must be a large quantity of invisible matter in galaxies—matter that must be present in the galaxy to supply the gravitational force, but that does not give off any light (or other electromagnetic radiation). To supply the required gravitational force, this *dark matter* must have a mass at least 10 times the mass of the visible matter in the galaxy. That is, more than 90 percent of the matter in the galaxy is in some unknown and invisible form. The dark matter may surround the galaxy in a spherical "halo" of radius several times the galaxy radius, as shown in Figure 16.11.

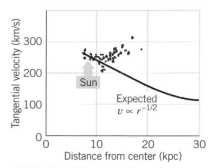

FIGURE 16.9 Tangential velocities of stars in our galaxy, determined from the Doppler shift of their light. The solid line is the prediction from Kepler's third law, Equation 16.17.

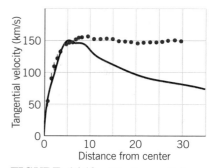

FIGURE 16.10 The tangential velocities of a distant galaxy as a function of the distance from its center. The solid line is the $r^{-1/2}$ dependence at distances beyond the central concentration of stars.

FIGURE 16.11 A representation of the suggested spherical halo of dark matter in our galaxy.

Galaxies are observed in gravitationally bound clusters of typically 100 members. The size of a cluster is perhaps 1 Mpc, about 100 times as large as a typical galaxy. Because these clusters rotate about their common center, we can perform similar measurements to compare the rotational speed of a galaxy with its distance from the center. As with individual galaxies, the clusters follow the dependence shown in Figure 16.10, suggesting that there is more matter in the clusters than we can account for from the galaxies alone. We conclude that dark matter also surrounds clusters of galaxies.

What kinds of objects make up this dark matter? Speculations about its nature are divided loosely into two categories: MACHOs (Massive Compact Halo Objects) and WIMPs (Weakly Interacting Massive Particles). Possible MACHOs include massive black holes, neutron stars, burnt-out white dwarf stars, or brown dwarf stars (Jupiter-sized objects of too small a mass to become a star). The WIMPs include neutrinos (if they have mass), magnetic monopoles, and other exotic types of stable elementary particles produced during the Big Bang but not yet observed on Earth.

Recently a gravitational lensing experiment has suggested the presence of MACHOs in our galactic halo. Gravitational lensing is an effect of general relativity, in which two objects line up with the observer, and the bending of light from the more distant object around the middle object produces an image of the distant object, just like the bending of light in passing through a lens produces an image of the object. Depending on the sizes and locations of the objects, the lens can form a double image of the distant object (see Problem 6 of Chapter 15) or it can focus the light from the distant object so that it appears to become brighter. Double imaging from a gravitational lens has been previously observed.

In 1993, there were several reported observations of brightening by a gravitational lens, in which, over the course of about a month, a star grew several times brighter and then returned to its previous brightness. The relative motion of the object, lens, and observer causes the faint image of the distant star to brighten and then dim as the lens passes between us and the star. The masses of the lens objects were estimated to be about $\frac{1}{10}$ the mass of the Sun, possibly suggesting white dwarf or brown dwarf stars (although the mass estimates are very uncertain). These experiments may be the first direct observation of dark matter in our galaxy, and they suggest that it may not be necessary to invent exotic kinds of elementary particles to account for the dark matter.

Later we discuss how dark matter may affect our theories about the evolution of the universe.

Vera Rubin (1928–, United States) An observational astronomer, she has made pioneering discoveries about the motion of stars and galaxies. By observing the Doppler shifts of stars in galaxies, she deduced that their rotational velocities are not consistent with attraction only by a large concentration of mass at the galactic center. Her work has been among the leading contributions to understanding the existence and amount of dark matter in the universe. She was awarded the National Medal of Science in 1993.

16.4 COSMOLOGY AND GENERAL RELATIVITY

General relativity can be applied to calculate the properties of the universe as a whole. For this case, the mass-energy density term in Equation 15.8 must describe the entire universe. We are not interested in the "local" variations in density on a scale of galactic size, but rather in the average density of the entire universe, evaluated over a distance that is large compared with the spacing between galaxies. (In a similar way, when we speak of the density of a solid we are interested not in the variations on the atomic scale but rather on the average density of the entire material, evaluated over a distance that is large compared with the spacing between atoms.) The density of the universe is not a constant; it changes with time as the universe expands.

Solving the equations of general relativity for the large-scale structure of the universe gives the following result, which is known as the Friedmann equation:

$$\left(\frac{dR}{dt}\right)^2 = \frac{8\pi}{3} G\rho R^2 - kc^2 \tag{16.18}$$

Here $R(t)$ represents the size or distance scale factor of the universe at time t, and ρ represents the total mass-energy density at the same time. (The density is expressed in mass units, such as kg/m^3, even if it represents radiation.)

The constant k that appears in Equation 16.18 specifies the overall geometrical structure of the universe: $k = 0$ if the universe is flat, like Figure 15.8; $k = +1$ if the universe is curved and closed, like Figure 15.9; $k = -1$ if the universe is curved and open, like Figure 15.10. When $k = +1$, the distance factor $R(t)$ is directly related to the size or "radius" of the universe, but its meaning is not so apparent when $k = 0$ or $k = -1$, since in both of the latter cases the universe is infinite in extent. In these cases $R(t)$ should be regarded as a scale factor that represents the expansion of the space; the absolute magnitude of R in this case is not significant, and only its variation with time is of interest, since any particular length (such as the distance between two galaxies) will vary with time just as R does.

To solve Equation 16.18, we must therefore specify the constant k. On the large scale, our universe seems quite close to being flat (as we discuss in Section 16.8), and we therefore take $k = 0$. This simplifies the mathematics and gives results that are not too far different from what we obtain with $k = \pm 1$, so for rough estimates our calculation should be acceptable.

The density ρ in Equation 16.18 must include both the matter and the radiation present in the universe. As we verify later, the present universe is dominated by matter; the contribution of radiation to the total density is negligible. As the universe expands, the amount of matter remains constant but the volume increases like R^3. Thus the matter density ρ_m decreases with increasing R according to $\rho_m \propto R^{-3}$. Putting this result into Equation 16.18 and integrating, we find

$$R(t) = At^{2/3} \tag{16.19}$$

where A is a constant. Using this result to eliminate R from Equation 16.18, we obtain

$$t = \frac{1}{\sqrt{6\pi G\rho_m}} \tag{16.20}$$ *Matter-dominated universe*

In contrast, the early universe was dominated by radiation; the mass density of the matter was negligible. From Equation 16.6 we see that the energy density of the radiation depends on $d\lambda/\lambda^5$. Since all wavelengths scale with R, we have $d\lambda \propto R$ and $\lambda^5 \propto R^5$. Thus the energy density of radiation ρ_r decreases with increasing R according to $\rho_r \propto R^{-4}$. Inserting this result into Equation 16.18 and integrating, we obtain

$$R(t) = At^{1/2} \tag{16.21}$$

and

Radiation-dominated universe

$$t = \sqrt{\frac{3}{32\pi G \rho_r}} \tag{16.22}$$

The Hubble parameter can be defined in terms of the time variation of the scale factor:

$$H = \frac{1}{R}\frac{dR}{dt} \tag{16.23}$$

If the universe has been expanding at a constant rate ($R \propto t$), then H^{-1} is the age of the universe. In the two cases we derived above, the age is less than H^{-1}. A matter-dominated universe expanding since $t = 0$ has an age of $\frac{2}{3}H^{-1}$, while a radiation-dominated universe has an age of $\frac{1}{2}H^{-1}$. In either case we can take H^{-1} as a rough measure of the present age.

Under the mutual gravitational interaction of the galaxies, the expansion rate should be slowing. The rate of change of the expansion depends on d^2R/dt^2, which is an acceleration. It is convenient to define the dimensionless *deceleration parameter q* as

Deceleration parameter

$$q = \frac{-R(d^2R/dt^2)}{(dR/dt)^2} \tag{16.24}$$

which depends on the deceleration d^2R/dt^2. For the universe expanding at a constant rate ($R \propto t$), $q = 0$. For the matter-dominated universe $q = \frac{1}{2}$ and for the radiation-dominated universe $q = 1$.

Differentiating Equation 16.18, keeping in mind that ρ is a function of t, and assuming that we are at present in a matter-dominated universe, we can find a relationship between the deceleration parameter and the density:

$$q = \frac{4\pi G \rho_m}{3H^2} \tag{16.25}$$

Notice that, since the term proportional to kc^2 vanishes when Equation 16.18 is differentiated, Equation 16.25 applies more generally than the special cases we have considered here.

We can therefore characterize the universe by several parameters: a shape parameter k, which describes whether it is flat or curved, open or closed; a radius or scale parameter $R(t)$, which measures the size of the universe as a function of time; the density ρ, which represents both matter and energy, and which is also a function of time; the Hubble parameter H, which is proportional to the rate of expansion; and the deceleration parameter q, which tells us the rate at which the expansion is slowing down.

The challenge to the observational astronomer is to obtain data on the distribution and motion of the stars and galaxies that can be analyzed to obtain values for these parameters.

16.5 THE BIG BANG COSMOLOGY

The present universe is characterized by a relatively low temperature and a low density of particles. Its structure and evolution are controlled by the gravitational force. Because the universe has been expanding and cooling, in the distant past it must have been characterized by a higher temperature and a greater density of particles. Let us imagine we could run the cosmic clock backward and examine the universe at earlier times, even before the formation of stars and galaxies. At some point in its history, the temperature of the universe must have been high enough to ionize atoms; at that time the universe consisted of a plasma of electrons and positive ions, and the electromagnetic force was important in determining the structure of the universe. At still earlier times, the temperature was hot enough that collisions between the ions would have knocked loose individual nucleons, so the universe consisted of electrons, protons, and neutrons, along with radiation. In this era the strong nuclear force was important in determining the evolution of the universe. At still earlier times the weak interaction played a significant role.

If we try to go back still further, we reach a time when the matter of the universe consisted only of quarks and leptons. Because we have never observed a free quark, we don't know much about their individual interactions, and so we can't describe this very early state of the universe. If someday we are able to understand the interactions of free quarks, we can penetrate this barrier and look to still earlier times. Eventually we reach a fundamental barrier when the universe had an age of only 10^{-43} s, which is known as the *Planck time* (see Problem 17). Beyond this time, quantum theory and gravity are hopelessly intertwined, and none of our present theories gives us any clue about the structure of the universe.

Later than the Planck time, but still before the condensation of bulk matter, the universe consisted of particles, antiparticles, and radiation in approximate thermal equilibrium at temperature T. The universe at this time was radiation-dominated: the energy density of the radiation exceeded the energy density of the matter. In a radiation-dominated universe, we can use Equation 16.22 to find a relationship between the temperature and the age. Inserting the radiation density from Equation 16.13, remembering to convert to mass units such that $\rho_r = U/c^2$, and evaluating all numerical factors, we obtain

$$T = \frac{1.5 \times 10^{10}\, \text{s}^{1/2} \cdot \text{K}}{t^{1/2}} \qquad (16.26)$$

where the temperature T is in K and the time t is in seconds. This equation relates the age of the early universe to its temperature.

The radiation of the early universe consisted of high-energy photons, whose average energy at the temperature T can be roughly estimated as kT, where k is the Boltzmann constant. The interactions between the radiation and the matter can be represented by two processes:

$$\text{photons} \rightarrow \text{particle} + \text{antiparticle}$$

$$\text{particle} + \text{antiparticle} \rightarrow \text{photons}$$

That is, photons can engage in pair production, in which their energy becomes the rest energy of a particle-antiparticle pair, or a particle and antiparticle can annihilate into photons. In each case, the energy of the photons must be at least as large as the rest energy of the particle and antiparticle.

EXAMPLE 16.2

(a) At what temperature is the thermal radiation in the universe energetic enough to produce nucleons and antinucleons? (b) What is the age of the universe when it cools to that temperature?

SOLUTION

(a) Let us consider the formation of proton-antiproton or neutron-antineutron pairs by photons:

$$\gamma + \gamma \longrightarrow p + \overline{p} \quad \text{and} \quad \gamma + \gamma \longrightarrow n + \overline{n}$$

To produce these reactions, the photons must have an energy at least as great as the nucleon rest energy, or about 940 MeV. The temperature of the photons must then be

$$T = \frac{E}{k} = \frac{mc^2}{k} = \frac{940\,\text{MeV}}{8.6 \times 10^{-5}\,\text{eV/K}} = 1.1 \times 10^{13}\,\text{K}$$

(b) From Equation 16.26 we can find the age of the universe when the photons have this temperature:

$$t = \left(\frac{1.5 \times 10^{10}\,\text{s}^{1/2} \cdot \text{K}}{T}\right)^2 = \left(\frac{1.5 \times 10^{10}\,\text{s}^{1/2} \cdot \text{K}}{1.1 \times 10^{13}\,\text{K}}\right)^2 = 2 \times 10^{-6}\,\text{s}$$

That is, at times earlier than 2 μs, the universe was hot enough for the photons to produce nucleon-antinucleon pairs, but after 2 μs the photons were not energetic enough to produce nucleon-antinucleon pairs. The annihilation reaction continues to occur, but after this time nucleon pair production ceases.

In this calculation we are using average photon energies as estimates. Photons in the tail of the blackbody spectrum are sufficiently energetic to produce nucleon-antinucleon pairs even after 2 μs, but *on the average* the photons have too little energy. More precisely, we could state that the rate of nucleon-antinucleon pair production drops rapidly at around 2 μs and becomes negligible at times much greater than 2 μs.

Let us now look at some of the major developments in the evolution of the universe.

$t = 10^{-6}$ s Let us begin the story at a time of 1 μs. From Equation 16.26 we find $T = 1.5 \times 10^{13}$ K or $kT = 1300$ MeV. The scale factor is smaller than that of the present universe by the red shift, 2.7 K/1.5 $\times 10^{13}$ K = 1.8×10^{-13}. If the universe is closed and finite, its radius is smaller than the present radius (10^{26} m) by this factor, so the universe is about the present size of the solar system (10^{13} m). At 1 μs, the universe consists of p, $\bar{p}$, n, $\bar{n}$, e$^-$, e$^+$, μ^-, μ^+, π^0, π^-, π^+, and perhaps other particles, plus photons, neutrinos, and antineutrinos. Because both pair production and annihilation can occur, the number of particles is roughly equal to the number of antiparticles for each species. Furthermore, the number of photons is roughly equal to the number of protons, which is in turn roughly equal to the number of electrons.

The number of protons is roughly equal to the number of neutrons. The relative number of neutrons and protons is determined by three factors:

1. *The Boltzmann factor* $e^{-\Delta E/kT}$.   Since protons have less rest energy than neutrons, there are more of them at any given temperature. The energy difference ΔE is $(m_n - m_p)c^2 = 1.3$ MeV, so the neutron-to-proton ratio can be expressed as $e^{-1.5 \times 10^{10}/T}$ for T in Kelvins. For $T \sim 10^{13}$ K, this ratio is very nearly 1, but it becomes different from 1 as T approaches 10^{10} K.

2. *Nuclear reactions.*   Reactions such as n + $\nu_e \rightleftarrows$ p + e$^-$ and n + e$^+ \rightleftarrows$ p + $\bar{\nu}_e$ can go in either direction and tend to make it easy for protons to turn into neutrons or neutrons into protons, as long as there are plenty of e$^-$, e$^+$, ν_e, and $\bar{\nu}_e$ around.

3. *Neutron decay.*   The neutron half-life is about 10 min, which is going to be important only at later times. For $t < 1$ s, there has not yet been enough time for an appreciable number of neutrons to decay.

At $t = 1$ μs, all three of these factors keep the neutron-to-proton ratio very close to 1.

$t = 10^{-2}$ s Between 10^{-6} s and 10^{-2} s, the temperature drops from 1.5×10^{13} K ($kT = 1300$ MeV) to 1.5×10^{11} K ($kT = 13$ MeV), and the distance scale factor increases by 100. Pions and muons are no longer being produced by the photons, and because their lifetimes are much shorter than 10^{-2} s, they have decayed into electrons, positrons, and neutrinos. Pair production of nucleons and antinucleons no longer occurs, but nucleon-antinucleon annihilation continues. As we discuss later, there is very slight imbalance of matter over antimatter of perhaps 1 part in 10^9. During this interval, all of the antimatter and most (99.9999999 percent) of the matter is annihilated. Pair production of electrons and positrons can still occur, so the universe consists of p, n, e⁻, e⁺, photons, and neutrinos. The neutron-to-proton ratio remains about 1.

t = **1** *s*   Between 10^{-2} s and 1 s, the temperature drops to 1.5×10^{10} K ($kT = 1.3$ MeV). In this interval, the Boltzmann factor which determines the neutron-to-proton ratio becomes different from 1; by $t = 1$ s, the nucleons consist of about 73 percent protons and 27 percent neutrons. During this period, the influence of the neutrinos has been decreasing; to convert a proton to a neutron by capturing an antineutrino ($\bar{\nu}_e + p \rightarrow n + e^+$) requires an antineutrino of at least 1.8 MeV, above the mean neutrino energy (1.3 MeV) at this temperature. This begins the time of "neutrino decoupling," when the interactions of matter and primordial neutrinos no longer occur. From this time on, the neutrinos continue to fill the universe, cooling along with the expansion of the universe. These primordial neutrinos presently have roughly the same density as the microwave photons, but a slightly lower temperature (about 2 K).

t = **6** *s*   Between 1 s and 6 s ($T = 6 \times 10^9$ K or $kT = 0.5$ MeV), the average photon energy decreases and becomes insufficient to produce electron-positron pairs. Electron-positron annihilation continues, and as a result all of the positrons and nearly all (99.9999999 percent) of the electrons are annihilated. The electrons have too little energy to convert protons to neutrons ($e^- + p \rightarrow n + \nu_e$ no longer occurs), and so the only remaining weak interaction process that influences the relative number of protons and neutrons is the radioactive decay of the neutron, which has a half-life of 10 minutes and so has not appreciably occurred by this time. The nucleons are now about 84 percent protons and 16 percent neutrons, or about 5 times as many protons as neutrons.

The composition of the universe after $t = 6$ s consists of some number N protons, the same number N electrons, and about $0.2N$ neutrons. There are no remaining positrons or antinucleons. Because particle-antiparticle annihilation has substantially reduced the number of nucleons while the number of photons remained stable, there are about $10^9 N$ photons (and about the same number of neutrinos).

EXAMPLE 16.3

Estimate the relative number of neutrons and protons among the nucleons at $t = 1$ s.

SOLUTION

At this time, the temperature is 1.5×10^{10} K. The neutron-to-proton ratio is determined by the Boltzmann factor, $e^{-\Delta E/kT}$, where ΔE is the neutron-proton rest energy difference. The exponent in the Boltzmann factor is

$$\frac{\Delta E}{kT} = \frac{1.3\,\text{MeV}}{(8.62 \times 10^{-5}\,\text{eV/K})(1.5 \times 10^{10}\,\text{K})} = 1.0$$

so the ratio of neutrons to protons is

$$\frac{N_n}{N_p} = e^{-\Delta E/kT} = e^{-1.0} = 0.37$$

The relative number of protons is then

$$\frac{N_p}{N_p + N_n} = \frac{1}{1 + N_n/N_p} = \frac{1}{1.37} = 0.73$$

The nucleons consist of 73 percent protons and 27 percent neutrons.

16.6 THE FORMATION OF NUCLEI AND ATOMS

Let us review developments in the Big Bang cosmology up to this point. (1) A hot, dense universe, full of photons and elementary particles of all varieties, has cooled to below 10^{10} K. (2) Most of the unstable particles have decayed away. (3) All of the original antimatter and most of the original matter annihilated one another, leaving a small number of protons, an equal number of electrons, and about one-fifth as many neutrons. (4) Neutrinos, which have about the same density as photons, decoupled at about 1 s and will continue cooling as the universe expands. This is the situation after $t = 6$ s.

As the neutrons and protons collide with one another, it is possible to form a deuteron (^{2}H nucleus):

$$n + p \longrightarrow d + \gamma$$

but the high density of photons can also produce the inverse reaction:

$$\gamma + d \longrightarrow n + p$$

We recall from Chapter 12 that the deuteron binding energy is 2.22 MeV. In order to have any appreciable buildup of deuterons, the photons present must first cool until their energies are below 2.22 MeV; otherwise the deuterons will be broken up as quickly as they can be formed. The energy 2.22 MeV corresponds to a temperature $T = 2.5 \times 10^{10}$ K, and we therefore might expect deuterons to be formed as soon as the temperature drops below 2.5×10^{10} K. However, this does not happen. The radiation does not have a single energy, but rather has a blackbody spectrum. A small fraction of the photons has energies *above* 2.22 MeV, and these photons continue to break apart the deuterons (Figure 16.12).

Before matter-antimatter annihilation occurred, there were about as many photons as nucleons and antinucleons, but after $t = 0.01$ s, the ratio of nucleons to photons is about 10^{-9}; about $\frac{1}{6}$ of the nucleons are neutrons.

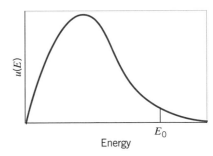

FIGURE 16.12 The blackbody radiation spectrum. The photons above $E_0 = 2.22$ MeV are energetic enough to break apart deuterium nuclei.

If the fraction of photons above 2.22 MeV is greater than $\frac{1}{6} \times 10^{-9}$, there will be at least one energetic photon per neutron, which effectively prevents deuteron formation. Our next job is to calculate to what temperature the photons must cool before fewer than $\frac{1}{6} \times 10^{-9}$ of them are above 2.22 MeV.

The Planck spectrum in terms of energy was given by Equation 16.9. We expect that the temperature must be much less than 2.5×10^{10} K, and so we are interested in the Planck distribution where $E \gg kT$, for which it is approximately

$$u(E)\, dE = \frac{8\pi E^3}{(hc)^3} e^{-E/kT}\, dE \tag{16.27}$$

That is, the distribution falls exponentially. Since each photon has energy E, the number density of photons at each energy is $u(E)dE/E$, or

$$n(E)\, dE = \frac{8\pi E^2}{(hc)^3} e^{-E/kT}\, dE \tag{16.28}$$

and the total number density above some energy E_0 is determined by integrating the number density from E_0 to ∞:

$$N_{E>E_0} = \int_{E_0}^{\infty} n(E)\, dE \tag{16.29}$$

which can be shown to be

$$N_{E>E_0} = \frac{8\pi}{(hc)^3} (kT)^3\, e^{-E_0/kT} \left[\left(\frac{E_0}{kT}\right)^2 + 2\left(\frac{E_0}{kT}\right) + 2 \right] \tag{16.30}$$

From Equation 16.11 we find the *total* number density of photons, and thus the fraction f above E_0 is $N_{E>E_0}/N$, which can be evaluated to be

$$f = 0.42 e^{-E_0/kT} \left[\left(\frac{E_0}{kT}\right)^2 + 2\left(\frac{E_0}{kT}\right) + 2 \right] \tag{16.31}$$

When $f = \frac{1}{6} \times 10^{-9}$, corresponding to the number needed to prevent deuteron formation, Equation 16.31 gives

$$\frac{E_0}{kT} \cong 28 \tag{16.32}$$

Deuteron formation

With $E_0 = 2.22$ MeV, the required temperature is about 9×10^8 K; when $T > 9 \times 10^8$ K, the number of photons with $E > 2.22$ MeV is greater than the number of neutrons, and deuteron (^{2}H) formation is prevented. When T drops below 9×10^8 K (which occurs at about $t = 250$ s), deuterons can be produced. From 6 s to 250 s, very little (except expansion and the corresponding temperature decrease) happens in the universe, but after $t = 250$ s things happen very quickly. Deuterons form and then react with the many protons and neutrons available to give

$$^2\text{H} + \text{p} \longrightarrow {}^3\text{He} + \gamma$$

and

$$^2\text{H} + \text{n} \longrightarrow {}^3\text{H} + \gamma$$

The energies of formation of these nuclei are, respectively, 5.49 MeV and 6.26 MeV, well above the 2.22 MeV threshold of the deuteron formation. If the photons are not energetic enough to break apart the deuterons, they are certainly not energetic enough to break apart ^{3}He and ^{3}H. The final steps in the formation of the heavier nuclei are

$$^3\text{He} + \text{n} \longrightarrow {}^4\text{He} + \gamma$$

and

$$^3\text{H} + \text{p} \longrightarrow {}^4\text{He} + \gamma$$

There are no stable nuclei with $A = 5$, so no further reactions of this sort are possible. Nor is it possible to have ^{4}He + ^{4}He reactions since ^{8}Be is highly unstable. (It would be possible to form stable ^{6}Li and ^{7}Li, but these are made in very small quantities relative to H and He; from Li further reactions are possible, such as ^{7}Li + ^{4}He → ^{11}B, and so forth, but these occur in still smaller quantities. The end products ^{2}H and He, along with the leftover original protons, make up about 99.9999 percent of the nuclei after the era of nuclear reactions.)

By $t = 250$ s, the original 16 percent neutrons present at $t = 6$ s had beta-decayed to about 12 percent, leaving 88 percent protons. Since most of the ^{2}H, ^{3}H, and ^{3}He were "cooked" into heavier nuclei, we can assume the universe to be composed mostly of ^{1}H and ^{4}He nuclei. Of the N nucleons present at $t = 250$ s, $0.12N$ were neutrons and $0.88N$ were protons. The $0.12N$ neutrons combined with $0.12N$ protons, forming $0.06N$ ^{4}He, and leaving $0.76N$ protons. The universe then consisted of $0.82N$ nuclei, of which 7.3 percent were ^{4}He and 92.7 percent were protons. Since helium is about four times as massive as hydrogen, by *mass* the universe is about 24 percent helium.

Helium abundance

At this point the universe begins a long and uneventful period of cooling, during which the *strong* interactions cease to be of importance.

The final step in the evolution of the primitive universe is the formation of neutral hydrogen and helium atoms from the ^{1}H, ^{2}H, ^{3}He, and ^{4}He nuclei and the free electrons. In the case of hydrogen, this takes place when the photon energy drops below 13.6 eV; otherwise any atoms that might happen to form will be immediately ionized by the radiation. There are still about 10^9 photons for every proton, and so we must wait for the radiation to cool until the fraction of photons above 13.6 eV is less than about 10^{-9}. We can use Equation 16.31 to find the value of E_0/kT for $f = 10^{-9}$, and the result is

$$\frac{E_0}{kT} = 26$$

With $E_0 = 13.6$ eV, the corresponding temperature is $T = 6070$ K, which occurs at time $t = 6.1 \times 10^{12}$ s $= 190{,}000$ y. These final estimates are actually not quite correct. We have been considering only the energy density of radiation present in the universe. As the universe cools, the contribution of the matter to the total energy density becomes more significant, and so the temperature drops more slowly than we would estimate. This contribution may increase this time by about a factor of 4 to about 700,000 y, and the radiation temperature is decreased by about a factor of 2, to $T \cong 3000 \ \mathrm{K}$.

Decoupling of electromagnetic radiation

After neutral atoms have formed, there are virtually no charged particles left in the universe, and the radiation field is not energetic enough to ionize the atoms. This is the time of the decoupling of the radiation field from the matter, and now electromagnetism, the third of the four basic forces, is no longer important in shaping the evolution of the universe. The large-scale development of the universe is from this point governed only by gravity.

The time after $t = 700,000$ y has been comparatively uneventful, at least from the point of view of cosmology. Fluctuations in the density of the hydrogen and helium triggered the condensation of galaxies, and then first-generation stars were born. Supernova explosions of the material from these stars permitted the formation of second generation systems, among which planets formed from the rocky debris.

Meanwhile, the decoupled radiation field, unaffected by the gravitational coming and going of matter, began the long journey that eventually took it, cooled again by a factor of 1000, to the radio telescopes of twentieth-century Earth.

The details of the Big Bang cosmology are summarized in Figure 16.13. It is a remarkable story, all the more so because we can understand most of its details, with the possible exception of the first instant, with nothing more than some basic theories of modern physics, most of which we can study (on a much smaller scale!) in our laboratories on Earth.

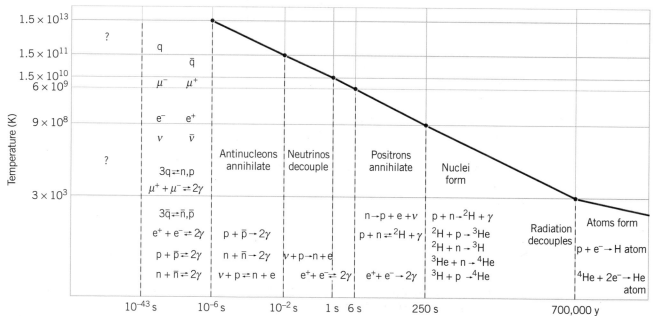

FIGURE 16.13 Evolution of the universe according to the Big Bang cosmology. The heavy solid line shows the temperature and time in the radiation-dominated era before decoupling. The most important reactions in each era are shown.

16.7 ECHOES OF THE BIG BANG

We have already discussed the 2.7-K background radiation from the Big Bang as the best single piece of experimental evidence for this particular cosmology. Yet the universe carries other memories of its birth, which can perhaps serve as other tests of the theory.

Neutrino Background   The neutrinos decoupled from matter much before the radiation, and the radiation was heated somewhat by particle-antiparticle annihilation. The neutrinos should be red shifted just like the photons, but should still be at a somewhat lower temperature than the photons, perhaps 2 K.

Neutrinos are extremely elusive particles, difficult to catch and detect, but the density of these early neutrinos ought to be about the same as the photon density, perhaps $10^8/m^3$. Detecting such neutrinos and measuring their energy spectrum and their temperature are beyond the realm of current technology, but would be a stringent test of the theory if such experiments could be done.

Helium Abundance   Much of the matter in the universe has been formed and reformed, and so has lost its "memory" of the Big Bang. There may, however, be "first generation" matter in stars and galaxies, which should show the roughly 24 percent helium abundance that characterized the formation of matter.

A variety of experiments suggests that the abundance of helium in the universe is 23 to 25 percent by mass, in excellent agreement with our rough estimate of 24 percent. These experiments include the emission of visible light from gas clouds near stars and the emission of radio waves by interstellar gas, both of which permit us to compare the amounts of hydrogen and helium present. In addition, the dynamics of stellar formation depends on the initial hydrogen and helium concentrations; present theories permit us to estimate their ratio from the observed properties of stars. The 24 percent abundance seems to be rather constant throughout the universe, as we would expect if it were predetermined by the Big Bang. (Not enough helium has been produced by nuclear fusion in stars in the last 15×10^9 y to change this ratio significantly.)

In fact (and here physics comes nearly full circle, from the very old and large to the very new and small), the early helium abundance is a function of the conditions before 10^{-6} s, when quarks and leptons filled the universe. The evolutionary rate in this era depends on the number of different kinds of quarks and leptons that can participate in reactions. It has been calculated that the helium abundance is probably not consistent with the existence of more than 3 generations of quarks and leptons. It seems remarkable that extrapolations to an unobservable state of the universe can yield such insight into the fundamental structure of matter.

Antimatter   The present model of the Big Bang assumes that there was a slight imbalance of antimatter and matter in the early universe, perhaps

1 part in 10^9. Our numerical estimates are based on *all* of the antimatter and 99.9999999 percent of the matter disappearing in mutual annihilation, with the leftover 0.0000001 percent of matter constituting the present universe. There is no evidence to indicate that there are still large quantities of antimatter present in our universe, but neither is the evidence *against* antimatter particularly strong. Whether there are antistars and antigalaxies we cannot say; we can study distant objects only by observing their light and their gravity, and since antimatter emits exactly the same light and experiences exactly the same gravity as matter, we can't tell the difference from our observations.

Where did the 10^{-9} excess of matter in the early universe come from? We really can't answer this question, but evidence gathered in particle physics experiments in 1964 may provide a clue. The decay of the neutral K meson shows interference effects between matter and antimatter, but only at the very low level of 1 part in 10^3. (J. W. Cronin and V. L. Fitch received the 1980 Nobel prize in physics for their work on this experiment.) The weak decay of the K^0 is so far the only case in which this asymmetry between matter and antimatter has been observed; all other experiments yield identical results when performed with antimatter as they do when performed with matter.

The distinction between matter and antimatter occurs at an early stage in the evolution of the universe, during the quark-antiquark era. The Grand Unified Theories (GUTs) include this asymmetry between quarks and antiquarks in a natural way, although there is as yet no accepted version of the GUTs that yields a convincing explanation for the K^0 experiment.

Mini Black Holes

Mini Black Holes   The enormous energy densities present at the earliest stages of the Big Bang could have compressed matter to a high enough density to have allowed tiny black holes to form, possibly with masses very much less than one gram. This is pure speculation, of course, since no black hole of *any* sort has been conclusively identified, but if such mini black holes exist, they might be one form of the dark matter (see Section 16.3) that is needed to explain the gravitational attraction in galaxies and clusters of galaxies, and they might also provide the "missing mass" that is necessary to "close" the universe (see Section 16.8).

Ripples in the Microwave Background

Ripples in the Microwave Background   Based on our model of the evolution of the universe according to the Big Bang theory, it is not clear why stars and galaxies form at all. A uniform mixture of radiation and particles could continue to expand indefinitely as a diffuse gas without allowing matter to achieve the densities necessary for stars to form.

When particles and antiparticles were undergoing pair production and annihilation in the early universe, there were small local fluctuations in the density, as particles disappeared from one place and appeared in another. One theory asserts that in an early era the universe experienced a period of very rapid expansion, which could have magnified these small fluctuations.

In 1992, the COBE satellite, which obtained the precise data on the spectrum of the cosmic microwave background radiation shown in Figure 16.6, observed ripples in the temperature of the background (Color Plate

18). These ripples are very small, on the order of 3×10^{-5} K, but they do suggest that in the early universe the distribution of matter and energy was nonuniform. The exact mechanism that connects the ripples with the formation of galaxies is not yet clear; one explanation is that primitive dark matter (of the WIMPs variety) collected in the "cold" locations and provided a large gravitational force that attracted the hydrogen and helium atoms after the radiation decoupling era. Whatever the correct explanation, the observation of these ripples in the temperature of the background radiation has provided us with a direct means to observe conditions in the early universe.

16.8 THE FUTURE OF THE UNIVERSE

What of the future? Does the universe go on expanding forever, or will it eventually stop its expansion and begin contraction? Will there be a Big Bang in reverse, a sort of Big Crunch, as all the matter in the universe rushes to a point, while the background radiation heats up again? Following the Big Crunch will there be another Big Bang, which will begin the evolution of a new universe? If so, this process may have been going on continuously, and the Big Bang is more like a Big Bounce.

According to Newtonian gravitation, the expansion of the universe is slowing under the influence of the gravitational attraction of its components. Whether there is enough deceleration to reverse the expansion is determined by how much mass is present. In a similar way, the curvature of spacetime in general relativity is determined by the density ρ_m of the universe; according to Equation 16.24 the deceleration parameter q, which is proportional to ρ_m, determined the rate of change of dR/dt and thus tells us how much the expansion is slowing.

The dependence of $R(t)$ on t for possible types of universe is illustrated in Figure 16.14. A relationship between k and q can be found by combining Equations 16.18 and 16.25:

$$\left(\frac{dR}{dt}\right)^2 (1 - 2q) = -kc^2 \qquad (16.33)$$

A flat universe with $k = 0$ must have $q = \frac{1}{2}$, and will continue to expand forever. The curved, open universe with $k = -1$ must have $0 < q < \frac{1}{2}$ and will likewise continue its expansion. The closed universe with $k = +1$ has $q > \frac{1}{2}$; this universe will reach a maximum radius and then begin to contract.

When we look at a distant galaxy, we see it as it existed when its light was emitted. The most distant galaxies, which are around 10 billion light-years away, emitted their light when the universe was much younger and therefore when their speed was much greater. Galaxies closer to us, on the other hand, emitted their light more recently. Comparing the light from near and distant galaxies should show evidence for deceleration. As a result of deceleration, distant galaxies should show relatively larger red shifts

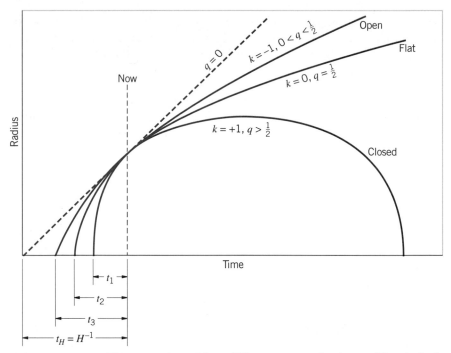

FIGURE 16.14 The expansion of four different types of universe. The dashed line represents a universe that has been expanding at a constant rate since $t = 0$, which occured a time H^{-1} before the present. The solid lines represent three different universes, all of which are younger than H^{-1}.

than we would expect from an extrapolation based on closer galaxies. Figure 16.15 shows an attempt to do this analysis; as you can see, the results are inconclusive—from these data, it is not possible to determine whether q is greater than or less than the critical value of $\frac{1}{2}$.

Another approach to this analysis is to try to determine the density of matter in the universe. From Equation 16.25, we can determine the critical value of the density corresponding to the critical value of the deceleration parameter:

Critical density
$$\rho_{cr} = \frac{3q_{cr}H^2}{4\pi G} \qquad (16.34)$$

Putting in the numbers, we obtain

$$\rho_{cr} \cong 1 \times 10^{-26} \, \text{kg/m}^3$$

If the density of the universe is less than ρ_{cr}, there is not enough matter to produce the deceleration to reverse the expansion, and the universe will continue to expand forever, like the open universe represented in Figure 16.14. If the density of the universe is greater than ρ_{cr}, there is enough matter present to reverse the expansion and close the universe. If the density equals ρ_{cr}, the universe is flat, and the expansion rate slows to zero just as the matter reaches infinite separation.

We define the ratio between the actual density of the universe and the critical value as Ω:

$$\Omega = \frac{\rho}{\rho_{cr}} \qquad (16.35)$$

The value of Ω determines the expansion properties of the universe:

$$\text{open: } \Omega < 1, \qquad \text{flat: } \Omega = 1, \qquad \text{closed: } \Omega > 1$$

From the visible galaxies, we estimate the density of matter in the universe to be

$$\rho_{gal} = 3 \times 10^{-28} \text{ kg/m}^3$$

so that from the visible matter alone we would estimate $\Omega = 0.03$, far below the value needed to close the universe. From our discussion in Section 16.3, we know that the dark matter in the universe may have a mass about 10 times that of the visible matter; including the dark matter we might have $\Omega = 0.3$. If the universe is flat or closed, as many astrophysicists believe, there is a considerable quantity of "missing mass" equivalent to at least 2 times the mass of the dark matter.

One possible form for this missing mass is neutrinos or other stable, weakly interacting particles that may have been produced in the Big Bang. Although we often assume the neutrinos to have zero mass, the experimental upper limits on the rest energies are not particularly small:

$$\nu_e: mc^2 < 10 \text{ eV}, \qquad \nu_\mu: mc^2 < 0.3 \text{ MeV}, \qquad \nu_\tau: mc^2 < 40 \text{ MeV}$$

If the density of neutrinos from the Big Bang is about the same as the density of photons ($4 \times 10^8/\text{m}^3$), then a neutrino mass of 10 eV/c^2 would lead to a density of about 7×10^{-27} kg/m^3. Combining the neutrinos with the galaxies and the dark matter, we obtain $\Omega \approx 1$.

This estimate for Ω is very rough, since we don't know the exact contributions of the dark matter or the neutrinos, or whether there are other relic particles from the Big Bang that may contribute (magnetic monopoles, mini black holes, WIMPs, and so forth). It seems safe to conclude that Ω is within an order of magnitude of 1 and may well be close to 1.

This conclusion poses a difficulty for astrophysicists called the *flatness problem*. Why is the universe so close to being flat? It seems coincidental that, of all the possible universes that might have evolved from the Big Bang, ours appears to be very nearly flat.

Actually, the flatness problem is more serious. By way of analogy, consider a projectile that is thrown upward from the surface of the Earth. The parameter Ω in effect measures the ratio between the gravitational potential energy and the kinetic energy: $\Omega = |U_{grav}|/K$. If the initial value of Ω is greater than 1, the gravitational energy exceeds the kinetic energy, so the projectile will rise to a maximum height and then fall back to Earth. When it reaches its maximum height, $K = 0$ and Ω becomes infinite. During the entire ascent, the value of Ω increases because the kinetic energy decreases more rapidly than the magnitude of the gravitational energy. If the projectile

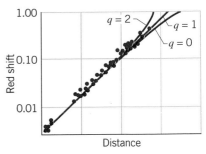

FIGURE 16.15 The Hubble law at large distances and red shifts. The red shift is $\Delta\lambda/\lambda$, which is linear in v only for small v.

is launched so that $\Omega < 1$, there is more than enough kinetic energy to overcome the Earth's gravity, and the projectile will escape the pull of the Earth. When it reaches infinite separation, $\Omega = 0$ because $U_{grav} = 0$. If we choose the initial velocity such that $\Omega = 1$, there is just enough energy to escape, and the projectile reaches infinite separation with $K = 0$. Throughout the entire journey, Ω remains exactly 1.

For the projectile as well as for the evolution of the universe, the conclusions are identical: If $\Omega = 1$ initially, it remains exactly 1 always, but if either $\Omega > 1$ or $\Omega < 1$, it grows further from 1. If the early universe had $\Omega = 1.000001$, after the passage of 15 billion years Ω would have grown very large; similarly, if the initial value of Ω were 0.999999, by now it would be very close to 0. It has been calculated that for Ω to be within an order of magnitude of 1 today (that is, between 0.1 and 10), it must have initially been in the range 1 ± 10^{-59}.

The explanation for the flatness of the universe is not clear. One possibility is the *inflationary model*, which suggests that the universe grew very rapidly (by perhaps 50 orders of magnitude) in a short interval of time between 10^{-35} s and 10^{-32} s. This is the instant at which the electroweak force and the strong force separated from their previous unification and became distinct forces. The sudden growth is due to a sort of phase transition that accompanies this separation. Just as the surface of a curved balloon suddenly inflated to 10^{50} times its size would appear flat, so the curved universe appeared very nearly flat after the inflation. Here again cosmology and particle physics have come together: When we understand how Grand Unified Theories (GUTs) affect the properties of elementary particles, we shall be better able to understand the early universe.

Although we have come very far in the twentieth century in our understanding of the basic physics of our universe, it is frustrating not to be able to determine which of two fates awaits us: (1) The universe expands forever; all stars and galaxies eventually use up their energy and become black dwarves or black holes. The universe becomes cold and dark, and all life ends. (2) The expansion slows and eventually stops; the galaxies begin moving closer together as the universe contracts. Eventually the universe collapses to a single point, possibly to be reborn again in a new Big Bang with a new set of physical laws. It is both frightening and inspiring to realize that we may be unique in the universe, and that no other generation of scientists may have this opportunity to live at a time when the fundamental laws of the universe begin to be explored and understood.

SUGGESTIONS FOR FURTHER READING

Many of the reference books on general relativity listed in Chapter 15 also include material on cosmology. Some nonmathematical introductory books on cosmology are:

T. Ferris, *The Red Limit* (New York, Wm. Morrow & Co., 1977). Certainly among the best-written works on *any* scientific topic; rich in historical and personal detail.

W. J. Kaufmann III, *Relativity and Cosmology* (New York, Harper and Row, 1977). Another well-written general work; includes general relativity and black holes.

L. Krauss, *The Fifth Essence* (New York, Basic Books, 1989). The story of dark matter.

A. Lightman, *Ancient Light* (Boston, Harvard University Press, 1991). An introduction to the people who have made the major discoveries.

S. Weinberg, *The First Three Minutes* (New York, Basic Books, 1977). A modern classic; if you read only one book on cosmology, choose this one.

Many of the above are available in paperback. There are several works on cosmology which include mostly calculus-level math (no tensors):

P. T. Landsberg and D. A. Evans, *Mathematical Cosmology* (Oxford, Clarendon Press, 1977). A superior nontensor mathematical treatment.

P. J. E. Peebles, *Principles of Physical Cosmology* (Princeton University Press, 1993). Emphasizes observational evidence for the Big Bang.

M. Rowan-Robinson, *Cosmology*, 2nd ed. (Oxford, Clarendon Press, 1981). Includes summaries of other models and discusses some of the unsolved problems of cosmology.

D. W. Sciama, *Modern Cosmology* (Cambridge, Cambridge University Press, 1971). Compact and highly readable; touches all aspects of cosmology.

E. R. Harrison, *Cosmology* (Cambridge, Cambridge University Press, 1981). A comprehensive and well organized survey.

A more challenging work, but still very readable and containing much information on observations:

J. V. Narlikar, *Introduction to Cosmology*, 2nd ed. (Cambridge, Cambridge University Press, 1993).

A few popular articles:

J. R. Gott III, J. E. Gunn, D. N. Schramm, and B. M. Tinsley, "Will the Universe Expand Forever?," *Scientific American* **234**, 62 (March 1976).

R. A. Muller, "The Cosmic Background Radiation and the New Aether Drift," *Scientific American* **238**, 64 (May 1978).

J. D. Barrow and J. Silk, "The Structure of the Early Universe," *Scientific American* **242**, 118 (April 1980).

W. L. Freedman, "The Expansion Rate and Size of the Universe," *Scientific American* **267**, 54 (November 1992).

D. E. Osterbrock, J. A. Gwinn, and R. S. Brashear, "Edwin Hubble and the Expanding Universe," *Scientific American* **269**, 84 (July 1993).

QUESTIONS

1. Can we look out into the distant universe without also looking back into time?

2. We have used throughout this chapter an age of the universe of about 15×10^9 y and a Hubble constant of about 67 km/s·Mpc. Are these numbers consistent?

3. Is Hubble's parameter a constant? Does it vary over large distances of space? Over long intervals of time?

4. Explain why the age of the universe must be less than $1/H$.

5. Why is it difficult to obtain precise values for the Hubble parameter and the deceleration parameter?

6. All natural processes are governed by the rule that the entropy must increase; the increase of entropy, as the universe "runs down," defines for us a direction of time. If the universe begins to contract and therefore to heat up, will the entropy of natural processes therefore decrease? Will the inhabitants of that universe observe time to be running backward?

7. The hydrogen in the universe contains a small fraction of deuterium. Assuming the deuterium originated in the Big Bang, what era of the Big Bang would we learn about by measuring the deuterium abundance? Can we accomplish this measurement using terrestrial hydrogen? What properties of deuterium could we use to determine its presence in distant regions of the galaxy?

8. Between $t = 1$ s and $t = 6$ s, the neutron fraction should drop from 27 to 8 percent; instead it drops only to about 16 percent. Why don't more neutrons turn into protons during this era? Is it as difficult for protons to turn into neutrons?

9. If we were able to observe the neutrinos from the early universe, would they have a spectrum determined by the Planck distribution?

PROBLEMS

1. Use Hubble's law to estimate the wavelength of the 590.0 nm sodium line as observed emitted from galaxies whose distance from us is (a) 1.0×10^6 light-years; (b) 1.0×10^8 light-years; (c) 1.0×10^{10} light-years.

2. Find the peak wavelength of the 2.7 K blackbody radiation.

3. (a) Differentiate Equation 16.9 to find the *energy* at which the maximum of the radiation spectrum occurs. (b) Evaluate the peak photon energy of the 2.7-K microwave background.

4. Evaluate the numerical constants in Equations 16.8 and 16.11 in order to obtain Equations 16.12, 16.13, and 16.14.

5. Photons of visible light have energies between about 2 and 3 eV. (a) Compute the number density of photons from the 2.7-K background radiation in that interval. (Use Equation 16.30 to find $N_{>2eV}$ and $N_{>3eV}$.) Would such photon densities be visible to the eye? (b) Assume the eye could detect about 100 photons/cm^3. At what temperature would the background radiation be visible? At what age of the universe would this have occurred?

6. The first rotational state of cyanogen is at an energy of 4.70×10^{-4} eV above the ground state. Compute the relative populations of the ground state and the first three rotational states at $T = 2.7$ K.

7. By differentiating Equation 16.18 with respect to t, derive Equation 16.25. (*Hint*: Don't forget that ρ and R are both functions of t.)

8. Expressing R as At^n, show that $q = n^{-1} - 1$. Evaluate q for the matter-dominated and radiation-dominated universes.

9. Derive Equation 16.26.

10. At what age did the universe cool below the threshold temperature for (a) nucleon production; (b) pi meson production?

11. (a) At what temperature was the universe hot enough to permit the photons to produce K mesons ($mc^2 = 500$ MeV)? (b) At what age did the universe have this temperature?

12. Derive Equations 16.30 and 16.31.

13. Consider the universe at a temperature of 5000 K. (a) At what age did this occur, and during which stage of the evolution of the universe? (b) Evaluate the average photon energy at that time. (c) If there are 10^9 photons per nucleon, evaluate the ratio between the radiation density and the mass density at that time.

14. The early universe was radiation dominated, and the present universe is matter dominated. (a) At what temperature were the radiation and matter densities equal? (b) What was the age of the universe when this occurred?

15. Suppose the *number density* of neutrinos from the Big Bang were the same as the present number density of photons. Find the rest energy of the neutrinos that would provide the critical density needed to close the universe.

16. Suppose the difference between matter and antimatter in the early universe were 1 part in 10^8 instead of 1 part in 10^9. (a) From Equation 16.31 evaluate the temperature at which deuterium begins to form. (b) At what age does this occur? (c) Evaluate the temperature and the corresponding time of radiation decoupling (when hydrogen atoms form).

17. Because we don't yet have a quantum theory of gravity, we cannot analyze the properties of the universe before the Planck time, about 10^{-43} s. If we assume that the properties of the universe during that era were determined by quantum theory, relativity, and gravity, the Planck time should be characterized by the fundamental constants of those three theories: h, c, and G. We can therefore write $t \propto h^i c^j G^k$, where i, j, and k are exponents to be determined. (a) Do a dimensional analysis to determine i, j, and k. (b) Assuming the proportionality parameter is of order unity, evaluate t.

18. The Hubble parameter could be as low as 50 km/s·Mpc or as high as 100 km/s·Mpc. Compute the critical density necessary to close the universe for these two extremes.

19. Derive Equation 16.33.

20. Suppose all the visible matter were distributed uniformly throughout the universe. On the average, how many hydrogen atoms would be found per cubic meter?

21. Suppose the universe were composed of uniformly distributed stars of the mass of the Sun (6.0×10^{30} kg). If the average density were the same as that of the visible matter in the present universe, what would be the spacing between the stars? Express your answer in light-years.

CONSTANTS
AND
CONVERSION
FACTORS

CONSTANTS*

Speed of light	c	2.99792458×10^8 m/s
Charge of electron	e	$1.6021773 \times 10^{-19}$ C
Boltzmann constant	k	1.38066×10^{-23} J/K
		8.6174×10^{-5} eV/K
Planck's constant	h	6.626076×10^{-34} J·s
		4.135670×10^{-15} eV·s
	$\hbar = h/2\pi$	1.054573×10^{-34} J·s
		6.582122×10^{-16} eV·s
Gravitational constant	G	6.6726×10^{-11} N·m^2/kg^2
Avogadro's constant	N_A	6.022137×10^{23} mole^{-1}
Universal gas constant	R	8.31451 J/mole·K
Stefan Boltzmann constant	σ	5.6705×10^{-8} W/m^2·K^4
Rydberg constant	R_∞	1.0973731571×10^7 m^{-1}
Hydrogen ionization energy		13.605698 eV
Bohr radius	a_0	$5.2917725 \times 10^{-11}$ m
Bohr magneton	μ_B	$9.2740154 \times 10^{-24}$ J/T
		5.7883826×10^{-5} eV/T
Nuclear magneton	μ_N	$5.0507865 \times 10^{-27}$ J/T
		3.1524517×10^{-8} eV/T
Fine structure constant	α	$1/137.035989$
	hc	1239.8424 eV·nm (MeV·fm)
	$e^2/4\pi\varepsilon_0$	1.439965 eV·nm (MeV·fm)

* The number of significant figures given for the numerical constants indicates the precision to which they have been determined; there is an experimental uncertainty, typically of a few parts in the last digit, except for the speed of light (which is exact).

SOME PARTICLE MASSES

	kg	u	MeV/c^2
Electron	$9.1093897 \times 10^{-31}$	5.485798×10^{-4}	0.5109991
Proton	$1.6726231 \times 10^{-27}$	1.007276470	938.2723
Neutron	1.674955×10^{-27}	1.008664924	939.5656
Deuteron	3.343586×10^{-27}	2.01355323	1875.6134
Alpha	6.644662×10^{-27}	4.00150618	3727.3803

CONVERSION FACTORS

$1 \text{ eV} = 1.6021773 \times 10^{-19} \text{ J}$

$1 \text{ u} = 931.4943 \text{ MeV}/c^2$
$\quad = 1.6605402 \times 10^{-27} \text{ kg}$

$1 \text{ y} = 3.156 \times 10^7 \text{ s} \cong \pi \times 10^7 \text{ s}$

$1 \text{ Å} = 10^{-10} \text{ m}$

$1 \text{ b} = 10^{-28} \text{ m}^2$

$1 \text{ Ci} = 3.7 \times 10^{10} \text{ decays/s}$

$1 \text{ light-year} = 9.46 \times 10^{15} \text{ m}$

$1 \text{ parsec} = 3.26 \text{ light-year}$

TABLE OF ATOMIC MASSES

The table gives the atomic masses of some isotopes of each element. All naturally occurring stable isotopes are included (with their natural abundances shown in italics in the last column). Some of the longer-lived radioactive isotopes of each element are also included, with their half-lives. Each element has many other radioactive isotopes that are not included in this table. More complete listings can be found in the sources from which this table was derived: *Table of Isotopes* (Seventh Edition), edited by C. M. Lederer and V. S. Shirley (New York, Wiley, 1978); G. Audi and H. Wapstra, "The 1993 Atomic Mass Evaluation," *Nuclear Physics* A**565**, 1 (1993).

In the half-life column, $My = 10^6$ y.

	Z	A	Atomic mass (u)	*Abundance or Half-life*		Z	A	Atomic mass (u)	*Abundance or Half-life*
H	1	1	1.007825	*99.985%*	Na	11	21	20.997655	22.5 s
		2	2.014102	*0.015%*			22	21.994437	2.61 y
		3	3.016049	12.3 y			23	22.989770	*100%*
He	2	3	3.016029	*1.38 × 10⁻⁴%*			24	23.990963	15.0 h
		4	4.002603	*99.99986%*			25	24.989954	59 s
Li	3	6	6.015122	*7.5%*			26	25.992590	1.1 s
		7	7.016004	*92.5%*	Mg	12	22	21.999574	3.86 s
		8	8.022486	0.84 s			23	22.994125	11.3 s
Be	4	7	7.016929	53.3 d			24	23.985042	*78.99%*
		8	8.005305	0.07 fs			25	24.985837	*10.00%*
		9	9.012182	*100%*			26	25.982593	*11.01%*
		10	10.013534	1.5 My			27	26.984341	9.46 m
		11	11.021658	13.8 s			28	27.983877	20.9 h
B	5	8	8.024607	0.77 s	Al	13	25	24.990429	7.18 s
		9	9.013329	0.85 as			26	25.986892	0.74 My
		10	10.012937	*19.9%*			27	26.981538	*100%*
		11	11.009306	*80.1%*			28	27.981910	2.24 m
		12	12.014352	20.2 ms			29	28.980445	6.56 m
C	6	10	10.016853	19.2 s	Si	14	26	25.992330	2.23 s
		11	11.011433	20.4 m			27	26.986704	4.16 s
		12	12.000000	*98.90%*			28	27.976926	*92.23%*
		13	13.003355	*1.10%*			29	28.976495	*4.67%*
		14	14.003242	5730 y			30	29.973770	*3.10%*
		15	15.010599	2.45 s			31	30.975363	2.62 h
N	7	13	13.005739	9.96 m			32	31.974148	172 y
		14	14.003074	*99.63%*	P	15	29	28.981801	4.14 s
		15	15.000109	*0.366%*			30	29.978314	2.50 m
		16	16.006100	7.1 s			31	30.973761	*100%*
		17	17.008450	4.2 s			32	31.973907	14.3 d
O	8	14	14.008595	71 s			33	32.971725	25.3 d
		15	15.003066	122 s	S	16	30	29.984903	1.18 s
		16	15.994915	*99.76%*			31	30.979555	2.57 s
		17	16.999132	*0.038%*			32	31.972071	*95.02%*
		18	17.999160	*0.204%*			33	32.971459	*0.75%*
		19	19.003577	26.9 s			34	33.967867	*4.21%*
		20	20.004076	13.6 s			35	34.969032	87.5 d
F	9	17	17.002095	64.5 s			36	35.967081	*0.017%*
		18	18.000938	110 m			37	36.971126	5.05 m
		19	18.998403	*100%*	Cl	17	33	32.977452	2.51 s
		20	19.999981	11 s			34	33.973763	1.53 s
		21	20.999949	4.2 s			35	34.968853	*75.77%*
Ne	10	18	18.005710	1.7 s			36	35.968307	0.30 My
		19	19.001880	17.2 s			37	36.965903	*24.23%*
		20	19.992440	*90.48%*			38	37.968011	37.2 m
		21	20.993847	*0.27%*			39	38.968009	55.6 m
		22	21.991386	*9.25%*					
		23	22.994467	37.2 s					
		24	23.993615	3.4 m					

	Z	A	Atomic mass (u)	Abundance or Half-life		Z	A	Atomic mass (u)	Abundance or Half-life
Ar	18	34	33.980270	0.844 s	Cr	24	48	47.954036	21.6 h
		35	34.975257	1.78 s			49	48.951341	42.3 m
		36	35.967546	*0.337%*			50	49.946050	*4.35%*
		37	36.966776	35.0 d			51	50.944772	27.7 d
		38	37.962732	*0.063%*			52	51.940512	*83.79%*
		39	38.964313	269 y			53	52.940653	*9.50%*
		40	39.962383	*99.60%*			54	53.938885	*2.36%*
		41	40.964501	1.82 h			55	54.940844	3.50 m
		42	41.963050	32.9 y			56	55.940645	5.94 m
K	19	37	36.973377	1.23 s	Mn	25	52	51.945570	5.59 d
		38	37.969080	7.64 m			53	52.941294	3.7 My
		39	38.963707	*93.26%*			54	53.940363	312 d
		40	39.963999	1.28 Gy			55	54.938049	*100%*
		41	40.961826	*6.73%*			56	55.938909	2.58 h
		42	41.962403	12.4 h			57	56.938287	87.2 s
		43	42.960716	22.3 h	Fe	26	52	51.948116	8.27 h
Ca	20	38	37.976319	0.44 s			53	52.945312	8.51 m
		39	38.970718	0.86 s			54	53.939615	*5.9%*
		40	39.962591	*96.94%*			55	54.938298	2.73 y
		41	40.962278	0.103 My			56	55.934942	*91.7%*
		42	41.958618	*0.647%*			57	56.935398	*2.15%*
		43	42.958767	*0.135%*			58	57.933280	*0.28%*
		44	43.955481	*2.09%*			59	58.934880	44.5 d
		45	44.956186	164 d			60	59.934077	1.5 My
		46	45.953693	*0.0035%*			61	60.936749	6.0 m
		47	46.954546	4.54 d	Co	27	57	56.936296	272 d
		48	47.952533	*0.187%*			58	57.935757	70.8 d
		49	48.955673	8.72 m			59	58.933200	*100%*
Sc	21	43	42.961151	3.89 h			60	59.933822	5.27 y
		44	43.959403	3.93 h			61	60.932479	1.65 h
		45	44.955910	*100%*	Ni	28	56	55.942136	6.10 d
		46	45.955170	83.8 d			57	56.939800	35.7 h
		47	46.952408	3.35 d			58	57.935348	*68.1%*
		48	47.952235	43.7 h			59	58.934351	0.075 My
Ti	22	44	43.959690	49 y			60	59.930790	*26.2%*
		45	44.958124	3.08 h			61	60.931060	*1.14%*
		46	45.952630	*8.0%*			62	61.928348	*3.63%*
		47	46.951764	*7.3%*			63	62.929673	100 y
		48	47.947947	*73.8%*			64	63.927969	*0.93%*
		49	48.947871	*5.5%*			65	64.930088	2.52 h
		50	49.944792	*5.4%*	Cu	29	61	60.933462	3.35 h
		51	50.946616	5.76 m			62	61.932587	9.74 m
		52	51.946898	1.7 m			63	62.929601	*69.2%*
V	23	48	47.952254	16.0 d			64	63.292768	12.7 h
		49	48.948517	338 d			65	64.927794	*30.8%*
		50	49.947163	*0.250%*			66	65.928873	5.10 m
		51	50.943964	*99.750%*			67	66.927750	61.9 h
		52	51.944780	3.75 m					
		53	52.944342	1.61 m					

	Z	A	Atomic mass (u)	*Abundance or Half-life*		Z	A	Atomic mass (u)	*Abundance or Half-life*
Zn	30	62	61.934334	9.19 h	Br	35	77	76.921380	57.0 h
		63	62.933215	38.5 m			78	77.921146	6.46 m
		64	63.929146	*48.6%*			79	78.918338	*50.69%*
		65	64.929245	244 d			80	79.918530	17.7 m
		66	65.926036	*27.9%*			81	80.916291	*49.31%*
		67	66.927131	*4.10%*			82	81.916805	35.3 h
		68	67.924847	*18.8%*			83	82.915181	2.40 h
		69	68.926553	56 m	Kr	36	76	75.925950	14.8 h
		70	69.925325	*0.62%*			77	76.924669	74.4 m
		71	70.927727	2.45 m			78	77.920388	*0.356%*
							79	78.920083	35.0 h
Ga	31	67	66.928205	3.26 d			80	79.916379	*2.25%*
		68	67.927983	67.6 m			81	80.916593	0.213 My
		69	68.925581	*60.1%*			82	81.913485	*11.6%*
		70	69.926027	21.1 m			83	82.914137	*11.5%*
		71	70.924707	*39.9%*			84	83.911508	*57.0%*
		72	71.926372	14.1 h			85	84.912530	10.8 y
		73	72.925170	4.86 h			86	85.910615	*17.3%*
							87	86.913359	76.3 m
Ge	32	68	67.928097	271 d	Rb	37	83	82.915114	86.2 d
		69	68.927972	39.0 h			84	83.914387	32.8 d
		70	69.924250	*21.2%*			85	84.911792	*72.17%*
		71	70.924954	11.4 d			86	85.911170	18.6 d
		72	71.922076	*27.7%*			87	86.909186	*27.83%*
		73	72.923460	*7.7%*			88	87.911323	17.8 m
		74	73.921178	*35.9%*	Sr	38	82	81.918404	25.6 d
		75	74.922860	82.8 m			83	82.917557	32.4 d
		76	75.921403	*7.4%*			84	83.913426	*0.56%*
		77	76.923549	11.3 h			85	84.912936	64.8 d
							86	85.909265	*9.8%*
As	33	73	72.923825	80.3 d			87	86.908882	*7.0%*
		74	73.923929	17.8 d			88	87.905617	*82.6%*
		75	74.921597	*100%*			89	88.907455	50.5 d
		76	75.922394	26.3 h			90	89.907738	29.1 y
		77	76.920648	38.8 h	Y	39	87	86.910880	79.8 h
							88	87.909506	106.6 d
Se	34	72	71.927112	8.4 d			89	88.905849	*100%*
		73	72.926767	7.1 h			90	89.907152	64.1 h
		74	73.922477	*0.89%*			91	90.907301	58.5 d
		75	74.922524	120 d	Zr	40	88	87.910225	83.4 d
		76	75.919214	*9.4%*			89	88.908889	78.4 h
		77	76.919915	*7.6%*			90	89.904702	*51.5%*
		78	77.917310	*23.8%*			91	90.905643	*11.2%*
		79	78.918500	<0.065 My			92	91.905039	*17.1%*
		80	79.916522	*49.6%*			93	92.906474	1.53 My
		81	80.917993	18.5 m			94	93.906314	*17.4%*
		82	81.916700	*8.7%*			95	94.908041	64.0 d
		83	82.919119	22.3 m			96	95.908275	*2.80%*
							97	96.910950	16.9 h

	Z	A	Atomic mass (u)	Abundance or Half-life		Z	A	Atomic mass (u)	Abundance or Half-life
Nb	41	91	90.906989	680 y	Pd	46	100	99.908505	3.63 d
		92	91.907192	35 My			101	100.908289	8.47 h
		93	92.906376	*100%*			102	101.905607	*1.0%*
		94	93.907282	0.020 My			103	102.906087	17.0 d
		95	94.906834	35.0 d			104	103.904034	*11.1%*
Mo	42	90	89.913935	5.67 h			105	104.905083	*22.3%*
		91	90.911749	15.5 m			106	105.903484	*27.3%*
		92	91.906810	*14.8%*			107	106.905129	6.5 My
		93	92.906811	3500 y			108	107.903895	*26.5%*
		94	93.905087	*9.3%*			109	108.905954	13.7 h
		95	94.905841	*15.9%*			110	109.905153	*11.7%*
		96	95.904678	*16.7%*			111	110.907640	23.4 m
		97	96.906020	*9.6%*					
		98	97.905407	*24.1%*					
		99	98.907711	65.9 h					
		100	99.907476	*9.6%*					
		101	100.910346	14.6 m					
Tc	43	95	94.907656	20.0 h	Ag	47	105	104.906528	41.3 d
		96	95.907870	4.3 d			106	105.906667	24.0 m
		97	96.906364	2.6 My			107	106.905093	*51.4%*
		98	97.907215	4.2 My			108	107.905954	2.37 m
		99	98.906254	0.211 My			109	108.904756	*48.16%*
		100	99.907657	15.8 s			110	109.906111	24.6 s
Ru	44	94	93.911365	51.8 m					
		95	94.910418	1.64 h					
		96	95.907604	*5.5%*					
		97	96.907560	2.88 d	Cd	48	104	103.909848	57.7 m
		98	97.905287	*1.86%*			105	104.909468	55.5 m
		99	98.905939	*12.7%*			106	105.906458	*1.25%*
		100	99.904219	*12.6%*			107	106.906614	6.50 h
		101	100.905582	*17.1%*			108	107.904183	*0.89%*
		102	101.904349	*31.6%*			109	108.904985	462 d
		103	102.906323	39.3 d			110	109.903006	*12.5%*
		104	103.905430	*18.6%*			111	110.904182	*12.8%*
		105	104.907750	4.44 h			112	111.902758	*24.1%*
							113	112.904401	*12.2%*
							114	113.903359	*28.7%*
							115	114.905431	53.5 h
							116	115.904756	*7.5%*
							117	116.907219	2.49 h
Rh	45	101	100.906613	3.3 y	In	49	111	110.905112	2.80 d
		102	101.906842	2.9 y			112	111.905534	15.0 m
		103	102.905504	*100%*			113	112.904062	*4.3%*
		104	103.906655	42.3 s			114	113.904918	71.9 s
		105	104.905692	35.4 h			115	114.903879	*95.7%*
							116	115.905261	14.1 s

	Z	A	Atomic mass (u)	Abundance or Half-life		Z	A	Atomic mass (u)	Abundance or Half-life
Sn	50	110	109.907854	4.11 h	Xe	54	122	121.908560	20.1 h
		111	110.907736	35.3 m			123	122.908477	2.08 h
		112	111.904822	*0.97%*			124	123.905895	*0.096%*
		113	112.905174	115.1 d			125	124.906398	16.9 h
		114	113.902783	*0.65%*			126	125.904268	*0.090%*
		115	114.903347	*0.36%*			127	126.905179	36.4 d
		116	115.901745	*14.5%*			128	127.903531	*1.91%*
		117	116.902955	*7.68%*			129	128.904780	*26.4%*
		118	117.901608	*24.2%*			130	129.903509	*4.1%*
		119	118.903311	*8.6%*			131	130.905083	*21.2%*
		120	119.902199	*32.6%*			132	131.904155	*26.9%*
		121	120.904239	27.1 h			133	132.905906	5.24 d
		122	121.903441	*4.63%*			134	133.905395	*10.4%*
		123	122.905723	129 d			135	134.907208	9.14 h
		124	123.905275	*5.79%*			136	135.907220	*8.9%*
		125	124.907785	9.64 d			137	136.911563	3.82 m
		126	125.907654	0.1 My					
Sb	51	119	118.903948	38.1 h	Cs	55	131	130.905460	9.69 d
		120	119.905076	15.9 m			132	131.906430	6.47 d
		121	120.903822	*57.4%*			133	132.905447	*100%*
		122	121.905180	2.70 d			134	133.906714	2.06 y
		123	122.904216	*42.6%*			135	134.905972	2.3 My
		124	123.905938	60.2 d			136	135.907307	13.2 d
		125	124.905247	2.73 y					
Te	52	118	117.905832	6.00 d	Ba	56	128	127.908309	2.43 d
		119	118.906410	16.1 h			129	128.908675	2.23 h
		120	119.904026	*0.095%*			130	129.906311	*0.106%*
		121	120.904935	16.8 d			131	130.906931	11.8 d
		122	121.903056	*2.59%*			132	131.905056	*0.101%*
		123	122.904271	*0.91%*			133	132.906003	10.5 y
		124	123.902819	*4.79%*			134	133.904504	*2.42%*
		125	124.904424	*7.12%*			135	134.905684	*6.59%*
		126	125.903305	*18.9%*			136	135.904571	*7.85%*
		127	126.905217	9.35 h			137	136.905822	*11.2%*
		128	127.904462	*31.7%*			138	137.905242	*71.7%*
		129	128.906596	69.6 m			139	138.908836	83.1 m
		130	129.906223	*33.9%*					
		131	130.908522	25.0 m					
I	53	125	124.904624	60.1 d	La	57	136	135.907650	9.87 m
		126	125.905619	13.0 d			137	136.906470	0.06 My
		127	126.904468	*100%*			138	137.907108	*0.090%*
		128	127.905805	25.0 m			139	138.906349	*99.910%*
		129	128.904988	15.7 My			140	139.909473	1.68 d
		130	129.906674	12.4 h			141	140.910958	3.92 h

	Z	A	Atomic mass (u)	Abundance or Half-life		Z	A	Atomic mass (u)	Abundance or Half-life
Ce	58	134	133.909030	76 h	Eu	63	149	148.917923	93.1 d
		135	134.909147	17.7 h			150	149.919699	35.8 y
		136	135.907650	*0.190%*			151	150.919846	*47.8%*
		137	136.907780	9.0 h			152	151.921741	13.5 y
		138	137.905986	*0.254%*			153	152.921227	*52.2%*
		139	138.906647	137.6 d			154	153.922976	8.59 y
		140	139.905435	*88.4%*			155	154.922890	4.68 y
		141	140.908272	32.5 d			156	155.924751	15.2 d
		142	141.909241	*11.1%*					
		143	142.912382	33.1 h	Gd	64	150	149.918656	1.79 My
		144	143.913643	285 d			151	150.920345	124 d
Pr	59	139	138.908933	4.4 h			152	151.919789	*0.20%*
		140	139.909072	3.39 m			153	152.921747	242 d
		141	140.907648	*100%*			154	153.920862	*2.2%*
		142	141.910041	19.1 h			155	154.922619	*14.8%*
		143	142.910813	13.6 d			156	155.922120	*20.5%*
Nd	60	140	139.909310	3.37 d			157	156.923957	*15.7%*
		141	140.909605	2.5 h			158	157.924101	*24.8%*
		142	141.907719	*27.1%*			159	158.926385	18.6 h
		143	142.909810	*12.2%*			160	159.927051	*21.9%*
		144	143.910083	*23.8%*			161	160.929666	3.66 m
		145	144.912569	*8.3%*					
		146	145.913113	*17.2%*	Tb	65	157	156.924021	99 y
		147	146.916096	11.0 d			158	157.925410	180 y
		148	147.916889	*5.7%*			159	158.925343	*100%*
		149	148.920145	1.72 h			160	159.927164	72.3 d
		150	149.920887	*5.6%*			161	160.927566	6.88 d
		151	150.923825	12.4 m					
Pm	61	143	142.910928	265 d	Dy	66	154	153.924423	3.0 My
		144	143.912586	363 d			155	154.925749	10.0 h
		145	144.912745	17.7 y			156	155.924278	*0.057%*
		146	145.914693	5.53 y			157	156.925461	8.1 h
		147	146.915134	2.62 y			158	157.924405	*0.100%*
		148	147.917468	5.37 d			159	158.925736	144.4 d
		149	148.918330	53.1 h			160	159.925194	*2.3%*
Sm	62	142	141.915204	72.5 m			161	160.926930	*18.9%*
		143	142.914624	8.83 m			162	161.926795	*25.5%*
		144	143.911996	*3.1%*			163	162.928728	*24.9%*
		145	144.913407	340 d			164	163.929171	*28.2%*
		146	145.913038	103 My			165	164.931700	2.33 h
		147	146.914894	*15.0%*					
		148	147.914818	*11.3%*	Ho	67	163	162.928730	4570 y
		149	148.917180	*13.8%*			164	163.930231	29.0 m
		150	149.917272	*7.4%*			165	164.930319	*100%*
		151	150.919929	90 y			166	165.932281	26.8 h
		152	151.919729	*26.7%*			167	166.933127	3.1 h
		153	152.922094	46.3 h					
		154	153.922206	*22.7%*					
		155	154.924636	22.3 m					

	Z	A	Atomic mass (u)	Abundance or Half-life		Z	A	Atomic mass (u)	Abundance or Half-life
Er	68	160	159.929080	28.6 h	W	74	178	177.945850	21.6 d
		161	160.930002	3.21 h			179	178.947072	37.5 m
		162	161.928775	*0.14%*			180	179.946706	*0.12%*
		163	162.930029	75.0 m			181	180.948198	121 d
		164	163.929197	*1.61%*			182	181.948205	*26.3%*
		165	164.930723	10.4 h			183	182.950224	*14.3%*
		166	165.930290	*33.6%*			184	183.950932	*30.7%*
		167	166.932046	*23.0%*			185	184.953420	75.1 d
		168	167.932368	*26.8%*			186	185.954362	*28.6%*
		169	168.934588	9.40 d			187	186.957158	23.7 h
		170	169.935461	*14.9%*	Re	75	183	182.950821	70.0 d
		171	170.938026	7.52 h			184	183.952524	38.0 d
Tm	69	167	166.932849	9.25 d			185	184.952955	*37.40%*
		168	167.934171	93.1 d			186	185.954986	90.6 h
		169	168.934211	*100%*			187	186.955750	*62.60%*
		170	169.935798	128.6 d			188	187.958112	17.0 h
		171	170.936426	1.92 y	Os	76	182	181.952186	22.1 h
Yb	70	166	165.933880	56.7 h			183	182.953110	13.0 h
		167	166.934947	17.5 m			184	183.952491	*0.018%*
		168	167.933895	*0.135%*			185	184.954043	93.6 d
		169	168.935187	32.0 d			186	185.953838	*1.6%*
		170	169.934759	*3.1%*			187	186.955748	*1.6%*
		171	170.936323	*14.3%*			188	187.955836	*13.3%*
		172	171.936378	*21.9%*			189	188.958145	*16.1%*
		173	172.938207	*16.1%*			190	189.958445	*26.4%*
		174	173.938858	*31.8%*			191	190.960928	15.4 d
		175	174.941273	4.19 d			192	191.961479	*41.0%*
		176	175.942569	*12.7%*			193	192.964148	30.5 h
		177	176.945257	1.9 h	Ir	77	189	188.958716	13.2 d
Lu	71	173	172.938927	1.37 y			190	189.960590	11.8 d
		174	173.940334	3.3 y			191	190.960591	*37.3%*
		175	174.940768	*97.41%*			192	191.962602	73.8 d
		176	175.942683	*2.59%*			193	192.962923	*62.7%*
		177	176.943755	6.71 d			194	193.965075	19.2 h
Hf	72	172	171.939460	1.87 y	Pt	78	188	187.959395	10.2 d
		173	172.940650	23.6 h			189	188.960832	10.9 h
		174	173.940042	*0.16%*			190	189.959930	*0.013%*
		175	174.941504	70 d			191	190.961684	2.9 d
		176	175.941403	*5.2%*			192	191.961035	*0.79%*
		177	176.943220	*18.6%*			193	192.962984	50 y
		178	177.943698	*27.3%*			194	193.962663	*32.9%*
		179	178.945815	*13.6%*			195	194.964774	*33.8%*
		180	179.946549	*35.1%*			196	195.964934	*25.3%*
		181	180.949099	42.4 d			197	196.967323	18.3 h
Ta	73	179	178.945934	1.79 y			198	197.967875	*7.2%*
		180	179.947466	*0.0123%*			199	198.970576	30.8 m
		181	180.947996	*99.9877%*					
		182	181.950152	114 d					

	Z	A	Atomic mass (u)	*Abundance* or Half-life		Z	A	Atomic mass (u)	*Abundance* or Half-life
Au	79	195	194.965017	186 d	Fr	87	212	211.996182	20.0 m
		196	195.966551	6.18 d			223	223.019731	21.8 m
		197	196.966551	*100%*	Ra	88	223	223.018497	11.43 d
		198	197.968225	2.694 d			224	224.020202	3.66 d
		199	198.968748	3.14 d			225	225.023603	14.9 d
Hg	80	194	193.965381	520 y			226	226.025402	1600 y
		195	194.966640	9.9 h	Ac	89	225	225.023220	10.0 d
		196	195.965814	*0.15%*			226	226.026089	29.4 h
		197	196.967195	64.1 h			227	227.027747	21.77 y
		198	197.966752	*10.0%*	Th	90	229	229.031754	7340 y
		199	198.968262	*16.9%*			230	230.033126	75,400 y
		200	199.968309	*23.1%*			231	231.036296	25.52 h
		201	200.970285	*13.1%*			232	232.038050	*100%*
		202	201.970625	*29.9%*			233	233.041576	22.3 m
		203	202.972857	46.6 d	Pa	91	230	230.034532	17.4 d
		204	203.973475	*6.9%*			231	231.035878	32,800 y
		205	204.976056	5.2 m			232	232.038581	1.31 d
Tl	81	201	200.970803	72.9 h	U	92	233	233.039627	0.1592 My
		202	201.972090	12.2 d			234	234.040945	0.245 My
		203	202.972329	*29.5%*			235	235.043922	*0.720%*
		204	203.973848	3.78 y			236	236.045561	23.42 My
		205	204.974412	*70.5%*			237	237.048723	6.75 d
		206	205.976095	4.20 m			238	238.050784	*99.275%*
Pb	82	202	201.972143	0.053 My			239	239.054289	23.5 m
		203	202.973375	51.9 h	Np	93	236	236.046570	0.115 My
		204	203.973028	*1.42%*			237	237.048166	2.14 My
		205	204.974467	15.2 My			238	238.050940	2.117 d
		206	205.974449	*24.1%*	Pu	94	238	238.049553	87.74 y
		207	206.975880	*22.1%*			239	239.052156	24,100 y
		208	207.976636	*52.4%*			240	240.053807	6563 y
		209	208.981075	3.25 h			241	241.056844	14.4 y
Bi	83	207	206.978456	32.2 y			242	242.058736	0.373 My
		208	207.979727	0.368 My	Am	95	241	241.056822	433 y
		209	208.980384	*100%*			242	242.059542	16.0 h
		210	209.984105	5.01 d			243	243.061372	7380 y
		211	210.987258	2.14 m	Cm	96	246	246.067217	4730 y
Po	84	207	206.981578	5.80 h			247	247.070346	15.6 My
		208	207.981231	2.90 y			248	248.072341	0.340 My
		209	208.982415	102 y	Bk	97	247	247.070298	1380 y
		210	209.982857	138.4 d	Cf	98	251	251.079579	898 y
At	85	209	208.986158	5.41 h	Es	99	252	252.082970	472 d
		210	209.987131	8.1 h					
		211	210.987481	7.21 h	Fm	100	257	257.095096	100.5 d
Rn	86	211	210.990585	14.6 h	Md	101	258	258.098427	55 d
		222	222.017570	3.82 d					

	Z	A	Atomic mass (u)	Abundance or Half-life
No	102	259	259.101050	58 m
Lr	103	260	260.105570	3.0 m
Rf	104	261	261.108910	65 s
Ha	105	262	262.114370	34 s

	Z	A	Atomic mass (u)	Abundance or Half-life
Sg	106	261	261.116360	0.23 s
Ns	107	262	262.123120	0.10 s
Hs	108	264	264.128630	0.08 ms
Mt	109	266	266.137830	3.4 ms
	110	267		0.5 ms
	111	272		4 ms

ANSWERS TO ODD-NUMBERED PROBLEMS

Chapter 1

1. mass m moves at speed $v/\sqrt{2}$
   mass $3m$ moves at $v/\sqrt{6}$
   in the direction $\theta = -35.3°$
3. $K = 46.0$ keV
   $v = 1.49 \times 10^6$ m/s
   $p = 9.90 \times 10^{-21}$ kg·m/s
7. (b) 1.1652×10^{-36} eV·nm
   (c) 8.0918×10^{-37}
9. $v = \alpha c$
13. (b) 2.426×10^{-3} nm
15. $v_1 = 2.47 \times 10^6$ m/s
    $v_2 = -0.508 \times 10^6$ m/s

Chapter 2

1. 70 km/s
3. 2.6×10^8 m/s
5. (a) 357.1 ns
   (b) 103 m
   (c) 28.8 m
9. $0.402c$
11. 5.0×10^7 m/s
15. $\Delta x' = 2.34$ km, $\Delta t' = 1.07$ μs
17. $v'_B = 0.72c$ at $\theta' = 146°$
    $v'_C = 0.85c$ at $\theta' = 180°$
    $v'_D = 0.82c$ at $\theta' = 164°$
23. $v > 0.990c$
27. $v > 0.0115c$
29. (a) 2.56×10^{-3} eV
    (b) 25.6 eV
    (c) 24.7 keV
    (d) 10.9 MeV
31. (a) 3.1 MeV
    (b) 7.8 MeV
33. 8.9×10^{-4} kg
35. $0.981c$

Chapter 3

1. 1.32 mm
3. (a) 0.388 nm
   (b) 7.2°; 97.2°
5. 4×10^{-9} eV to 4×10^{-7} eV
7. 1.3×10^{18}
9. $h = 6.57 \times 10^{-34}$ J·s,
   $\phi = 2.28$ eV
11. $V + 2.42$ V
13. (a) 288 nm
    (b) 1.33 V
19. 483 nm
21. (a) 2.52 μm
    (b) 0.405
23. 0.33 W
25. (a) 10.33 keV
    (b) 0.06 keV
27. $2E^2/(2E + mc^2)$
29. (a) 0.402 MeV
    (b) 0.260 MeV
31. 6.4×10^3 eV/c, 3.9×10^{-4} eV
33. $E = 4mc^2$, $K_e = 2mc^2/3$

Chapter 4

1. (a) 0.0279 nm
   (b) 12 fm

(c) 0.025 fm
(d) 0.73 nm

3. 0.01 V; 100 V; 10^9 V

5. (a) 89 MeV
(b) 4.2 MeV
(c) 1.1 MeV

7. (a) 0.10 nm
(b) 0.11 nm

9. 33 nm

11. $15°$ ($n = 1$), $32°$ ($n = 2$), $52°$ ($n = 3$)

15. (a) 2000 eV/c
(b) 4 eV

17. 5.5×10^{-24} s

19. (a) 0.71 MeV
(b) $\Delta E = 0.66$ MeV

21. (a) 2×10^{-5} eV/c
(b) 1.2×10^{-15} eV

23. (a) 990 eV/c
(b) 2.4×10^{-5} eV

27. (a) 1.0×10^2 cm
(b) 0.11 cm

Chapter 5

3. 1.1 eV

5. (a) 89.6 eV
(b) 41.8 eV, 71.6 eV, etc.

7. (a) 0.1955
(b) 0.6090
(c) 0.1955

9. (a) 2160 eV
(b) 4.70×10^4 eV/c
(c) 4.20×10^{-3} nm

11. 2.0 MeV

17. $5.00E_0$, $10.00E_0$

25. $a = m\omega_0/2\hbar$
$E = 3\hbar\omega/2$

27. 4.96 eV; 9.92 eV

Chapter 6

3. (a) 45.7 nm
(b) 86.2 nm

7. 33 MeV

9. (a) 3.37×10^{-5}
(b) 4.41×10^{-3}
(c) 1.33×10^{-2}
(d) 0.982

11. 2.84

13. 4.39×10^{-3} MeV

15. 59/s

17. 91.13 nm (Lyman),
820.1 nm (Paschen)

19. $5 \rightarrow 4$: $\Delta E = 0.306$ eV
$5 \rightarrow 3$: $\Delta E = 0.97$ eV
$5 \rightarrow 2$: $\Delta E = 2.86$ eV
$5 \rightarrow 1$: $\Delta E = 13.1$ eV

23. (a) 1.51 eV
(b) 13.6 eV
(c) 7.65 eV

25. 0.178 nm

27. $E_1 = -54.4$ eV, $E_2 = -13.6$ eV
$E_3 = -6.04$ eV, $E_4 = -3.40$ eV

29. 7×10^{-8} eV

33. $a_0 = 1.19 \times 10^{29}$ m
$E_2 - E_1 = 2.0 \times 10^{-78}$ eV

35. 0.440 nm

39. $n = 745$, $r = 29$ μm

Chapter 7

3. 50

7. $35°$, $66°$, etc.

13. 0.0490

15. 5.4×10^{-3}

17. $6a_0$ ($2s$), $5a_0$ ($2p$)

25. 121.50254 nm, 121.50200 nm

Chapter 8

1. (a) N, P, As, Sb, Bi
(b) Co, Rh, Ir

3. Na: $Z_{eff} = 1.8$; K: $Z_{eff} = 2.3$

5. Fe

9. 11 kV

11. (a) $L = 5, S = 1$
(b) $L = 2, S = 4$
(c) $L = 2, S = 1$

13. $S = 0, 1$
$L = 0, 1, 2, 3, 4$

15. Na: 0.67 eV; Li: 0.66 eV

17. $3p$: 2.1×10^{-3} eV

19. (a) $(2, 1, +1, +\frac{1}{2})$, etc.
(b) 36
(c) 30

21. (a) 670.8 nm
(b) 58.4 nm
(c) Li: 230 nm; He: 50.6 nm

Chapter 9

1. 15.4 eV

3. 1.13 nm

5. (a) 30.9×10^{-30} C·m
   (b) 88.0%

7. 42.7%

11. 4.00×10^{-6} eV

13. HD: 1.14×10^{14} Hz, 0.074 nm,
    0.0114 eV

15. 23.1 mm, 11.6 mm, 7.71 mm

21. (b) 2.37×10^{21} eV/m^2
    (c) 2.1×10^{-3} eV

23. (a) 5.15×10^{-10} eV
    (b) 2.65×10^{-19} eV

25. (a) $p(L = 1)/p(L = 0) = 1.64$
    (b) $p(L = 1)/p(L = 0) = 8.38 \times 10^{-3}$

27. 3780 K

Chapter 10

5. $[(kT/m)(3 - 8/\pi)]^{1/2}$

7. (a) $e^{-E/kT}$
   (b) $E/(1 + e^{E/kT})$
   (c) $NE/(1 + e^{E/kT})$
   (d) $R(E/kT)^2[e^{E/kT}/(1 + e^{E/kT})^2]$

9. (a) $f(+1) = 0.3295$, $f(0) = 0.3333$
   $f(-1) = 0.3372$

11. (a) 0.0379 eV
    (b) 2.79×10^{21}

15. (a) $U = 1.84 \times 10^{17}$ eV/m^3
    (b) 0.0346
    (c) 2.26×10^{-17}

17. $E_F = 7.04$ eV, $E_{av} = 4.22$ eV

19. 0.12 eV

21. (a) 4.37×10^{-36}
    (b) 6.00×10^{-6}

23. 244 keV

Chapter 11

1. (a) $\frac{1}{8}$
   (b) $a = 2r$
   (c) 0.5236

5. $-\dfrac{e^2}{4\pi\varepsilon_0}\dfrac{1}{R}\left(8 - \dfrac{6}{2/\sqrt{3}} + \dfrac{24}{\sqrt{11/3}}\right)$

7. (a) 6.82 eV
   (b) 6.45 eV
   (c) 3.27 eV

9. $U_C = -8.96$ eV, $U_R = 1.00$ eV

13. 0.255 nm

15. (a) 7.09 eV
    (b) $2.20kT$

17. 0.23×10^{25} m^{-3}

19. 7076 K

23. (a) -0.094 eV
    (b) 7.62 nm

25. (a) $1 - 3.46 \times 10^{-10}$
    (b) 3.46×10^{-10}

Chapter 12

1. (a) $^{19}_{9}\text{F}_{10}$
   (b) $^{199}_{79}\text{Au}_{120}$
   (c) $^{107}_{47}\text{Ag}_{60}$

3. (a) 15 MeV
   (b) 824 MeV

5. 12.6 fm

7. (a) 1636.5 MeV,
       7.868 MeV/nucleon
   (b) 1118.5 MeV,
       8.410 MeV/nucleon

9. ^{3}He: 7.718 MeV
   ^{3}H: 8.482

11. ^{4}He: 19.814 MeV

13. 2.6×10^{-3} fm

15. (a) 35 min
    (b) 0.020 min^{-1}
    (c) 46 s^{-1}

17. 0.0598

19. (a) 7.40×10^7 s^{-1}
    (b) 6.62×10^6 s^{-1}

21. 37.5 μCi

23. (a) $e^+ + \nu$
    (c) $^{224}_{90}\text{Th}_{134}$

25. ^{210}Bi and ^{211}At

27. 1.93×10^{-3} W

29. ^{232}Th: 1.04×10^{-20} s^{-1}

33. 11.506 MeV

35. 0.864 MeV

37. (a) 6
    (b) 4
    (c) 42.658 MeV
    (d) 27.8 μW

39. (a) 4.67×10^{-9} eV
    (b) 1.95×10^{-3} eV
    (c) 0.097 mm/s

Chapter 13

1. (a) $^1_1H_0$
   (c) $^{30}_{15}P_{15}$

3. (a) $\frac{1}{2}$
   (b) $\frac{3}{4}$
   (c) $\frac{15}{16}$

5. 8.6×10^{-4} b

7. $1.58\ \mu Ci$

9. 4.49 b (^{63}Cu), 2.20 b (^{65}Cu)

13. (a) -10.313 MeV
    (b) 2.314 MeV

15. $K_p = 12.201$ MeV,
    $K_\alpha = 11.152$ MeV
    $K_p = 22.706$ MeV,
    $K_\alpha = 0.647$ MeV

17. 8.660 MeV

19. 9.042 MeV

21. (a) 6.545 MeV
    (b) 4.807 MeV

23. 180.02 MeV

25. 1.943 MeV, 1.199 MeV,
    7.551 MeV, 7.296 MeV,
    1.732 MeV, 4.966 MeV

27. 14.1 MeV

29. 7.274 MeV

33. (a) 5.594 MeV
    (b) 0.56 W

Chapter 14

1. (a) strong
   (c) weak
   (e) strong

3. (a) L_e
   (b) S
   (c) B

5. (a) ν_e
   (c) π^0

9. $0.999635c$, 40.2 GeV

11. (a) 820 eV
    (c) 8.9 keV

13. (a) 34 MeV
    (c) 218 MeV
    (e) 76 MeV

15. (a) K^+
    (c) K^0
    (e) μ^+

19. (a) 2205 MeV
    (c) 1800 MeV

21. $K_\pi = 57$ MeV
    $K_n = 310$ MeV
    $\theta_n = 9.7°$

23. (a) $s \rightarrow d + u + \bar{u}$
    (c) $u + u \rightarrow$ energy
    (e) energy $\rightarrow d + \bar{d}$

25. (a) weak interaction
    (b) $D_s^+ \rightarrow \phi + \pi^+$
         $C = +1 \rightarrow 0 + 0$
         $S = +1 \rightarrow 0 + 0$
    (c) $c \rightarrow s + W^+$
        and $W^+ \rightarrow u + \bar{d}$

Chapter 15

1. $6.49 \times 10^{-6}\ \lambda$

5. 1.8×10^{-5} s

9. -0.230 MeV

19. (a) 1.8721×10^{-64} J·m
    (b) 4.1443×10^{-29} MeV·m^2
    (c) 3.5468×10^{22} m
    (d) 3.2944×10^{-74} MeV

21. (a) 140 MeV
    (b) 85 MeV
    (c) 2.0×10^{59} MeV
    (d) -4.0×10^{59} MeV

23. 2.58 fm

25. (a) 6.21×10^{-15} W

Chapter 16

1. (a) 590.1 nm
   (b) 594.5 nm
   (c) 1625 nm

3. (a) $(2.431 \times 10^{-4}$ eV/K$)T$
   (b) 6.56×10^{-4} eV

5. (a) 1.2×10^{-3719} m^{-3}
   (b) 1000 K, 7×10^6 y

11. (a) 5.9×10^{12} K
    (b) 6.7×10^{-6} s

13. (a) 9×10^{12} s $= 3 \times 10^5$ y
    (b) 1.17 eV
    (c) 1.24

15. 14 eV

17. (a) $i = \frac{1}{2}, j = -\frac{5}{2}, k = \frac{1}{2}$
    (b) 1.3×10^{-43} s

21. 2.9×10^3 light-years

PHOTO CREDITS

Chapter 1
Opener: From *Galileo's Dialogue Concerning the Two Chief World Systems*, 1632.

Chapter 2
Opener: FAR SIDE ©FARWORKS, Inc. Distributed by Universal Press Syndicate. Reprinted with permission. All rights reserved. Page 25: Courtesy AIP Niels Bohr Library. Figure 2.7: Courtesy Education Development Center. Page 26: The New York Times.

Chapter 3
Opener: Courtesy U.S. Department of Energy. Figure 3.2: From *Atlas of Optical Phenomena* by M. Cagnet, et al. Springer-Verlag, Prentice Hall, 1962. Figure 3.7: Amoros, Buerger, and de Amoros, *The Laue Method*, Academic Press, 1975. Figure 3.9: Courtesy Education Development Center. Figure 3.8: Amoros, Buerger, and de Amoros, *The Laue Method*, Academic Press, 1975. Page 75: Courtesy AIP Niels Bohr Library. Page 82: Courtesy AIP, W. F. Meggers Collection, Page 85: Courtesy AIP Niels Bohr Library.

Chapter 4
Opener: Philippe Plailly/Photo Researchers. Page 101: Courtesy AIP Niels Bohr Library. Figure 4.1: Cagnet, et al., *Atlas of Optical Phenomena*, Springer-Verlag, 1971. Figure 4.2: Courtesy Sumio Iijima, Arizona State University. Figure 4.3: Courtesy Education Development Center. Figure 4.7: From E. O. Wollan, C. G. Shull, and M. C. Marney, *Physical Review* 73, 527 (1948). Figure 4.10: Courtesy C. Jorisson, Institut für Angewande Physik der Universitat Tübingen. Page 115: Courtesy AIP Niels Bohr Library. Figure 4.25: Courtesy A. Tonomura, T. Matsuda, and T. Kawasaki, Advanced Research Laboratory. From *American Journal of Physics* 57, 117 (1989).

Chapter 5
Opener: Revised for *Scientific American* Article, October 1981, "Accoustics of Violins," Page 139: Courtesy AIP Niels Bohr Library. Figure 5.23: Courtesy Education Development Center.

Chapter 6
Opener: Michael Gilbert/Photo Researchers. Page 179: Courtesy AIP Niels Bohr Library. Page 189: Courtesy AIP Niels Bohr Library.

Chapter 7
Opener: Courtesy William P. Spencer, Massachusetts Institute of Technology, from *Scientific American*, May 1981, p. 132.

Chapter 8
Opener: Kenneth Eward/BioGrafx-Science Source/Photo Researchers. Page 237: Courtesy AIP Niels Bohr Library. Page 249: Courtesy University of Oxford, Museum of the History of Science, AIP Niels Bohr Library.

Chapter 9
Opener: Ken Eward/Photo Researchers. Page 287: Courtesy The Nobel Institute.

Chapter 10
Opener: Gregory Sams/Photo Researchers. Figure 10.22: From K. Mendelssohn, *The Quest for Absolute Zero*, McGraw-Hill, New York, 1966.

Chapter 11
Opener: D. Jeremy Burgess/Photo Researchers.

Chapter 12
Opener: Elscint/Science Photo Library/Photo Researchers. Page 378: Courtesy Los Alamos Scientific Laboratory, University of California. Page 386: Courtesy AIP Niels Bohr Library.

Chapter 13
Opener: Courtesy U.S. Navy. Figure 13.1b: Courtesy Argonne National Laboratory. Figure 13.2b: Courtesy Purdue University. Page 423: Courtesy AIP Emilio Segre Visual Archives, Herzfeld Collection. Figure 13.20: Courtesy D. Bruce Sodee, M.D. Page 438: Courtesy AIP, Meggers Gallery of Nobel Laureates.

Chapter 14
Opener: Dan McCoy/Rainbow. Page 456: Courtesy Bryn Mawr College. Page 461: Courtesy AIP Emilio Segre Visual Archives, Physics Today Collection. Figure 14.2: Courtesy Lawrence Berkeley Laboratory. Figure 14.18: Courtesy William R. Kropp, University of California, Irvine, and the IMB Collaboration.

Chapter 15
Opener: Georg Gerster/Photo Researchers. Figure 15.19: Courtesy Brookhaven National Laboratory. Page 508: Courtesy Library of Congress. Page 509: Courtesy Royal Observatory, Edinburgh. Figure 15.24: Courtesy Kitt Peak National Observatory. Figure 15.25: Courtesy Hale Observatories, California Institute of Technology.

Chapter 16
Opener: Painting by Vincent van Gogh, "The Starry Night" (1889). Photo courtesy of the Museum of Modern Art, NY. Page 522. Courtesy Hale Observatories, AIP Neils Bohr Library, Figure 16.1: Courtesy California Institute of Technology. Page 527: Courtesy AIP Neils Bohr Library. Figure 16.8: Courtesy Mount Wilson and Palomar Observatories. Page 532: Photo by Mark Godfrey, courtesy of Vera Rubin.

INDEX